Bioenergetics

Bioenergetics

Edited by Zoe Hooper

SYRAWOOD
PUBLISHING HOUSE

New York

Published by Syrawood Publishing House,
750 Third Avenue, 9th Floor,
New York, NY 10017, USA
www.syrawoodpublishinghouse.com

Bioenergetics
Edited by Zoe Hooper

International Standard Book Number: 978-1-68286-404-3 (Hardback)

Cataloging-in-publication Data

Bioenergetics / edited by Zoe Hooper.
    p. cm.
Includes bibliographical references and index.
ISBN 978-1-68286-404-3
1. Bioenergetics. 2. Energy metabolism. 3. Mitochondria. 4. Biochemistry. I. Hooper, Zoe.
QH510 .B56 2017
572.43--dc23

Printed in the United States of America.

TABLE OF CONTENTS

PREFACE

Bioenergetics is characterized by the energy involved in making and breaking of chemical bonds in the molecules found in organisms. It is a sub discipline of biochemistry. This book is compiled in such a manner, that it will provide in-depth knowledge about the theory and practice of bioenergetics and its related fields. It aims to provide a comprehensive study of the living organisms by examining components like growth, development and metabolism which are essential to understand the field of bioenergetics. From theories to research to practical applications, case studies related to all contemporary topics of relevance to this field have been included in this book. Coherent flow of topics, student-friendly language and extensive use of examples make this book an invaluable source of knowledge for all researching and studying the field of bioenergetics.

The information shared in this book is based on empirical researches made by veterans in this field of study. The elaborative information provided in this book will help the readers further their scope of knowledge leading to advancements in this field.

Finally, I would like to thank my fellow researchers who gave constructive feedback and my family members who supported me at every step of my research.

Editor

Synergistic Effect of High Charge and Energy Particle Radiation and Chronological Age on Biomarkers of Oxidative Stress and Tissue Degeneration: A Ground-Based Study Using the Vertebrate Laboratory Model Organism *Oryzias latipes*

Xuan Zheng[1,2], Xinyan Zhang[3], Lingling Ding[4], Jeffrey R. Lee[5], Paul M. Weinberger[6], William S. Dynan[1,7]*

1 Department of Neuroscience and Regenerative Medicine, Georgia Regents University, Augusta, Georgia, United States of America, 2 Center for Gene Diagnosis, Zhongnan Hospital, Wuhan University, Wuhan, China, 3 Department of Biostatistics, University of Alabama at Birmingham, Birmingham, Alabama, United States of America, 4 Department of Anatomy and Embryology, Wuhan University School of Medicine, Wuhan, China, 5 Department of Pathology, Georgia Regents University, Augusta, Georgia, United States of America, 6 Department of Otolaryngology and Center for Biotechnology & Genomic Medicine, Georgia Regents University, Augusta, Georgia, United States of America, 7 Departments of Radiation Oncology and Biochemistry, Emory University, Atlanta, Georgia, United States of America

Abstract

High charge and energy (HZE) particles are a main hazard of the space radiation environment. Uncertainty regarding their health effects is a limiting factor in the design of human exploration-class space missions, that is, missions beyond low earth orbit. Previous work has shown that HZE exposure increases cancer risk and elicits other aging-like phenomena in animal models. Here, we investigate how a single exposure to HZE particle radiation, early in life, influences the subsequent age-dependent evolution of oxidative stress and appearance of degenerative tissue changes. Embryos of the laboratory model organism, *Oryzias latipes* (Japanese medaka fish), were exposed to HZE particle radiation at doses overlapping the range of anticipated human exposure. A separate cohort was exposed to reference γ-radiation. Survival was monitored for 750 days, well beyond the median lifespan. The population was also sampled at intervals and liver tissue was subjected to histological and molecular analysis. HZE particle radiation dose and aging contributed synergistically to accumulation of lipid peroxidation products, which are a marker of chronic oxidative stress. This was mirrored by a decline in PPARGC1A mRNA, which encodes a transcriptional co-activator required for expression of oxidative stress defense genes and for mitochondrial maintenance. Consistent with chronic oxidative stress, mitochondria had an elongated and enlarged ultrastructure. Livers also had distinctive, cystic lesions. Depending on the endpoint, effects of γ-rays in the same dose range were either lesser or not detected. Results provide a quantitative and qualitative framework for understanding relative contributions of HZE particle radiation exposure and aging to chronic oxidative stress and tissue degeneration.

Editor: Andrea Motta, National Research Council of Italy, Italy

Funding: This work was supported by the US Department of Energy Low Dose Radiation Research Program, award number DE-SC0002343 to WSD and the National Aeronautics and Space Administration, award number NNX11AC30G to WSD. The funders had no role in study design, data collection and analysis, decision to publish, or preparation of the manuscript.

Competing Interests: The authors have declared that no competing interests exist.

* Email: wdynan@emory.edu

Introduction

Exposure to fast-moving atomic nuclei, known as high charge and energy (HZE) particles, is a limiting factor in the design of exploration-class human space missions, defined as those that venture beyond low earth orbit and thus beyond the partial protection afforded by the earth's magnetic field [1]. Prior work has shown that HZE particle radiation is associated with cancer and a number of other aging-like phenomena, including atherosclerosis, bone loss, and cognitive or behavioral impairment (for examples see [2–10]; reviewed in [11]).

Persistent oxidative stress is suspected to be an underlying factor in many of these degenerative effects observed at the organ level (reviewed in [12,13]. Elevated levels of reactive oxygen species or their reaction products have been measured following HZE particle radiation exposure of both cells and animals [14–18]. Although HZE particles appear to be particularly effective, other DNA damaging agents also induce oxidative stress. Possible mechanisms leading to generation or release of reactive oxygen species include mitochondria injury [19,20], DNA damage-mediated activation of NADPH oxidases [21,22], and p53-mediated repression of PGC-1α, a master regulator of mitochondrial function and antioxidant gene expression [23]. Oxidative stress propagates from cell to cell, in part by signaling mechanisms [16,24], and is a target for countermeasure development [14,25].

The primary objective of the present study was to evaluate the increment in persistent oxidative stress due to radiation exposure, relative to the natural increase that occurs during aging. This is particularly challenging at the low doses representative of anticipated human exposure. Recent data from the Mars Science Laboratory suggest that a human round-trip mission to Mars would incur a physical radiation dose of about 0.16 Gy and an equivalent dose of about 0.66 Sv [26]. The large-scale, lifetime study described here was designed to address the challenges of measuring the potentially small effect of radiation doses in this range against a substantial age-dependent background. Secondary objectives of the study were to investigate potential mechanisms underlying increased oxidative stress through characterization of radiation-associated changes in gene expression, histology, and mitochondrial ultrastructure.

We performed studies using a vertebrate model organism, *Oryzias latipes* (the Japanese medaka fish). Medaka and other teleost models, such as the zebrafish, occupy a unique experimental niche. They have many vertebrate-specific organs of radiobiological interest and homologs of most or all of the DNA damage response and repair genes found in humans and other mammals. Most laboratory fish species also undergo regular, time-dependent aging similar to that in mammals. Importantly for the present study, it is feasible to maintain large numbers of individuals for lifetime studies at a cost much lower than for rodents.

The medaka species has a nearly 50-year history of use in radiation research [27–29]. The embryos tolerate a wide range of temperatures and can be chilled to delay development [30], which facilitates the logistics of beam-line studies [31]. We have previously investigated markers of normal aging in the medaka and shown that the liver is one of the first organs to show age-dependent degenerative changes [32]. The liver is also target of HZE carcinogenesis in the mouse [3], making it a logical focus for the study of HZE effects.

Our results show that a single exposure to HZE particle radiation, in a dose range overlapping that of anticipated human exposure, significantly elevates the levels of lipid peroxidation products in liver, when measured months or years following initial irradiation. The effect of radiation is synergistic with normal aging. A single developmental exposure to HZE particle radiation exposure is also associated with abnormal mitochondrial ultrastructure, cystic degeneration, and persistently decreased levels of expression of PPARGC1A, a master regulator of mitochondrial and antioxidant gene expression. Effects on the various endpoints were either lesser or not measurable with γ-rays in the same dose range, indicating that many of these endpoints are radiation quality dependent.

Materials and Methods

Animal maintenance

This study was carried out in strict accordance with the recommendations in the Guide for the Care and Use of Laboratory Animals of the National Institutes of Health. The protocol (number BR09-10-259) was approved by the Institutional Animal Care and Use Committee at Georgia Regents University. (Office of Laboratory Animal Welfare Assurance number A3307-1). The experiments did not involve procedures accompanied by pain, distress, or discomfort for which analgesics or anesthetics would be indicated.

Breeding stocks of CAB wild type Japanese medaka (*Oryzias latipes*) were maintained as described [32]. Embryos were collected and incubated for 24 h in aerated 0.67X seawater (3.5% NaCl) containing 0.0001% methylene blue. Viable (unstained) embryos were transferred to embryo rearing medium (0.1% NaCl, 0.003% KCl, 0.004% $CaCl_2$ dihydrate, 0.016% Mg_2SO_4 heptahydrate, 0.001%, $NaHCO_3$, pH 7.4) and held at 18°C with daily medium exchanges. Embryos collected over a 5 d period were pooled and shipped at ambient temperature to the NASA Space Radiation Laboratory in Brookhaven, New York. Embryos were re-staged according to established criteria [33] and stage 22–34 embryos were pooled and re-distributed randomly among six dose groups, including one mock-irradiated control group (n≈250 per group). Based on our prior study of normal aging [32], this group size provides 80% power to detect a 10% difference in mean survival assuming a two-sided α of 0.05.

Embryos were exposed, at mid-day, to 1 GeV/nucleon ^{56}Fe ions at 0.1–0.4 Gy/min in T-25 flasks filled with embryo rearing medium. The primary method of calibrating the dose at NSRL is via a calibration ion chamber that is periodically calibrated using a standard gamma ray source. Before the exposures, the calibration ion chamber was used to measure the dose at the same time as an in-line secondary ion chamber was read. This reading served to transfer the calibration to the secondary ion chamber, which remained in the beam during the exposures. The secondary ion chamber was used to measure the integrated dose delivered and to cut off the beam when the desired total dose was reached. To facilitate experimental logistics, groups were exposed in order from highest to lowest dose, which allowed time for decay of activation products in the higher dose groups prior to returning embryo-containing flasks from the beam line facility to the biology laboratory.

After 48 h, embryos were shipped from Brookhaven to the home laboratory. They were maintained with a 12 h:12 h light:dark photoperiod at ambient temperature in ~200 ml aerated embryo rearing medium, which was changed daily. Every 2–3 d, newly hatched fry were transferred to containers containing continuously aerated tank water. They were fed twice daily with Zeigler Aquatox feed (Aquatic Ecosystems) passed through a 106 μm mesh filter. After two months, fry were transferred to a rack habitat system (Aquatic Habitats, Apopka, FL). Fish were maintained in conditioned water with quality parameters as follows: pH 7.5–8.3; conductivity, 500–560 μS; alkalinity, 80–100 mg/L as $CaCO_3$; hardness, 100–120 mg/L as $CaCO_3$; and dissolved oxygen, 5–7 mg/L. Fish were fed freely until they reached satiation twice daily, once with brine shrimp plus flake food in the morning and once with flake food in the afternoon.

A parallel γ-ray exposure arm of the experiment was performed at the home laboratory using a Gammacell Exactor (MDS Nordion, Ottawa, ON, Canada) at 0.8 Gy/min. Timed exposures were performed, with radiation dose estimated based on the manufacturer's calibration as adjusted for decay. We previously verified the accuracy of the calibration of this instrument using thermoluminescence dosimetry devices (Landauer Inc., Glenwood, IL, USA). [34]. There were seven dose groups, including one mock-irradiated control group. Embryos were maintained thereafter as for the HZE-irradiated cohort.

Survival was scored prior to hatching based on exclusion of methylene blue dye and after hatching by direct observation of motility. The condition of the fish was monitored at least once daily. Fish were humanely euthanized if they displayed an inability to ambulate (swim) properly or weight loss in excess of 15% of body weight.

Tissue processing and staining

At 250, 400, 500, and 600 d post-irradiation, eight male fish were randomly chosen from each dose group (except for the 27 Gy γ-ray group, where only four male survivors remained). As in previous work, we evaluated livers in male fish only [32], because the female liver, which produces vitellogenin as part of oogenesis, is highly variable between individuals. Fish were euthanized with 400 mg/L Tricaine (MS-222) and livers were divided into portions for histologic fixation and RNA extraction. Fixation was performed in Bouin's solution for 48 h. Samples were rinsed in 50% ethanol and kept in 70% ethanol at 4°C until they lost their yellow color. After paraffin embedding, 5 μm sections were stained with hematoxylin and eosin and evaluated blindly by a pathologist. For anti-4-hydroxy-2-nonenal (4-HNE) staining, sections were deparaffinized, rehydrated, and stained as described [32] using 1.1 ng/μl anti-4-HNE primary antibody (HNE-J2, Japan Institute for the Control of Aging, Fukuroi, Japan) and 1:500 biotinylated goat anti-mouse IgG secondary antibody (Vector Laboratories). Immune complexes were visualized using VECTASTAIN ABC standard and ImmPACT DAB kits (Vector Laboratories) with hematoxylin counterstaining. Where indicated, primary antibody was neutralized with 4-HNE-BSA (Cell Biolabs, San Diego, CA, USA). DAB staining was quantified using a Nuance multispectral imaging system 3.0.0 (Cambridge Research and Instrumentation Inc. Woburn, MA, USA). Spectral profiles were defined using a slide without primary antibody to define hematoxylin and a slide with high DAB intensity to define DAB plus hematoxylin. Subsequently a pure DAB spectral profile was calculated from the hematoxylin and the DAB plus hematoxylin profiles. Images were captured at 20 nM intervals from 480 to 720 nM, and spectrally unmixed. The hematoxylin component image was used to create an investigator-defined region of interest (ROI) mask, which was then applied to the DAB-specific component image. Nuance values (in optical density, OD) were recorded for the DAB component only, as average signal per ROI as a continuous scalar variable.

Polymerase chain reaction

For mRNA analysis, total liver RNA was extracted using Trizol and PureLink RNA Mini Kit (Ambion by Life Technologies, Grand Island, NY). Reverse transcription using a QuantiTect Reverse Transcription Kit (Qiagen, Valencia, CA) was performed to obtain cDNA. We identified medaka orthologs of various genes of radiobiological interest based on their annotation in the ENSEMBL genome browser (www.ensembl.org) and verified them based on alignment to other full-length vertebrate homologs. Primers were designed such that at least one member of each pair spanned an mRNA splice junction (Table S1). The qPCR reactions were performed using a QuantiFast SYBR Green PCR kit (Qiagen, Valencia, CA). The inability to amplify genomic DNA, which may be present as a contaminant in RNA samples, was verified experimentally. Specificity of the PCR products was verified by gel electrophoresis and melting curve analysis. The $\Delta\Delta C_t$ method was used to determine fold change between treatment groups. Normalization was performed using the geometric average of two reference genes, ACTB (β-actin) and RPL7 [35,36].

Electron microscopy

Liver tissue was fixed in 2% glutaraldehyde in 0.1 M sodium cacodylate (NaCac) buffer, pH 7.4, postfixed in 2% osmium tetroxide in NaCac, stained en bloc with 2% uranyl acetate, dehydrated with a graded ethanol series and embedded in Epon-Araldite resin. Thin sections were cut with a diamond knife on a Leica EM UC6 ultramicrotome (Leica Microsystems, Inc., Bannockburn, IL), collected on copper grids and stained with uranyl acetate and lead citrate. Cells were observed in a JEM 1230 transmission electron microscope (JEOL USA Inc., Peabody, MA) at 110 kV and imaged with an Ultras can 4000 CCD camera & First Light Digital Camera Controller (Gating Inc., Pleasanton, CA).

Statistical methods

For 4-hydroxynonenal (4-HNE), we attempted to fit to a general linear model using age and dose as predictors and 4-HNE scores as the outcome. The homogeneity assumption was not met, and a logarithmic transformation of the outcome was therefore applied. HZE particle radiation and γ-ray cohorts were analyzed separately. For RNA analysis, we again fitted to a general linear model using age and dose as predictors and $\Delta\Delta C_t$ values as the outcome. For the HZE cohort, assumptions were met for the PPARGC1A, SOD2, CDKN1A, SIRT3, and PTGES genes. For the γ-ray cohort, assumptions were met for the PPARGC1A, CAT, CDKN1A, and PTGES genes. For analysis of mitochondrial ultrastructure, differences in area were evaluated by ANOVA and then by a Dunnett's test comparing different dose groups with the non-irradiated control. The fraction of abnormal mitochondria was evaluated by a Chi-square test. For necrotic cysts, data from different time points were pooled and evaluated by ordinal logistic regression. In all cases the unit of analysis was single animals or tissue samples derived from single animals.

Results

Experimental design

The experimental design was inspired by an early study of X-ray effects [37] and is summarized in Figure 1A. Medaka embryos were staged and randomized among experimental groups as described in Materials and Methods. Pools of embryos received a single exposure to HZE particle radiation or to reference γ-rays (n≈250 per dose group). Irradiation with HZE particles (1000 MeV/u ^{56}Fe nuclei) was performed at the National Space Radiation Laboratory in Brookhaven, New York, using a range of doses from 0.1 Gy to 9.0 Gy in 3-fold increments. A separate cohort was irradiated with γ-rays using a ^{137}Cs irradiator, using a range of doses from 0.1 Gy to 27 Gy in 3-fold increments. Populations were then observed and sampled at intervals until the median lifespan was exceeded. Measured endpoints included survival, growth, and histological and molecular changes in the liver.

Effect of radiation exposure on growth and survival

Following irradiation, embryos were allowed to hatch out and survival was monitored for a minimum of 750 days, until about 80% mortality had been reached in all groups, including the non-irradiated controls. There was acute mortality in the 9 Gy HZE group and in the 27 Gy γ-ray group. All of the lower dose groups showed mortality curves similar to those for the non-irradiated controls. We observed some normal peri-hatching mortality, a plateau in survival that lasted about one year, and a period of steady decline in the number of survivors that lasted until the end of the experiment (Figures 1B and 1C). The median lifespan for individuals that survived the peri-hatching period was in the 500–550 day range, which is similar to that observed previously for medaka reared in our laboratory [32].

We did not observe gross developmental abnormalities or changes in swimming or feeding behavior in any of the sublethally irradiated groups. There were also no statistically significant effects

A

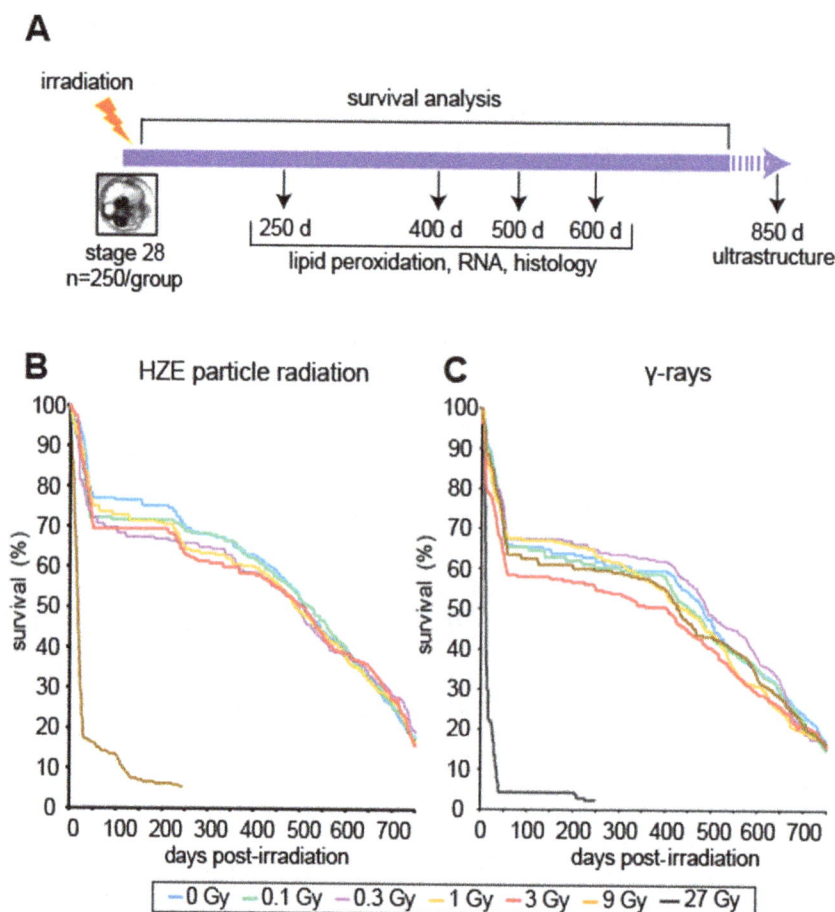

B HZE particle radiation **C** γ-rays

0 Gy — 0.1 Gy — 0.3 Gy — 1 Gy — 3 Gy — 9 Gy — 27 Gy

Figure 1. Experimental design and radiation survival. A. Embryos were irradiated with 1 GeV/u ^{56}Fe ions or with γ-rays, reared, and scored for survival as described in Materials and Methods. At the indicated times, male fish from each dose group were sacrificed and the livers analyzed. B, C. Kaplan-Meier survival curves for fish exposed as embryos to indicated doses and types of radiation.

of sublethal irradiation on weight or length of individuals sampled at intervals from the population (Figure S1).

Effect of age and radiation exposure on a marker of persistent oxidative stress

Given the prior findings that HZE particle exposure is associated with oxidative stress, we hypothesized that outwardly normal individuals might nevertheless exhibit changes in oxidative stress-related biomarkers. At 250, 400, 500, and 600 days post-irradiation, we sampled eight individuals from each dose group and harvested livers for histologic and molecular analysis.

We evaluated levels of 4-hydroxy-nonenal, an end product of lipid peroxidation and a biomarker of chronic oxidative stress. We performed quantitative immunohistochemistry as described in Materials and Methods. Representative primary images of liver sections from control and irradiated individuals are shown in Figure 2A. The staining was specific for 4-HNE, as evidenced by the absence of staining when the primary antibody was omitted or when the primary antibody was neutralized with 4-HNE-BSA (Figure S2). For quantification, images were analyzed by Nuance multispectral analysis to separate anti-4-HNE staining from the hematoxylin counterstain. False-color renderings generated from the Nuance analysis are shown below the corresponding bright field images (Figure 2B).

Values obtained from the Nuance analysis are plotted in Figure 2C. Some trends are apparent in the raw data. Mean 4-HNE levels increased with age to a similar extent in both the HZE and γ-ray cohorts. In addition, mean 4-HNE levels were higher in all of the HZE-irradiated groups than in the corresponding age-matched control groups. To more rigorously evaluate age and radiation dose as predictors of 4-HNE levels, we performed multiple regression analysis. Residual values were not normally distributed and we therefore applied a logarithmic transformation. For the HZE particle radiation cohort, age and radiation dose were significant predictors. The best fit was obtained with a model of the form:

$$\ln(4HNE) = a_1(age) + a_2(dose) + a_3(dose)^2$$

where a_1, a_2, and a_3 are empirically determined. Parameter estimates, uncertainties, and P values are given in Table 1. To examine the goodness of fit, we analyzed the correlation between actual values of ln(4-HNE) versus those predicted by the model (Figure 2D). Based on the coefficient of determination (R^2), the model explains 26% of the variation in the HZE cohort, and the residuals are randomly distributed.

The form of the model implies that the effects of age and dose are synergistic. That is, additive contributions of the age and dose terms to the logarithmically transformed 4-HNE values translate

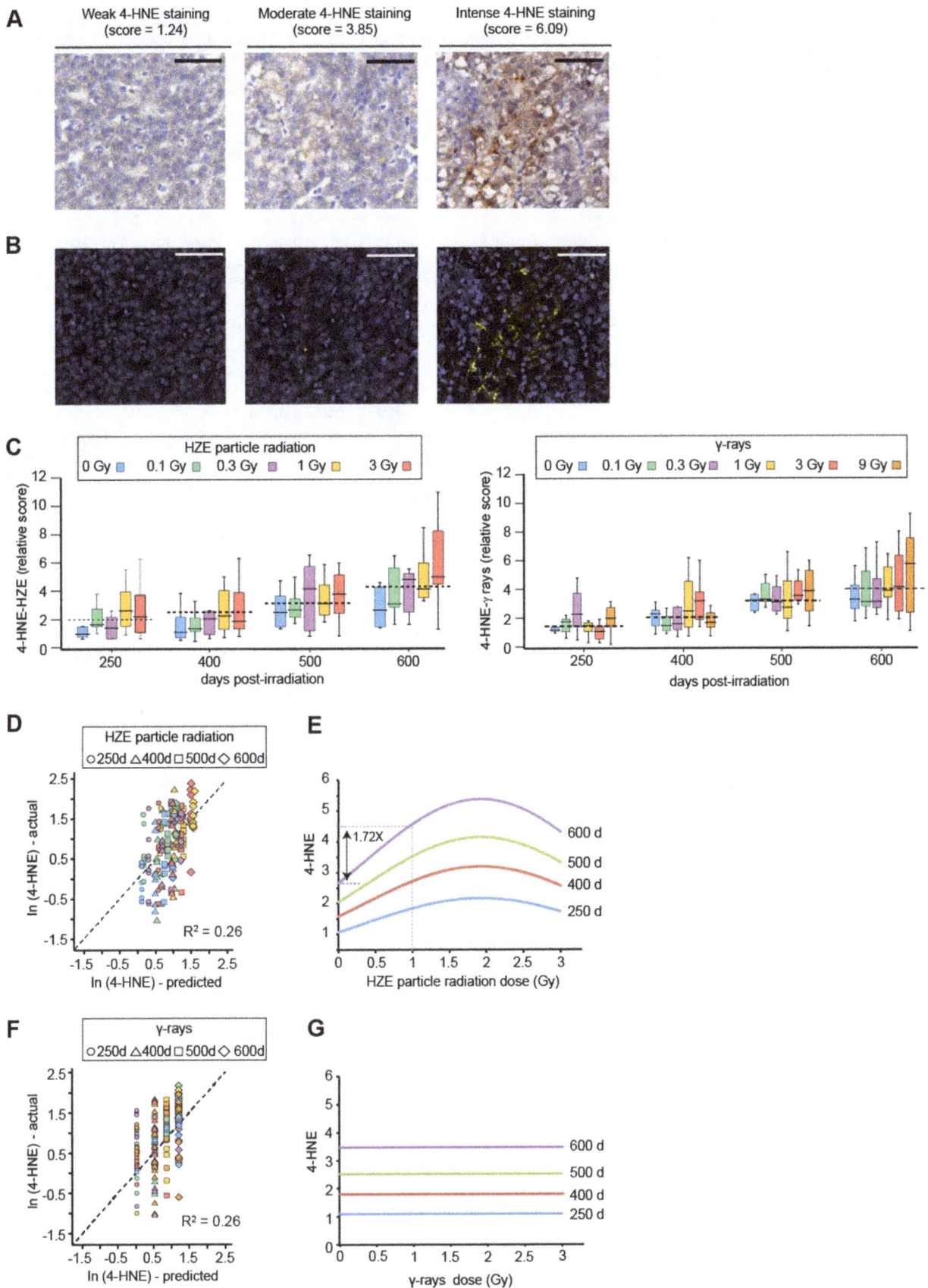

Figure 2. Age and dose-dependent accumulation of 4-hydroxynonenal in liver tissue. Liver sections were stained with anti-4-hydroxy-2-nonenal (4-HNE) and hematoxylin counterstain as described in Materials and Methods. A. Bright-field images showing weak, moderate, or intense anti-4-HNE immunohistochemical staining. Scale bars = 20 μm. B. Nuance renderings of images from Panel A (yellow, 4-HNE; blue, cell nuclei). C.

Quantification of 4-HNE staining based on Nuance analysis. Box plot shows mean and interquartile ranges. Color denotes dose group. Values are normalized to 0 Gy, 250 day group. D. Plot showing correlation between actual and predicted natural logarithm (ln)-transformed 4-HNE values for HZE cohort. Predicted values are based on regression model incorporating age, dose, and dose-squared parameters. Each symbol represents one individual. Shape denotes age group; color denotes dose group using same values as in Panel B. E. Plot showing predicted 4-HNE values as a function of age and HZE dose. F, G. Same as Panels D, E for γ-ray cohorts. Regression model incorporates an age parameter only; dose parameters were non-significant.

into multiplicative contributions to the real (non-transformed) 4-HNE values. To illustrate the relative contribution of age and dose in practical terms, we plotted 4-HNE values, as predicted by the model, as a function of dose (Figure 2E). In the example depicted in the figure, exposure to 1.0 Gy of HZE particle radiation increases 4-HNE levels by 1.72-fold. Note that because of the multiplicative relationship between age and radiation effects, the fold increment attributable to radiation is the same regardless of age.

We performed a similar analysis the γ-ray cohort. Age was a significant predictor but radiation dose was not. The best fit was thus obtained with a model of the form:

$$\ln(4HNE) = a_1(age)$$

Parameter estimates, uncertainties, and P values are again given in Table 1. To examine the goodness of fit, we again analyzed the correlation between actual values of ln(4-HNE) versus those predicted by the model (Figure 2F). Based on the coefficient of determination (R^2), the model explains 26% of the variation. To illustrate the model in practical terms, we again plotted 4-HNE levels versus dose (Figure 2G). Note that in this representation, the γ-ray model reduces to a series of parallel lines representing different age groups, as the radiation dose term is non-significant.

Effect of age and radiation exposure on PPARGC1A and other mRNAs

It was of interest to investigate potential mechanisms underlying the elevated 4-HNE values in the HZE-exposed cohort. As one approach, we measured changes in mRNA levels for candidate genes that are involved in mitochondrial maintenance, antioxidant defense, growth control, or radiation-induced intercellular signaling. One of these was the medaka ortholog of PPARGC1A, which encodes a transcriptional coactivator, PGC-1α. In mammals, PGC-1α provides a unique link between the DNA damage response, metabolism, and oxidative stress [23]. It promotes expression of genes that are required for mitochondrial biogenesis and maintenance, and it coordinates expression of anti-oxidant defense genes. Declines in expression have been linked to aging and various disease states, reviewed in [38,39]. Expression is also

repressed by p53 as part of the chronic DNA damage response [23].

We hypothesized that a decline in PGC-1α expression might be a factor in radiation-induced oxidative stress. We designed primers specific for the medaka ortholog of PPARGC1A and performed qPCR using medaka ACTB (β-actin) and RPL7 as internal reference genes. Primer sequences and ENSEMBL transcript IDs for these and other genes in the study are shown in Table S1. All primer pairs yielded reverse-transcriptase PCR products showing a single sharp peak (Figure S3) on melting curve analysis.

Relative PPARGC1A mRNA levels for HZE particle radiation and γ-ray cohorts are shown in Figure 3A. There was much more stochastic variation than with the 4-HNE data. However, multiple regression analysis showed that age and radiation dose were significant predictors of $\Delta\Delta C_t$ values for both the HZE and γ-ray cohorts. A model of the form:

$$\Delta\Delta C_t = a_1(age) + a_2(dose)$$

provided the best fit. As $\Delta\Delta C_t$ is related to logarithmically transformed mRNA abundance values, the model again implies a multiplicative, synergistic effect on real (non-transformed) mRNA abundance. Figure 3B and 3D show correlation plots for actual and predicted values for the HZE particle-irradiated and γ-ray cohorts, respectively. The models accounted for 11.5% and 17.1% of the experimental variation in the two cohorts. Parameters, uncertainties, and P values for the model are given in Table 2. As with the 4-HNE data, independently determined age parameters were similar for the HZE particle radiation and γ-ray cohorts. Because the dose parameters were significant for both cohorts, it was possible to estimate a nominal relative biological effectiveness (RBE) value of 2.1. This should, however, be interpreted with caution because of the large uncertainties in the parameter estimates.

To illustrate the relative contributions of age and radiation dose, we plotted the predicted PPARGC1A mRNA levels as a function of dose (Figure 3C and 3E). The magnitude of the age effect was similar to that with 4-HNE, that is, a 2 to 4-fold change over the period from 250 to 600 days post exposure. The magnitude of the dose effect was much smaller, however. A 1 Gy dose of HZE or γ-rays leads to only a 12.3% or 6.1% decline in PPARGC1A

Table 1. Age and dose dependence for 4-HNE.

Predictor	HZE			γ-ray		
	Parameter estimate	Parameter uncertainty	P value	Parameter estimate	Parameter uncertainty	P value
Intercept	−0.582	0.201	0.0044	−0.734	0.184	<0.0001
Age	0.0026	0.0004	<0.0001	0.0033	0.0004	<0.0001
Dose	0.735	0.228	0.0015	NS[a]	NS	NS
Dose squared	−0.191	0.072	0.0091	NS	NS	NS

[a]Not significant.
Units for age are days; units for dose are Gy. Parameters are for ln-transformed data.

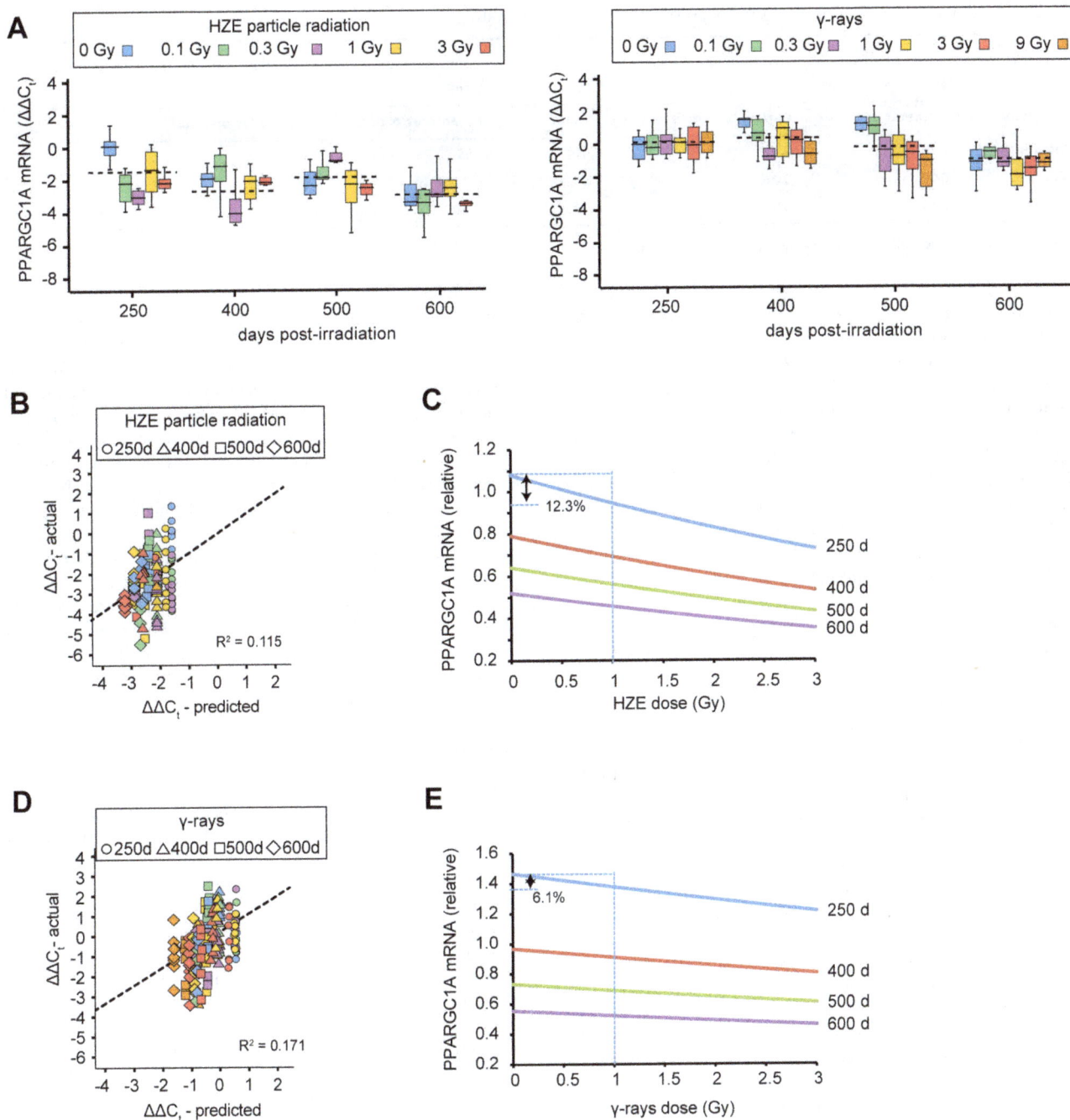

Figure 3. Age and dose-dependent decline in PPARGC1A mRNA in liver tissue. A. Quantification of PPARGC1A mRNA. Box plot shows mean and interquartile ranges. Color denotes dose group. Values are normalized to 0 Gy, 250 day group. B. Plot showing correlation between actual and predicted PPARGC1A values. Each symbol represents one individual. Shape denotes age group, color denotes dose group using same values as in Panel A. Plot showing predicted PPARGC1A values as a function of age and HZE dose. Note that vertical axis shows relative mRNA amounts (i.e., back-transformed from $\Delta\Delta C_t$ values). D, E, same as Panels B, C for γ-ray cohorts.

expression, respectively. Thus, although a decrease in PPARGC1A may contribute to increased oxidative stress, it seems likely there may be other, unidentified genes to account for the magnitude of the oxidative stress phenotype.

In addition to PPARGC1A, we investigated the effects of age and radiation dose on several other genes that are involved in mitochondrial maintenance, anti-oxidant defense, growth control, or radiation-induced intercellular signaling (Figure S4). The largest effects of radiation dose were seen with CDKN1A, which encodes

the p21 (WAF1/CIP1) protein, a potent cyclin-dependent kinase inhibitor. CDKN1A is a prototypical target of p53 activation and a marker of cell senescence (reviewed in [40,41]). Expression of CDKN1A decreased (rather than increased) with dose in both the HZE particle radiation and γ-ray cohorts, with a nominal RBE of 3.7 (Figure S5). The results suggest that developmental exposure to radiation is associated with neither a chronic p53-dependent DNA damage response nor with widespread senescence. In contrast to

Table 2. Age and dose dependence for expression of PPARGC1A mRNA.

	HZE			γ-ray		
Predictor	Parameter estimate	Parameter uncertainty	P value	Parameter estimate	Parameter uncertainty	P value
Intercept	−0.86	0.35	0.014	1.55	0.30	0.0001
Age	−0.003	0.0007	<0.0001	−0.004	0.0006	<0.0001
Dose	−0.19	0.09	0.03	−0.09	0.03	0.0005

Units for age are days; units for dose are Gy. Parameters are for $\Delta\Delta C_t$.

PPARGC1A, the effects of age on CDKN1A expression were small and inconsistent between cohorts,

Other genes tested included SOD2 (superoxide dismutase) and CAT (catalase), which are oxidative stress defense genes; SIRT3 (Sirtuin 3), a mitochondrial histone deacetylase that is protective against aging and cellular damage (reviewed in [42]); and PTGES (prostaglandin E synthase), which catalyzes a step in biosynthesis of prostaglandin E2, a mediator of radiation bystander signaling that is induced at the protein level by HZE irradiation in mice [43]. In some cases, regression analysis showed significant effects of age or radiation dose (Table S2) although the effect sizes were small and the biological significance is uncertain.

Change in mitochondrial morphology

The age and radiation-related increases in persistent oxidative stress seen in Figure 2, and the declines in PPARGC1A expression seen in Figure 3, led us to hypothesize that mitochondria might be compromised in HZE-irradiated individuals. To test this, we sampled the few long-term HZE radiation survivors that remained after other experiments were complete and prepared liver samples for ultrastructural analysis.

Three male fish each from the 0 Gy, 1 Gy, and 3 Gy groups were analyzed. Representative electron micrographs of liver sections are shown in Figure 4. Whereas samples from the 0 Gy groups showed predominantly normal mitochondrial size and shape (panel A), most of the samples from the 1 Gy and 3 Gy groups showed grossly enlarged or elongated mitochondria, sometimes with disorganized, circular cristae (panels B and C). Mitochondrial area was about 1.4-fold higher in the 1 Gy group and nearly 2-fold higher in the 3 Gy group, relative to non-irradiated controls (panel D). As an alternative method of analysis, we also determined the fraction of mitochondria with an area of more than 25 μm^2. The 1 Gy and 3 Gy groups had a significantly higher fraction of these severely enlarged mitochondria (panel D).

Presence of distinctive, necrotic cysts in livers from HZE cohort

We anticipated that radiation exposure might lead to liver cancer, particularly as the medaka liver is an established model for chemical carcinogenesis [44]. However, there were no histologically verified liver tumors in approximately 350 specimens examined. Indeed, among the 2000+ individuals observed in the lifetime study, only 2–3 animals were seen with grossly evident tumors. These sporadic tumors, which in most cases occurred near the eyes, were not analyzed further.

We did observe that about one third of the fish sampled for histology from the HZE-irradiated groups showed distinctive, necrotic cysts. Examples of these are shown in Figure 5A. Whereas normal liver is composed of a regular pattern of hepatocytes, separated by sinusoidal spaces, the cystic regions consisted of empty areas, up to 100 μm in diameter, surrounded by necrotic cells. A summary showing the incidence and severity of lesions in each HZE dose and age group are shown in Figure 5B. For statistical analysis, we pooled the data for different time points and for different severities and performed an ordinal logistic regression (Figure 5C). There was a significant increase over the non-irradiated controls at all doses ≥0.3 Gy. Above this threshold, incidence was dose-independent over a 10-fold range. Necrotic cysts were not present at significant levels in the γ-ray exposed groups (Fig. 5D and 5E).

Spongiosis hepatis is another type of cystic lesion that occurs spontaneously in adult medaka [32,45]. It differs from the necrotic cysts in that the hepatic skeleton remains intact and the area surrounding the lesion remains normal. Some incidence of spongiosis hepatis was seen in most experimental groups, and in only one instance was the incidence significantly elevated relative to controls (0.3 Gy γ-rays) (Figure S5).

Discussion

We report here the acute and long-term effects of HZE particle and γ-ray exposure in a laboratory model organism, the Japanese medaka fish. Acute mortality was seen only at the highest doses tested (9 Gy HZE particle radiation and 27 Gy γ-rays). Acute mortality following exposure to 27 Gy of γ-rays is consistent with prior reports of LD_{50} values of 20 and 26 Gy for low-linear energy transfer radiation in the medaka [31,46].

In contrast to the acute effects, we found that long-term effects of HZE particle radiation were seen at lower doses, where a single exposure, early in life, led to a persistent increase in oxidative stress. Elevated levels of a quantifiable biomarker, 4-HNE, indicated that oxidative damage was present. Multiple regression analysis indicated synergy between radiation exposure and chronological age as predictors of 4-HNE levels. An additional, qualitative indicator of persistent oxidative stress was the abnormal mitochondrial ultrastructure observed in aged, HZE-exposed individuals. Mitochondrial homeostasis is maintained by a cycle of fission and fusion. In mammals, oxidative stress has been shown to perturb this cycle, leading to enlarged and elongated morphology [47]. In our HZE-exposed specimens, we observed similar, bizarrely elongated and enlarged mitochondria, evident more than two years after the original exposure.

The synergistic age and radiation-dependent decline in PPARGC1A mRNA may be one cause of the observed oxidative stress. PPARGC1A is a master regulator of genes involved in mitochondrial maintenance and defense against oxidative stress [48]. The decreased levels are suggestive of decreased mitochondrial function and inability to effectively detoxify reactive oxygen species. Of the candidate genes tested, the only other one to show such a marked, dose-dependent decline was CDKN1A, a classically TP53-inducible gene and a senescence marker.

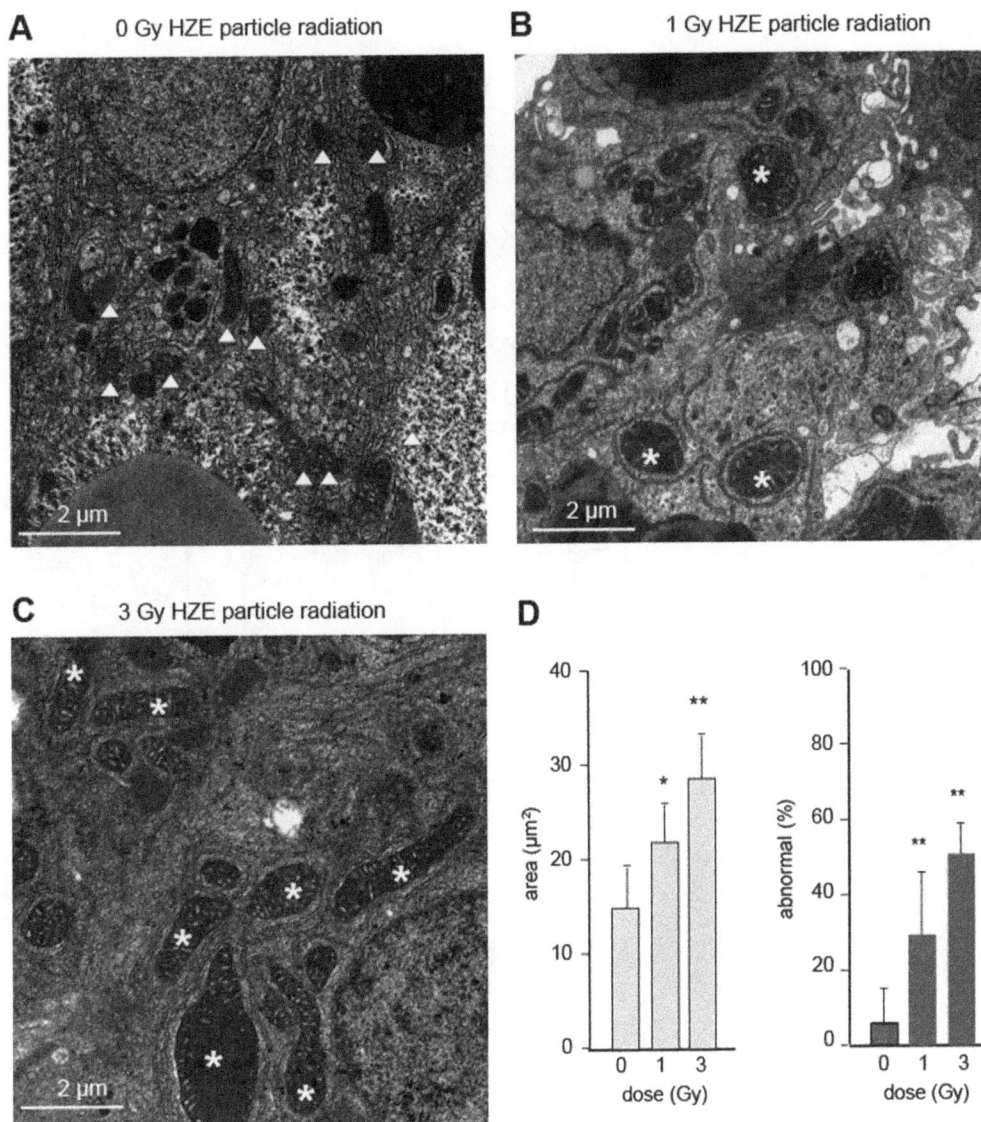

Figure 4. Analysis of mitochondrial ultrastructure. Panels A–C show relatively normal mitochondria (triangles) from a non-irradiated individual and two examples of enlarged and elongate mitochondria (*) from HZE-exposed groups. D. Mitochondrial area as a function of HZE dose (left panel). Percent of elongated mitochondria as a function of dose (right panel).

Although we did not have access to species cross-reactive antibodies that could be used to measure TP53 protein levels directly, the decline (rather than increase) in CDKN1A expression argues against the presence of activated p53 in the aging, irradiated populations. There were some other genes that showed significant radiation responses, although effect sizes were small and biological significance is uncertain.

Our best-fit model for the effect of HZE radiation exposure on 4-HNE levels assumed a nonlinear dose-response curve, that is, a curve that bends over at the highest doses, whereas the best-fit models for the PPARGC1A RNA data assumed a linear dose-response relationship. This apparent difference may or may not be biologically meaningful, as the radiation effect on PPARGC1A was smaller than the effect on 4-HNE, and we may not have had sufficient power to distinguish between linear and nonlinear models. The nonlinearity in the 4-HNE data primarily affects the predicted response at doses above the range of anticipated human exposure, and so may not be of practical significance.

We also observed distinctive necrotic cysts, but not liver cancer, in the HZE-exposed fish. They occurred in all HZE-exposed groups and did not show a significant dose response relationship. We did not see these lesions in non-irradiated fish or in the γ-ray cohort, or in our previous study of aging in the medaka [32]. While the cystic degeneration was a very notable feature of the HZE-exposed group, we do not at this time have a clear understanding of its etiology; perhaps it is an emergent phenomenon reflecting the various kinds of stress imposed by HZE exposure.

To investigate differences between the effects of HZE particle radiation and γ-rays, we exposed a separate cohort of medaka embryos to γ-radiation. It was necessary to establish the two cohorts sequentially, rather than simultaneously, because of the logistical difficulty in handling large numbers of embryos simultaneously. The γ-ray cohort was established about six weeks after the HZE cohort and was drawn from the same stocks. The two cohorts had approximately the same median lifespan and

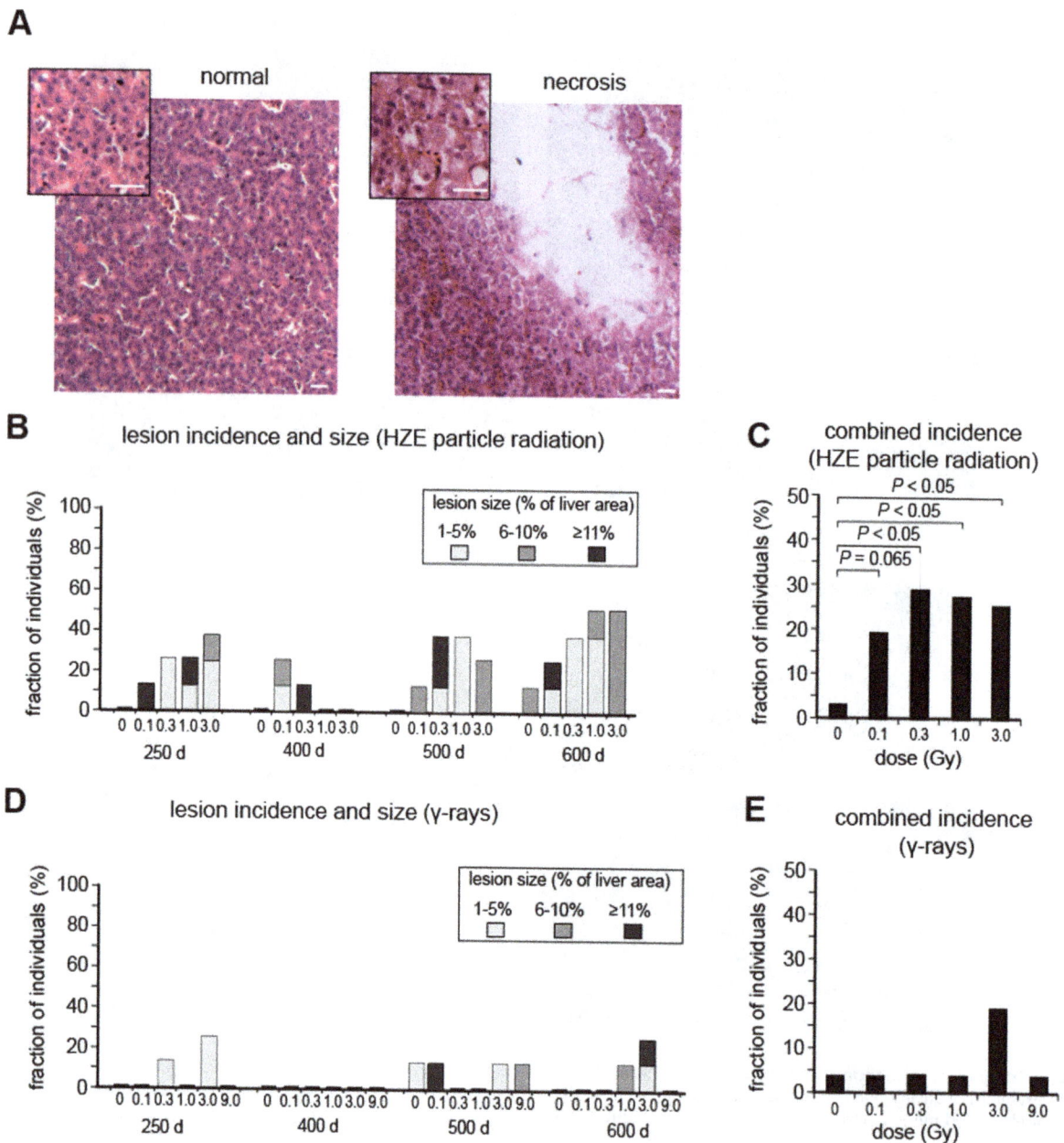

Figure 5. Necrotic cysts in livers of radiation-exposed individuals. A. Representative hematoxylin and eosin stained sections. The left panel shows normal liver, and the right panel shows a necrotic cyst. Insets show regions of each section at higher magnification. Scale bars are 20 μm. B. Stacked column graph showing the incidence of necrotic cysts, classified according to the percentage area of the liver that was affected. C. Pooled data showing incidence of necrotic cysts at different doses of HZE particle radiation. Lesions of different severity were combined and classified as abnormal. Different age groups were also combined. P values are shown based on ordinal logistic regression. D. Stacked column graph, as in Panel B but for γ-ray exposed groups. E. Pooled data showing incidence of necrotic cysts at different doses of γ-rays.

showed a similar age-dependent increase in 4-HNE levels and age-dependent decline in PPARGC1A mRNA. Thus, we regard them as biologically comparable. There was some radiation dependent increase in 4-HNE levels in the γ-ray cohort, but it was not statistically significant in the regression analysis. There were also radiation-dependent declines in PPARGC1A and CDKN1A, although these were smaller than for the HZE radiation-exposed cohort.

Quantitative and qualitative differences in HZE particle radiation and γ-rays presumably reflect the distinctive physics of tissue interaction. HZE particles produce a dense burst of reactive oxygen species along a nanometer-scale core track, whereas γ-rays, at the same dose, deposit energy along more numerous but less densely ionizing tracks. Dense ionization along the core HZE track leads to potentially irreparable DNA damage [12,49]. At the same time, radial propagation of secondary electrons produces damage at sites elsewhere in the target cell or in neighboring cells. Thus, a single encounter with an HZE particle thus creates damage that is simultaneously denser and more widespread, potentially affecting DNA and the mitochondria simultaneously and in different ways.

We speculate that this may produce a sufficient burst of damage from which cells never fully recover – an initial burst of reactive oxygen species leads to a self-perpetuating cycle of mitochondrial injury, leakage of endogenous reactive oxygen species, and further damage to mitochondria or other cellular components.

The medaka model has some limitations, notably the paucity of species cross-reactive antibodies. We were thus not able to measure levels of TP53 or phosphorylated ATM proteins, which would have provided direct information regarding the presence of a chronic DNA damage response. We were not able to measure 8-oxodeoxyguanosine, a major base oxidation product, because of high background staining. We were not able to detect γ-H2AX, a marker of unrepaired DNA double-strand breaks. It may be that examination of mRNA expression profiles in greater depth, for example by deep sequencing, will provide further information about the activation of DNA damage response pathways, but results of such studies are not available at the present time.

Conclusions

A main rationale for animal studies of HZE particle radiation effects is the need to better understand human risk. Results here show that effects of HZE particle radiation can persist over the lifetime of an organism. These persistent effects occur at doses representative of the range of anticipated human exposure.

Supporting Information

Figure S1 Comparison of growth rates in different experimental groups. Eight male fish were sampled from each dose group and time point. Colors denote dose as shown in key. A. HZE particle cohort, body length and body weight as indicated. B. γ-ray cohort, body length and weight as indicated.

Figure S2 Controls for 4-HNE staining. Liver sections were stained with anti-4-HNE and hematoxylin counterstain as described in Materials and Methods. Bright-field images are shown. Scale bars = 20 μm. Left panel, staining under normal conditions; center panel, primary antibody omitted; right panel, primary antibody pre-incubated with 2 ng/μl 4-HNE BSA. Note absence of staining when primary antibody was omitted or pre-incubated with antigen.

Figure S3 Melting curves for PCR products derived from primer pairs used in this study. Melting curve analysis was performed for PCR products of each gene were analyzed, as indicated. Panels show fluorescence as a function of temperature. The first derivative of the melting curve is superimposed. Colors denote results with two independent samples. Note the sharp melting transition seen with each product. Genes are grouped by functional category as indicated.

Figure S4 Quantification of candidate mRNAs other than PPARGC1A. Box plot shows mean and interquartile ranges. Color denotes dose group as shown in key. Values are normalized to 0 Gy, 250 day group. Genes are grouped by functional category as indicated. Left panels, HZE particle radiation, right panels, γ-rays. Note that CDKN1A shows a decline in HZE particle-irradiated individuals (in almost all cases, irradiated groups show lower mean expression than age-matched control groups; see Table S2 for regression analysis). There was a smaller, but significant dose-dependent decline for γ-rays. Although age or dose were statistically significant predictors for some other genes, the magnitude of the effects were small and in some cases inconsistent between HZE radiation and γ-ray cohorts.

Figure S5 Spongiosis hepatis in in livers of radiation-exposed individuals. A. Representative hematoxylin and eosin stained section showing spongiosis hepatis; compare with normal and with necrotic cysts in Fig. 5 of main text. Inset shows region at higher magnification. Scale bars are 20 μm. B. Stacked column graph showing the incidence and size of regions of spongiosis in HZE particle radiation-exposed cohort. C. Pooled data showing incidence of spongiosis at different doses of HZE particle radiation. Lesions of different severity were combined and classified as abnormal. Different age groups were also combined. P values are shown based on ordinal logistic regression. D, E. Same as Panels B, C for γ-ray cohort.

Table S1 Primer pairs used in this study. Functional category, gene symbol, forward and reverse primer sequences, and ENSEMBL transcript identifiers are shown.

Table S2 Age and dose dependence for expression of select mRNAs. Table provides results of regression models based on quantification of mRNAs shown in Fig. S5. Category, gene symbol, parameter values and uncertainties, and P values are indicated. Parameter values are omitted where assumptions are violated for a univariate model.

Author Contributions

Conceived and designed the experiments: XZ LD JRL PMW WSD. Performed the experiments: XZ JRL PMW. Analyzed the data: XZ XYZ LD JRL PMW WSD. Contributed reagents/materials/analysis tools: XYZ PMW. Wrote the paper: XZ XYZ WSD.

References

1. National Research Council (2008) Managing Space Radiation Risk in the New Era of Space Exploration: The National Academies Press.
2. Shukitt-Hale B, Casadesus G, McEwen JJ, Rabin BM, Joseph JA (2000) Spatial learning and memory deficits induced by exposure to iron-56-particle radiation. Radiat Res 154: 28–33.
3. Weil MM, Bedford JS, Bielefeldt-Ohmann H, Ray FA, Genik PC, et al. (2009) Incidence of acute myeloid leukemia and hepatocellular carcinoma in mice irradiated with 1 GeV/nucleon (56)Fe ions. Radiat Res 172: 213–219.
4. Villasana L, Rosenberg J, Raber J (2010) Sex-dependent effects of 56Fe irradiation on contextual fear conditioning in C57BL/6J mice. Hippocampus 20: 19–23.
5. Alwood JS, Yumoto K, Mojarrab R, Limoli CL, Almeida EA, et al. (2010) Heavy ion irradiation and unloading effects on mouse lumbar vertebral microarchitecture, mechanical properties and tissue stresses. Bone 47: 248–255.
6. Yumoto K, Globus RK, Mojarrab R, Arakaki J, Wang A, et al. (2010) Short-term effects of whole-body exposure to (56)fe ions in combination with musculoskeletal disuse on bone cells. Radiat Res 173: 494–504.
7. Yu T, Parks BW, Yu S, Srivastava R, Gupta K, et al. (2011) Iron-ion radiation accelerates atherosclerosis in apolipoprotein E-deficient mice. Radiat Res 175: 766–773.
8. Cherry JD, Liu B, Frost JL, Lemere CA, Williams JP, et al. (2012) Galactic cosmic radiation leads to cognitive impairment and increased A beta plaque accumulaton in a mouse model of Alzheimer's disease. PLoS One 7: e53275.
9. Lonart G, Parris B, Johnson AM, Miles S, Sanford LD, et al. (2012) Executive function in rats is impaired by low (20 cGy) doses of 1 GeV/u (56)Fe particles. Radiat Res 178: 289–294.
10. Britten RA, Davis LK, Johnson AM, Keeney S, Siegel A, et al. (2012) Low (20 cGy) doses of 1 GeV/u (56)Fe–particle radiation lead to a persistent reduction in the spatial learning ability of rats. Radiat Res 177: 146–151.

11. Bielefeldt-Ohmann H, Genik PC, Fallgren CM, Ullrich RL, Weil MM (2012) Animal studies of charged particle-induced carcinogenesis. Health Phys 103: 568–576.

12. Azzam EI, Jay-Gerin JP, Pain D (2012) Ionizing radiation-induced metabolic oxidative stress and prolonged cell injury. Cancer Lett 327: 48–60.

13. Li M, Gonon G, Buonanno M, Autsavapromporn N, de Toledo SM, et al. (2013) Health Risks of Space Exploration: Targeted and Nontargeted Oxidative Injury by High-Charge and High-Energy Particles. Antioxid Redox Signal.

14. Limoli CL, Giedzinski E, Baure J, Rola R, Fike JR (2007) Redox changes induced in hippocampal precursor cells by heavy ion irradiation. Radiat Environ Biophys 46: 167–172.

15. Poulose SM, Bielinski DF, Carrihill-Knoll K, Rabin BM, Shukitt-Hale B (2011) Exposure to 16O-particle radiation causes aging-like decrements in rats through increased oxidative stress, inflammation and loss of autophagy. Radiat Res 176: 761–769.

16. Gonon G, Groetz JE, de Toledo SM, Howell RW, Fromm M, et al. (2013) Nontargeted stressful effects in normal human fibroblast cultures exposed to low fluences of high charge, high energy (HZE) particles: kinetics of biologic responses and significance of secondary radiations. Radiat Res 179: 444–457.

17. Datta K, Suman S, Kallakury BV, Fornace AJ Jr. (2012) Exposure to heavy ion radiation induces persistent oxidative stress in mouse intestine. PLoS One 7: e42224.

18. Tseng BP, Giedzinski E, Izadi A, Suarez T, Lan ML, et al. (2013) Functional Consequences of Radiation-Induced Oxidative Stress in Cultured Neural Stem Cells and the Brain Exposed to Charged Particle Irradiation. Antioxid Redox Signal.

19. Kim GJ, Fiskum GM, Morgan WF (2006) A role for mitochondrial dysfunction in perpetuating radiation-induced genomic instability. Cancer Res 66: 10377–10383.

20. Limoli CL, Giedzinski E, Morgan WF, Swarts SG, Jones GD, et al. (2003) Persistent oxidative stress in chromosomally unstable cells. Cancer Res 63: 3107–3111.

21. Kang MA, So EY, Simons AL, Spitz DR, Ouchi T (2012) DNA damage induces reactive oxygen species generation through the H2AX-Nox1/Rac1 pathway. Cell Death Dis 3: e249.

22. Sun C, Wang Z, Liu Y, Liu Y, Li H, et al. (2014) Carbon ion beams induce hepatoma cell death by NADPH oxidase-mediated mitochondrial damage. J Cell Physiol 229: 100–107.

23. Sahin E, Colla S, Liesa M, Moslehi J, Muller FL, et al. (2011) Telomere dysfunction induces metabolic and mitochondrial compromise. Nature 470: 359–365.

24. Buonanno M, de Toledo SM, Pain D, Azzam EI (2011) Long-term consequences of radiation-induced bystander effects depend on radiation quality and dose and correlate with oxidative stress. Radiat Res 175: 405–415.

25. Villasana LE, Rosenthal RA, Doctrow SR, Pfankuch T, Zuloaga DG, et al. (2013) Effects of alpha-lipoic acid on associative and spatial memory of sham-irradiated and 56Fe-irradiated C57BL/6J male mice. Pharmacol Biochem Behav 103: 487–493.

26. Zeitlin C, Hassler DM, Cucinotta FA, Ehresmann B, Wimmer-Schweingruber RF, et al. (2013) Measurements of energetic particle radiation in transit to Mars on the Mars Science Laboratory. Science 340: 1080–1084.

27. Wittbrodt J, Shima A, Schartl M (2002) Medaka–a model organism from the far East. Nat Rev Genet 3: 53–64.

28. Furutani-Seiki M, Wittbrodt J (2004) Medaka and zebrafish, an evolutionary twin study. Mech Dev 121: 629–637.

29. Takeda H, Shimada A (2010) The art of medaka genetics and genomics: what makes them so unique? Annu Rev Genet 44: 217–241.

30. Egami N, Etoh H (1966) Effect of temperature on the rate of recovery from radiation-induced damage in the fish Oryzias latipes. Radiat Res 27: 630–637.

31. Kuhne WW, Gersey BB, Wilkins R, Wu H, Wender SA, et al. (2009) Biological effects of high-energy neutrons measured in vivo using a vertebrate model. Radiat Res 172: 473–480.

32. Ding L, Kuhne WW, Hinton DE, Song J, Dynan WS (2010) Quantifiable biomarkers of normal aging in the Japanese medaka fish (Oryzias latipes). PLoS One 5: e13287.

33. Iwamatsu T (2004) Stages of normal development in the medaka Oryzias latipes. Mech Dev 121: 605–618.

34. Bladen CL, Lam WK, Dynan WS, Kozlowski DJ (2005) DNA damage response and Ku80 function in the vertebrate embryo. Nucleic Acids Res 33: 3002–3010.

35. Zhang Z, Hu J (2007) Development and validation of endogenous reference genes for expression profiling of medaka (Oryzias latipes) exposed to endocrine disrupting chemicals by quantitative real-time RT-PCR. Toxicol Sci 95: 356–368.

36. Vandesompele J, De Preter K, Pattyn F, Poppe B, Van Roy N, et al. (2002) Accurate normalization of real-time quantitative RT-PCR data by geometric averaging of multiple internal control genes. Genome Biol 3: RESEARCH0034.

37. Egami N, Eto H (1973) Effect of x-irradiation during embryonic stage on life span in the fish, Oryzias latipes. Exp Gerontol 8: 219–222.

38. Austin S, St-Pierre J (2012) PGC1alpha and mitochondrial metabolism–emerging concepts and relevance in ageing and neurodegenerative disorders. J Cell Sci 125: 4963–4971.

39. Johri A, Chandra A, Beal MF (2013) PGC-1alpha, mitochondrial dysfunction, and Huntington's disease. Free Radic Biol Med 62: 37–46.

40. Jung YS, Qian Y, Chen X (2010) Examination of the expanding pathways for the regulation of p21 expression and activity. Cell Signal 22: 1003–1012.

41. Sperka T, Wang J, Rudolph KL (2012) DNA damage checkpoints in stem cells, ageing and cancer. Nat Rev Mol Cell Biol 13: 579–590.

42. Tao R, Vassilopoulos A, Parisiadou L, Yan Y, Gius D (2014) Regulation of MnSOD enzymatic activity by Sirt3 connects the mitochondrial acetylome signaling networks to aging and carcinogenesis. Antioxid Redox Signal 20: 1646–1654.

43. Cheema AK, Suman S, Kaur P, Singh R, Fornace AJ Jr, et al. (2014) Long-Term Differential Changes in Mouse Intestinal Metabolomics after gamma and Heavy Ion Radiation Exposure. PLoS One 9: e87079.

44. Okihiro MS, Hinton DE (1999) Progression of hepatic neoplasia in medaka (Oryzias latipes) exposed to diethylnitrosamine. Carcinogenesis 20: 933–940.

45. Boorman GA, Botts S, Bunton TE, Fournie JW, Harshbarger JC, et al. (1997) Diagnostic criteria for degenerative, inflammatory, proliferative nonneoplastic and neoplastic liver lesions in medaka (Oryzias latipes): consensus of a National Toxicology Program Pathology Working Group. Toxicol Pathol 25: 202–210.

46. Egami N, Eto H (1962) Dose-survival time relationship and protective action of reserpine against X-irradiation in the fish, Oryzias latipes. Ann Zool Japan 35: 188–198.

47. Chan DC (2012) Fusion and fission: interlinked processes critical for mitochondrial health. Annu Rev Genet 46: 265–287.

48. St-Pierre J, Drori S, Uldry M, Silvaggi JM, Rhee J, et al. (2006) Suppression of reactive oxygen species and neurodegeneration by the PGC-1 transcriptional coactivators. Cell 127: 397–408.

49. Autsavapromporn N, De Toledo SM, Buonanno M, Jay-Gerin JP, Harris AL, et al. (2011) Intercellular communication amplifies stressful effects in high-charge, high-energy (HZE) particle-irradiated human cells. J Radiat Res 52: 408–414.

2

The Impact of Curtin University's Activity, Food and Attitudes Program on Physical Activity, Sedentary Time and Fruit, Vegetable and Junk Food Consumption among Overweight and Obese Adolescents: A Waitlist Controlled Trial

Leon M. Straker[1]*, Erin K. Howie[2], Kyla L. Smith[2], Ashley A. Fenner[3], Deborah A. Kerr[4], Tim S. Olds[5], Rebecca A. Abbott[2], Anne J. Smith[2]

1 School of Physiotherapy and Exercise Science, Curtin University, GPO Box U1987, Perth, Western Australia, 6845, Australia, 2 School of Physiotherapy and Exercise Science, Curtin University, Perth, Australia, 3 School of Psychology and Speech Pathology, Curtin University, Perth, Australia, 4 School of Public Health, Curtin University, Perth, Australia, 5 Health and Use of Time (HUT) Group, University of South Australia, Adelaide, Australia

Abstract

Background: To determine the effects of participation in Curtin University's Activity, Food and Attitudes Program (CAFAP), a community-based, family-centered behavioural intervention, on the physical activity, sedentary time, and healthy eating behaviours of overweight and obese adolescents.

Methods: In this waitlist controlled clinical trial in Western Australia, adolescents (n = 69, 71% female, mean age 14.1 (SD 1.6) years) and parents completed an 8-week intervention followed by 12 months of telephone and text message support. Assessments were completed at baseline, before beginning the intervention, immediately following the intervention, and at 3-, 6-, and 12- months follow-up. The primary outcomes were physical activity and sedentary time assessed by accelerometers and servings of fruit, vegetables and junk food assessed by 3-day food records.

Results: During the intensive 8-week intervention sedentary time decreased by −5.1 min/day/month (95% CI: −11.0, 0.8) which was significantly greater than the rate of change during the waitlist period (p = .014). Moderate physical activity increased by 1.8 min/day/month (95% CI: −0.04, 3.6) during the intervention period, which was significantly greater than the rate of change during the waitlist period (p = .041). Fruit consumption increased during the intervention period (monthly incidence rate ratio (IRR) 1.3, 95% CI: 1.10, 1.56) and junk food consumption decreased (monthly IRR 0.8, 95% CI: 0.74, 0.94) and these changes were different to those seen during the waitlist period (p = .004 and p = .020 respectively).

Conclusions: Participating in CAFAP appeared to have a positive influence on the physical activity, sedentary and healthy eating behaviours of overweight and obese adolescents and many of these changes were maintained for one year following the intensive intervention.

Trial Registration: Australia and New Zealand Clinical Trials Registry ACTRN12611001187932

Editor: Fiona Gillison, University of Bath, United Kingdom

Funding: This trial was funded by a Healthway Health Promotion Research Project Grant #19938. LS was supported by a National Health and Medical Research Council senior research fellowship #APP1019980. KS was supported by an APA/CRS scholarship. The funders had no role in study design, data collection and analysis, decision to publish, or preparation of the manuscript.

Competing Interests: The authors have declared that no competing interests exist.

* Email: L.Straker@curtin.edu.au

Background

Adolescents who are overweight or obese are at greater risk of physical and mental health problems both during adolescence and subsequent adulthood [1,2]. Physical activity, sedentary behaviour and healthy eating behaviours not only contribute to obesity but also have important independent health implications [3]. The primary outcome focus of most interventions for overweight and obese adolescents has been adiposity rather than activity and healthy eating behaviours [4]. Aside from the importance of these behaviours to multiple health issues, including adiposity [3], a focus on weight-related outcomes may have unintended negative psychological consequences [5]. Evidence also suggests that interventions targeting both activity and healthy eating behaviours should be multi-disciplinary and involve families in community settings for sustained change [4,6,7]. Whilst the few studies focussed on behavioural outcomes for overweight and obese adolescents have reported some encouraging findings [8,9,10], they have either lacked assessments of sustained change beyond immediately post-intervention [11,12,13,14] or lacked objective measures of activity [8,9,15]. Additionally, few studies have used detailed dietary assessment methods, such as food records, to describe changes in healthy eating outcomes for adolescents [16].

Curtin University's Activity, Food and Attitudes Program (CAFAP) was a community-based, family-centered behavioural intervention implemented by a multi-disciplinary team of health practitioners. The current study determined the effects of participation in CAFAP on activity and healthy eating behaviours of overweight and obese adolescents and how behaviours were maintained for one year following the intervention. Specific hypotheses tested were:

- Time in sedentary, light, moderate and vigorous physical activity and intake of fruit, vegetables and 'junk food' would change following participation in the CAFAP program and changes would be maintained for up to 12 months

- The rate of change of time in sedentary, light, moderate and vigorous physical activity and intake of fruit, vegetables and 'junk food' over the wait-list control period would differ from the rate of change over the intervention and maintenance periods.

Methods

Study Design

This study was a staggered entry, within-subject, waitlist controlled clinical trial conducted in Western Australia. This design was selected to minimize ethical concerns with withholding treatment in a no treatment control for 18 months [17] and the likely unacceptably high drop-out over multiple assessments over an extended study duration for a no-treatment or minimal standard care control group. Additionally, the within-subject design reduces error variance from individual differences, thus increasing the power with the expected high dropout rates and reduced sample sizes found in previous studies [4,10,13,18]. The trial was registered (Australia and New Zealand Clinical Trials Registry # ACTRN12611001187932) and the protocol published [19]. The protocol for this trial and supporting CONSORT checklist are available as supporting information; see Checklist S1 and Protocol S1. Participants were assessed at six time-points including baseline, 3-months after baseline prior to beginning the intervention (end of waitlist period), immediately following the 8-week intervention (end of intervention period), and at 3-, 6-, and 12-months following the intervention (maintenance periods). Entry into the program was staggered into three waves, beginning in February, May and August 2012 to control for bias from external events and seasonal changes. Follow-up assessments were completed by December 2013. The program was conducted at three sites (two urban and one rural) with a high proportion of low socio-economic status residents. Curtin University Human Research Ethics Committee approved all procedures (HR105/2011). Written informed assent was obtained from adolescents and consent was obtained from parents prior to commencing the study.

Recruitment and Participants

One hundred and twenty three participants enquired about the program from the community following information provided by health professionals, community newspapers and radio media, and distribution of flyers, of which 69 entered the study. Recruitment began in late November 2011 and continued through to August 2012. To be included in the study, participants had BMI-for-age and sex greater than the 85th percentile on the Centers for Disease Control BMI-for-age growth charts [20] and passed a medical screening prior to participation. Participants were excluded if their obesity was related to a diagnosed metabolic, genetic, or endocrine disease, were receiving current treatment for a psychological disorder, or were unable to attend sessions twice weekly at the community locations. Sample size estimates were reported in the protocol paper [21]. Seven cohorts ranging in size from 6 to 13 adolescents and the same number of parents were conducted in three waves. Recruitment ceased after the planned three waves despite smaller than anticipated numbers due to grant funding constraints. Figure 1 shows the flow of participants through the study.

Intervention

This intervention was adapted to be conducted in the community from a previous program conducted in healthcare and university settings. Additional detailed formative work with adolescents, researchers and practitioners guided the development and refinement of the curriculum [22]. In summary, a team of 13 multidisciplinary community practitioners (psychologist, physio-therapist/exercise physiologist, dietitian) were trained to implement the program across the seven cohorts, with one facilitator from each domain working with each cohort typically. The theoretical framework for the intervention was self-determination theory [23] and goal setting theory [24] and is described in detail elsewhere [25]. Both instructors and parents were encouraged to provide a need-supportive environment to increase adolescents' autonomous motivation for physical activity and healthy eating behaviours.

The initial 8-week intervention included 2-hour group sessions twice per week in community locations. Both adolescents and parents participated in all sessions, with each session having separate and joint activities. An outline of the sessions and which facilitators supported each session is provided in Table 1. Phone contact was made with participants who missed a session to increase attendance adherence. At every session, adolescents participated in 45-minutes of enjoyable physical activity and a behavioural component on healthy eating, activity and overcoming barriers (see Table 1). During the CAFAP activity sessions, participants were taught to self-monitor their heart rate and self-perceived level of exertion. The sessions began with a warm up, usually 1–2 group games, to increase the heart rate and body temperature in preparation for the circuit exercises. During the circuit training, participants moved through alternate 'huff and puff' and 'strength' stations. 'Huff and puff' stations, like boxing, were designed to increase the participants' heart rate and increase their level of cardiovascular exertion. 'Strength' stations included a focus on different parts of the body, specifically trunk (e.g. oblique crunches), upper limb (e.g. bicep curls) and lower limb (e.g. squats). Sessions usually ended with another group game and cool-down stretches. Adolescents were encouraged to bring their own music to play for the group during the activity sessions. Adolescents set weekly, manageable goals and parents were guided to set their own goals to support these adolescent goals.

During the 12-month follow-up, participants received structured telephone and text message contact at a decreasing

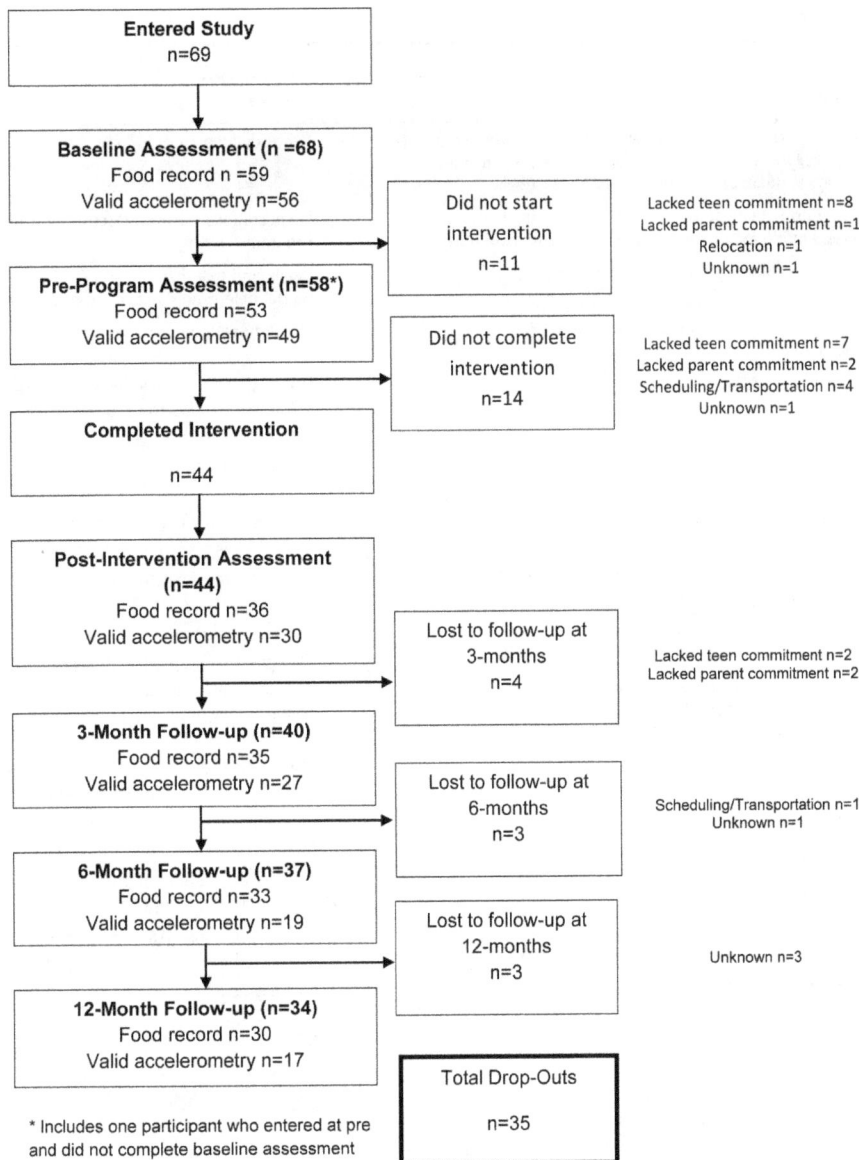

Figure 1. Participant flow diagram for the waitlist controlled trial of Curtin University's Activity, Food, and Attitudes Program.

frequency based on the same theoretical principles and key messages as during the intensive face-to-face contact period. Contact was based on self-determination theory and goal setting theory and focused on eating more fruits and vegetables, eating less junk food, being less sedentary and being more active. The structuring of text messages is described in detail elsewhere [26]. The phone coaching was completed by members of the facilitation/assessment team who were well known to participants and aimed to provide structure, support attempts at change, and promote adolescents' sense of autonomy. A protocol for adverse events was in place, but no adverse events were reported. Program fidelity was assessed on several occasions for each site by independent observation.

Further details about the intervention, including resources for health professionals wishing to conduct similar programs, are provided at the study website: http://cafap.curtin.edu.au/.

Measures

The primary outcomes were time in sedentary, light, moderate and vigorous activity and serves of fruit, vegetables and 'junk food' [21].

Activity. Adolescents were instructed to wear Actical monitors (Respironics; Bend, Oregon, USA) on an elastic band on their right hip for all waking hours for 7 days except for during water-related activities. Data were collected in 15-second epochs. Diary information and visual inspection of the data were used to determine individual daily wear-time. Periods where participants attended CAFAP sessions or assessments were excluded from the data. Accelerometry data were processed using a customised LabView V7 (National Instruments, Austin, TX, USA) program to determine daily minutes of sedentary, light, moderate and vigorous physical activity using intensity cut-points for children [27]. Participants were included in the primary analysis at each assessment period if they had at least three days of at least 8 hours of wear-time [28].

Table 1. Overview of CAFAP sessions.

Session#	Adolescents	Parents
1	[All]Introduction to program – overview of program, introductions, group rules, key messages: be more active, be less inactive, eat more fruit and vegetables, eat less junk food, set goals	[All]Introduction to program – joint session
	[PE]Introduction to Activity sessions – types of activity, benefits, how to assess moderate intensity, warm up, circuit, cool down	[D]Expectations – Parent discussion of their expectations of the program
2	[PE]Activity session – adding stations, review heart rate	[D]Walk and talk (parents and facilitators go for a walk together to model being more active) –topics – get to know each other
	[D]Healthy eating – energy balance, basic nutrition principles: variety and nutrients.	[D]Healthy eating - joint session
3	[PE]Activity session – adding new circuit stations	[P]Understanding adolescence –teens need choice, competence, belonging Parents can provide structure, be involved, support teenager choices
	[PE]Healthy activity – be more active, be less inactive, benefits of being active, activity and inactivity, energy out balance	[PE]Health activity - joint session
4	[PE]Activity session – adding stations	[P]Providing structure – Setting up house rules, monitoring behaviour and observance of rules, consequences for breaking rules.
	[D]Portion size –portion size guidelines, food group intake, eat more fruit and vegetables, eat less junk food	[D]Portion size - joint session
5	[PE]Activity session – increasing speed	Parent introduction to goal setting – setting goals to support teenager goals.
	[P]Teens introduction to goal setting – how to set goals, feedback on current activity and eating behaviours (using pre-intervention assessment information), start to set goals	Walk and talk – topics – get to know each other, how it's going, review house rules
6	[PE]Activity session – increasing speed (30 mins only)	[D]Fast food and dinner- fast food (takeaway) and parent planning for 'fast' dinners at home (4:30–5:30pm)
	[All]Teens setting goals (1.5 hours)	[All]Parents setting goals to support teens (last ½ hour only- with teens)
7	[PE]Activity session (30 mins only)	[P]Review and debrief of progress – supporting teenager activity and food goals and competence
	[PE]Goals- (30 mins) - teens reflect on goal progress, write new weekly goals (20 mins) and share with parents (10 mins)	[P]Goals - (20 mins) - 10 mins to review weekly progress then join with teens for 10 mins
	[PE]Family activity– review of key activity messages, benefits of activity, identification of positive and less positive activity, inactivity & sleep habits of family, opportunities for goals	[PE]Family activity – joint session
8	[PE]Activity session	[D]Walk and talk – topics – parenting issues
	[D]Family food – review of key food messages, identification of positive and less positive food habits of family members, encourage positive eating behaviours to use in goal setting	[D]Family Food – joint session
9	[PE]Activity session – (30 mins)	[P]Parent-teen relationships – parenting styles to provide structure, be involved, support teenager choices, maintaining good relationships.
	[PE]Goals – (30 mins) - teens reflect on goal progress, write new weekly goals (20 mins) and share with parents (10 mins)	[P]Goals - (15 mins) - 5 mins to review weekly progress then join with teens for 10 mins
	[P]Overcoming barriers to achieve goals – things that help, influence of mood	[P]Overcoming barriers to achieve goals – joint session
10	[PE]Activity session 45 minutes then teens complete checklist and knowledge and skills mastery check	[D]Walk and talk
	[D]Snacks- problem solving, snacks that help you eat more fruit and vegetables and eat less junk food	[D]Snacks – joint session
11	[PE]Activity session- 30 minutes	[D]Food budgeting – money spent on food groups, planning to get more good food for your money
	[PE]Goals – (30 mins)- teens reflect on goal progress, write new weekly goals (20 mins) and share with parents (10 mins)	[D]Goals - (15 mins) - 5 mins to review weekly progress then join with teens for 10 mins
	[D]Food labelling – understanding labels, sugar in drinks, eat less junk	[D]Food labelling – joint session
12	[PE]Activity session	[P]Community opportunity – ideas for healthy activity and good food in your community using family findings from resource homewor
	[P]Togetherness – creative reflection on program involvement using paint colours to symbolise what participants have gained	[P]Togetherness – joint session
13	[PE]Activity session	[D]Supermarket visit – nutrition and cost of foods, reviewing food labelling skills and budgeting skills

Table 1. Cont.

Session#	Adolescents	Parents
	[P]Teen problem solving to achieve goals – (35 mins) - dealing with setbacks	
	[P]Goals – (25 mins) - teens reflect on goal progress, write new weekly goals (15 mins) and share with parents (10 mins)	[D]Goals - (15 mins) - 5 mins to review weekly progress then join with teens for 10 mins
14	[PE]Activity session	[D]Parents in the kitchen - cooking healthy snacks with fruits and vegetables
	[D]Teens in the kitchen - cooking healthy snacks with fruits and vegetables	[P]Parent problem solving to achieve goals - dealing with setbacks
15	[PE]Activity session – (55 mins)	[D]Recipe ideas – modifying recipes to be tasty and healthy, feedback on program to date
	[PE]Goal review – (5 mins) - Reflect back on program goal set at the beginning	[D]Goals- (10 mins) - 5 mins to review weekly progress and prep for 3month goals for 5 mins
	[All]3 month goal setting. Long term goals. (15 mins just teens)	[All]3 month goal setting – (15 mins) - Follow up/support requirements for post program, Joint session (join after 15 mins)
16	[PE]Activity session – confirm follow up arrangements	[D]Cooking Celebration Preparation- prepare celebration food, confirm follow up arrangements
	[All]Cooking Celebration - cooking healthy party foods	[All]Cooking Celebration – joint session

Session facilitated by: All, physiotherapist/exercise physiologist (PE), dietitian (D) or psychologist (P).
Shading indicates adolescent and parent joint sessions.

Food Intake. Dietary intake was assessed using a 3-day food record including one weekend day. Adolescents were trained in completing the record and estimating portion size using household measures. The completed food record was reviewed by the research dietitian and details were clarified with the adolescent. The records were analysed for the number of serves of fruits and vegetables, determined according to the Australian Guide to Healthy Eating serve sizes [29], and junk food according to published criteria [29,30]. Serves of fruit (150 g fresh fruit and 30 g dried fruit) were calculated excluding fruit juice, given the difficulty in identifying juice that had been sweetened and the propensity to consume excessive amounts of juice. Vegetables (75 g) included all vegetables other than fried potato, which were included in the junk food serves. The term 'junk food' was used to describe foods that are considered energy-dense, nutrient-poor foods [29,30] and do not belong to the core food groups. To account for the typically high prevalence of underreporting in self-reported food intake [31], underreporting ratios were calculated. Total energy expenditure was estimated from resting energy expenditure calculations [32], and activity energy expenditure from individual participant accelerometer data [33]. When accelerometer data was missing, energy activity energy expenditure was calculated for a sedentary individual [33]. A ratio of reported energy intake from diaries and total energy expenditure (rEI:TEE) was calculated and included as a continuous variable in the analyses [34,35].

Other Measures. Participants completed surveys of basic demographic information and anthropometric measurements were taken at each assessment at the community facility where the program was delivered. Weight and height were measured and used to calculate age and sex adjusted BMI z-scores [36]. Staff assessing outcomes were not blind to the participant's intervention stage, but did not have access to participant prior scores. Participants were not able to be blinded to the intervention. Data on adolescent fitness, food behaviours, perceived parent need-supportive behaviours, autonomous motivation, mental health and quality of life; parent mental health, autonomous motivation to support adolescents, demonstration of need-supportive behaviours, and family functioning; adolescent and parent perceptions of facilitator support; program fidelity; and adolescent, parent and facilitator perceptions of the program were collected and are being reported in detail elsewhere.

Statistical Analyses

Data were visually inspected for potential outliers. Outliers were checked for data entry errors and corrected where applicable, and biologically implausible values were removed. Data were screened for normality using histograms and multiple measures of location.

Descriptive statistics were calculated (means and standard deviations) at each time point. A comparison of baseline values between those participants completing both the intervention and maintenance periods were compared with those not completing either the intervention or maintenance periods using ANOVA and X^2.

To assess within-subject change, changes in physical activity were assessed using separate linear mixed models with random intercepts for sedentary, light, moderate and vigorous physical activity and adjusted for accelerometer wear-time. Participants who participated in at least 2 assessments were included in statistical models (total used in analysis, n = 56). Missing values were accounted for in the linear mixed models, which uses a likelihood-based estimation procedure resulting in non-biased estimates by imputation of missing responses based upon the surrounding responses and modelled covariance structure. Count data of servings of fruit, vegetables, and junk food were analysed using negative binomial regression utilising generalized estimating equations, with an exchangeable correlation structure and robust estimates of standard errors. Due to the bias that is likely to result from excluding underreporters [35], the ratio of estimated energy expenditure and reported energy intake was included in the models for food servings as a time-varying covariate [34]. Linear contrasts are presented as incidence rate ratios (IRR). Model fits were assessed through residual plots and diagnostics.

In all models, a priori linear contrasts compared the mean point estimate at each time point to pre-intervention point estimates. Further, the monthly rates of change over the following periods were estimated: the waitlist period between baseline and pre-intervention, the intervention period between pre-intervention and post-intervention, and the maintenance period between post-intervention and 12-months. Due to the waitlist control design, changes over the intervention and maintenance periods were compared to changes during the waitlist period. The change over periods was adjusted for varying time between assessments by expressing changes as rate of change per month rather than absolute change across the period. No explicit control for multiple comparisons was performed, rather 95% confidence intervals for all parameter estimates are presented together with actual p-values where appropriate. All analyses were conducted using Stata/IC 13.0 for Windows (StataCorp LP, College Station TX, USA).

Results

Baseline characteristics

Adolescents who completed the study were comparable to those who dropped out before completing the intensive intervention or during the maintenance phase except for the average number of days wearing the accelerometer and daily minutes of moderate physical activity (Table 2).

Changes in Physical Activity and Sedentary Time

The changes in physical activity at each time point and the rate of monthly change over each period can be seen in Table 3. From baseline to pre-intervention (the waitlist period), there was a significant increase in the point estimate of sedentary time and a decrease in light physical activity as seen in Figure 2. Moderate and vigorous physical activity did not significantly change during the waitlist period.

Following the intensive 8-week intervention there was a small but non-significant reduction in the point estimate of sedentary time but the rate of change in sedentary time over the intervention period was significantly different to the rate of change over the waitlist period, indicating a decrease in sedentary behaviour during the intervention period. There were small but non-significant differences in light, moderate and vigorous point estimates after the intervention. The rate of change in moderate activity over the intervention period was significantly different to the rate of change during the waitlist period, indicating an increase in moderate activity during the intervention.

During the entire 12-month maintenance period there were only small changes in physical activity and sedentary time.

Changes in Servings of Fruit, Vegetables, and Junk Food

Servings of fruits, vegetables and junk food did not change significantly from baseline to pre-intervention as seen in Table 3.

Following the intensive intervention, there was a significant increase in the point estimate of servings of fruit and the rate of change in fruit consumption during the intervention period was significantly different to the rate of change over the waitlist period, indicating an increase in fruit consumption (see Figure 3). There was no significant change in vegetable consumption during the intensive intervention period. Following the intervention, the point estimate of servings of junk food decreased and the rate of change

Table 2. Baseline characteristics of adolescents in Curtin University's Activity, Food, and Attitudes Program in total sample and those who did and did not complete the program (n, % or Mean (SD)).

	Total	Did not complete intervention	Did not complete maintenance	Completed	P-Value
n	68*	25	10	34	
Gender (% female)	71.0	64.0	60.0	79.4	0.31
Age (years)	14.1 (1.6)	14.6 (1.7)	13.4 (1.4)	13.9 (1.5)	0.10
Height (cm)	162.9 (8.6)	164.4 (9.4)	161.4 (7.4)	162.2 (8.5)	0.54
Weight (kg)	87.8 (20.4)	89.0 (22.5)	81.5 (14.5)	88.7 (20.5)	0.59
BMI (kg/m2)	32.8 (6.3)	32.6 (6.5)	31.2 (4.7)	33.5 (6.6)	0.59
BMI z-score	2.1 (0.4)	2.0 (0.5)	2.1 (0.3)	2.2 (0.4)	0.60
n (valid accelerometry)	56	16	8	32	
Mean days worn	**6.1 (1.3)**	**5.6 (1.3)**	**5.4 (1.2)**	**6.6 (1.1)**	**0.01**
Mean weartime (min/day)	779.0 (74.3)	758.3 (69.5)	759.8 (94.3)	794.2 (70.0)	0.21
Sedentary (min/day)	547.4 (91.7)	537.1 (90.1)	510.1 (111.6)	561.8 (86.8)	0.32
Light PA (min/day)	197.7 (54.0)	181.2 (43.4)	221.2 (61.2)	200.0 (53.5)	0.22
Moderate PA (min/day)	**32.6 (18.4)**	**38.7 (21.5)**	**27.8 (11.6)**	**30.7 (17.9)**	**0.02**
Vigorous PA (min/day)	1.5 (2.6)	1.4 (6.3)	0.7 (0.9)	1.7 (3.1)	0.11
N (valid food record record)	58	16	10	32	
Fruit (servings/day)	0.8 (0.8)	0.8 (1.0)	0.4 (0.3)	0.8 (0.8)	0.29
Vegetables (servings/day)	1.2 (1.0)	1.3 (1.1)	1.2 (1.2)	1.2 (1.0)	0.96
Junk food (servings/day)	5.8 (3.5)	6.4 (4.6)	5.0 (1.8)	5.7 (3.4)	0.60
Energy (kJ/day)	8101.1 (2503.0)	8405.8 (2807.0)	7236.8 (1854.6)	8215.3 (2530.1)	0.48

*One participant did not complete baseline testing but entered the study during the waitlist period.
P-values for ANOVA comparison between three groups: Did not complete intervention, Did not complete maintenance, Completed.
Variables with significant group differences highlighted in bold.

Table 3. Mean (95%CI) physical activity and healthy eating point estimates and rates of change across the study.

	Point Estimates						Rate of Change		
	Baseline	Pre	Post	3- months	6- months	12- Months	Waitlist Period (Baseline to Pre)	Intervention Period (Pre to Post)	Maintenance Period (Post to 12-months)
	Minutes per day (SE)						Mean Δ per month, min per day (95% CI)		
Sedentary[a]	532.3 (3.3)	548.2 (3.7)	537.9 (4.2)	544.8 (5.3)	540.9 (9.1)	545.8 (8.3)	5.3 (1.8, 8.8)	**-5.1** (-11.0, 0.8) [†]	0.7 (-0.8, 2.2)
Light	199.7 (2.7)*	186.4 (3.5)	192.8 (4.1)	190.9 (4.9)	191.9 (8.7)	185.5 (8.1)	-4.4 (-7.6, -1.2)	3.2 (-2.5, 8.9)	-0.6 (-2.0, 0.8)
Moderate	31.1 (1.3)	29.4 (1.1)	33.0 (1.5)	29.0 (1.7)	31.4 (2.0)	32.1 (2.4)	-0.9 (-2.1, 0.3)	**1.8** (-0.04, 3.6) [†]	-0.1 (-0.5, 0.4)
Vigorous	1.5 (0.2)	1.3 (0.2)	1.6 (0.2)	0.8 (0.3)	1.0 (0.3)	2.0 (0.4)	-0.1 (-0.3, 0.1)	0.1 (-0.1, 0.4)	0.04 (-0.04, 0.1)
	Servings per day (SE)						Monthly incidence rate ratio (95% CI)		
Fruit[b]	0.8 (0.1)	0.6 (0.1)	**1.1** (0.2)*	**1.1** (0.1)*	**0.9** (0.1)*	**1.0** (0.2)*	0.94 (0.86, 1.03)	**1.33** (1.11, 1.60) [†]	0.99 (0.97, 1.02)
Vegetables	1.3 (0.2)	1.3 (0.1)	1.3 (0.2)	1.4 (0.2)	**1.7** (0.2)*	1.4 (0.2)	1.00 (0.91, 1.10)	1.00 (0.85, 1.18)	1.01 (0.98, 1.03)
Junk Food	4.6 (0.3)	4.6 (0.4)	**3.2** (0.3)*	**3.4** (0.3)*	**3.3** (0.4)*	4.3 (0.5)	1.00 (0.95, 1.06)	**0.83** (0.74, 0.94) [†]	1.02 (1.00, 1.05)

[a] Activity variables estimated from mixed models with random intercepts adjusted for wear-time
[b] Healthy eating variables estimated from negative binomial regression using general estimating equations with random intercepts adjusted for underreporting ratio
*significant difference from Pre point estimate (p<.05)
[†] significant difference in the rate of change compared to Waitlist Period (p<.05)

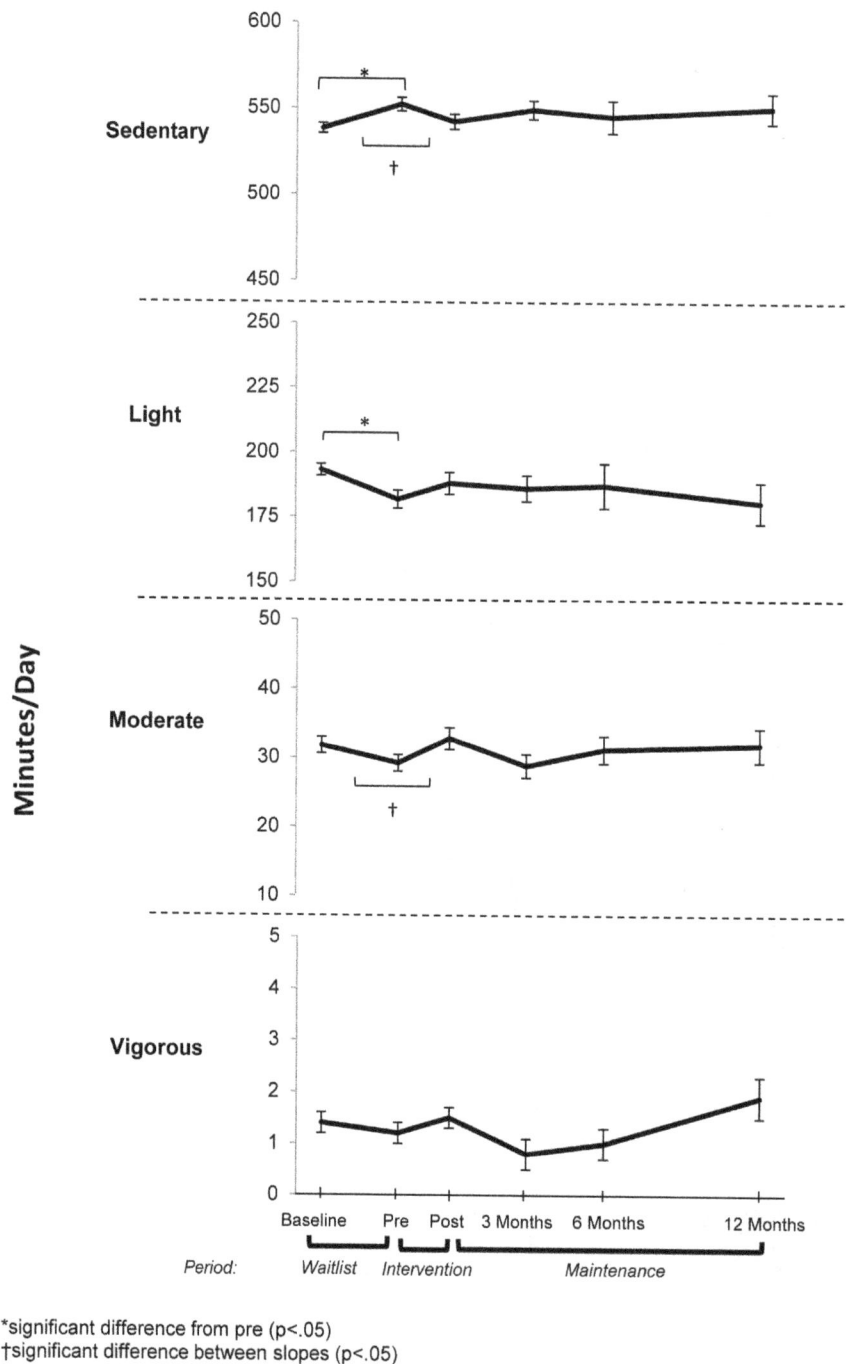

Figure 2. Mean (± standard error) changes in activity by intensity (results from mixed models).

during the intervention period was significantly different to the rate of change over the waitlist period.

During the entire 12-month maintenance period, there were minimal changes in servings of fruit, vegetables and junk food, except the point estimate of vegetable consumption was higher at 6-months compared to pre-intervention.

Secondary Outcome: BMI z-scores

BMI z-scores did not change significantly during the waitlist or intervention periods but the point estimates at 3-months (2.05, SE 0.02), 6-months (2.03, SE 0.02), and 12-months (2.03, SE.04) were

lower (p = .035, p = .042, p = .060 respectively) than at pre-intervention (2.11, SE 0.02). The rates of change in BMI z-scores were not different between the waitlist period (−0.004 per month, 95% CI: −0.02,.01), intervention period (−0.01 per month, 95% CI: −0.04, 0.01) and maintenance period (−0.005 per month, 95% CI: −0.01, 0.002).

Discussion

Overweight and obese adolescents participating in CAFAP were able to make small, statistically significant and potentially clinically

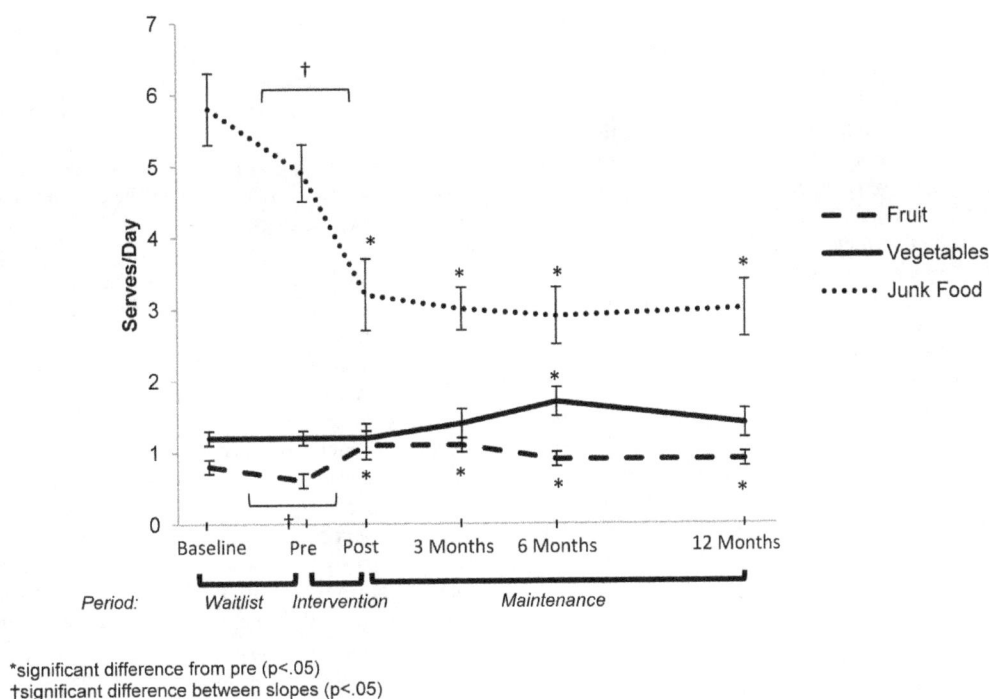

Figure 3. Mean (± standard error) changes in servings of fruit, vegetables and junk food (results from negative binomial regression).

useful improvements in physical activity, sedentary time and healthy eating behaviour trajectories and maintained many of these changes for one year following the intensive intervention. This was the first objective assessment of a community-based family-centered intervention for overweight and obese adolescents focussed on behaviours as the primary outcomes.

Physical Activity and Sedentary Time

Physical activity decreased during the 3 month waitlist period whilst sedentary time increased. This trend was in the same direction but of greater magnitude than previous findings on the typical activity trajectory during adolescence [37,38] and may indicate measurement reactivity or anticipation of the impending intervention. A review of longitudinal studies on physical activity across adolescence found physical activity to decrease an average of 7 percent per year [38]. In a time-use survey of over 6,000 Australian adolescents, moderate-to-vigorous physical activity decreased an average of 13–17 minutes per day per year [37]. In light of these findings, simply maintaining levels of physical activity and sedentary time during adolescence should be considered a positive outcome for interventions [38].

The CAFAP intervention was effective at halting and potentially reversing the negative developmental trajectories of increasing sedentary time and decreasing moderate physical activity. Indeed the effect size of 4 minutes a day more moderate to vigorous activity when comparing the waitlist and intervention periods is consistent with activity interventions across child and adolescent samples [39]. Of the few physical activity interventions for overweight and obese adolescents that have measured physical activity by accelerometry, most have found no significant changes [12,14,18]. Two similar lifestyle interventions, used self-report to assess physical activity and found limited changes in specific activities [8,13,40,41]. While the change observed in the current study is comparable to that found in other successful interventions,

and of similar magnitude to the 10% change anticipated in the *a priori* power calculations [19], there is insufficient evidence from both adult and child samples regarding the specific health implications of a change in this magnitude. Even following the intervention, on average participants were only achieving half of the recommended 60 minutes of MVPA per day [42], with just six percent of participants achieving an average of 60 minutes per day at post-intervention and the 12-month follow-up. However, as previous researchers have suggested, preventing a further decline in already low activity levels may be a key target during this age transition [38], and participating in 30 minutes of activity per day or 150 minutes per week in adulthood has been shown to have widespread health effects [43].

This was the first study to objectively measure physical activity up to 12 months following a lifestyle intervention for overweight and obese adolescents. The maintenance period was less effective at improving physical activity and sedentary time. While the text messaging during this 12 month period to support this behaviour was based on the limited available evidence, little is known on how adolescents actually change their behaviour and respond to such messages [26].

Healthy Eating

Compared to research on physical activity levels during adolescence, even less is known on how food behaviours typically change during this life stage [44], partially due to the difficulty in obtaining high quality food record data from adolescents.

During the intervention, servings of junk food decreased and servings of fruit increased. There were no changes in vegetable consumption. Evidence from studies in children suggests that fruit consumption may increase more in response to intervention than vegetable consumption [15]. While other studies have reported increases in selected measures of diet or macronutrients [8,41], only one previous study has used food recalls to measure changes in fruit, vegetable, or junk food intake in overweight and obese adolescents. One program [10] increased combined fruit and vegetable consumption by 0.6 servings per day after 6 months of intervention, similar to the 0.5 servings per day increase in fruit in CAFAP. Using a 15-item food frequency questionnaire, the Loozit study found an improved proportion of participants meeting fruit and vegetable guidelines after a two-month intervention [41]. While the consumption of fruits and vegetables following participation in CAFAP still fell short of Australian guidelines for 2 serves of fruit and 5 serves of vegetables each day, [29] the increase of half a serve of fruit per day almost doubled the amount of fruit consumed by participants at baseline. The magnitude of change was also much greater than the 10% modelled *a priori* [19] and the changes observed in activity. Evidence on the precise health impact of a change of this magnitude given baseline levels of fruit, vegetables and junk food is lacking and thus an important topic for future research.

In the CAFAP intervention, positive changes in fruit and junk food consumption were successfully maintained up to 12 months following the intervention, however, junk food had increased at 12 months. Additionally, vegetable consumption increased 6 months after the intervention. The Loozit study found no changes in fruit, vegetable, or junk food consumption after 12 or 24 months following the intervention [8,40]. CAFAP focused on behaviours instead of weight loss, which may have contributed to the maintained changes in healthy eating seen in the current study.

Limitations

The assessment utilised objective measures of physical activity and detailed 3-day food records. While underreporting is known to be a problem of food records, particularly in overweight and obese adolescents [31], they remain the best available method for detecting short-term changes in diet and intake patterns [45]. Additionally, the energy expenditure estimates from accelerometry were compared to reported energy intake, underreporting was consistent throughout the study and the underreporting ratio was included in the statistical models [34,35].

The study assessments were taken at multiple time-points during a year-long maintenance period, addressing a noted paucity of studies with follow-up of sustained behaviours beyond immediately post-intervention [4,15]. As is common with many adolescent interventions [4,10,13,18], there was high attrition through the study but analyses were performed on all participants with two time points of data, and sensitivity analyses confirmed findings in those who completed the study. A process evaluation of the intervention is currently underway to explore barriers to successful completion.

While this quasi-experimental study did not have a concurrent control group, the waitlist period provided a within-subject control period comparison for changes seen in the outcomes across the intervention and maintenance periods. This design was selected as providing the best evidence enabling high external validity (staggered entry) and high internal validity (within person control period) whilst providing best practice health behaviour interventions for high-risk adolescents with a 12 months follow-up. For the physical activity and sedentary behaviour outcomes, the changes during the waitlist period were greater than previously published [37,38], but were in the anticipated direction. The lack of literature on dietary trends across adolescence does not provide a reference for changes in healthy eating outcomes. Whilst there

may have been some reactivity during the waitlist period, changes during the intervention were greater than those seen during the waitlist period, suggesting positive effects of the intervention on moderate physical activity, sedentary time and servings of fruit and junk food.

While behaviours are critical outcomes, further research is needed to determine whether such changes in behaviours translate into changes in fitness, mental and physical health status and quality of life. Further research to understand the patterns for both healthy eating behaviour (such as which meals or specific foods are best to target) and physical activity (such as which day of the week or types of physical activities are best to target) could inform more effective interventions, as could a process evaluation of the current study. Research should also explore whether the effects observed were due to the theoretical mechanism proposed [25], that is, whether training parents in need-supportive behaviours can enhance adolescent autonomous motivation and subsequent physical activity and healthy eating behaviours.

The delivery of the program by community health professionals in community settings, whilst challenging, provided high external validity for the findings to be replicable. However, the small sample size and delivery of the intervention in just three sites of similar populations suggest caution in extrapolation to other samples with different social and other characteristics.

Conclusions

This study found that a community-based, family focussed multi-disciplinary physical activity and healthy eating intervention can have a positive influence on behaviours in overweight and obese adolescents and many of these changes can be maintained for up to a year following the intervention. This is encouraging, especially when contrasted with the common pessimistic trajectories of physical activity and healthy eating during adolescence. Improving physical activity and healthy eating behaviours during adolescence is important for current and future physical and mental health and thus successful programs should be made widely available to benefit as many adolescents as possible.

Acknowledgments

The authors would like to thank the adolescents and parents who participated in CAFAP; the CAFAP facilitators and the research staff and Curtin University colleagues.

The authors confirm that all ongoing and related trials for this drug/intervention are registered in the Australia and New Zealand Clinical Trials Registry, # ACTRN12611001187932.

Author Contributions

Conceived and designed the experiments: LS AF DK TO RA AS. Performed the experiments: KS AF. Analyzed the data: EH AS. Wrote the paper: LS EH KS.

References

1. Herman KM, Craig CL, Gauvin L, Katzmarzyk PT (2009) Tracking of obesity and physical activity from childhood to adulthood: The Physical Activity Longitudinal Study. Int J Pediatr Obes 4: 281–288.
2. Soric M, Jembrek Gostovic M, Gostovic M, Hocevar M, Misigoj-Durakovic M (2013) Tracking of BMI, fatness and cardiorespiratory fitness from adolescence to middle adulthood: the Zagreb Growth and Development Longitudinal Study. Ann Hum Biol.
3. Iannotti RJ, Wang J (2013) Patterns of physical activity, sedentary behavior, and diet in U.S. adolescents. J Adolesc Health 53: 280–286.
4. Luttikhuis HO, Baur L, Jansen H, Shrewsbury VA, O'Malley C, et al. (2009) Interventions for treating obesity in children. Cochrane Database of Syst Rev: 187.
5. Bacon L, Aphramor L (2011) Weight science: evaluating the evidence for a paradigm shift. Nutr J 10: 9.
6. Hoelscher DM, Kirk S, Ritchie L, Cunningham-Sabo L (2013) Position of the academy of nutrition and dietetics: interventions for the prevention and treatment of pediatric overweight and obesity. J Acad Nutr Diet 113: 1375–1394.
7. Zook KR, Saksvig BI, Wu TT, Young DR (2014) Physical activity trajectories and multilevel factors among adolescent girls. J Adolesc Health 54: 74–80.
8. Nguyen B, Shrewsbury VA, O'Connor J, Steinbeck KS, Hill AJ, et al. (2013) Two-year outcomes of an adjunctive telephone coaching and electronic contact intervention for adolescent weight-loss maintenance: the Loozit randomized controlled trial. Int J Obes (Lond) 37: 468–472.
9. Davis JN, Ventura EE, Tung A, Munevar MA, Hasson RE, et al. (2012) Effects of a randomized maintenance intervention on adiposity and metabolic risk factors in overweight minority adolescents. Pediatr Obes 7: 16–27.
10. Bean MK, Mazzeo SE, Stern M, Evans RK, Bryan D, et al. (2011) Six-month dietary changes in ethnically diverse, obese adolescents participating in a multidisciplinary weight management program. Clin Pediatr (Phila) 50: 408–416.
11. Davis JN, Tung A, Chak SS, Ventura EE, Byrd-Williams CE, et al. (2009) Aerobic and strength training reduces adiposity in overweight Latina adolescents. Med Sci Sport Exer 41: 1494–1503.
12. Wengle JG, Hamilton JK, Manlhiot C, Bradley TJ, Katzman DK, et al. (2011) The 'Golden Keys' to health - a healthy lifestyle intervention with randomized individual mentorship for overweight and obesity in adolescents. Paediatr Child Health 16: 473–478.
13. Evans RK, Franco RL, Stern M, Wickham EP, Bryan DL, et al. (2009) Evaluation of a 6-month multi-disciplinary healthy weight management program targeting urban, overweight adolescents: Effects on physical fitness, physical activity, and blood lipid profiles. International Journal of Pediatr Obes 4: 130–133.
14. Kong AS, Sussman AL, Yahne C, Skipper BJ, Burge MR, et al. (2013) School-based health center intervention improves body mass index in overweight and obese adolescents. J Obes 2013: 575016.
15. Evans CE, Christian MS, Cleghorn CL, Greenwood DC, Cade JE (2012) Systematic review and meta-analysis of school-based interventions to improve daily fruit and vegetable intake in children aged 5 to 12 y. Am J Clin Nutr 96: 889–901.
16. Collins CE, Warren J, Neve M, McCoy P, Stokes BJ (2006) Measuring effectiveness of dietetic interventions in child obesity: a systematic review of randomized trials. Arch Pediatr Adolesc Med 160: 906–922.
17. Warren JM, Golley RK, Collins CE, Okely AD, Jones RA, et al. (2007) Randomised controlled trials in overweight children: practicalities and realities. Int J Pediatr Obes 2: 73–85.
18. Robbins LB, Pfeiffer KA, Maier KS, Lo YJ, Wesolek Ladrig SM (2012) Pilot intervention to increase physical activity among sedentary urban middle school girls: a two-group pretest-posttest quasi-experimental design. J Sch Nurs 28: 302–315.
19. Straker LM, Smith KL, Fenner AA, Kerr DA, McManus A, et al. (2012) Rationale, design and methods for a staggered-entry, waitlist controlled clinical trial of the impact of a community-based, family-centred, multidisciplinary program focussed on activity, food and attitude habits (Curtin University's Activity, Food and Attitudes Program–CAFAP) among overweight adolescents. BMC Public Health 12: 471.
20. Kuczmarski RJ, Ogden CL, Grummer-Strawn LM, Flegal KM, Guo SS, et al. (2000) CDC growth charts: United States. Advance data: 1–27.
21. Straker L, McManus A, Kerr D, Smith K, Davis M, et al. (2010) CAFAP: A multi-disciplinary family-centred community-based intervention for overweight/obese adolescents. J Sci Med Sport 13, Supplement 1: e11–e12.
22. Smith K, Straker L, McManus A, Fenner A (2014) Barriers and enablers for participation in healthy lifestyle programs by adolescents who are overweight: a qualitative study of the opinions of adolescents, their parents and community stakeholders. BMC Pediatrics 14: 53.
23. Deci EL, Ryan RM (2000) The "what" and "why" of goal pursuits: Human needs and the self-determination of behavior. Psychological Inquiry 11: 227–268.
24. Locke EA, Latham GP (1990) A theory of goal setting and task performance. Englewood Cliffs, New Jersey: Prentice Hall.
25. Fenner AA, Straker LM, Davis MC, Hagger MS (2013) Theoretical underpinnings of a need-supportive intervention to address sustained healthy lifestyle changes in overweight and obese adolescents. Psychol Sport Exerc 14: 819–829.
26. Smith KL, Kerr DA, Fenner AA, Straker LM (2014) Adolescents just do not know what they want: a qualitative study to describe obese adolescents' experiences of text messaging to support behavior change maintenance post intervention. J Med Internet Res 16: e103.
27. Colley RC, Tremblay MS (2011) Moderate and vigorous physical activity intensity cut-points for the Actical accelerometer. J Sports Sci 29: 783–789.
28. Rich C, Geraci M, Griffiths L, Sera F, Dezateux C, et al. (2013) Quality control methods in accelerometer data processing: defining minimum wear time. PLoS One 8: e67206.
29. National Health and Medical Research Council (2013) Educator Guide. Canberra: National Health and Medical Research Council.
30. Rangan AM, Randall D, Hector DJ, Gill TP, Webb KL (2008) Consumption of 'extra' foods by Australian children: types, quantities and contribution to energy and nutrient intakes. Eur J Clin Nutr 62: 356–364.
31. Singh R, Martin BR, Hickey Y, Teegarden D, Campbell WW, et al. (2009) Comparison of self-reported and measured metabolizable energy intake with total energy expenditure in overweight teens. Am J Clin Nutr 89: 1744–1750.
32. Henes ST, Cummings DM, Hickner RC, Houmard JA, Kolasa KM, et al. (2013) Comparison of Predictive Equations and Measured Resting Energy Expenditure Among Obese Youth Attending a Pediatric Healthy Weight Clinic: One Size Does Not Fit All. Nutrition in Clinical Practice 28: 617–624.
33. Puyau MR, Adolph AL, Vohra FA, Zakeri I, Butte NF (2004) Prediction of activity energy expenditure using accelerometers in children. Med Sci Sport Exerc 36: 1625–1631.
34. Jennings A, Cassidy A, van Sluijs EM, Griffin SJ, Welch AA (2012) Associations between eating frequency, adiposity, diet, and activity in 9–10 year old healthy-weight and centrally obese children. Obesity 20: 1462–1468.
35. Rennie KL, Coward A, Jebb SA (2007) Estimating under-reporting of energy intake in dietary surveys using an individualised method. The British Journal of Nutrition 97: 1169–1176.
36. USDA/ARS Children's Nutrition Center (2003) Children's BMI-percentile-for-age Calculator.
37. Olds T, Wake M, Patton G, Ridley K, Waters E, et al. (2009) How do school-day activity patterns differ with age and gender across adolescence? J Adolesc Health 44: 64–72.
38. Dumith SC, Gigante DP, Domingues MR, Kohl HW, 3rd (2011) Physical activity change during adolescence: a systematic review and a pooled analysis. Int J Epidemiol 40: 685–698.
39. Metcalf B, Henley W, Wilkin T (2012) Effectiveness of intervention on physical activity of children: systematic review and meta-analysis of controlled trials with objectively measured outcomes (EarlyBird 54). BMJ 345: e5888.
40. Nguyen B, Shrewsbury VA, O'Connor J, Steinbeck KS, Lee A, et al. (2012) Twelve-month outcomes of the loozit randomized controlled trial: a community-based healthy lifestyle program for overweight and obese adolescents. Arch Pediatr Adolesc Med 166: 170–177.
41. Shrewsbury VA, Nguyen B, O'Connor J, Steinbeck KS, Lee A, et al. (2011) Short-term outcomes of community-based adolescent weight management: The Loozit(R) Study. BMC Pediatr 11: 13.
42. Department of Health and Aging (2014) National Physical Activity Guidelines for Australians. Canberra: Commonwealth of Australia.
43. Brown WJ, Bauman AE, Bull FC, Burton NW (2012) Development of Evidence-based Physical Activity Recommendations for Adults (18-64 years). Australian Gorvement Department of Health.
44. Smith KL, Straker LM, Kerr DA, Smith AJ (2014) Overweight adolescents eat what? And when? Analysis of consumption patterns to guide dietary message development for intervention. J Hum Nutr Diet.
45. Kirkpatrick SI, Reedy J, Butler EN, Dodd KW, Subar AF, et al. (2014) Dietary assessment in food environment research: a systematic review. Am J Prev Med 46: 94–102.

3

Arabidopsis mTERF15 Is Required for Mitochondrial *nad2* Intron 3 Splicing and Functional Complex I Activity

Ya-Wen Hsu[1,2], Huei-Jing Wang[2], Ming-Hsiun Hsieh[2], Hsu-Liang Hsieh[1], Guang-Yuh Jauh[2,3]*

1 Institute of Plant Biology, National Taiwan University, Taipei, 116, Taiwan, ROC, 2 Institute of Plant and Microbial Biology, Academia Sinica, Nankang, Taipei, 11529, Taiwan, ROC, 3 Biotechnology Center, National Chung-Hsing University, Taichung, 402, Taiwan, ROC

Abstract

Mitochondria play a pivotal role in most eukaryotic cells, as they are responsible for the generation of energy and diverse metabolic intermediates for many cellular events. During endosymbiosis, approximately 99% of the genes encoded by the mitochondrial genome were transferred into the host nucleus, and mitochondria import more than 1000 nuclear-encoded proteins from the cytosol to maintain structural integrity and fundamental functions, including DNA replication, mRNA transcription and RNA metabolism of dozens of mitochondrial genes. In metazoans, a family of nuclear-encoded proteins called the mitochondrial transcription termination factors (mTERFs) regulates mitochondrial transcription, including transcriptional termination and initiation, via their DNA-binding activities, and the dysfunction of individual mTERF members causes severe developmental defects. *Arabidopsis thaliana* and *Oryza sativa* contain 35 and 48 mTERFs, respectively, but the biological functions of only a few of these proteins have been explored. Here, we investigated the biological role and molecular mechanism of Arabidopsis mTERF15 in plant organelle metabolism using molecular genetics, cytological and biochemical approaches. The null homozygous T-DNA mutant of *mTERF15*, *mterf15*, was found to result in substantial retardation of both vegetative and reproductive development, which was fully complemented by the wild-type genomic sequence. Surprisingly, mitochondria-localized mTERF15 lacks obvious DNA-binding activity but processes mitochondrial *nad2* intron 3 splicing through its RNA-binding ability. Impairment of this splicing event not only disrupted mitochondrial structure but also abolished the activity of mitochondrial respiratory chain complex I. These effects are in agreement with the severe phenotype of the *mterf15* homozygous mutant. Our study suggests that Arabidopsis mTERF15 functions as a splicing factor for *nad2* intron 3 splicing in mitochondria, which is essential for normal plant growth and development.

Editor: Emanuele Buratti, International Centre for Genetic Engineering and Biotechnology, Italy

Funding: This work was supported by research grants from Academia Sinica (Taiwan; http://ipmb.sinica.edu.tw/index.html/), the National Science and Technology Program for Agricultural Biotechnology (NSTP/AB, 098S0030055-AA, Taiwan) and the Ministry of Science and Technology (99-2321-B-001-036-MY3 and 102-2321-B-001-040-MY3; http://www.most.gov.tw/) to G.-Y. Jauh. The funders had no role in study design, data collection and analysis, decision to publish, or preparation of the manuscript.

Competing Interests: The authors have declared that no competing interests exist.

* Email: jauh@gate.sinica.edu.tw

Introduction

Mitochondria, which originated through the endosymbiosis of α-proteobacteria into ancestral host cells, are the cellular powerhouses and play vital roles in diverse eukaryotic cell processes through the production of ATP and various metabolic intermediates [1,2]. Recent studies also suggest that dysfunctional mitochondria are involved in many neurodegenerative diseases such as aging and cognitive decline in a wide range of metazoans, including humans [3]. Maintaining the structural and metabolic integrity of this semi-autonomous organelle is essential for the normal function of eukaryotic cells. Nevertheless, over the course of symbiotic evolution, the majority of mitochondrial genes migrated into the nuclear genome of the original host, leaving an incomplete set of essential genes in the mitochondrial genomes of most organisms, including plants [4–6]. Complicated and dynamic communication and coordination between the nucleus and mitochondria greatly impact many fundamental cellular

processes in, and even the lives of, most eukaryotes [7]. Indeed, based on the complete sequence of the Arabidopsis mitochondrial genome, it has been reported that 57 mitochondrial genes encode the subunits of multiprotein complexes that are required for the respiratory chain, heme and cytochrome assembly, and mitochondrial ribosomes [4]. Additionally, plant mitochondria are more complex than those found in other kingdoms and exhibit unique RNA metabolism, including RNA transcription, splicing, editing, degradation, and translation [8–10]. The proteins involved in these processes are predominantly encoded by the nuclear genome and are imported into mitochondria after protein synthesis. For example, recent studies suggested that one protein family, called the mitochondrial transcription termination factors (mTERFs), plays important roles in regulating the organellar transcription machinery.

The mTERF proteins were first identified two decades ago as regulators of transcription termination in human mitochondria [11]. Phylogenetic analyses of mTERF homologs in metazoans and plants revealed the presence of 4 subfamilies, mTERF1 to mTERF4 [12]. These proteins share a common 30-amino-acid repeat module called the mTERF motif [13]. The proteins within this family possess diverse numbers and arrangements of these motifs, yet the folding patterns of these proteins are similar. Moreover, crystal structure studies of mTERF1, mTERF3 and mTERF4 suggest that the helical structure of the mTERF motifs may be essential for their nucleic acid-binding activities [14,15]. Human mTERF1 binds specific sites located at the 3′-end of the 16S rRNA and tRNA$^{Ler(UUR)}$ genes to terminate mitochondrial transcription [16]. Additionally, mTERF1 binds to the mitochondrial transcription initiation site to create a DNA loop that allows for the recycling of the transcriptional machinery [16]. This simultaneous link between mitochondrial transcriptional initiation and termination sites may explain the high rate of mitochondrial rRNA biogenesis. mTERF2 binds to mitochondrial DNA (mtDNA) in a non-specific manner and associates with nucleoids. mTERF2 loss-of-function mice exhibit myopathies, memory deficits, and impaired respiratory function due to a reduction in the number of mitochondrial transcripts [17,18]. mTERF3 is essential, and mTERF3-knockout mice die during the early stages of embryogenesis; mTERF3 non-specifically interacts with mtDNA promoter regions and mediates the repression of mitochondrial transcription initiation [19]. Mouse NSUN4 (NOL1/NOP2/Sun domain family, member 4) is a methyltransferase involved in the methylation of ribosomes. Mouse mTERF4 first binds 16S and 12S rRNAs, then forms a stoichiometric complex with NSUN4, which is essential for proper mitochondrial ribosomal assembly and translation [14,20,21].

The genome of Arabidopsis contains 35 identified mTERFs, which vary in the number and arrangement of their mTERF motifs [22]. Several recently characterized Arabidopsis mTERF proteins displayed diverse and complicated functions during mitochondrial and plastid transcription, such as SINGLET OXYGEN-LINKED DEATH ACTIVATOR10 (SOLDAT10) [23], BELAYA SMERT/RUGOSA2 (BSM/RUG2) [22,24], MTERF DEFECTIVE IN ARABIDOPSIS 1 (MDA1) [25] and SUPPRESSOR of hot1-4 1 (SHOT1) [26]. SOLDAT10, the first Arabidopsis mTERF to be characterized, is targeted to plastids and regulates plastid homeostasis. Arabidopsis soldat10 mutants suppressed the $^1O_2$-mediated cell-death and growth-retardation phenotypes of the conditional fluorescent (flu) mutant, in which excess levels of the reactive oxygen species (ROS) $^1O_2$ were generated within chloroplasts after the dark-to-light shift [23]. BSM/RUG2 is essential for embryogenesis and postembryonic development in Arabidopsis and displays DNA-binding activity towards plastid DNA. Additionally, BSM/RUG2 localizes to both plastids and mitochondria, and the perturbation of BSM/RUG2 affects the processing and steady-state levels of plastid transcripts [22] and decreases the expression of mitochondrial transcripts [24]. Arabidopsis MDA1 is a plastid-localized protein involved in abiotic stress tolerance. MDA1 mutation leads to significant defects in plant growth and chloroplast development but enhanced tolerance to salt and osmotic stress, perhaps via reduced sensitivity to abscisic acid [25]. Mutating SHOT1 suppresses the heat-hypersensitive phenotype caused by the loss of heat shock protein 101 in Arabidopsis [26]. However, SHOT1 is a mitochondrial-resident mTERF protein, and the thermotolerance phenotype of shot1 may be caused by reduced ROS-mediated oxidative damage during heat stress. Additionally, SHOT1 plays an important role in regulating the expression of mitochondrial-encoded genes and mitochondrial genome copy number. Although it is known that many mTERF mutations cause defective development and stress

responses, studies of the precise molecular functions of these mTERF proteins are limited in Arabidopsis organelles.

In Arabidopsis, several protein families that participate in organellar RNA splicing, including those containing the chloroplast RNA splicing and ribosome maturation (CRM) domain [27–29], the pentatricopeptide repeat (PPR) [30–35], the plant organellar RNA recognition (PORR) domain [36,37] and the ACCUMULATION OF PHOTOSYSTEM ONE (APO) motif [38]. These domains are recognized as RNA binding domains and interact with specific elements within introns. Although several splicing factors have been identified in plastids and mitochondria, the detailed mechanism of RNA splicing in plant organelles is still being investigated. Recently, maize mTERF4 was identified as a splicing factor associated with several introns and splicing factors in chloroplasts [39].

In the present study, we investigated another Arabidopsis mTERF protein, mTERF15, and identified its role in post-transcriptional modification of mitochondrial RNA processing. Mutating mTERF15 significantly disturbed normal plant vegetative and reproductive growth, which may be the result of defective mitochondrial development and/or function. Interestingly, mTERF15 is a mitochondria-localized protein that binds to RNA but not to double-stranded DNA (dsDNA). Moreover, an analysis of mitochondrial intron splicing in mterf15 mutants revealed defective RNA splicing of nad2 intron 3, which led to a deficiency in complex I activity. The possible molecular function of mTERF15 in regulating Arabidopsis mitochondrial RNA metabolism is discussed.

Materials and Methods

Plant material and growth conditions

The mTERF15 T-DNA insertion mutant (SALK_134099) was obtained from the Arabidopsis Biological Resource Center (Ohio State University, USA). Seeds from Arabidopsis thaliana ecotype Columbia-0 and mterf15/+ plants were surface-sterilized by chloride gas after stratification for 3 d at 4°C and then either germinated on half-strength Murashige and Skoog medium (Duchefa Biocheme, The Netherlands) containing 1% sucrose and 0.7% phytoagar (Duchefa) or grown in soil inside a growth chamber on a 16-h light/8-h dark cycle at 22°C.

Complementation of the mterf15 mutant

Approximately 3.3 kb of the genomic sequence corresponding to locus At1g74120, including a 2-kb putative promoter and the 1.3-kb coding region, was amplified using Phusion polymerase (Finnzymes, http://www.finnzymes.com) and then ligated into a modified pPZP221 vector containing a C-terminal GFP sequence followed by the Nopaline synthase (NOS) terminator. The primers used for cloning are in Table S1. After verification of the DNA sequence, the construct was introduced into Agrobacterium tumefaciens strain GV3101. Arabidopsis plants heterozygous for mTERF15 were transfected with Agrobacterium by the vacuum infiltration method [40]. Seeds corresponding to the T$_1$ generation were collected and grown on solid half-strength Murashige and Skoog medium (Duchefa) containing 1% sucrose, 0.7% phytoagar (Duchefa) and the antibiotic G418 at 100 μg/μl. Leaves from T$_1$ plants were harvested for genotyping. Plants with an mterf15 homozygous background harboring mTEFR15p::mTERF15-GFP were collected for further studies.

Ultrastructural sample preparation and tetramethylrhodamine methyl ester perchlorate (TMRM) staining

For transmission electron microscopy, samples were frozen in a high-pressure freezer (Leica EMPACT2, 2000–2050 bar). Freeze

Table 1. The phenotype of progeny from SALK 134099 heterozygotes (*mterf15/+*) with abnormal morphology.

	Normal seedlings	Seedlings with retarded growth	Non-germinated seeds	Ratio (normal: abnormal)	*p*-value (χ^2 value, 3:1)
Wild-type	308 (99.68)	0	1 (0.32)	*NA	NA
SALK_134099 heterozygotes	246 (74.77)	27 (8.21)	56 (17.02)	2.96:1	0.924

Phenotypes were recorded after seeds were germinated on half-strength MS solid medium and then incubated for 7 d with a 16-h light/8-h dark cycle at 22°C.
*NA, Not available.
Data indicates total number of plants observed and numbers in brackets indicate their corresponding percentage.

substitution involved the use of anhydrous acetone (containing 1% OsO_4 and 0.1% UA) with a Leica EM AFS2. The samples were kept at $-85°C$ for 3 d, at $-60°C$ for 1 d, at $-20°C$ for 1 d, and at $0°C$ for 1 d and then brought to room temperature. The samples were infiltrated and embedded in Spurr resin. Ultrathin sections (70–90 nm) were cut using a Reichert Ultracut S or Leica EM UC6 (Leica, Vienna, Austria) and collected on 100 mesh copper grids. After staining with 5% uranyl acetate in 50% methanol for 10 min and 0.4% lead citrate for 4 min, sections were observed under a Philips CM 100 Transmission Electron Microscope at 80 KV; images were captured using a Gatan Orius CCD camera.

TMRM staining (Molecular Probes, Invitrogen) was performed to examine the mitochondrial membrane potential in protoplasts. Isolated protoplasts were stained with 20 nM TMRM and immediately observed under a laser scanning confocal microscope (Zeiss, LSM510, Carl Zeiss, http://www.zeiss.com). TMRM fluorescence and chlorophyll autofluorescence were induced by excitation with a 543-nm HeNe laser and a 488-nm Argon laser. The emission signals from TMRM and from chlorophyll were recorded with using 565- to 615- and 650-nm band-pass filters, respectively. The fluorescence intensity and area were quantified using LSM510 software (Carl Zeiss, http://www.zeiss.com).

Northwestern blot analysis

Northwestern blot analysis was performed to test the RNA-binding ability of mTERF15 as described previously [41]. Briefly, *mTERF15* cDNA lacking the 34 N-terminal amino acids was ligated into the pET52b vector to produce the mTERF15-strep recombinant protein. The recombinant proteins were expressed in *E. coli* strain BL21 and incubated with StrepTactin-Sepharose resins (GE Healthcare) according to the manufacturer's instructions. The resins were washed extensively with incubation buffer before elution of the recombinant proteins. The mTERF15-strep recombinant protein and the MBP-strep control protein were transferred from 12% SDS-polyacrylamide gels onto a PVDF membrane and then renatured overnight in renaturation buffer (0.1 M Tris-HCl, pH 7.5 and 0.1% (v/v) NP-40) at 4°C. After 4 washes with renaturation buffer (15 min), blots were blocked with blocking buffer [(10 mM Tris-HCl, pH 7.5, 5 mM Mg acetate, 2 mM DTT, 5% (w/v) BSA and 0.01% (v/v) Triton X-100)] at room temperature for 5 min. Full length of *nad2* intron 3 was *in vitro* synthesized using SP6 polymerase (Promega). Total RNA and *in vitro* synthesized *nad2* intron3 were labeled with $\gamma^{32}P$ ATP using T4 polynucleotide kinase (Fermentas). Hybridization was performed at 4°C in the presence of 5 μg labeled RNA. Blots were washed 4 times (5 min each) in a washing buffer (10 mM Tris-HCl, pH 7.5, 5 mM Mg acetate, 2 mM DTT) to remove unbound RNA. The radioisotope signals were detected via autoradiography on X-ray film (Kodak).

RNA isolation and qRT-PCR analysis

Total RNA was isolated using the Qiagen RNeasy Mini Kit (QIAGEN) and then treated with TURBO DNA-free DNase (Ambion) to remove contaminating genomic DNA. Reverse transcription of RNA was performed using the M-MLV transcriptase system (Invitrogen) and random hexamer primers.

Gene expression analysis via qRT-PCR was carried out using Power SYBR Green Supermix (Applied Biosystems) running on an ABI Prism 7000 sequence detection system (Applied Biosystems). Primer sets were designed as described previously [31] (Table S2) to monitor the expression of genes involved in the splicing of junctions within mitochondria.

Northern blot analysis

A total amount of 5 μg RNA was denatured at 65°C for 10 min and then separated on 2% formaldehyde-agarose gels in MOPS buffer and transferred to a nylon membrane (PerkinElmer). The *nad2* probes were labeled using the Rediprime II DNA labeling system (GE Healthcare). Probe primers are listed in Table S1. The prehybridization, hybridization and washing steps were performed following the manufacturer's instructions.

Crude mitochondria isolation and complex I Activity assay

Crude mitochondria were prepared as described previously [36,42]. Approximately 200 mg fresh tissue was extracted in 2 ml extraction buffer (75 mM MOPS-KOH, pH 7.6, 0.6 M sucrose, 4 mM EDTA, 0.2% polyvinylpyrrolidone 40, 8 mM cysteine, 0.2% bovine serum albumin) at 4°C. The lysate was centrifuged first at 1300×g for 5 min (twice), after which the supernatant was centrifuged at 22,000×g for 20 min at 4°C. The resulting pellet (which usually contains most of the thylakoid and mitochondrial membranes) was resuspended in buffer (10 mM MOPS-KOH, pH 7.2, 0.3 M sucrose) and stored at −80°C.

Clear native electrophoresis was performed as described previously [43]. Crude mitochondria were washed with distilled water, resuspended with buffer (50 mM NaCl, 50 mM imidazole/HCl, 2 mM 6-aminohexanoic acid, 1 mM EDTA, pH 7.0) and solubilized by adding DDM (10%) at a DDM-to-protein ratio of 2.5 (g/g). After centrifugation at 100,000×g for 15 min, 10% glycerol and 0.02% Ponceau S were added to the supernatant for 4–16% native PAGE with anode buffer (25 mM imidazole/HCl, pH 7.0) and cathode buffer (50 mM Tricine, 7.5 mM imidazole, pH 7.0, 0.01% DDM and 0.05% DOC). Clear native gels were washed 3 times with distilled water for 5 min and incubated with 1 mM nitroblue tetrazolium and 0.14 mM NADH in 0.1 M Tris, pH 7.4. The reaction was stopped with 40% methanol and 10% acetic acid after a dark blue signal appeared on the gel.

Results

Disruption of the Arabidopsis *mTERF15* gene results in significant growth and developmental retardation

A screen of Arabidopsis T-DNA insertional mutants from the SALK collection allowed us to identify several mutants that were defective in post-embryonic development and/or seed germination. We found one T-DNA mutant showing abnormal phenotype in progenies segregated from SALK_134099 heterozygotes. The progeny of SALK_134099 heterozygotes showed growth retardation at a ratio of 3:1 (normal to defective) (χ^2, $p = 0.924$; Table 1). Most abnormal progenies derived from SALK_134099 heterozygous plants were unable to germinate on solid medium (17.02%), although a few grew slowly (8.21%, Table 1). The SALK_134099 homozygous mutant containing a T-DNA insertion in the coding region of Arabidopsis *mTERF15* (At1g74120; Figure 1A) and showing significant growth and development retardation was

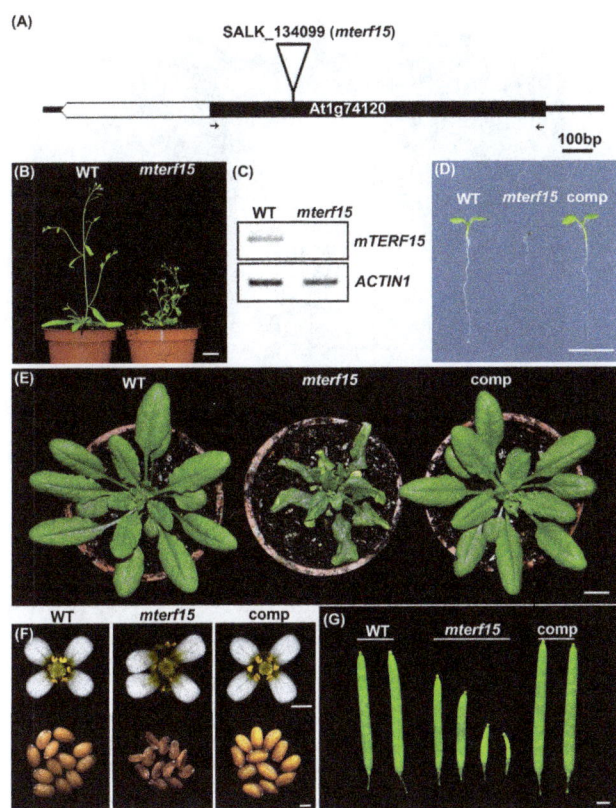

Figure 1. Phenotypes associated with *mterf15* mutants. (A) Schematic diagram of the Arabidopsis *mTERF15* gene (At1g74120) and the single T-DNA insertion within the coding region for *mterf15* (SALK_134099). Primers used to analyze *mTERF15* transcripts are indicated by black arrows. **(B)** Comparison of whole-plant morphology between a 1.5-month-old wild-type plant and a 3-month-old *mterf15* mutant. Scale bar = 1 cm. **(C)** RT-PCR analysis of *mTERF15* expression in 7-d-old wild-type (WT) and *mterf15* seedlings. The housekeeping gene *ACTIN1* was used as an internal control. **(D)** Morphology of 7-d-old seedlings for wild-type and *mterf15* lines and a complemented line transformed with *mTERF15p::mTERF15-GFP*. Scale bar = 1 cm. **(E)** Morphology and arrangement of leaves for 1-month-old wild-type, 2-month-old *mterf15* and 1-month-old *mterf15* complementation lines. Scale bars = 1 cm **(F)** Morphology of flowers and seeds from the wild-type, *mterf15* and complementation lines. Scale bars = 1 mm (flowers) and 200 μm (seeds). **(G)** Morphology of siliques from the wild-type, *mterf15* and complementation lines. Scale bars = 1 mm.

named *mterf15* (Figure 1B). The full-length *mTERF15* transcript was not detected in the homozygous *mterf15* plant (Figure 1C), suggesting that all of the observed phenotypes resulted from the null mutation of the *mTERF15* allele (SALK_134099). Developmental defects were widespread in homozygous *mterf15* plants. Compared with wild-type seedlings, *mterf15* seedlings were small in stature, had small cotyledons and displayed reduced growth of primary roots (Figure 1D). The rosette leaves of *mterf15* were smaller, wrinkled and twisted whereas the wild type plant showed larger and smooth round-shaped leaves. Moreover, the extended petiole region of *mterf15* was shorter than the wild type plant. Compared with 1.5-month-old wild-type plants that were ready for blossoming, 3-month-old adult *mterf15* plants were dwarfed and just about ready for flowering (Figure 1E). The flowers of *mterf15* plants were abnormal, with misplaced petals, extremely defective anthers and a limited amount of pollen (Figure 1F, upper panel). Unlike wild-type siliques, mature *mterf15* siliques were much shorter (Figure 1G), and no viable seeds were found (Figure 1F, lower panel).

To further verify that these defects were indeed caused by the disruption of the *mTERF15* allele, *mterf15* heterozygotes were complemented with the *mTERF15* native promoter-driven *mTERF15* coding sequence tagged with *GFP* (*mTERF15p::mTERF15-GFP*). The T_2 progeny of homozygotes for the *mTERF15* mutation which harbors at least one copy of the transgene *mTERF15p::mTERF15-GFP* showed similar phenotypes as 7-d-old wild-type seedlings and adult plants (Figure 1 and Figure S1). All of the above results show that functional *mTERF15* is essential for normal plant growth and development.

mTERF15, a plant-specific and mitochondria-localized protein, is important for mitochondrial biogenesis and membrane potential

The Arabidopsis *mTERF15* locus contains a single exon and appears to encode a 51-kDa protein. Architecture analysis with SMART (http://smart.embl-heidelberg.de) revealed that mTERF15 contains 5 typical mTERF motifs (Figure S2A). A search of the NCBI database (http://www.ncbi.nlm.nih.gov) revealed mTERF15 homologs present in *Vitis vinifera*, *Populus trichocarpa*, *Brachypodium distachyon*, *Sorghum bicolor* and *Zea mays*. Comparing protein similarities and identities between flowering plant mTERF15 orthologs and metazoan mTERF1, Arabidopsis mTERF15 is evolutionarily closer to plant mTERF15 orthologs than metazoan mTERF1. This result suggests that mTERF15 is a plant-specific mTERF protein (Table S3). The phylogenetic tree branched into 2 groups, monocots and dicots (Figure S2B). Amino acid sequence alignments of all 6 homologs revealed stronger conservation of the C terminus compared with the N terminus. All of these homologs appeared to contain 5 mTERF motifs with a conserved proline residue at position 8, except for the first motif (Figure S2A).

In silico analysis of the BIO-array (http://bar.utoronto.ca/efp_arabidopsis/cgi-bin/efpWeb.cgi) and Genevestigator (https://www.genevestigator.com/gv/plant.jsp) microarray databases suggested various expression levels of *mTERF15* in different tissues, with the highest and lowest expression levels in mature seeds and pollen, respectively. However, in the Genevestigator microarray database, *mTERF15* showed the highest expression in sperm cells and the lowest in leaf phloem protoplasts. Quantitative real-time PCR spatiotemporal analysis in Arabidopsis revealed *mTERF15* expression in all organs, with high levels of accumulation in seedlings, flowers and siliques (Figure S3).

Transient protoplast assays and stable transgenic studies suggested that mTERF15 is a mitochondrial protein [22]. To confirm the subcellular localization of mTERF15, we observed the root hairs of 7-d-old *mTERF15p::mTERF15- GFP* transgenic seedlings via confocal microscopy after staining with the mitochondrial indicator MitoTracker orange. The co-localization of the mTERF15-GFP and MitoTracker signals demonstrated that mTERF15 is a mitochondrial protein (Figure S4).

We further investigated the morphology of chloroplasts and mitochondria from the leaves of 1-month-old wild-type and 2-month-old *mterf15* plants by transmission electron microscopy. The sizes and features of *mterf15* chloroplasts were similar to those of wild-type chloroplasts (Figure 2A, D). Although the sizes of the mitochondria did not differ between *mterf15* and wild-type plants, the inner membrane systems were aberrant in *mterf15* mitochondria; wild-type mitochondria (Figure 2B–2C) showed normal cristae structures with clear inner spaces, whereas *mterf15* mitochondria (Figure 2E–2F) lacked any obvious cristae structure. Therefore, the integrity of the mitochondrial inner membrane

system may have been compromised, potentially resulting in decreased mitochondrial membrane potential.

To assess mitochondrial membrane integrity, we treated protoplasts from leaves of the wild-type, complementation line and *mterf15* plants with tetramethylrhodamine methyl ester (TMRM), a lipophilic, cationic, red-orange fluorescent dye that accumulates within mitochondria in living cells. The intensity of fluorescence is associated with the mitochondrial membrane potential [44,45]. We observed large and strong fluorescent TMRM signals in protoplasts from the wild-type and complementation lines, which suggests that mitochondria were active in these lines (Figure 2G, I); however, TMRM signals were slightly weaker in *mterf15* protoplasts (Figure 2H). The mean fluorescence intensity for *mterf15* protoplasts was reduced by ~80% compared with wild-type protoplasts (Figure 2J). These results suggest that mTERF15 is crucial for maintaining mitochondrial biogenesis and membrane integrity.

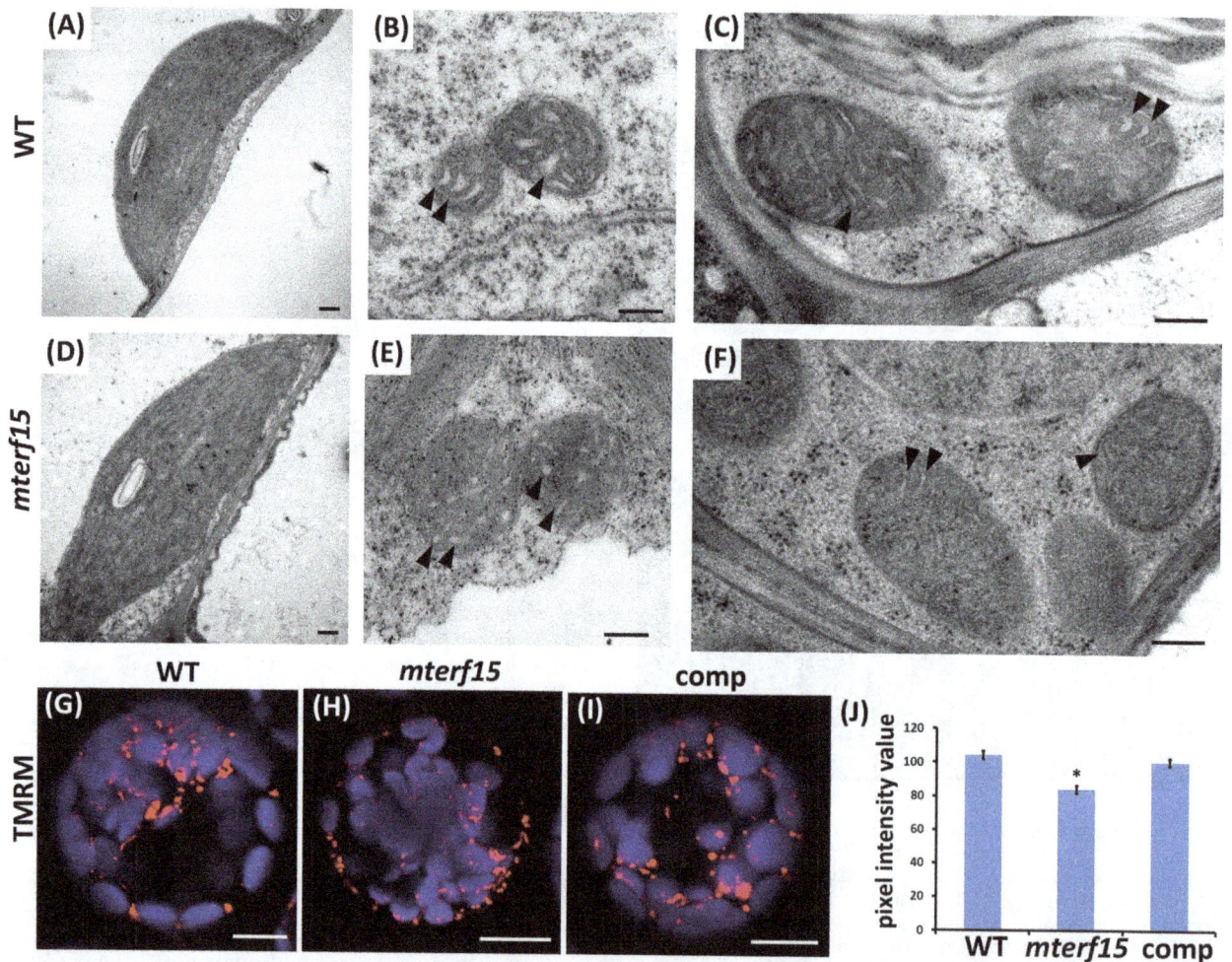

Figure 2. Morphological analysis of mitochondria in wild-type (WT) and *mterf15* plants. (A and D) Transmission electron microscopy of thin sections corresponding to chloroplasts from 1-month-old wild-type (A) and 2-month-old *mterf15* plants (D). Sections were processed by chemical fixation. Scale bar = 2 μm. (B, C, E, F) Transmission electron microscopy of thin sections corresponding to mitochondria from wild-type (B, C) and *mterf15* (E, F) plants. Images B and E correspond to chemically fixed leaves. Images C and F correspond to leaves fixed by high-pressure freezing. Scale bar = 1 μm. Arrowheads pointed the inner membrane spaces in mitochondria. (G, H, I) Overlay of signals from chlorophyll autofluorescence (blue) with signals from the mitochondrial membrane potential marker tetramethylrhodamine methyl ester perchlorate (TMRM; red) in protoplasts from the wild-type (G), *mterf15* (H) and complementation lines (I). Scale bar = 10 μm. (J) Quantification of the mean fluorescent intensity after TMRM staining. Data are mean ± SD of 15 individual protoplasts from the wild-type, *mterf15* and complementation lines. Asterisks represent significant differences (*P<0.5, **P<0.01, ***P<0.001; Student's *t* test) relative to wild type.

mTERF15 is an RNA-binding protein that is involved in mitochondrial *nad2* intron 3 splicing and normal complex I activity

Because most eukaryotic mTERFs function as nucleic acid-binding proteins, we performed *in vitro* binding studies with dsDNA-cellulose and northwestern blot analyses to determine the dsDNA- and RNA-binding abilities of mTERF15. After incubation with commercial dsDNA-cellulose resin, most of the mTERF15 was present in the unbound fraction (Figure S5A), which implies a lack of dsDNA-binding activity for mTERF15. Additionally, southwestern blot analysis was employed to validate the DNA-binding ability of mTERF15. As shown in Figure S5B, there is no DNA-binding signal for the mTERF15 protein in southwestern blots. The above results imply that mTERF15 may not be a DNA-binding protein. Therefore, we investigated the RNA-binding ability of mTERF15 by northwestern blot analysis (Figure 3). Strep-tagged mTERF15 lacking the first N-terminal 34 amino acids, which contain a putative targeting signal (MW = 48 kDa), was separated by SDS-PAGE, transferred to a PVDF membrane and probed with a specific antibody against strep-tagged or radiolabeled total RNA prepared from 7-d-old Arabidopsis seedlings. The presence of the mTERF15-strep and MBP-strep fusion protein were confirmed by immunoblot analysis using an antibody against the strep-tag, and 48-kDa bands corresponding to mTERF15-strep were detected in a dose-dependent manner in the RNA-binding assay but not in the control. These results suggest that Arabidopsis mTERF15 functions as an RNA-binding protein.

Previously, Babiychuk *et al*. (2011) showed that one mTERF protein, BSM, is required for embryogenesis and plays a role in chloroplast intron splicing. Therefore, we examined 23 intron-splicing events in mitochondria by quantitative RT-PCR (qRT-PCR). In mitochondria, 9 genes require intron splicing for the complete maturation of their respective transcripts, and 4 of these contain a single intron: *rps3*, *cox2*, *ccmFc* and *rpl2*. The others (*nad1*, *nad2*, *nad4*, *nad5* and *nad7*) contain multiple introns and are required for the normal function of mitochondrial complex I. Using the specific primer sets designed by Longevialle *et al*. (2007), we examined individual splicing events in mitochondria obtained from wild-type, *mterf15* and complemented plants. All splicing events showed similar results in both the wild-type and complemented plants; however, the RNA splicing of *nad2* intron 3 (*nad2-3*) was significantly reduced in *mterf15* plants (Figure 4A).

In Arabidopsis, the mature *nad2* mRNA is formed by merging exons from 2 different transcripts, *nad2a* and *nad2b*, through one *trans*-splicing and 3 *cis*-splicing events. The genomic fragment of *nad2a* carries exons 1 and 2 of *nad2* separated by *nad2* intron 1. The other 3 exons of *nad2*, exons 3 through 5, are from *nad2b* transcripts (Figure 4B). To validate the results of the qRT-PCR analysis, we used RT-PCR to examine the accumulation of un-spliced *nad2b* transcripts in *mterf15* plants; the amplification of *nad2* exon 5 by primers 3F and 3R was used for normalization (lower panel in Figure 4B). RT-PCR with primers 1F and 2R revealed no mature transcripts from the splicing of *nad2* intron 3 in *mterf15* plants (Figure 4B). Additionally, the transcript levels were greater in *mterf15* plants than in wild-type plants upon amplification of the exon–intron junction with the primer sets 1F&1R and 2F&2R. This *nad2* intron 3 splicing defect was further confirmed by northern blot analysis. The *nad2b* transcripts revealed by probes for *nad2* exon 3 and 5 showed no mature *nad2* mRNAs in *mterf15* plants (m in Figure 4C) and was accompanied by the overaccumulation of un-spliced transcripts (u in Figure 4C). A similar accumulation of un-spliced transcripts in *mterf15* was confirmed using probes specific to *nad2* intron 3 (u in Figure 4C). This splicing defect was fully rescued in complement-

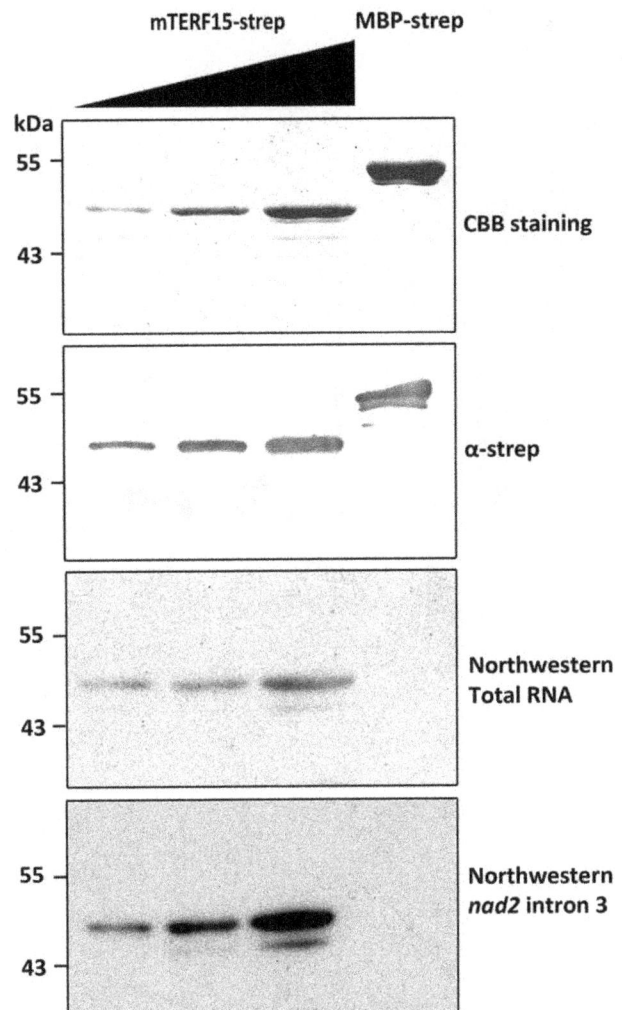

Figure 3. Northwestern blot assay of RNA-binding by mTERF15 recombinant protein. The first panel shows Coomassie brilliant blue (CBB) staining of MBP-strep and mTERF15-strep lacking the first 34 N-terminal amino acids. The second panel shows immunoblot analysis with anti-strep antibody to recognize the strep-tag. The third panel shows the northwestern blot analysis using radioisotope-labeled total RNA isolated from 7-d-old seedlings. The final panel shows the northwestern blot analysis using radioisotope-labeled *nad2* intron 3; signals are observed on the PVDF membrane corresponding to the RNAs bound to mTERF15.

ed plants. Another *cis*-splicing transcript, *nad1* exon 3, was used as a positive control, with comparable results to those observed in qRT-PCR: the splicing was normal, but the accumulation of *nad1-3* was greater in *mterf15* plants than in wild-type and complemented plants (Figure 4C).

Because mTERF15 functions as an RNA-binding protein and is involved in *nad2* intron 3 splicing, mTERF15 may directly interact with *nad2* intron 3. First, we confirmed the interaction between mTERF15 and full-length *nad2* intron 3 (~2.6 kb) by northwestern blot analysis (Figure 3, lowest panel). The interaction between mTERF15 and *nad2* intron 3 was stronger than the interaction with total RNA. We first examined the interaction between mTERF15 and full-length *nad2* intron 3 (~2.6 kb) by northwestern blot analysis (Figure 3, lowest panel). The interaction between mTERF15 and *nad2* intron 3 was stronger than the interaction with total RNA. Then RNA-EMSA (RNA-Electro-

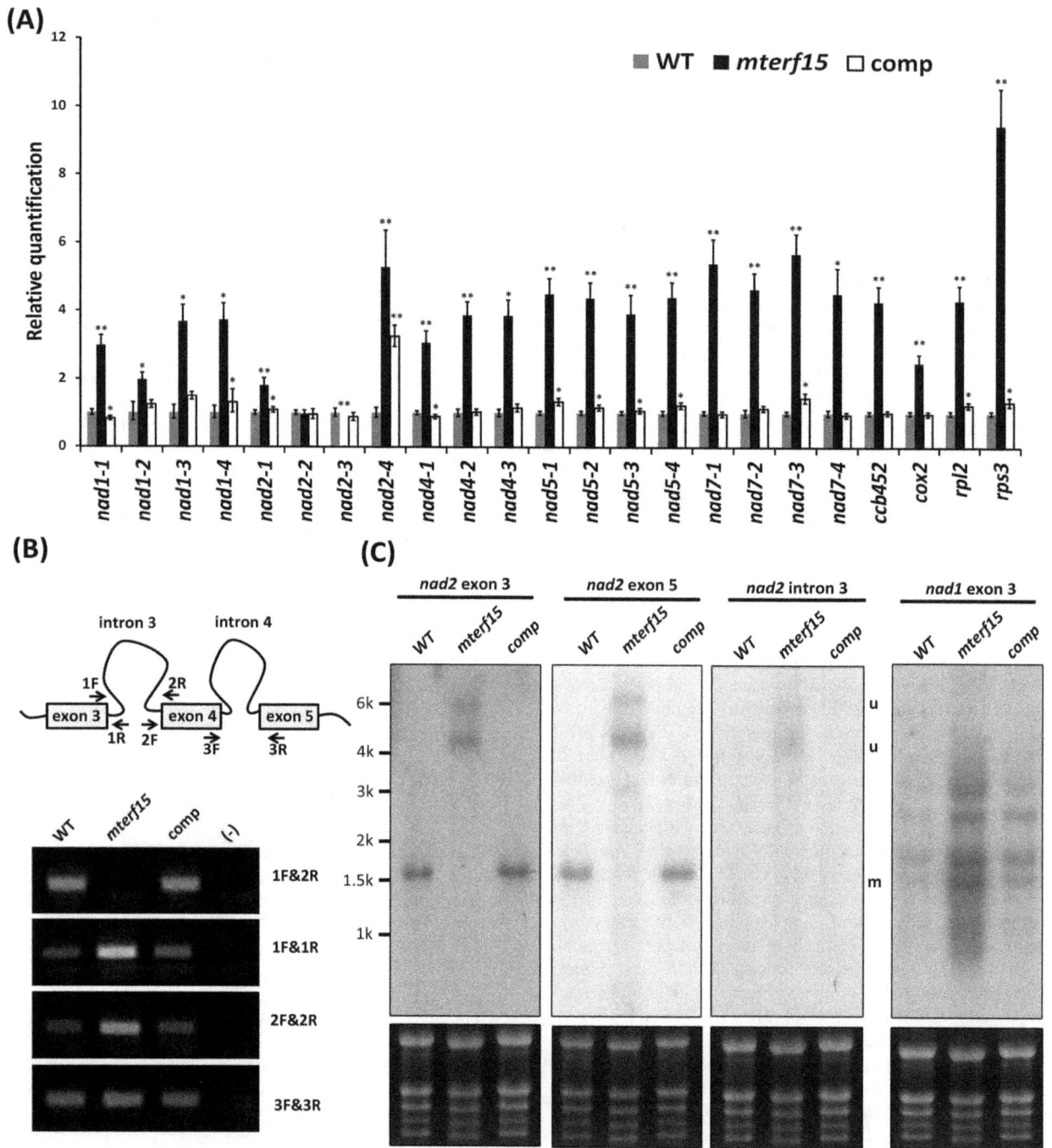

Figure 4. Mitochondrial splicing efficiency in *mterf15* mutants. (A) qRT-PCR analysis of mitochondrial RNA splicing in 7-d-old seedlings from the wild-type, *mterf15* and complementation lines. Data were normalized to the expression of 5 housekeeping genes (*YSL8, RPL5B, UBC, ACTIN2/ ACTIN8* and *TUB6*). Asterisks represent significant differences (*$P<0.5$, **$P<0.01$; Student's t test) relative to wild type. **(B)** Diagram of *nad2b* transcripts including exon 3 to 5 and RT-PCR analysis of transcript levels from the *nad2* exon–intron junction. Primers used for RT-PCR analysis are indicated with black arrows. **(C)** Northern blot analysis using probes recognizing *nad2b* transcripts (*nad2* exon 3 and *nad2* exon 5), *nad2* intron 3, and *nad1* exon 3. The radioisotope signal is shown in the top panel, and the rRNA signals shown in the lower panel were used as loading controls. Marker sizes are indicated. m, mature transcript; u, un-spliced transcript.

phoresis Mobility Shift Assay) was used to confirm the RNA-binding activity of mTERF15 (Figure S6). Additionally, two unrelated mitochondrial intron sequences, *ccmF$_C$* and *rpl2*, were used as the control probes in RNA-EMSA. Since the full length of *nad2* intron 3 is too large to be used as the probe for RNA-EMSA, we divided it into five fragments (approximately 500 bp each),

named Fragment 1-5 (Figure S6A). Fragment 1, part of domain 1, is the largest and substantially varies in sequence and secondary structure among characterized group II introns. Fragment 5 contains phylogenetically conserved and typical RNA structure found in all mitochondrial introns [51]. As shown in Figure S6B, only Fragment 1 could be recognized by mTERF15 in a dosage-dependent manner. Neither Fragments 2–5 nor *ccmFc* and *rpl2* could be recognized by mTERF15. Above results suggest that mTERF15 is an RNA-binding protein and is involved in the proper splicing of *nad2* intron 3.

In Arabidopsis, the *nad2* gene encodes a subunit of mitochondrial complex I (an NADH:ubiquinone oxidoreductase), which is required to translocate protons from the mitochondrial matrix into the intermembrane space. Complex I activity was therefore investigated in *mterf15*, in which the splicing of *nad2* intron 3 is defective. Isolated crude mitochondria were subjected to complex separation on clear native gels to visualize protein complexes by silver staining (Figure 5A) or to complex I activity assays (Figure 5B). The amount and activity of complex I were greatly reduced in *mterf15*, whereas the wild-type and complementation lines showed normal activity. Therefore, intron-splicing defects in *nad2* intron 3 led to reduced complex I formation and activity. Disturbing mitochondrial complex I impairs energy generation and leads to the extreme retardation of diverse aspects of growth and development in *mterf15* plants (Figure 1).

The steady-state levels of several mitochondrial proteins, including Nad9, Cox2, Cox3, AtpA, alternative oxidase (AOX) and the channel-forming protein Porin, were examined by immunoblot analysis of crude mitochondrial extracts from wild-type and *mterf15* plants (Figure 6). The steady-state levels of 3 proteins, Cox2, Cox3 and AtpA, were greater in *mterf15* plants than in wild-type plants. The level of alternative oxidase (AOX) was significantly increased in *mterf15* plants, which is consistent with a defect in NADH oxidation [46]. The level of another complex I subunit, Nad9, was decreased in *mterf15* plants, which may be due to an imbalance in the stoichiometry of multi-subunit complexes [47].

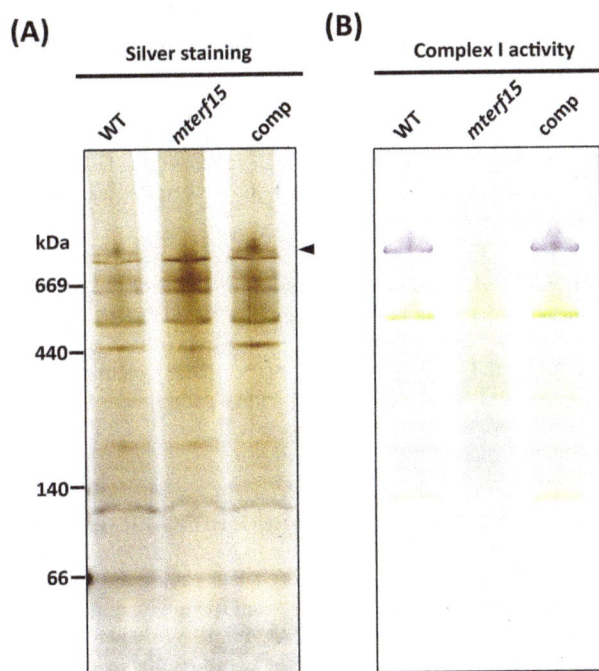

Figure 6. Relative accumulation of mitochondrial proteins in *mterf15* mutants. Crude mitochondria were isolated from the wild-type, *mterf15* and complementation lines. Following protein quantification based on the Bradford assay, equal amounts of protein were subjected to immunoblot analysis using various antibodies recognizing Nad9 [55], Cox2, Cox3, AtpA [56], and AOX [57], as well as Porin.

Discussion

Our genetic, molecular and biochemical results indicate that mTERF15 is a unique, mitochondria-localized protein (Figures S2 and S4) with RNA-binding activity (Fig. 3). This protein is critical for post-transcriptional modification in the RNA splicing of mitochondrial *nad2* intron 3 (Figure 4). Mutating *mTERF15* impaired the normal activity of mitochondrial respiratory chain complex I (Figure 5), thereby resulting in abnormal mitochondrial development (Figure 2) and the widespread retardation of plant growth and development (Figure 1).

mTERFs constitute a broad family of eukaryotic proteins that are essential for the initiation and termination of organellar transcription and the translation and replication of the organellar transcription machinery [12]; however, mounting evidence suggests an emerging role for these proteins in RNA splicing in plants [39]. The human genome contains only 4 genes encoding mTERF paralogs, whereas the Arabidopsis genome contains at least 35 genes for mTERF proteins with a diverse arrangement and number of mTERF motifs. Using transient assays and stable transgenic plants expressing GFP-tagged mTERF proteins, Babiychuk *et al.* (2011) identified 11 and 17 mTERF proteins that localize to chloroplasts and mitochondria, respectively. However, only a few mTERFs have been well studied and are essential for vegetative growth [24–26] and embryogenesis [22,23]. Whether Arabidopsis mTERF proteins share similar conserved molecular functions with their mammalian counterparts or have evolved additional regulatory mechanisms is unclear. In contrast to animal cells, plant cells harbor 2 types of nucleoid-containing organelles – chloroplasts and mitochondria. Moreover, plant mitochondrial genomes are much larger and more complex, requiring intron splicing for proper gene expression [48]. The biological functions of mTERF proteins may be complicated in plant cells because recent co-expression analyses of the 35 Arabidopsis mTERF members indicated the association of mTERF proteins with DNA and RNA metabolism [49]. Recently,

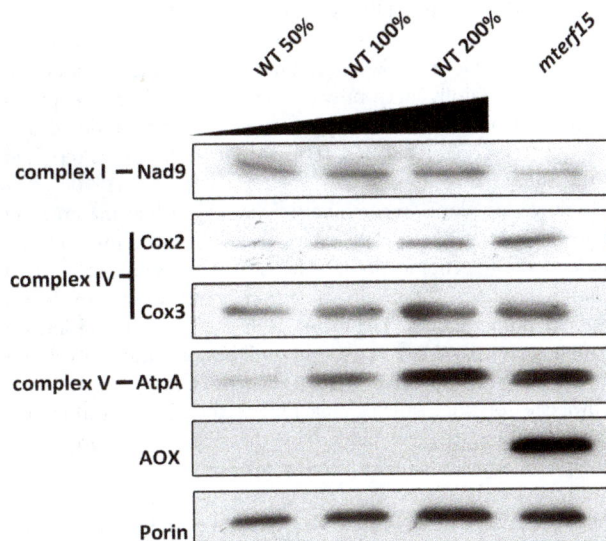

Figure 5. Clear native PAGE analysis of mitochondrial complex I activity. (A) Silver staining of the protein amount corresponding to crude mitochondria. **(B)** Complex I activity in crude mitochondria. Activity was visualized as a purple-blue color resulting from the utilization of the substrate (NADH) and the electron acceptor (nitroblue tetrazolium). Arrowhead indicated the band corresponding to Complex I.

it was found that Zm-mTERF4, an ortholog of Arabidopsis BSM/RUG2, is required for the splicing of several RNAs necessary for plastid translation in maize [39]. Therefore, an understanding of the mTERF family may provide new insights into the plant-specific functions of mTERF proteins in the transcriptional and post-transcriptional regulation of organellar nucleoids, as reported in this study.

In flowering plants, recombinogenic events such as the creation of intron-split genes greatly affect the dynamic nature of the mitochondrial genome [50]. The Arabidopsis mitochondrial genome contains 23 group II introns, and most are found dispersed in *nad* genes. The exons, including the flanking introns of these genes dispersed among the mtDNA, are transcribed and are the mRNAs generated by the splicing machinery. Until now, knowledge of the splicing machinery found in Arabidopsis mitochondria had been limited by the difficulty of organellar genome manipulation. However, a growing number of studies have revealed the involvement of a nuclear-encoded splicing factor in the excision of these introns [51]. Several proteins are involved in *nad2* intron splicing. For example, 2 proteins, ABA overly-sensitive 5 (ABO5) and RCC1/UVR8/GEF-like 3 (RUG3), were identified as splicing factors regulating the *cis*-splicing of *nad2* intron 3 [35,52]. ABO5, which encodes a PPR protein, was isolated from a mutational screen for ABA sensitivity and is required for *cis*-splicing of *nad2* intron 3 in mitochondria [35]. Another protein, *RUG3*, which encodes an RCC1/UVR8-like protein, is responsible for the efficient splicing of *nad2* introns 2 and 3 in mitochondria [52]. The mTERF15 protein, identified in this study, is required for *nad2* intron 3 RNA splicing, as demonstrated by RT-PCR and by northern blotting showing the accumulation of *nad2* intron 3 in *mterf15* plants (Figure 4B–C). Moreover, RNA-binding assays also demonstrated the direct interaction between mTERF15 and *nad2* intron 3. Our results suggest that mTERF15 is a new splicing factor in mitochondria. The next steps are to identify the specific elements of *nad2* intron 3 that are required for the binding of mTERF15 and to explore the relationships and/or potential interactions between ABO5, RUG3 and mTERF15 in *nad2* intron 3 splicing.

The *nad2* intron 3 splicing defect led to the loss of mitochondrial complex I activity in *mterf15*. Similar results were observed with the *RUG3* mutant [52]. Both mutants showed reduced root growth, small stature, and growth retardation. Another complex I mutant, *ndufs4* (NADH dehydrogenase [ubiquinone] fragment S subunit 4), also exhibited a similar phenotype [53]. Although approximately one-third of mitochondrial ATP production is associated with complex I, the presence of alternative NADH dehydrogenases allows the electron transport chain to bypass complex I when complex I efficiency is reduced or when plants encounter stress [54]. Interestingly, *abo5*, *rug3* and *ndufs4* plants produced viable seeds and propagated to the next generation, whereas *mterf15* showed a more severe phenotype. Homozygous *mterf15* plants showed a similar phenotype as these mutants in terms of complex I deficiency; however, seeds from homozygous *mterf15* plants could not germinate, even in solid medium over three months. It is possible that mTERF15 in maternal tissues of heterozygotic *mterf15* embryo may partially contribute to the developing homozygous *mterf15* seeds, which is completely absent in the developing seeds of homozygous *mterf15* plants. This suggests that the germination defect of *mterf15* is not only due to embryo lethality but also caused by aberrant reproductive organs, especially maternal tissues. The defective maternal tissues may reduce the flow of nutrient into the endosperm and eventually cause seed lethality [58]. Therefore,

mTERF15 may have additional roles in mitochondria, as observed in other eukaryotes, during embryogenesis and seed development.

Here, we identify mTERF15 as an RNA-binding protein that is required for *nad2* intron 3 splicing in mitochondria. The disruption of this splicing event leads to decreased complex I activity as well as growth and developmental retardation in *mterf15* mutants. Further studies of the detailed mechanism of mTERF15 are required to reveal the molecular mechanisms involved in post-transcription regulation in Arabidopsis mitochondria.

Supporting Information

Figure S1 *mTERF15p::mTERF15-GFP* **fully complements developmental defects found in** *mterf15* **homozygous mutant.** 7-day-old seedlings from 6 independent complementation lines, homozygotes for *mTERF15* mutation and harbouring at least one copy of the transgenes *mTERF15p::mTERF15-GFP*. Seeds were surface-sterilized and germinated on half MS medium under 22°C with light period control (16-h light/8-h dark).

Figure S2 **Protein alignment and phylogenetic tree of mTERF15 and homologs.** (A) Protein sequence alignment of mTERF15 and its homologs in *Vitis vinifera* (Vv), *Populus trichocarpa* (Pt), *Brachypodium distachyon* (Bd), *Sorghum bicolor* (Sb) and *Zea mays* (Zm). The alignment was generated by use of CLUSTALW, which creates pairwise alignments to calculate the divergence between pairs of sequences. Five mTERF motifs present in all homologs are underlined in black. Asterisks indicate the location of highly conserved proline residues. (B) Phylogenetic tree of mTERF15 and its homologs. The phylogenetic tree was created with use of Molecular Evolutionary Genetics Analysis (MEGA) by the Unweighted Pair Group Method with Arithmatic Mean (UPGMA) method.

Figure S3 **Spatial expression profile of** *mTERF15*. qRT-PCR analysis of the expression of *mTERF15* in seedlings, roots, stems, rosette leaves, flowers, siliques and pollen. Primers used in this experiment are in Table S1.

Figure S4 **Subcellular localization of mTERF15 protein.** Confocal microscopy of the localization of the mTERF15 protein in root hairs of the *mterf15* complementation line harboring mTERF15p::mTERF15-GFP. (A) Signal corresponding to the mTERF15-GFP fusion protein; (B) signal corresponding to the MitoTracker marker; and (C) merged signals from GFP and the MitoTracker marker. Scale bar = 10 μm.

Figure S5 **DNA-binding assay with mTERF15 recombinant protein.** (A) *In vitro* binding studies of dsDNA-cellulose with mTERF15-GST and GST. mTERF15-GST and GST were incubated with dsDNA-cellulose for 2 h at room temperature. The resin was washed with washing buffer 3 times and after different salt concentrations to elute bound protein. The detailed experimental procedure was previously described (Wobbe and Nixon, 2013). (B) Southwestern blot assay of DNA-binding ability by mTERF15-strep. The upper panel is Coomassie brilliant blue (CBB) staining of MBP-strep and mTERF15-strep without the first N-terminal 34 amino acid. The lower panel is southwestern blot analysis with radioisotope-labeled HindIII-digested mtDNA. **Methods of Figure S5: DNA-binding assay with dsDNA-cellulose resin.** This experimental procedure has been described previously (Wobbe & Nixon, 2013). Briefly, 1.8 μM of the mTERF15-GST fusion protein and GST control were incubated with 60 mg dsDNA-cellulose resin (Sigma) in DCBB buffer [50 mM HEPES, pH 8.0, 50 mM NaCl, 1 mM EDTA, 1 mM beta-mercaptoethanol and protease inhibitor cocktail (Roche)] at room temperature for 2 hr and under gentle rotation. The resin was washed 3 times with DCBB buffer, then bound protein was eluted at different salt concentrations (100, 300, 600 and 1000 mM NaCl in DCBB buffer). The samples corresponding to each step were collected and analyzed on SDS-PAGE. **Southwestern blot assay.** The recombinant protein were separated on 12% SDS-polyacrylamide gels and transferred onto a PVDF membrane and then renatured overnight in renaturation buffer (0.1 M Tris-HCl, pH 7.5 and 0.1% (v/v) NP-40) at 4°C. After 4 washes with renaturation buffer (15 min), blots were blocked with blocking buffer [(10 mM Tris-HCl, pH 7.5, 2 mM DTT, 5% (w/v) BSA and 0.01% (v/v) Triton X-100] at room temperature for 5 min. Mitochondrial DNA was isolated from crude mitochondria fraction, digested with HindIII and labeled with radioisotope using T4 polynucleotide kinase (Fermentas). Hybridization was performed at 4°C overnight in the presence of 5 μg labeled mtDNA. Blots were washed 4 times (5 min) in washing buffer (10 mM Tris-HCl, pH 7.5, 2 mM DTT) to remove unbound DNA. Radioisotope signals from PVDF membrane were detected via autoradiography on X-ray film (Kodak).

Figure S6　RNA-EMSA of mTERF15 recombinant protein with different mitochondrial transcribed intron fragments. (A) The diagram of *nad2* intron 3 and the regions and sizes of 5 fragments in *nad2* intron 3 used for assays. (B) Different concentration mTERF15 recombinant proteins (20, 10, 5 and 0 nM) were incubated with 100 nM radioisotope labeled indicated RNA probes. **Methods of Figure S6:** mTERF15 fusion protein were purified as mentioned in Northwestern blot analysis. For *in vitro* transcription, mitochondrial introns were amplified and cloned to pJET vector. Primer sets are nad2int3-1F "ACGCCAAGCTATTTAGGTGACACTATAGAATACGGGCGGCTGTAGGACGGAC" and nad2int3-1R "CTGTTCACCGTTGGATCTCGCC" for *nad2* intron 3 Fragment 1; nad2int3-5F "ACGCCAAGCTATTTAGGTGACACTATAGAATACGCGTGTTATCTGAAGGGAGCACG" and nad2int3-5R "GGGGGAGGGGGTTTTCTTCG" for *nad2* intron 3 Fragment 5; rpl2-F "ACGCCAAGCTATTTAGGTGACACTATAGAATA-

CATGAGACCAGGGAGAGCAAGAGCAC" and rpl2-mid-intron-R "CGTTGCTAAGCCAAGGTCCC" for *rpl2* intron; ccmFc-F "ACGCCAAGCTATTTAGGTGACACTATAGAATACATGGTCCAACTACATAACTTTTTC" and ccmFc-mid-intron-R "GCTTTGCCAACACAACATTAGG" for *ccmF_C* intron. Then, RNA probes were *in vitro* synthesized using SP6 polymerase (Promega) and labeled with $\gamma^{32}P$ ATP using T4 polynucleotide kinase (Fermentas). Different concentration of mTERF15 recombinant proteins (20, 10, 5 and 0 nM) were incubated with 100 nM *in vitro* synthesized RNAs for 30 min at room temperature in binding buffer (20 mM Tris/HCl pH 7.5, 180 mM NaCl, 2 mM dithiothreitol, 17 μg/μl BSA, 0.5 mM EDTA, and 20 μg/ml heparin). The mixtures were separated by 5% polyacrylamide gel and gel was dried. Signals were detected via autoradiography on X-ray film (Kodak).

Table S1　Primer sequences used in mTERF125 cloning, gene expression, and probe preparation for northern blotting and RT-PCR for the *nad2* exon–intron junction.

Table S2　Primer sequences for determining RNA splicing efficiency in mitochondria.

Table S3　Protein identity and similarity of mTERF15 with its plant homologs and mTERF1 in metazoans.

Acknowledgments

We thank Dr. Géraldine Bonnard (Institut de Biologie Moléculaire des Plantes, Centre National de la Recherche Scientifique, Strasbourg, France) for generously sharing the anti-NAD9 antibody, Ms. Mei-Jane Fang for her assistance with qPCR (DNA Analysis Core Laboratory, Institute of Plant and Microbial Biology, Academia Sinica) and Dr. Wann-Neng Jane for assistance with transmission and scanning electron microscopy (Plant Cell Biology Core Laboratory, Institute of Plant and Microbial Biology, Academia Sinica).

Author Contributions

Conceived and designed the experiments: YWH HJW GYJ. Performed the experiments: YWH HJW. Analyzed the data: YWH MHH HLH GYJ. Contributed reagents/materials/analysis tools: YWH MHH HLH GYJ. Wrote the paper: YWH MHH GYJ.

References

1. Andersson SG, Zomorodipour A, Andersson JO, Sicheritz-Ponten T, Alsmark UC, et al. (1998) The genome sequence of *Rickettsia prowazekii* and the origin of mitochondria. Nature 396: 133–140.
2. Gray MW, Burger G, Lang BF (1999) Mitochondrial evolution. Science 283: 1476–1481.
3. Gkikas I, Petratou D, Tavernarakis N (2014) Longevity pathways and memory aging. Front Genet 5: 155.
4. Unseld M, Marienfeld JR, Brandt P, Brennicke A (1997) The mitochondrial genome of *Arabidopsis thaliana* contains 57 genes in 366,924 nucleotides. Nat Genet 15: 57–61.
5. Martin W, Herrmann RG (1998) Gene transfer from organelles to the nucleus: how much, what happens, and why? Plant Physiol 118: 9–17.
6. Martin W, Stoebe B, Goremykin V, Hapsmann S, Hasegawa M, et al. (1998) Gene transfer to the nucleus and the evolution of chloroplasts. Nature 393: 162–165.
7. Welchen E, Garcia L, Mansilla N, Gonzalez DH (2014) Coordination of plant mitochondrial biogenesis: keeping pace with cellular requirements. Front Plant Sci 4: 551.
8. Liere K, Weihe A, Borner T (2011) The transcription machineries of plant mitochondria and chloroplasts: composition, function, and regulation. J Plant Physiol 168: 1345–1360.
9. Binder S, Brennicke A (2003) Gene expression in plant mitochondria: transcriptional and post-transcriptional control. Philos Trans R Soc Lond B - Biol Sci 358: 181–188; discussion 188–189.
10. Hammani K, Giege P (2014) RNA metabolism in plant mitochondria. Trends Plant Sci 19: 380–389.
11. Kruse B, Narasimhan N, Attardi G (1989) Termination of transcription in human mitochondria: identification and purification of a DNA binding protein factor that promotes termination. Cell 58: 391–397.
12. Linder T, Park CB, Asin-Cayuela J, Pellegrini M, Larsson NG, et al. (2005) A family of putative transcription termination factors shared amongst metazoans and plants. Curr Genet 48: 265–269.
13. Roberti M, Polosa PL, Bruni F, Manzari C, Deceglie S, et al. (2009) The MTERF family proteins: mitochondrial transcription regulators and beyond. Biochim Biophys Acta 1787: 303–311.
14. Yakubovskaya E, Guja KE, Mejia E, Castano S, Hambardjieva E, et al. (2012) Structure of the essential MTERF4:NSUN4 protein complex reveals how an MTERF protein collaborates to facilitate rRNA modification. Structure 20: 1940–1947.
15. Yakubovskaya E, Mejia E, Byrnes J, Hambardjieva E, Garcia-Diaz M (2010) Helix unwinding and base flipping enable human MTERF1 to terminate mitochondrial transcription. Cell 141: 982–993.
16. Martin M, Cho J, Cesare AJ, Griffith JD, Attardi G (2005) Termination factor-mediated DNA loop between termination and initiation sites drives mitochondrial rRNA synthesis. Cell 123: 1227–1240.
17. Pellegrini M, Asin-Cayuela J, Erdjument-Bromage H, Tempst P, Larsson NG, et al. (2009) MTERF2 is a nucleoid component in mammalian mitochondria. Biochim Biophys Acta 1787: 296–302.

18. Wenz T, Luca C, Torraco A, Moraes CT (2009) mTERF2 regulates oxidative phosphorylation by modulating mtDNA transcription. Cell Metab 9: 499–511.

19. Park CB, Asin-Cayuela J, Camara Y, Shi Y, Pellegrini M, et al. (2007) MTERF3 is a negative regulator of mammalian mtDNA transcription. Cell 130: 273–285.

20. Camara Y, Asin-Cayuela J, Park CB, Metodiev MD, Shi Y, et al. (2011) MTERF4 regulates translation by targeting the methyltransferase NSUN4 to the mammalian mitochondrial ribosome. Cell Metab 13: 527–539.

21. Spahr H, Habermann B, Gustafsson CM, Larsson NG, Hallberg BM (2012) Structure of the human MTERF4-NSUN4 protein complex that regulates mitochondrial ribosome biogenesis. Proc Natl Acad Sci U S A 109: 15253–15258.

22. Babiychuk E, Vandepoele K, Wissing J, Garcia-Diaz M, De Rycke R, et al. (2011) Plastid gene expression and plant development require a plastidic protein of the mitochondrial transcription termination factor family. Proc Natl Acad Sci U S A 108: 6674–6679.

23. Meskauskiene R, Wursch M, Laloi C, Vidi PA, Coll NS, et al. (2009) A mutation in the Arabidopsis mTERF-related plastid protein SOLDAT10 activates retrograde signaling and suppresses $^1O_2$-induced cell death. Plant J 60: 399–410.

24. Quesada V, Sarmiento-Manus R, Gonzalez-Bayon R, Hricova A, Perez-Marcos R, et al. (2011) Arabidopsis *RUGOSA2* encodes an mTERF family member required for mitochondrion, chloroplast and leaf development. Plant J 68: 738–753.

25. Robles P, Micol JL, Quesada V (2012) Arabidopsis MDA1, a nuclear-encoded protein, functions in chloroplast development and abiotic stress responses. PLoS One 7: e42924.

26. Kim M, Lee U, Small I, des Francs-Small CC, Vierling E (2012) Mutations in an Arabidopsis mitochondrial transcription termination factor-related protein enhance thermotolerance in the absence of the major molecular chaperone HSP101. Plant Cell 24: 3349–3365.

27. Asakura Y, Barkan A (2007) A CRM domain protein functions dually in group I and group II intron splicing in land plant chloroplasts. Plant Cell 19: 3864–3875.

28. Asakura Y, Bayraktar OA, Barkan A (2008) Two CRM protein subfamilies cooperate in the splicing of group IIB introns in chloroplasts. RNA 14: 2319–2332.

29. Zmudjak M, Colas des Francs-Small C, Keren I, Shaya F, Belausov E, et al. (2013) mCSF1, a nucleus-encoded CRM protein required for the processing of many mitochondrial introns, is involved in the biogenesis of respiratory complexes I and IV in Arabidopsis. New Phytol 199: 379–394.

30. de Longevialle AF, Hendrickson L, Taylor NL, Delannoy E, Lurin C, et al. (2008) The pentatricopeptide repeat gene *OTP51* with two LAGLIDADG motifs is required for the *cis*-splicing of plastid *ycf3* intron 2 in *Arabidopsis thaliana*. Plant J 56: 157–168.

31. de Longevialle AF, Meyer EH, Andres C, Taylor NL, Lurin C, et al. (2007) The pentatricopeptide repeat gene *OTP43* is required for *trans*-splicing of the mitochondrial *nad1* Intron 1 in *Arabidopsis thaliana*. Plant Cell 19: 3256–3265.

32. Koprivova A, des Francs-Small CC, Calder G, Mugford ST, Tanz S, et al. (2010) Identification of a pentatricopeptide repeat protein implicated in splicing of intron 1 of mitochondrial *nad7* transcripts. J Biol Chem 285: 32192–32199.

33. Chateigner-Boutin AL, des Francs-Small CC, Delannoy E, Kahlau S, Tanz SK, et al. (2011) OTP70 is a pentatricopeptide repeat protein of the E subgroup involved in splicing of the plastid transcript *rpoC1*. Plant J 65: 532–542.

34. Khrouchtchova A, Monde RA, Barkan A (2012) A short PPR protein required for the splicing of specific group II introns in angiosperm chloroplasts. RNA 18: 1197–1209.

35. Liu Y, He J, Chen Z, Ren X, Hong X, et al. (2010) *ABA overly-sensitive 5 (ABO5)*, encoding a pentatricopeptide repeat protein required for *cis*-splicing of mitochondrial *nad2* intron 3, is involved in the abscisic acid response in Arabidopsis. Plant J 63: 749–765.

36. Francs-Small CC, Kroeger T, Zmudjak M, Ostersetzer-Biran O, Rahimi N, et al. (2012) A PORR domain protein required for *rpl2* and *ccmF_c* intron

37. Kroeger TS, Watkins KP, Friso G, van Wijk KJ, Barkan A (2009) A plant-specific RNA-binding domain revealed through analysis of chloroplast group II intron splicing. Proc Natl Acad Sci U S A 106: 4537–4542.

38. Watkins KP, Rojas M, Friso G, van Wijk KJ, Meurer J, et al. (2011) APO1 promotes the splicing of chloroplast group II introns and harbors a plant-specific zinc-dependent RNA binding domain. Plant Cell 23: 1082–1092.

39. Hammani K, Barkan A (2014) An mTERF domain protein functions in group II intron splicing in maize chloroplasts. Nucleic Acids Res doi: 10.1093/nar/gku112.

40. Bechtold N, Pelletier G (1998) In planta Agrobacterium-mediated transformation of adult *Arabidopsis thaliana* plants by vacuum infiltration. Methods Mol Biol 82: 259–266.

41. Thangasamy S, Chen PW, Lai MH, Chen J, Jauh GY (2012) Rice LGD1 containing RNA binding activity affects growth and development through alternative promoters. Plant J 71: 288–302.

42. Pineau B, Layoune O, Danon A, De Paepe R (2008) L-galactono-1,4-lactone dehydrogenase is required for the accumulation of plant respiratory complex I. J Biol Chem 283: 32500–32505.

43. Wittig I, Karas M, Schagger H (2007) High resolution clear native electrophoresis for in-gel functional assays and fluorescence studies of membrane protein complexes. Mol Cell Proteomics 6: 1215–1225.

44. Cupp JD, Nielsen BL (2012) *Arabidopsis thaliana* organellar DNA polymerase IB mutants exhibit reduced mtDNA levels with a decrease in mitochondrial area density. Physiol Plant 49: 91–103.

45. Udy DB, Belcher S, Williams-Carrier R, Gualberto JM, Barkan A (2012) Effects of reduced chloroplast gene copy number on chloroplast gene expression in maize. Plant Physiol 160: 1420–1431.

46. Rasmusson AG, Geisler DA, Moller IM (2008) The multiplicity of dehydrogenases in the electron transport chain of plant mitochondria. Mitochondrion 8: 47–60.

47. Adam Z, Ostersetzer O (2001) Degradation of unassembled and damaged thylakoid proteins. Biochem Soc Trans 29: 427–430.

48. Gray MW (1989) Origin and evolution of mitochondrial DNA. Annu Rev Cell Biol 5: 25–50.

49. Kleine T (2012) *Arabidopsis thaliana* mTERF proteins: evolution and functional classification. Front Plant Sci 3: 233.

50. Bonen L (2008) *Cis*- and *trans*-splicing of group II introns in plant mitochondria. Mitochondrion 8: 26–34.

51. de Longevialle AF, Small ID, Lurin C (2010) Nuclearly encoded splicing factors implicated in RNA splicing in higher plant organelles. Mol Plant 3: 691–705.

52. Kuhn K, Carrie C, Giraud E, Wang Y, Meyer EH, et al. (2011) The RCC1 family protein RUG3 is required for splicing of *nad2* and complex I biogenesis in mitochondria of *Arabidopsis thaliana*. Plant J 67: 1067–1080.

53. Meyer EH, Tomaz T, Carroll AJ, Estavillo G, Delannoy E, et al. (2009) Remodeled respiration in *ndufs4* with low phosphorylation efficiency suppresses Arabidopsis germination and growth and alters control of metabolism at night. Plant Physiol 151: 603–619.

54. Millar AH, Whelan J, Soole KL, Day DA (2011) Organization and regulation of mitochondrial respiration in plants. Annu Rev Plant Biol 62: 79–104.

55. Lamattina L, Gonzalez D, Gualberto J, Grienenberger JM (1993) Higher plant mitochondria encode an homologue of the nuclear-encoded 30-kDa subunit of bovine mitochondrial complex I. Eur J Biochem 217: 831–838.

56. Luethy MH, Horak A, Elthon TE (1993) Monoclonal Antibodies to the α- and β-Subunits of the Plant Mitochondrial F1-ATPase. Plant Physiol 101: 931–937.

57. Elthon TE, Nickels RL, McIntosh L (1989) Monoclonal antibodies to the alternative oxidase of higher plant mitochondria. Plant Physiol 89: 1311–1317.

58. Chaudhury AM, Berger F (2001) Maternal control of seed development. Semin Cell Dev Biol 12: 381–386.

4

ZNF667/Mipu1 Is a Novel Anti-Apoptotic Factor That Directly Regulates the Expression of the Rat Bax Gene in H9c2 Cells

Lei Jiang, Hao Wang, Chunli Shi, Ke Liu, Meidong Liu, Nian Wang, Kangkai Wang, Huali Zhang, Guiliang Wang, Xianzhong Xiao*

Department of Pathophysiology, Xiangya School of Medicine, Central South University, 110 Xiangya Road, Changsha, Hunan 410078, P. R. China

Abstract

ZNF667/Mipu1, a C_2H_2-type zinc finger transcription factor, was suggested to play an important role in oxidative stress. However, none of the target genes or potential roles of ZNF667 in cardiomyocytes have been elucidated. Here, we investigated the functional role of ZNF667 in H9c2 cell lines focusing on its molecular mechanism by which it protects the cells from apoptosis. We found that ZNF667 inhibited the expression and the promoter activity of the rat proapoptotic gene Bax gene, and at the same time prevented apoptosis of H9c2 cells, induced by H_2O_2 and Dox. Western immunoblotting analysis revealed that ZNF667 also inhibited Bax protein expression, accompanied by attenuation of the mitochondrial translocation of Bax protein, induced by H_2O_2. EMSA and target detection assay showed that the purified ZNF667 fusion proteins could interact with the Bax promoter sequence in vitro, and this interaction was dependent upon the ZNF667 DNA binding sequences or its core sequence in the promoter. Furthermore, ChIP assay demonstrated that a stimulus H_2O_2 could enhance the ability of ZNF667 protein binding to the promoter. Finally, a reporter gene assay showed that ZNF667 could repress the activity of the Bax gene promoter, and the repression was dependent upon its binding to the specific DNA sequence in the promoter. Our work demonstrates that ZNF667 that confers cytoprotection is a novel regulator of the rat Bax gene, mediating the inhibition of the Bax mRNA and protein expression in H9c2 cardiomyocytes in response to H_2O_2 treatment.

Editor: Guo-Chang Fan, University of Cincinnati, College of Medicine, United States of America

Funding: This work was supported by the Major National Basic Research Program of China (2007CB512007 for X.X.) and the National Natural Science Foundation of China (30971205 for L.J.). The funders had no role in study design, data collection and analysis, decision to publish, or preparation of the manuscript.

Competing Interests: The authors have declared that no competing interests exist.

* Email: xianzhongxiao@xysm.net

Introduction

Zinc finger proteins are a superfamily of transcription factors. The rat zinc finger protein 667, ZNF667, provisionally named myocardial ischemic preconditioning upregulated protein 1 (Mipu1) in our lab due to its upregulation during myocardial ischemia/reperfusion, belongs to the KRAB/C_2H_2 zinc finger proteins that contains a KRAB domain at its N-terminus and 14 zinc fingers at its C-terminus. Both the ZNF667 mRNA and protein are expressed abundantly and predominantly in the brain and heart [1,2]. It has also been shown that ZNF667 is a nuclear protein that is localized to the nucleus through its KRAB domain or the linker adjacent to its zinc finger region, unlike most of the KRAB/C_2H_2 zinc finger proteins where their zinc finger motifs are required for nuclear targeting. Like other KRAB/C_2H_2 zinc finger proteins, ZNF667 is a DNA binding protein and binds to the specific core sequence 5′-CTTA-3′, acting as a transcriptional repressor [3], suggesting a role in the regulation of downstream genes. Studies have shown that H_2O_2 induces ZNF667 expression in rat heart-derived H9c2 cells [4,5], but to date the target gene(s) of ZNF667 is unknown. Based on informatic analysis, we have

shown certain genes, including several Bcl-2 family members, contain the potential DNA sequence in their promoter regions; one of them is the Bax gene from the rat which contains up to six sites [6]. Therefore, we hypothesized that ZNF667 might be involved in Bax regulation and interfere with the apoptotic pathway in rat heart-derived H9c2 cells in response to oxidative stress.

Bcl-2 family proteins play an important role in cardiomyocyte apoptosis during oxidative stress [7-9]. The Bcl-2 family of proteins consists of both antiapoptotic (such as Bcl-2 and Bcl-$_{XL}$) and proapoptotic (such as Bax and Bak) proteins [10,11]. It has been documented that the mitochondrial apoptotic pathway is controlled by the members of Bcl-2 family [10,11]. Bax together with Bak is a requisite gateway to the mitochondrial apoptotic machinery because cells that are doubly deficient in Bax and Bak fail to release cytochrome c and are resistant to all apoptotic stimuli that activate the intrinsic pathway [12,13]. In healthy cells, Bax is predominantly cytosolic and binds to its antiapoptotic partners (for example Bcl-2, Bcl-$_{XL}$), which prevents full Bax activation and apoptosis [14]. Upon cellular stress, when the capacity of its partners can be overwhelmed (for example Bax upregulation), Bax begins to oligomerize and translocates to the mitochondrial membrane. When Bax punctures the outer mitochondrial membrane (OMM), the mPTP is open and the mitochondrial contents release into cytoplasm [15-17]. Therefore, Bax is a major final mediator of the OMM permeabilization.

Because ZNF667 is a transcription repressor, we focused on its repressive roles in this study. Using H_2O_2 and Dox (doxorubicin) as stimuli, we have demonstrated that ZNF667 confers cytoprotection. We show for the first time that ZNF667 is a novel transcriptional regulator of the rat Bax gene, negatively regulating the expression of Bax by directly binding to the functional binding elements in the Bax promoter sequence in H9c2 cells.

Materials and Methods

Cell culture, transfection and treatment

The rat embryonic heart-derived H9c2 cells were obtained from the American Tissue Culture Collection (ATCC), and were grown in Dulbecco's modified Eagle's medium (DMEM) supplemented with 10% fetal calf serum (Gibco). Murine macrophage-like Raw264.7 cells, obtained from the Shanghai Type Culture Collection (Shanghai, China), were also grown in DMEM with 10% fetal calf serum. All cell lines were kept in an incubator at 37°C with 5% CO_2 and 95% air in a humidified atmosphere. Transient transfection was done with Lipofectamine 2000 (Invitrogen) according to the manufacture's instructions. H_2O_2 was diluted in PBS, and then diluted in culture medium. Dox was directly added to the cell culture medium at the described concentrations.

Total RNA and quantitative real-time PCR

Total RNA was extracted from H9c2 cells using the Trizol method (Invitrogen) according to manufacture's instructions. Total RNA (1μg) was reverse-transcribed using AMV Reverse Transcriptase (TaKaRa). Quantitative RT-PCR (qRT-PCR) was performed in triplicates in an ABI PRISM 7500HT System with 50 ng cDNA product, 125 nM of primers, and Fast SYBR Green Master Mix (Applied Biosystems). The primers used were listed as follows: forward 5'-CAGGAGGAATGGGAATGGC-3' and reverse 5'- TTAGACCCTAATTCAGGACCCC-3' for ZNF667 (194 bp), forward 5'- TGGTTGCCCTCTTCTACTTTG -3' and reverse 5' GTCACTGTCTGCCATGTGGG 3' for Bax (193 bp) and forward 5'-ACTCCCATTCTTCCACCTTTG-3' and 5'- CCCTGTTGCTGTAGCCATATT-3' for GAPDH (105 bp). Relative expression of target genes was calculated by the $2^{-\triangle\triangle CT}$ method as previously described [6]. Final data were presented as fold changes against control.

Construction of plasmids

The plasmids pcDNA3.1-ZNF667, pGEX-ZF, pEGFP-ZNF667, pEGFP-ZF, pEGFP-KRAB and pRNA-U6.1-ZNF667 (shRNA) have been constructed as previously described [3,18]. The full-length ZNF667 (pcDNA3.1-ZNF667) DNA was used as a template and Pyrobest (Takara) was used as the DNA polymerase for the PCR amplification of the truncated ZNF667. For the construction of the reporter plasmid, the DNA sequence corresponding to the bases -812 to -53 of the rat Bax promoter (Genbank accession number: AB046392) (Fig. 1) was amplified by PCR using the following primers: 5'-GGGGTACCAAAGTAA-GAGGATAAGAGAG-3' and 5'-AAAAAGCTTTCACG TGACTCC CCGCAGAC-3', and the PCR product was double digested with Kpn I and Hind III and inserted into pre-digested pGL3 basic vector (Promega) to produce the luciferase reporter plasmid pBa-luc, since both ends of the promoter sequence

contained the sites of the two restriction endonucleases (underlined). The mutation of the putative ZNF667 binding site or its core sequence was performed by changing the core sequence CTTA to GCGC in the promoter using fusion PCR to produce the mutant luciferase reporter plasmid pBM-luc. All of the constructs were verified by DNA sequencing analysis (Invitrogen, Shanghai, China).

Recombinant protein expression and purification

All fusion proteins were produced in Escherichia coli strain DH5α and expression was induced with 1.0 mM isopropyl β-D-thiogalactoside for 4 h according to the supplier's instructions (Amersham Pharmacia Biotech). All fusion proteins were expressed and purified as described previously [3].

Target detection assays

Target detection assay was done essentially as described in Ref [3]. To biotin-label Bax promoter, the chemically-synthesized biotin-labeled primer pairs (See "Constructions of plasmids" described above) were used for PCR amplification of Bax promoter using the luciferase reporter plasmid pBa-luc constructed above as a template.

EMSA

EMSA was performed as described in Ref [3]. For mutation assay, the chemically-synthesized biotin-labeled primer pairs (See "Constructions of plasmids" described above) were used for the PCR amplification of the mutant promoter sequence using the mutant luciferase reporter plasmid pBM-luc constructed above as a template to biotin-label the mutant Bax promoter sequence.

Cell mitochondrial extraction

In order to estimate the mitochondrial Bax protein, the mitochondrial fraction was prepared using a test kit (Pierce, cat#89874) according to the manufacturer's instructions. The purity of the protein fraction was confirmed by immunoblotting with enolase (Santa Cruz Biotechnology) and VDAC (Thermo Fisher Scientific) antibodies for cytosol and for mitochondria, respectively.

Chromatin immunoprecipitation (ChIP)

The ChIP assay was performed using a ChIP assay kit (Millipore, cat#17-295) according to the manufacturer's instructions with some modifications. Briefly, logarithmically growing H9c2 cells (1×10^8 cells) pre-treated with 0.5 mM H_2O_2 for 6 h or untreated were cross-linked using formaldehyde (final concentration 1% vol/vol) in DMEM medium for 10 min on ice. After stopping the cross-linking and sonicating on ice to make soluble

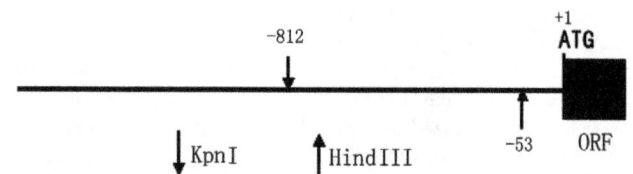

Figure 1. The scheme of the rat Bax gene promoter sequence. The Bax promoter sequence from -53 to -812 which contains six ZNF667 binding sites can be produced by PCR using the forward primer containing a Kpnã site and the reverse primer containing a Hind ãÂ site, and the PCR products can be inserted into pre-digested pGL3 basic vector to produce the luciferase reporter gene vector after digested with Kpnã/Hind ãÂ. ATG, translation start; ORF, open reading frame.

chromatin using an Ultrasonic homogenizer 4710(Cole-Parmer Instrument Co.), the cell lysates were diluted, precleared by incubation with protein A/G-Sepharose beads(Santa Cruz Biotechnology) and then incubated with ZNF667 polyclonal antibody we had prepared previously [4] and negative control anti-mouse-IgG antibody (Boster Biotechnologies, China) overnight at 4°C, with shaking to allow immunocomplexes to form. DNA–protein complexes were collected using protein A/G-Sepharose beads followed by several rounds of washing. Bound DNA–protein complexes were eluted from the antibodies with two incubations in elution buffer (100 mM NaHCO$_3$, 1% SDS) at room temperature for 15 min. Cross-links were reversed by the addition of NaCl followed by incubation at 65°C for 4 h. RNase A and proteinase K were sequentially added and incubated for 1 h at 37°C. DNA fragments were purified using a QIAquick PCR purification kit (Qiagen, cat#28106), and used as templates for PCR amplifications. The PCR products were fractionated on 2% agarose gels, and stained with ethidium bromide. The PCR primer pairs for the ChIP were: 5′-GGGGTACCAAAGTAAGAGGATAAGAGAG-3′ and 5′-TTGCTAAGGAGTTT GAGGCAAGCC-3′, and the target sequence was 310 bp in length.

Western blotting

After transfection, cells were cultured for 24 h. Cell lysates were collected by scraping in lysis buffer (50 mM Tris-HCl, pH 7.4, 150 mM NaCl, 1 mM EDTA, 1% Triton X-100) containing a cocktail of protease inhibitors (1× complete inhibitor). Cell debris was removed by centrifugation. Protein concentrations were determined by the Bradford assay. 20-100 µg of the protein extract was separated on 12% SDS polyacrylamide gels and electroblotted onto PVDF membranes (Millipore). The membranes were blocked with 2% bovine serum albumin (BSA) in TBS [50 mM Tris-HCl, 150 mM NaCl, pH 7.5] at 4°C for 16 h or overnight. The blocked membranes were incubated with primary antibodies in 2% BSA in TTBS [0.05% Tween 20 in TBS] for 1 to 4 h at room temperature. ZNF667, Myc tag, Bax, VDAC and β-actin were detected using rabbit polyclonal ZNF667 antibody generated in our previous study [4], rabbit monoclonal antibody against Myc tag (Cell signaling technology), rabbit polyclonal antibody against human Bax (BD Pharminogen), rabbit polyclonal antibody against VDAC (voltage-dependent anion channel protein) (Thermo Fisher Scientific) and rabbit monoclonal antibody against human β-actin (Cell signaling technology), respectively. The secondary antibodies were donkey anti-rabbit Ig-HRP (Boster Biotechnologies). Secondary antibodies were prepared in 5% milk TTBS and incubated with blots for 1 h at room temperature. The detection was performed using ECL-plus detection reagent according to the manufacturer's instructions (Amersham) or with DAB (diaminobenzidine) staining kit (Boster Biotechnologies, cat#AR1021) according to the manufacturer's instructions.

Apoptosis detection

H9c2 cells were transfected and treated as described. For flow cytometry analysis, cells were harvested by trypsinization, washed twice with ice-cold PBS, resuspended in ice-cold PBS, and fixed with 70% ethanol. Then the cells were incubated with 1 mL of PI/Triton X-100 staining solution (0.1% Triton X-100 in PBS, 0.2 mg/mL RNase A, and 10 µg/mL propidium iodide) for 30 min at room temperature. The stained cells were analyzed using a FACScan flow cytometry in combination with BD Lysis II Software (Becton Dickinson). Ten thousand events were analyzed for each sample. To analyze morphological changes in nuclei, the cells were fixed in 4% paraformaldehyde-phosphate-buffered

saline, stained with Hoechst 33258 (10 µg/ml) for 10 min, and washed three times with phosphate-buffered saline. The stained nuclei were visualized under a fluorescent inverted microscope. Caspase-3 activity was measured using a kit (Beyotime, cat#C1115) according to the manufacturer's instructions.

Reporter gene assay

For reporter gene analysis, 1.5×10^5 Raw264.7 cells were plated on 24-well plates and 0.5µg of pBa-luc or pBM-luc, 0.1-1µg of ZNF667 (or its truncated plasmids) and 20 ng of PRL-TK control plasmid (Promega) were co-transfected into cells by Lipofectamine 2000 reagent (Invitrogen) according to manufacturer's directions. Cells were collected 24 h after transfection, and firefly and Renilla luciferase activities were measured consecutively with a Dual-Luciferase Reporter assay reagents kit (Promega, cat#E1910) according to manufacture's recommendations using Lumat LB 9507 (Berthold Technologies). Firefly luciferase activity was normalized to Renilla luciferase activity. All assays shown were repeated at least four times.

Statistical analysis

Data are expressed as the means ± SEM. Each experiment was performed at least three times, and the difference among three or more groups was analyzed with ANOVA and Student–Newman–Keuls post hoc test. P<0.05 was considered significant.

Results

ZNF667 negatively regulates the expression of the Bax gene

Our previous studies have reported that H$_2$O$_2$ could increase the Bax expression [17,19,20]. In this study, we performed qRT-PCR with ZNF667-overexpressed or –knockdowned cells to determine if ZNF667 has any effect on the Bax gene expression. As shown in Fig. 2A and B, there was a significant increase or a significant reduction in ZNF667 protein levels after ZNF667 gene transfection or transfection with ZNF667 siRNA (#1) compared to the respective control. In cells transfected with ZNF667 plasmid (pcDNA3.1-ZNF667) or ZNF667 siRNA (pRNA-U6.1-ZNF667), there was a significant decrease or increase in Bax mRNA levels compared to control plasmid (pcDNA3.1) (Fig. 2C, bar 1 vs.2) or control siRNA (pRNA-U6.1) cells (Fig. 2C, bar 3 vs.4), respectively, which turns out contrary to ZNF667 levels. H$_2$O$_2$ largely induced Bax mRNA expression (Fig. 2C, bar1 vs. 5). However, the expression of Bax in H$_2$O$_2$-treated cells transfected with pcDNA3.1-ZNF667 was lower than that in the treated cells transfected with pcDNA3.1 (Fig. 2C, bar 5 vs.6). In contrast, Bax in the H$_2$O$_2$-treated cells transfected with ZNF667 siRNA (#1) was further induced to near 200% of that in the treated cells transfected with control siRNA (Fig. 2C, bar 7 vs.8). These results indicate that ZNF667 could inhibit the Bax gene expression induced with or without H$_2$O$_2$ in H9c2 cells.

To further determine the possibility that ZNF667 regulates Bax transcription, we cloned the rat Bax gene promoter (Genbank accession number: AB046392, -812 to -53) and performed a reporter gene assay. As shown in Fig. 2D, the Bax promoter activity was inhibited by ZNF667 overexpression, but it was promoted by ZNF667 knockdown. The Bax promoter activity was induced by H$_2$O$_2$ treatment, but the induction was inhibited by ZNF667 overexpression, and promoted by ZNF667 knockdown (Fig. 2D). These data are similar to those of Bax mRNA levels as described above. Taken together, all these results suggest that ZNF667 could inhibit Bax transcription perhaps via repressing its promoter activity.

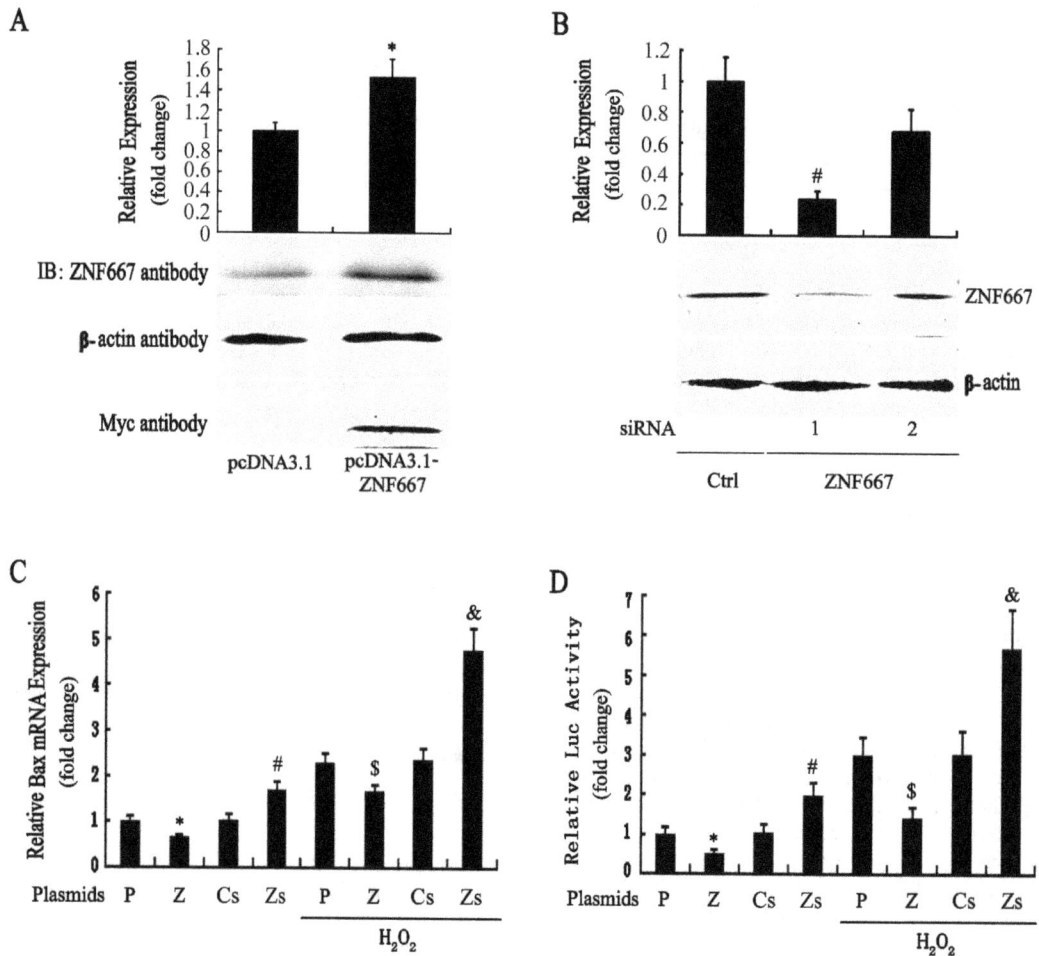

Figure 2. ZNF667 represses the Bax gene expression in H9c2 cardiomyocytes. Cells were transfected with plasmid pcDNA3.1 or pcDNA3.1-ZNF667, and transfected with pRNA-U6.1 or pRNA-U6.1-ZNF667 (shRNA), as indicated. Immunoblotting showing the effect of ZNF667 gene transfection (**A**) or ZNF667 siRNA (**B**) on expression of ZNF667 protein using the anti-ZNF667 and/or anti-myc tag antibodies (n = 3). Upper panel showing densitometry analysis of ZNF667 band against β-actin band. *P<0.05 vs. pcDNA3.1, #P<0.05 vs. control siRNA. After transfected with the indicated vector, or co-transfected with the indicated vector and the reporter construct pBa-luc plus pRL-TK, and serum starved in DMEM overnight, the cells were treated with 0.5 mM H_2O_2 for 12 h or untreated. qRT-PCR was performed to quantify Bax mRNA expression levels, and shown as the relative difference from the pcDNA3.1 or control siRNA normalized to GAPDH expression levels (n = 4, triplicate for each sample) (**C**). The reporter activity is shown as the relative luciferase activity normalized to the pRL-TK activity (n = 8) (**D**). *P<0.05 vs. pcDNA3.1; #P<0.05 vs. ctrl siRNA; $P<0.05 vs. pcDNA3.1+ H_2O_2; &P<0.05 vs. ctrl siRNA + H_2O_2. IB, immunoblot; P, pcDNA3.1; Z, pcDNA3.1-ZNF667; Cs, control siRNA; Zs, ZNF667 siRNA; Luc, luciferase.

Besides, we performed experiments with Dox-treated H9c2 cells. After 1 µM Dox was used to treat H9c2 cells, Bax mRNA levels were induced, accompanied with an increase in Bax promoter activity, but pre-transfection of pcDNA3.1-ZNF667 inhibited this induced expression and the activation of the promoter (Fig. S1). In contrast, ZNF667 siRNA increased Bax expression and its promoter activation, induced or non-induced (Fig. S1), which is similar to those of H_2O_2 treatment.

ZNF667 inhibits Bax protein expression and mitochondrial translocation induced by H_2O_2

The above results support the hypothesis that ZNF667 may directly inhibit the expression of the Bax gene. To observe the effect of ZNF667 on Bax protein expression, Bax protein was detected by Western blotting assay. In cells transfected with pcDNA3.1-ZNF667, Bax protein was reduced compared to control plasmid (pcDNA3.1) cells (Fig. 3A, lane1 vs. 3, p<0.05) as its mRNA did. H_2O_2 significantly induced the expression of Bax

protein (Fig. 3A, lane1 vs. 2). However, ZNF667 overexpression in part attenuated the inducible expression compared with the control (Fig. 3A, lane 2 vs 4). By contrast, Bax protein levels were largely upregulated in the cells transfected with ZNF667 siRNA (#1) (Fig. 3B, bar 1 vs.3) or in the cells transfected with ZNF667 siRNA (#1) plus H_2O_2 (Fig. 3B, bar 1 vs. 4) compared to the control siRNA, but Bax protein expression in cells treated with the combination of ZNF667 siRNA and H_2O_2 was much more than that treated with the control siRNA plus H_2O_2 or ZNF667 siRNA alone (Fig. 3B, bar 4 vs. 2 or 3). These results indicate that ZNF667 inhibits Bax protein expression induced or non-induced by H_2O_2.

To further observe the effect of ZNF667 on Bax protein expression, differing amounts of ZNF667 expression plasmid were used to transfect cells. As shown in Fig. 3C, Bax protein levels were decreasing with the increasing plasmid concentrations in the range used here. These results suggest that ZNF667 can inhibit the

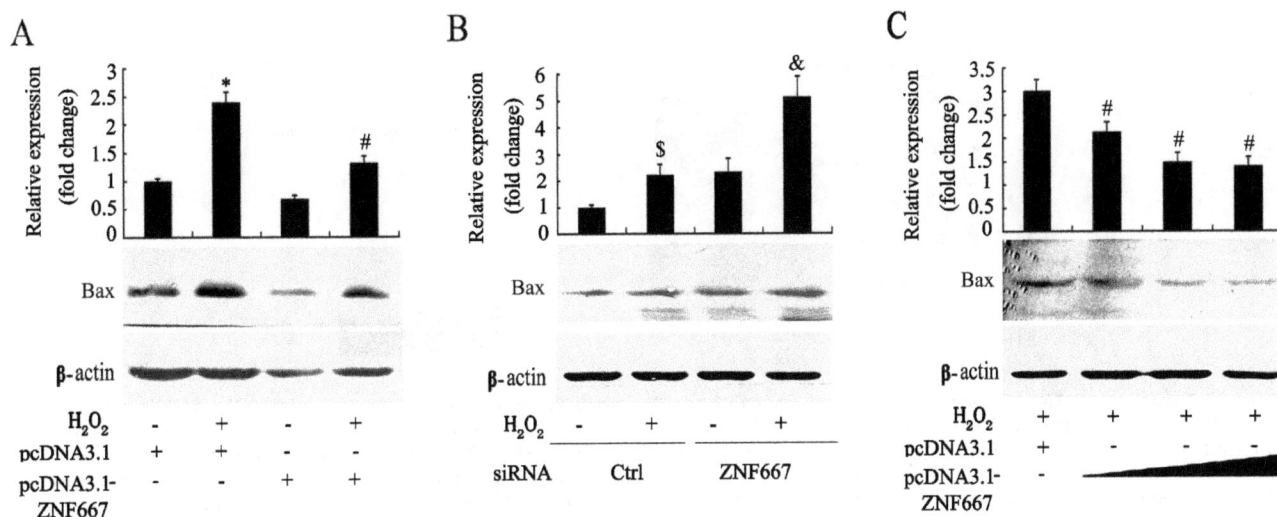

Figure 3. ZNF667 inhibits Bax protein expression in H9c2 cells. After cells were transfected with pcDNA3.1 or pcDNA3.1-ZNF667 (**A**), transfected with pRNA-U6.1 or pRNA-U6.1-ZNF667 (shRNA) (**B**), and transfected with different doses of pcDNA3.1-ZNF667 plasmid (4-24 μg) (**C**), respectively, for 24 h, the cells were treated with or without 0.5 mM H_2O_2 for another 24 h. Total proteins were extracted and subjected to SDS-PAGE followed by Western blotting analysis (n = 3). For A, B and C, shown in bottom panel are representative results, and the semi-quantitative analysis of Bax band against β-actin shown in top panel. *$P<0.05$ vs. pcDNA3.1; #$P<0.05$ vs. pcDNA3.1+ H_2O_2. $$P<0.05$ vs. ctrl siRNA; &$P<0.05$ vs. ctrl siRNA + H_2O_2.

expression of Bax protein in a dose-dependent manner in the H_2O_2-treated H9c2 cells.

The experiments were also performed with Dox-treated cells. ZNF667 overexpression inhibited in part Bax expression, and ZNF667 siRNA increased the Bax protein levels (data not shown), similar to the results mentioned above in H_2O_2-treated cells.

Other researchers reported that H_2O_2/Dox could mediate the translocation of Bax protein to mitochondria [21,22]. To investigate if ZNF667 affects the mitochondrial translocation of Bax protein, a cytosol-free mitochondrial fraction was prepared. As shown in Fig. S2, after H9c2 cells were treated with H_2O_2, there was a marked increase in the mitochondrial Bax protein. However, the increase was significantly attenuated by ZNF667 upregulation, and promoted by ZNF667 downregulation. In healthy cells, predominantly as a cytosolic monomer, Bax protein binds to prosurvival relatives such as Bcl-2, which blocks Bax activation and its mitochondrial translocation [13]. Upon oxidative stress, Bax expression increases, which may overwhelm the capacity of its antiapoptotic partners to bind it, leading to its comformational change and mitochondrial translocation [9,10,13]. Thus, ZNF667 inhibits the mitochondrial translocation of Bax protein induced by H_2O_2 at least in part through suppressing the expression of Bax protein in H9c2 cardiomyocytes.

Recombinant ZNF667 protein can interact with the Bax promoter in vitro

Being a transcriptional repressor, ZNF667 could inhibit the expression of the rat Bax gene and protein, which enticed us to investigate whether the Bax gene is one of its downstream genes and regulated directly. Based on bioinformatic analysis, we found that the promoter sequence of the Bax pro-apoptotic gene contains up to six putative ZNF667 binding sites or its core sequence CTTA, and the transcription factor ZNF667 regulates the expression of the Bax gene perhaps by directly binding to the promoter. In order to test this possibility, we first investigated if the ZNF667 protein can bind to the Bax promoter. We expressed and purified ZNF667 truncated fusion proteins. Using the purified ZF

and ZF2 fusion proteins, the target detection assay was performed to test whether or not they could bind the Bax promoter. After the membrane was visualized with Pierce Lightshift chemiluminescent EMSA kit (Pierce, cat#20148), we were able to compare the membrane and the SDS-PAGE gel stained with Coomassie blue. As shown in Fig. 4A, there was a single distinct binding band present in either the lane containing the ZF fusion protein corresponding to 80 kDa or the lane containing the ZF2 fusion protein corresponding to 46 kDa, suggesting that both ZF and ZF2 fusions could form a protein-DNA complex with the Bax promoter in vitro, indicating that ZNF667 could bind to the promoter perhaps through its binding sequence.

To further investigate if formation of the protein-DNA complex is dependent on the core sequence, a mutated Bax promoter sequence in which CTTA was substituted with GCGC was used for the binding analysis by EMSA. As shown in Fig. 4B, although the wild type promoter sequence could interact with the ZF fusion protein, displaying a distinct band, the mutant promoter sequence and the ZF fusion could not form a protein-DNA complex, and no band was observed (Fig. 4B, lanes 3 vs. 4), indicating the binding of ZNF667 protein and the Bax promoter via the core sequences.

The results showed that ZF region of ZNF667 is sufficient for DNA binding (Fig. 4), consistent with our previous report [3]. The ZF was fractioned into two parts, namely, ZF1 and ZF2. Similarly, using the target detection assay with the purified ZF truncated fusions (ZF1 and ZF2) and the DNA probe, we demonstrated that the last six ZFs (ZF2) of ZNF667 protein at its C-terminus is necessary and sufficient for its DNA binding while the other eight ZFs (ZF1) is not required for the binding (Fig. S3).

ZNF667 protein interacts with the Bax promoter within cells and the interaction is enhanced by H_2O_2

Protein-DNA binding in vitro does not necessarily mean binding within cells. In order to investigate the probability of ZNF667 binding to the Bax promoter within cells, ChIP was performed using the anti-ZNF667 antibody to pull down all the DNA that physically interacted with the ZNF667 protein, from

A

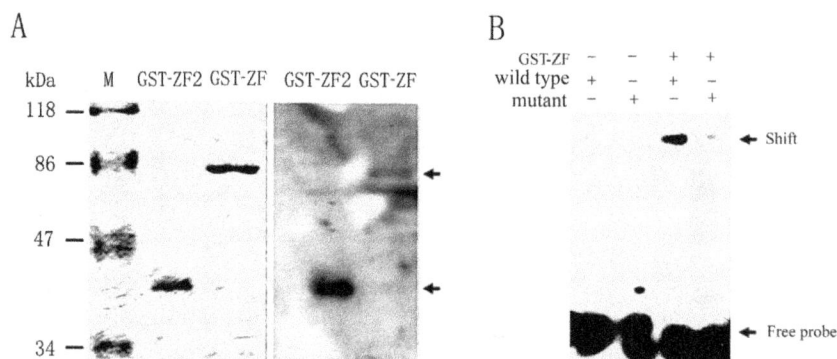

B

Figure 4. Purified GST-ZNF667 fusion proteins interact with the Bax gene promoter in vitro. (A) Purified GST-ZF, GST-ZF2 were separated on 10% SDS-PAGE and stained with Coomasie blue (left panel). A target detection assay was performed after the same proteins in A were transfected onto a nitrocellulose membrane. Biotin-labeled DNAs were hybridized to ZF and ZF2 of ZNF667. Arrows represent ZF-DNA complex (upper) and ZF2-DNA complex (lower), respectively (right panel). (B) Purified GST-ZF was added into the binding buffer containing the Bax promoter sequence or the mutant Bax sequence. After incubated at 30°C for 30 min, the mixtures were loaded onto a non-denatured electrophoresis. For both A and B, the biotin-labeled DNA-protein complexes were detected by a lightshift chemiluminescent EMSA kit. EMSA, electrophoretic mobility shift assay.

H_2O_2-treated or untreated H9c2 cells. DNA was extracted from the precipitated complex, and PCR was performed to detect the presence of the Bax promoter sequence from -53 to -812, which contains six binding sites. The result showed that, in the H_2O_2-treated H9c2 cells, the Bax promoter was found in the immune-complex pulled down by the anti-ZNF667 antibody, but not present in the precipitation that was pulled down by the control IgG or when no antibody was added (Fig. 5, right part), indicating that ZNF667 is interacting with the Bax promoter. Similarly, Bax was not observed in all the other precipitated complexes from the H_2O_2-untreated H9c2 cells besides the complex pulled down by the anti-ZNF667 antibody (Fig. 5, left part). However, there was an increase in ZNF667 bound to the Bax promoter in response to H_2O_2 treatment compared to untreated cells (Fig. 5, lanes 4 vs.7). Therefore, the above experiments demonstrated that, although ZNF667 binds to Bax promoter sequence, a stimulus, such as H_2O_2 enhanced its binding within cells.

ZNF667 directly represses the rat Bax promoter activity

In order to quantitatively analyze the effect of ZNF667 overexpression on the Bax promoter, we constructed a luciferase reporter gene vector (pBa-luc) by inserting the Bax promoter into

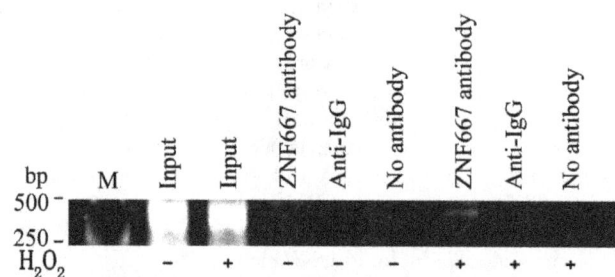

Figure 5. ZNF667 protein binds to the Bax gene promoter within H9c2 cells. ChIP assays were performed with H9c2 cells with indicated antibodies or without antibody as a negative control. The precipitated DNA was amplified by PCR, and the PCR products were separated on agarose gels, stained with ethidium bromide and visualized under an ultraviolet light using a gel imaging system. Input lanes show products after PCR amplification of chromatin DNA prior to immunoprecipitation. Anti-IgG, secondary antibody.

the pGL3 basic vector, in which the Bax promoter was located upstream of the firefly luciferase gene driving expression of firefly luciferase. We then constructed a mutation vector (pBM-luc), in which all the core sequences were mutated by nucleotide substitution from 5'-CTTA-3' to 5'-GCGC-3', as reported previously [3]. We performed studies to determine whether ZNF667 was transcriptionally repressive in a DNA binding-dependent manner. We examined the effects of ZNF667 expression on the activity of the Bax promoter containing either the ZNF667 DNA or the mutant DNA binding sequence. The construction scheme of the Bax promoter luciferase reporter vector used in these assays is shown in Fig. 1. As seen in Fig. 6, the co-transfection of RAW264.7 cells with a ZNF667 expressing plasmid (pcDNA3.1-ZNF667 or pEGFP-ZNF667) and the Bax promoter construct, which contained six ZNF667 core sequences (pBa-luc) could repress the promoter activity in a dose-dependent manner (Fig. 6, bars 2-5), whereas the co-transfection of the cells with pcDNA3.1 empty vector and the same reporter could not repress the promoter activity (Fig. 6, bar 1 vs. 4). Co-transfection of RAW264.7 cells with either pEGFP or pEGFP-KRAB and the same reporter construct failed to reduce the activity from the reporter gene promoter construct (Fig. 6, bars 7 and 8 vs. 6), suggesting that ZNF667 inhibited the activity of the firefly luciferase gene by inhibiting the Bax promoter, and this inhibition requires its intact structure. However, co-transfection of RAW264.7 cells with pcDNA3.1-ZNF667 and an all-binding-site-mutant ZNF667 binding sequence-reporter construct (pBM-luc) also failed to reduce the activity from the reporter gene promoter construct in both high and low doses (Fig. 6, bar 9 vs.2 and bar 10 vs.5), suggesting that ZNF667 inhibited the activity of the firefly luciferase gene by direct binding to the Bax promoter through its binding sequence. These results suggest that ZNF667 has transcription repressor activity on the Bax promoter, and this repression is dependent upon its binding to the promoter via the specific DNA sequence.

ZNF667 prevents H9c2 cells from H_2O_2/Dox-induced apoptosis

It has been demonstrated that H_2O_2-induced apoptosis in cardiomyocytes is involved in the mitochondrial intrinsic pathway, in which Bax is an important pro-apoptotic gene associated with the mitochondrial apoptotic signal pathway. In our above results,

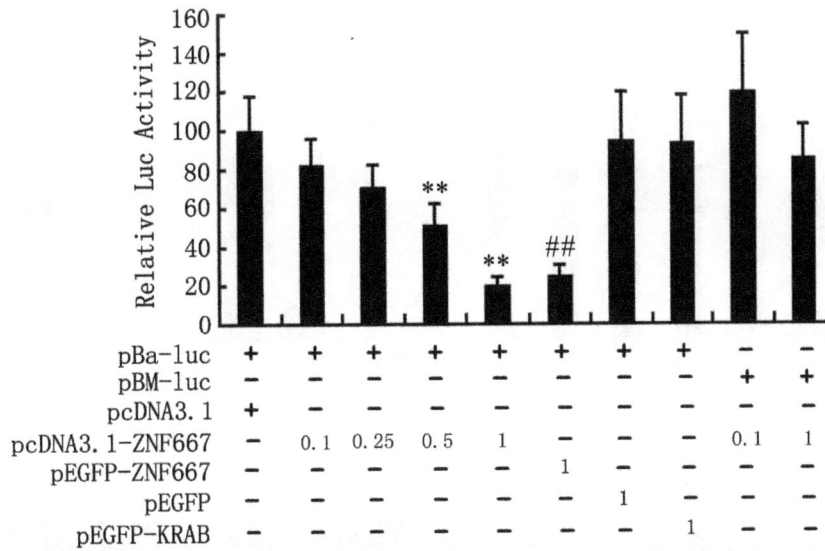

Figure 6. ZNF667 represses the activity of firefly luciferase driven by the Bax promoter. Bax promoter-driven reporter gene (pBa-luc) was co-transfected into Raw264.7 cells with increasing amounts (0.1-1μg) of pcDNA3.1-ZNF667, or 1μg of ZNF667-pEGFP or truncated ZNF667(ZF-pEGFP and KRAB-pEGFP), or pcDNA3.1 empty vector. Site-mutated reporter gene (pBM-luc) was also co-transfected into Raw264.7 cells with 0.1 or 1μg of pcDNA3.1-ZNF667 plasmid. Resultant luciferase activities are expressed as relative luciferase activities normalized to the pRL-TK activity. **P<0.01 vs. pcDNA3.1; ##P<0.01 vs. pEGFP. Luc, luciferase.

we have demonstrated that ZNF667 could inhibit the expression of Bax at both the mRNA and protein levels, blocking Bax translocation from cytoplasm to mitochondria, indicating that ZNF667 may confer cytoprotection against oxidative stress-mediated apoptosis. To determine this possibility, we detected cell apoptosis using flow cytometry, Hoechst staining and caspase-3 activity measurement by a kit. As shown in Fig. 7 A, B and C, H_2O_2 treatment led to a significant increase in the apoptosis rates, a paralleled increase in caspase-3 activities and more apoptotic nuclei (the condensed, fragmented and degraded nuclei), but upregulation of ZNF667 attenuated in part the apoptosis induction and the apoptotic enzyme activation. However, downregulation of ZNF667 increased the apoptosis rates and the enzyme activity, reversing the effects of upregulation.

In addition, we also tested the possibility that ZNF667 inhibits Dox-mediated apoptosis because it inhibits Bax expression induced by Dox. As expected, ZNF667 upregulation/downregulation inhibited/promoted Dox-mediated apoptosis and caspase-3 activation in H9c2 cells (Fig. S4), which is similar to those from H_2O_2-treated H9c2 cells. Taken together, all these results suggest that ZNF667 could protect H9c2 cells from H_2O_2/Dox-induced apoptosis, and the molecular mechanism may be involved in its inhibiting Bax expression.

Discussion

During myocardial ischemia or reperfusion, the expression of many genes such as c-fos, c-jun, junB, Egr-1, and HSP70 is upregulated [23,24], and some of them are considered to be involved in the endogenous cardioprotection against myocardial ischemia/reperfusion injury. ZNF667 gene, then called Mipu1 (GenBank Accession No. AY221750), was isolated and cloned by Yuan and colleagues at our lab as a novel gene, which was characterized by a KRAB domain at the N-terminus and 14 successive C_2H_2-type zinc-finger domains at the C-terminus and was up-regulated in rat heart after a transient I/R procedure [25]. Further important observations support a central role for

ZNF667/Mipu1 in the survival of C2C12 myogenic cells, because ZNF667 overexpression could reduce the growth arrest induced by serum withdrawal [26]. It was also shown that ZNF667 protein was localized to the nucleus of H9c2 cardiomyocytes and was upregulated after the cells were treated with H_2O_2 [2-5]. These observations indicated that ZNF667 might play a role in maintaining cell homeostasis and protecting the cells from being injured by oxidative stress. Previously, we have described that ZNF667 acts as a transcriptional repressor [3], even using the GAL4 system [18]. However, its downstream gene(s) which is regulated directly and could help us to explain its physiological/pathophysiological functions is unknown, and little is known about whether ZNF667 plays a role in Dox-treated H9c2 cardiomyocytes.

Because ZNF667 is a transcription repressor, we focused on its transcriptional repression in this study. Because ZNF667 is an antiapoptotic factor and because Bax is an important pro-apoptotic gene associated with the mitochondrial apoptotic signal pathway the promoter of which contains ZNF667 binding sites, Bax became our first target. In our sequence-based analysis of the rat Bax gene promoter, there was six potential ZNF667 binding sites located on the promoter sequence ranging from -813 to -52 (Fig. 1). In order to investigate the relationship between ZNF667 and Bax, we cloned the Bax gene promoter and, using the H9c2 cardiomyocytes or Raw 264.7 cells, we identified the repressive effect of ZNF667 on the promoter. The Bax expression and its promoter activity were inhibited by ZNF667 overexpression but promoted by ZNF667 knockdown (Fig. 2 and 3). Using the purified ZNF667 fusion proteins (GST-ZF and GST-ZF2), we showed that the ZNF667 could interact with the Bax promoter in vitro (Fig. 4), and that nucleotide substitutions in the ZNF667 binding sequences could abolish the in vitro protein-DNA interaction (Fig. 4B), indicating the specific interaction between the ZNF667 protein and the Bax promoter. H_2O_2 could enhance the affinity between the ZNF667 protein and the Bax promoter within the cells (Fig. 5). Moreover, nucleotide substitutions in ZNF667 binding sites of the Bax promoter directly affected the

A

B

C

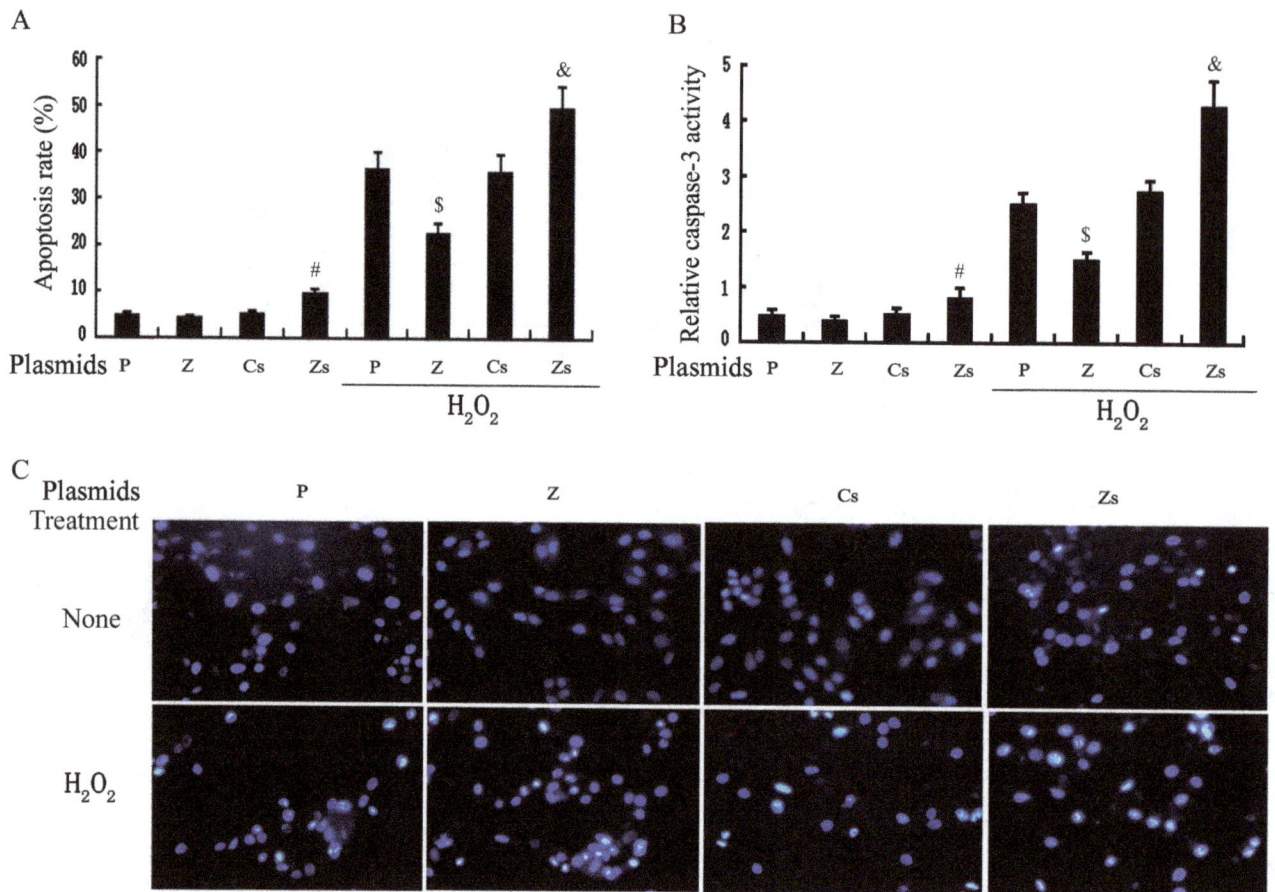

Figure 7. ZNF667 inhibits H$_2$O$_2$-mediated apoptosis and caspase-3 activation in H9c2 cells. Cells were transfected with the indicated plasmids for 24 h and followed by treatment with or without H$_2$O$_2$ for 6 h as indicated. The cells were used for apoptosis analysis by flow cytometry (A) or for caspase-3 activity analysis by a kit (B). #$P<0.05$ vs. siRNA Ctrl; $$P<0.05 vs. pcDNA3.1+ H$_2$O$_2$; &$P<0.05$ vs. Ctrl siRNA + H$_2$O$_2$. (C) Treated as described above, the cells were stained with Hoechst 33258, and observed by fluorescence microscopy. Original magnifications: ×200. P, pcDNA3.1; Z, pcDNA3.1-ZNF667; Cs, control siRNA; Zs, ZNF667 siRNA.

inhibitory effect of ZNF667 on the promoter activity (Fig. 6). To our knowledge, this study is the first to identify a specific downstream target gene of this novel KRAB/C$_2$H$_2$ zinc finger protein.

To observe which of the fourteen zinc fingers of ZNF667 are required for its DNA binding, we also expressed and purified the ZF1 fusion containing the eight ZFs adjacent to the N-terminus and the ZF2 fusion containing the six ZFs at the C-terminus (Fig. S3A). Using the two fusion proteins, we performed the target detection assay. The results showed that ZF2 is required and sufficient for ZNF667 binding to the DNA sequence (Fig. S3). Generally, only three, or even two zinc fingers, are sufficient for binding to the DNA sequence [27,28]. However, ZNF667 is localized in the nucleus not by its zinc fingers but by its KRAB domain, unlike other KZNFs [3]. Therefore, more than eight zinc fingers can be used for protein-protein interactions and/or RNA-protein interactions [29-33]. If this is true, ZNF667 may play many other roles by its interactions with other molecules in addition to its transcriptional functions. Like other KZNFs [33,34], ZNF667 plays a repressive role by recruiting the co-repressor KAP-1 through its KRAB domain (data not shown).

In the present study, Dox/H$_2$O$_2$ was used as our research tools because ZNF667 responds to H$_2$O$_2$ treatment as reported in our previous study and because Dox is an inducer of cardiomyocyte

apoptosis. Bax is a proapoptotic factor and overexpression of Bax increases apoptosis in cardiomyocytes and other cells [35,36]. Since ZNF667 directly contacts the Bax promoter, repressing the promoter activity and inhibiting Bax expression during Dox/H$_2$O$_2$ treatment, it was inferred that ZNF667 should block H$_2$O$_2$-induced translocation of Bax protein to mitochondria, which was then demonstrated by the experiments. Because ZNF667 has inhibitory effects on the induced expression and translocation of Bax protein, it should have inhibitory effects on Dox/H$_2$O$_2$-induced apoptosis, which was also verified by our experiments as expected. Therefore, all the evidence obtained here supports ZNF667 is a protective factor against Dox/H$_2$O$_2$ stress in cardiomyocytes. The molecular mechanism is involved in its inhibiting Bax expression. Although Our previous studies have implied that Mipu1 may confer cytoprotection [1,5,6,26], it is in this study that we for the first time shows that ZNF667/Mipu1 is a cytoprotective protein in H9c2 cardiomyocytes at least during H$_2$O$_2$/Dox stress.

The results discussed in this research suggest that ZNF667 is a transcription repressor of the rat Bax gene, protecting cardiomy-ocytes against Dox/H$_2$O$_2$-induced apoptosis. However, Bax and apoptosis are associated with cancer [37-40]. In cancer cells, an increase in Bax expression will promote apoptosis. As the last six zinc fingers of the rat ZNF667 are sufficient for its DNA binding, we compared the fingers of ZNF667 from rat with that from human, it was found that the identities are 93%, indicating that human ZNF667 may also have the same DNA binding sequence. Bio- informatic analysis shows that the human Bax promoter contains a ZNF667 DNA binding sequence, which implies that human ZNF667 may also repress human Bax expression. If this is true, ZNF667 may be involved in cancer. Interestingly, it was also found that human ZNF667 was indeed upregulated in human brain astrocytomas [41]. Of course, the roles of ZNF667 in cancer need to be deeply elucidated in the future.

Collectively, our present study demonstrated that ZNF667 is a novel direct regulator of the rat Bax gene and a protective protein in cardiomyocytes, which has the potential to protect the rat heart against ischemia/reperfusion injury. Our findings also provided new clues for ZNF667 functions in other processes or diseases .

Supporting Information

Figure S1   ZNF667 inhibits Dox-induced expression of the rat Bax gene. Cells were treated with different doses of Dox for 12 h (A) or with 1 μM Dox for the indicated times (B) or transfected with the indicated vector for 24 h and treated with 1 μM Dox for 12 h (C), or co-transfected with the reporter construct pBa-luc (0.5 μg) and pRL-TK (0.02 μg) plus pcDNA3.1 (0.5 μg) or pcDNA3.1-ZNF667 (0.5μg), or pRNA-U6.1 (sC, 0.5μg) or pRNA-U6.1-ZNF667 (0.5μg), and serum starved in DMEM overnight followed by treatment with 1 μM Dox for 12 h (D). Total RNA was extracted from cells of each group, and qRT-PCR was performed to quantify Bax mRNA expression levels, shown as the relative difference from the control or pcDNA3.1 or control siRNA normalized to GAPDH expression levels (n = 3), respectively. The reporter activities are shown as the relative luciferase activity normalized to the pRL-TK activity (n = 8). $*P<0.05$ vs. pcDNA3.1; $\#P<0.05$ vs. ctrl siRNA; $\$P<0.05$ vs. pcDNA3.1+ Dox; $\&P<0.05$ vs. ctrl siRNA + Dox. P, pcDNA3.1; Z, pcDNA3.1-ZNF667; Cs, control siRNA; Zs, ZNF667 siRNA; Dox, doxorubicin; Luc, luciferase.

Figure S2   ZNF667 inhibits Bax mitochondrial translocation induced by H_2O_2. Cells were transfected with pcDNA3.1 or pcDNA3.1-ZNF667 (A), and transfected with pRNA-U6.1 or pRNA-U6.1-ZNF667 (B), respectively, for 24 h. The cells were then treated with H_2O_2 for 6 h or untreated. The cell mitochondria was prepared using a kit. The mitochondrial proteins were subjected to Western blot analyses using the antibodies indicated. For both A and B, shown in the bottom panel are representative results, and shown in the top panel are the

means ±SEM of three independent experiments. $*P<0.05$ vs. pcDNA3.1; $\#P<0.05$ vs. pcDNA3.1+ H_2O_2; $\$P<0.05$ vs. siRNA Ctrl; $\&P<0.05$ vs. siRNA Ctrl + H_2O_2. VDAC, voltage-dependent anion channel protein.

Figure S3   Target detection assay revealed the last zinc fingers of ZNF667 is required and sufficient for its DNA binding. A. Scheme of ZNF667 and its truncated fusions. GST was fused to their N-termini. **B and D.** Purified GST-ZF, GST-ZF2, GST-ZF1 and GST were separated on 10% SDS–PAGE, as indicated on the top and stained with Coomassie blue. **C and E.** A target detection assay was performed in the presence of zinc after the same proteins in B and D, respectively, were transferred to two nitrocellulose membranes. After the membranes were incubated with biotin-labeled probes and washed, the bound DNA was visualized using a Lightshift chemiluminescent EMSA kit (Pierce). The target detection assay revealed zinc-dependent binding of the ZF and the ZF2 to the probe (indicated by arrows). This shows the last six zinc fingers of ZNF667 (ZF2) is sufficient and necessary for its binding to the DNA binding sites. GST, glutathione S-transferase.

Figure S4   ZNF667 prevents apoptosis and caspase-3 activation mediated by Dox in H9c2 cells. The cells transfected with the indicated plasmids were treated or untreated with Dox for 6 h. The cells were harvested by trypsinization, and then used for apoptosis analysis by flow cytometry (A) or for caspase-3 activity analysis by a kit (B). $\#P<0.05$ vs. Ctrl siRNA; $\$P<0.05$ vs. pcDNA3.1+ Dox; $\&P<0.05$ vs. Ctrl siRNA + Dox. P, pcDNA3.1; Z, pcDNA3.1-ZNF667; Cs, control siRNA; Zs, ZNF667 siRNA.

Acknowledgments

We thank Drs. Zhenyu Zhao and Yansheng Feng for assistance in analysis of luc activity. This work was supported by the National Natural Science Foundation of China (30971205 for L.J.) and the Major Basic Research Program of China (2007CB512007 for X.X.).

Author Contributions

Conceived and designed the experiments: LJ XX. Performed the experiments: LJ HW CS ML GW. Analyzed the data: LJ KL NW KW HZ. Contributed reagents/materials/analysis tools: LJ KL ML NW HZ. Wrote the paper: LJ. Contributions of others were mentioned in the acknowledgment.

References

1. Wang G, Zuo X, Liu J, Jiang L, Liu Y, et al. (2009) Expression of Mipu1 in response to myocardial infarction in rats.. Int. J. Mol. Sci.10: 492-506.
2. Wang G, Zuo X, Jiang L, Wang K, Wei X, et al. (2010) Tissue expression and subcellular localization of Mipu1, a novel myocardial ischemia-related gene. Braz J Med Biol Res. 43: 43-51.
3. Jiang L, Tang D, Wang K, Zhang H, Yuan C, et al. (2007) Functional analysis of a novel KRAB/C2H2 zinc finger protein Mipu1. Biochem. Biophy. Res Commun. 356: 829-835.
4. Jiang L, Zhang B, Wang G, Wang K, Xiao X (2009) Expression, purification and characterization of rat zinc finger protein Mipu1 in Escherichia coli. Mol.Cel.Biochem. 328: 137–144.
5. Qu S, Zhu H, Wei X, Zhang C, Jiang L, et al. (2010) Oxidative stress-mediated up-regulation of myocardial ischemic preconditioning up-regulated protein 1 gene expression in H9c2 cardiomyocytes is regulated by cyclic AMP-response element binding protein. Free Rad.Biol. Med. 49: 580–586.
6. Wang K, Lei J, Zou J, Xiao H, Chen A, et al. (2013) Mipu1, a novel direct target gene, is involved in hypoxia inducible factor 1-mediated cytoprotection. PLoS One. 8(12): e82827.
7. Zhou X, Sheng Y, Yang R, Kong X (2010) Nicotine promotes cardiomyocyte apoptosis via oxidative stress and altered apoptosis-related gene expression. Cardiology. 115(4):243-50.
8. Nishikawa S, Tatsumi T, Shiraishi J, Matsunaga S, Takeda M, et al. (2006) Nicorandil regulates Bcl-2 family proteins and protects cardiac myocytes against hypoxia-induced apoptosis. J Mol Cell Cardiol. 40(4):510-9.
9. Jin HJ, Xie XL, Ye JM, Li CG (2013) TanshinoneIIA and cryptotanshinone protect against hypoxia-induced mitochondrial apoptosis in H9c2 cells. PLoS One. 8(1): e51720.
10. Youle RJ, Straser A (2008) The BCL-2 protein family: opposing activities that mediate cell death. Nat Rev Mol Cell Biol. 9: 47-59.
11. Danial NN (2007) BCL-2 family proteins: critical checkpoints of apoptotic cell death. Clin Cancer Res. 13(24): 7254-63.
12. Lindsten T, Ross AJ, King A, Zong WX, Rathmell JC, et al. (2000) The combined functions of proapoptotic Bcl-2 family members bak and bax are essential for normal development of multiple tissues. Mol Cell. 6(6): 1389-99.
13. Wei MC, Zong WX, Cheng EH, Lindsten T, Panoutsakopoulou V, et al. (2001) Proapoptotic BAX and BAK: a requisite gateway to mitochondrial dysfunction and death. Science. 292(5517): 727-30.

14. Fletcher JI, Meusburger S, Hawkins CJ, Riglar DT, Lee EF, et al. (2008) Apoptosis is triggered when prosurvival Bcl-2 proteins cannot restrain Bax. Proc Natl Acad Sci USA. 105(47): 18081-7.

15. Lalier L, Cartron PF, Juin P, Nedelkina S, Manon S et al. (2007) Bax activation and mitochondrial insertion during apoptosis. Apoptosis 12: 887-896.

16. Tait SW, Green DR (2010) Mitochondria and cell death: outer membrane permeabilization and beyond. Nat Rev Mol Cell Biol 11:621-632.

17. Dewson G, Ma S, Frederick P, Hockings C, Tan I, et al. (2012) Bax dimerizes via a symmetric BH3:groove interface during apoptosis. Cell Death Differ 19: 661-670.

18. Wang G, Zuo X, Yuan C, Zheng Y, Jiang L, et al. (2009) Mipu1, a novel rat zinc-finger protein, inhibits transcriptional activities of AP-1 and SRE in mitogen-activated protein kinase signaling pathway. Mol. Cell Biochem. 322: 93-102.

19. Zhang B, Wang H, Jiang B, Liang P, Liu M, et al. (2010) Nucleolin/C23 is a negative regulator of hydrogen peroxide-induced apoptosis in HUVECs. Cell Stress Chaperones 15: 249-257.

20. Liu S, Li J, Tao Y, Xiao X (2007) Small heat shock protein alpha B crystallin binds to p53 to sequester its translocation to mitochondria during hydrogen peroxide-induced apoptosis. Biochem Biophys Res Commun 354: 109-114.

21. Xiang SY, Ouyang K, Yung BS, Miyamoto S, Smrcka AV, et al. (2013) PLCε, PKD1, and SSH1L transduce RhoA signaling to protect mitochondria from oxidative stress in the heart. Sci Signal. 6(306): ra108.

22. Sardão VA, Oliveira PJ, Holy J, Oliveira CR, Wallace KB (2009) Doxorubicin-induced mitochondrial dysfunction is secondary to nuclear p53 activation in H9c2 cardiomyoblasts. Cancer Chemother Pharmacol. 64: 811–827.

23. Plumier JC, Robertson HA, Currie RW (1996) Differential accumulation of mRNA for immediate early genes and heat shock genes in heart after ischaemic injury. J. Mol. Cell. Cardiol. 28:1251–1260.

24. Nelson DP, Wechsler SB, Miura T, Stagg A, Newburger JW, et al. (2002) Myocardial immediate early gene activation after cardiopulmonary bypass with cardiac ischemia–reperfusion. Ann. Thorac. Surg. 73:156–162.

25. Yuan C, Zhang H, Liu Y, Wang Q, Xiao X (2004) Cloning and characterization of a new gene Mipu1 up-regulated during myocardial ischemia-reperfusion. Prog. Biochem. Biophys. 31: 231–236.

26. Yuan C, Liu Y, Wang Q, Xiao X (2004) Effects of new gene Mipu1 over-expression on the cell growth cycle of C2C12 cells cultured under condition of serum withdrawal. J. Clin Res. 21(8):837-839.

27. Zweidler-Mckay PA, Grimes HL, Flubacher MM, Tsichlis PN (1996) Gfi-1 encodes a nuclear zinc finger protein that binds DNA and functions as a transcriptional repressor. Mol. Cell. Biol. 16: 4024-4034.

28. Nunez N, Clifton MM, Funnell AP, Artuz C, Hallal S, et al. (2011) The multi-zinc finger protein ZNF217 contacts DNA through a two-finger domain. J. Biol. Chem. 286: 38190-38201.

29. Jheon AH, Ganss B, Cheifetz S, Sodek J (2001) Characterization of a novel KRAB/C2H2 zinc finger transcription factor involved in bone development. J. Biol. Chem. 276: 18282-18289.

30. Gou D, Wang J, Gao L, Sun Y, Peng X, et al. (2004) Identification and functional analysis of a novel human KRAB/C2H2 zinc finger gene ZNF300. Biochem.Biophys. Acta 1676: 203-209.

31. Razin SV, Borunova VV, Maksimenko OG, Kantidze OL (2012) Cys2His2 zinc finger protein family: classification, functions, and major members. Biochemistry (Mosc) 77: 217-226.

32. Burdach J, O'Connell MR, Mackay JP, Crossley M (2012) Two-timing zinc finger transcription factors liaising with RNA. Trends Biochem Sci. 37: 199-205.

33. Tian C, Xing G, Xie P, Lu K, Nie J, et al. (2009) KRAB-type zinc-finger protein Apak specifically regulates p53-dependent apoptosis. Nat.Cell Biol. 11: 580-591.

34. Urrutia R (2003) KRAB-containing zinc-finger repressor proteins. Genome Biol. 4: 231.

35. Fortuno MA, Zalba G, Ravassa S, D'Elom E, Beaumont FJ, et al. (1999) p53-mediated upregulation of BAX gene transcripti on is not involved in Bax-alpha protein overexpression in the left ventricle of spontaneously hypertensive rats. Hypertension. 33: 1348-1352.

36. Jin KL, Graham SH, Mao XO, He X, Nagayama T, et al. (2001) Bax kappa, a novel Bax splice variant from ischemic rat brain lacking an ART domain, promotes neuronal cell death. J. Neurochem 77: 1508-1519.

37. Lowe SW, Lin AW (2000) Apoptosis in cancer. Carcinogenesis 21 (3): 485-495.

38. Shigematsu S, Fukuda S, Nakayama H, Inoue H, Hiasa Y, et al. (2011) ZNF689 suppresses apoptosis of hepatocellular carcinoma cells through the down-regulation of Bcl-2 family members. Exp Cell Res. 317(13):1851-9.

39. González-Gironès DM, Moncunill-Massaguer C, Iglesias-Serret D, Cosialls AM, Pérez-Perarnau A, et al. (2013) AICAR induces Bax/Bak-dependent apoptosis through upregulation of the BH3-only proteins Bim and Noxa in mouse embryonic fibroblasts. Apoptosis. 18: 1008-1016.

40. Liu K, Lou J, Wen T, Yin J, Xu B, et al. (2013) Depending on the stage of hepatosteatosis, p53 causes apoptosis primarily through either DRAM-induced autophagy or BAX. Liver Int. 33(10):1566-74.

41. Wang Z, Gu L, Liu H, Wang D, Tu Z, et al. (2013) Expression of Mipu1 in human brain astrocytomas and its clinical significance. Chin J Neurosurg Dis Res. 12: 37-40.

5

The Circadian Clock Maintains Cardiac Function by Regulating Mitochondrial Metabolism in Mice

Akira Kohsaka[1]*, Partha Das[1], Izumi Hashimoto[1], Tomomi Nakao[1], Yoko Deguchi[1], Sabine S. Gouraud[1], Hidefumi Waki[1], Yasuteru Muragaki[2], Masanobu Maeda[1]

1 Department of Physiology, Wakayama Medical University School of Medicine, Wakayama, Japan, 2 First Department of Pathology, Wakayama Medical University School of Medicine, Wakayama, Japan

Abstract

Cardiac function is highly dependent on oxidative energy, which is produced by mitochondrial respiration. Defects in mitochondrial function are associated with both structural and functional abnormalities in the heart. Here, we show that heart-specific ablation of the circadian clock gene *Bmal1* results in cardiac mitochondrial defects that include morphological changes and functional abnormalities, such as reduced enzymatic activities within the respiratory complex. Mice without cardiac *Bmal1* function show a significant decrease in the expression of genes associated with the fatty acid oxidative pathway, the tricarboxylic acid cycle, and the mitochondrial respiratory chain in the heart and develop severe progressive heart failure with age. Importantly, similar changes in gene expression related to mitochondrial oxidative metabolism are also observed in C57BL/6J mice subjected to chronic reversal of the light-dark cycle; thus, they show disrupted circadian rhythmicity. These findings indicate that the circadian clock system plays an important role in regulating mitochondrial metabolism and thereby maintains cardiac function.

Editor: Nicholas S. Foulkes, Karlsruhe Institute of Technology, Germany

Funding: This work was supported by JSPS KAKENHI Grant Number 25460316 (to A. K.) and the Takeda Science Foundation (to A. K.). The funders had no role in study design, data collection and analysis, decision to publish, or preparation of the manuscript.

Competing Interests: The authors have declared that no competing interests exist.

* Email: kohsaka@wakayama-med.ac.jp

Introduction

Heart function relies on oxidative energy that is supplied continuously by mitochondria. Because energy demand in the heart varies across the sleep/wake cycle, timely supply of energy may benefit the heart and optimize its function. This time-dependent energy metabolism is subject to the circadian clock system, in which a set of core clock genes play a major role. Without clock gene function, the daily rhythms in heart energy metabolism are impaired [1,2]. Interestingly, previous studies have demonstrated that defects in clock gene function not only dampen the metabolic rhythms in the heart but also alter cardiac function [3], indicating a close relationship between clock gene function and heart energy homeostasis with respect to the regulation of cardiac function. However, little is known about how the circadian clock system coordinates the heart energy balance, which directly impacts cardiac function.

As the molecular machinery of the circadian clock was identified, it became clear that the roles of the clock genes were not limited to the regulation of physiological rhythms. Clock genes also participate in fundamental physiological processes, such as the immune response [4,5], gastrointestinal digestion/absorption [6,7], and renal function [8,9]. Perhaps most striking was the observation that even fuel and energy metabolism is under the control of clock genes. Using mouse models of clock gene dysfunction, clock genes were demonstrated to regulate carbohy-

drate and lipid metabolism, such as hepatic gluconeogenesis [10,11], pancreatic insulin secretion [12], and fat cell differentiation [13,14]. In addition to these tissue-specific roles of clock genes in energy metabolism, more ubiquitous metabolic processes, such as the control of cellular redox, are associated with clock-gene function [15,16,17]. More importantly, recent studies have revealed that mitochondrial oxidative respiration is also influenced by the function of clock genes in liver and muscle [18,19]. Because functional clock genes are widespread throughout the body [20,21], mitochondrial function may be impaired in almost every cell if clock genes do not function appropriately, and this mitochondrial dysfunction may be more apparent in energy-demanding tissues, such as the heart and muscles.

Defects in mitochondrial function have been implicated in the development of heart failure in humans and rodents [22,23]. Mitochondrial function is highly dependent on the number and/or structure of these organelles and on their capacity for metabolic processes, including fatty acid oxidation (FAO), the tricarboxylic acid (TCA) cycle, and mitochondrial respiration. A substantial portion of the literature has focused on the role of mitochondrial metabolism in the maintenance of cardiac function and has uncovered an integrated regulation of mitochondrial dynamics and bioenergetics by transcription factors and their related molecules [24,25,26]. These factors include peroxisome proliferator-activated receptors (PPARs) [27,28], estrogen-related receptors (ERRs) [29,30], and PPARγ coactivator-1 (PGC-1) [31], and transcript levels of these genes influence the contractility of cardiac muscle. Importantly, some of these transcription factors interact with protein products of clock genes. Specifically, clock genes and genes encoding PPARα and PGC-1α, key players in mitochondrial dynamics and bioenergetics, interact to regulate their own transcription [32,33,34], indicating a close association between the clock and mitochondrial function at the molecular level.

In this study, we hypothesized that cardiac mitochondria may require normal circadian clock function to maintain cardiac function. To demonstrate that dysregulation of the circadian clock leads to cardiac mitochondrial defects, which then leads to reduced cardiac function, we used two mouse models of circadian clock dysregulation, one in which the clock gene *Bmal1* was knocked out in the heart and another that was exposed to continuous disturbance of the external light-dark (LD) cycle, which disrupts behavioral and physiological rhythms.

Results

Heart-specific *Bmal1* knockout mice develop congestive heart failure with age

To test the hypothesis that the circadian clock is involved in the maintenance of cardiac function by regulating mitochondrial metabolism, we first examined the effect of heart-specific disruption of the clock on both cardiac morphology and function. We generated heart-specific *Bmal1* knockout (H-Bmal1$^{-/-}$) mice by crossing mice bearing a floxed *Bmal1* gene with mice expressing the Cre recombinase under the control of the αMHC promoter. We compared the cardiac phenotypes of 12-week-old H-Bmal1$^{-/-}$ mice with those of littermate control animals with a floxed *Bmal1* gene but no *Cre* transgene. As shown in previous reports using mice with heart-specific disruption of clock genes [35], H-Bmal1$^{-/-}$ mice showed an increase in heart size and heart weight (HW)/body weight (BW) ratio compared with control animals (Figure 1, A and B). Histological analysis revealed that the increase in heart size observed in the H-Bmal1$^{-/-}$ mice was due to thickening of the left ventricular (LV) wall (Figure 1, C and D). In addition to these morphological changes, heart function was also altered in H-Bmal1$^{-/-}$ mice. Although neither the LV internal diameter at diastole (LVIDd) nor systole (LVIDs), as assessed by echocardiogram, was significantly different, LVIDs tended to be higher in H-Bmal1$^{-/-}$ mice compared with control animals, which led to a significant decrease in fractional shortening (Figure 1E), indicating that H-Bmal1$^{-/-}$ mice developed heart failure. Consistent with these functional analyses, the expression levels of markers of heart failure (*ANP* and *BNP*) were increased in heart tissue from H-Bmal1$^{-/-}$ mice (Figure 1F). Although cardiac function may have been affected by disrupted physiological rhythmicity, such as behavioral and cardiovascular rhythms, the daily rhythms in locomotor activity, blood pressure, and heart rate were not different between control and H-Bmal1$^{-/-}$ mice (Figure S1, A and B).

We further explored the impact of age on heart failure in H-Bmal1$^{-/-}$ mice by examining 24-week-old or older animals. As was observed in the 12-week-old animals, the HW/BW ratio was increased in 24-week-old H-Bmal1$^{-/-}$ mice, although the degree of the difference between H-Bmal1$^{-/-}$ and control animals was smaller at 24 weeks of age than at 12 weeks of age (Figure 1B and Figure S2A). In contrast, functional analyses showed that H-Bmal1$^{-/-}$ animals at 24 weeks of age exhibited a more profound decrease in heart function than at 12 weeks of age. Although not observed in 12-week-old animals, both LVIDd and LVIDs were significantly increased in 24 week-old H-Bmal1$^{-/-}$ mice com-

pared with control animals (Figure 1E and Figure S2B). Furthermore, systolic contractility (fractional shortening) was more severely decreased in 24-week-old than 12 week-old H-Bmal1$^{-/-}$ mice when compared with age-matched control animals (16% vs. 20% decrease at 12 and 24 weeks of age, respectively; Figure 1E and Figure S2B).

The progressive reduction in heart function with age resulted in early mortality in H-Bmal1$^{-/-}$ mice. Although almost all control animals survived beyond the study period (96 weeks), the majority of H-Bmal1$^{-/-}$ mice died at approximately 40 weeks of age (average life span, control 95.8 weeks vs. H-Bmal1$^{-/-}$ 41.3 weeks; log-rank test, $\chi^2 = 62.0$, df = 1, p<0.001; Figure 1G). The existence of quite severe heart failure was indicated in histological analyses from 33-week-old H-Bmal1$^{-/-}$ mice, which showed ventricular dilation, thinned myocardial walls, thrombosis in the ventricle, and myocardial fibrosis (Figure 1H). In addition to the heart tissue, histological changes that indicated severely decreased cardiac function were observed in other organs of 33-week-old H-Bmal1$^{-/-}$ mice. The lung tissue of these animals showed vascular wall thickness, congestion, and infiltration of inflammatory cells, which are all often observed with severe congestive heart failure (Figure 1I). Signs of congestion in H-Bmal1$^{-/-}$ animals were also observed in the liver, including dilatation of hepatic sinusoids and central veins (Figure 1I). Collectively, our data show that defects in the function of the *Bmal1* gene in heart lead to progressive heart failure, which results in congestion of multiple organs and a short life span.

Bmal1 deficiency alters cardiac energy metabolism

To examine whether the heart failure observed in H-Bmal1$^{-/-}$ mice was due to alterations in cardiac energy metabolism, we performed a DNA microarray analysis to detect changes in expression of the genes responsible for this failure. As approximately 8% of the genes expressed in the heart show circadian rhythmicity [36], and because we only used samples dissected at ZT 2 (ZT 0 defined as lights on), the limitations of our microarray analysis are that the detected changes might simply be due to shifted peaks or troughs of gene expression and that the analysis might not detect all gene expression changes. Nevertheless, we found that broad classes of genes regulating cellular energy metabolism were upregulated or downregulated in the heart tissues of 12-week-old H-Bmal1$^{-/-}$ mice compared with those of control animals (Figure 2A). Because fatty acids are the preferred metabolic substrate for adult cardiomyocytes, we first investigated the expression of genes regulating fatty acid metabolism. Although a few genes were upregulated, the majority of genes associated with fatty acid transport were downregulated in Bmal1$^{-/-}$ hearts (Figure 2A). In addition to fatty acid transporters, the mRNA expression levels of a variety of enzymatic components that regulate mitochondrial FAO (e.g., *Ehhadh* and *Hadha*) were also downregulated in Bmal1$^{-/-}$ hearts (Figure 2A). Importantly, despite the decrease in mitochondrial FAO genes, genes for peroxisomal FAO (*Acox 2* and *Acox3*) were upregulated in Bmal1$^{-/-}$ hearts (Figure 2A), indicating some compensation from mitochondrial to peroxisomal FAO. Validation analyses using quantitative PCR also revealed a significant decrease in the expression levels of *Cpt2*, *Acsl1* (the rate-limiting enzyme for the activation of fatty acids), *Fabp3*, *Ehhadh*, and *Hadha* (the rate-limiting enzyme for mitochondrial FAO) in Bmal1$^{-/-}$ hearts (Figure 2B).

Because glucose metabolism in cardiomyocytes takes on greater importance under pathologic conditions including myocardial ischemia and hypertrophy, we next investigated the possibility of compensatory induction of the glycolytic pathway in Bmal1$^{-/-}$

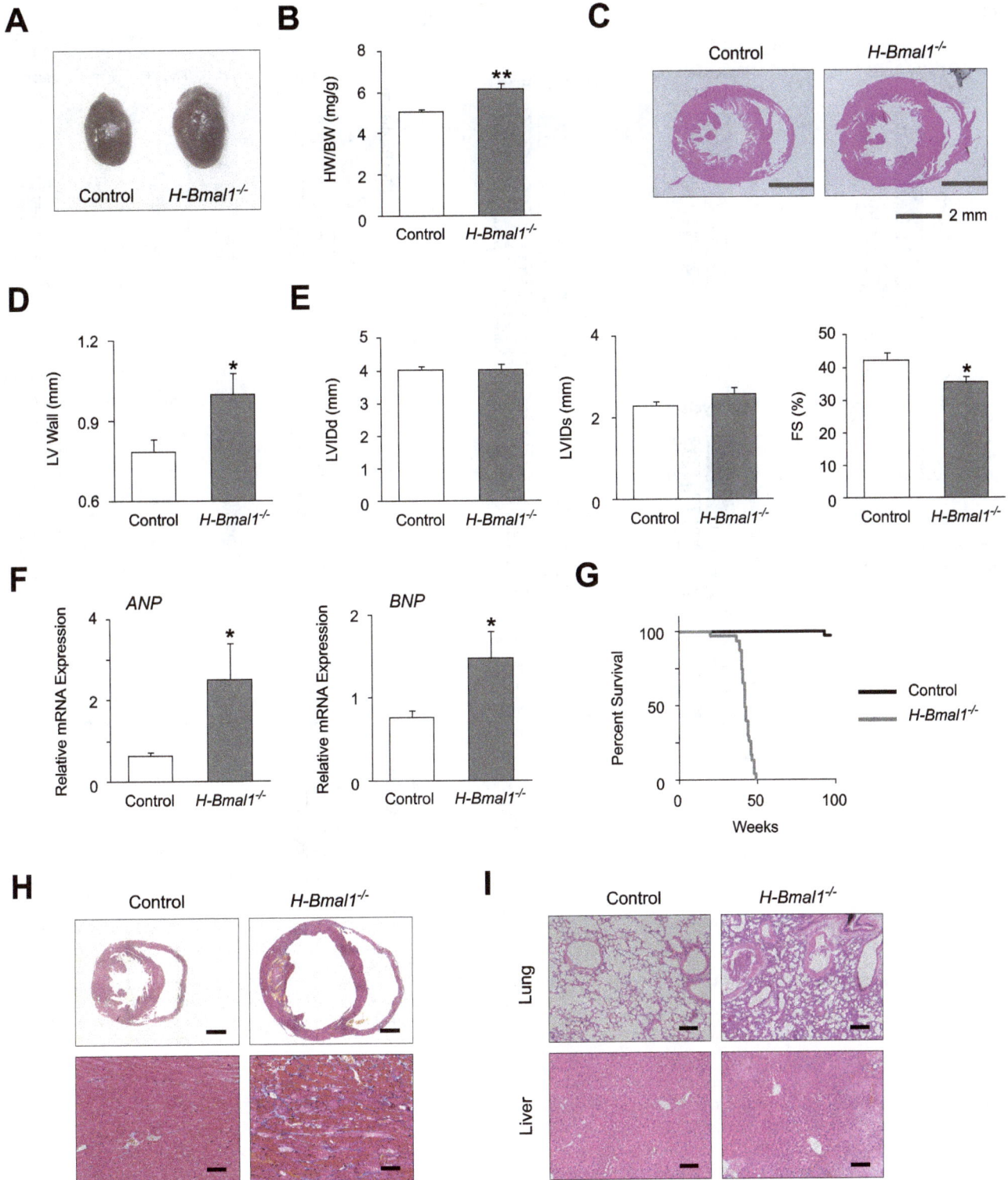

Figure 1. _H-Bmal1_^{−/−} mice develop progressive congestive heart failure with age. (A) Representative gross morphology of hearts from 12-week-old control and _H-Bmal1_^{−/−} mice. (B) Ratios of heart weight to body weight (HW/BW) at 12 weeks of age (n = 6 per group). (C) Low-power views after H&E staining of transverse sections from control and _Bmal1_^{−/−} hearts at 12 weeks of age. (D) LV lateral wall thickness determined in histological images used in (C) (n = 5 per group). (E) Echocardiographic analysis in 12-week-old control and _H-Bmal1_^{−/−} animals (n = 8–9 per group). LV internal diameter at diastole (LVIDd) and at systole (LVIDs) and fractional shortening (FS) are shown in bar graph format. (F) Transcript expression levels of _ANP_ and _BNP_ in control and _H-Bmal1_^{−/−} mice at 12 weeks of age (n = 6 per group). (G) Kaplan-Meier survival curves of control and _H-Bmal1_^{−/−} animals (n = 31 per group). (H) Low-power views (top panels, scale bar: 1 mm) and high-magnification views (bottom panels, scale bar: 25 μm) of Masson's trichrome staining of heart sections from 33-week-old control and _H-Bmal1_^{−/−} mice. (I) High-magnification views of H&E staining of lung (top panels) and liver (bottom panels) sections from the same animals used in (H). Scale bar: 50 μm. Data are the mean ± SEM. *$P < 0.05$, **$P < 0.01$, unpaired two-tailed Student's t-test.

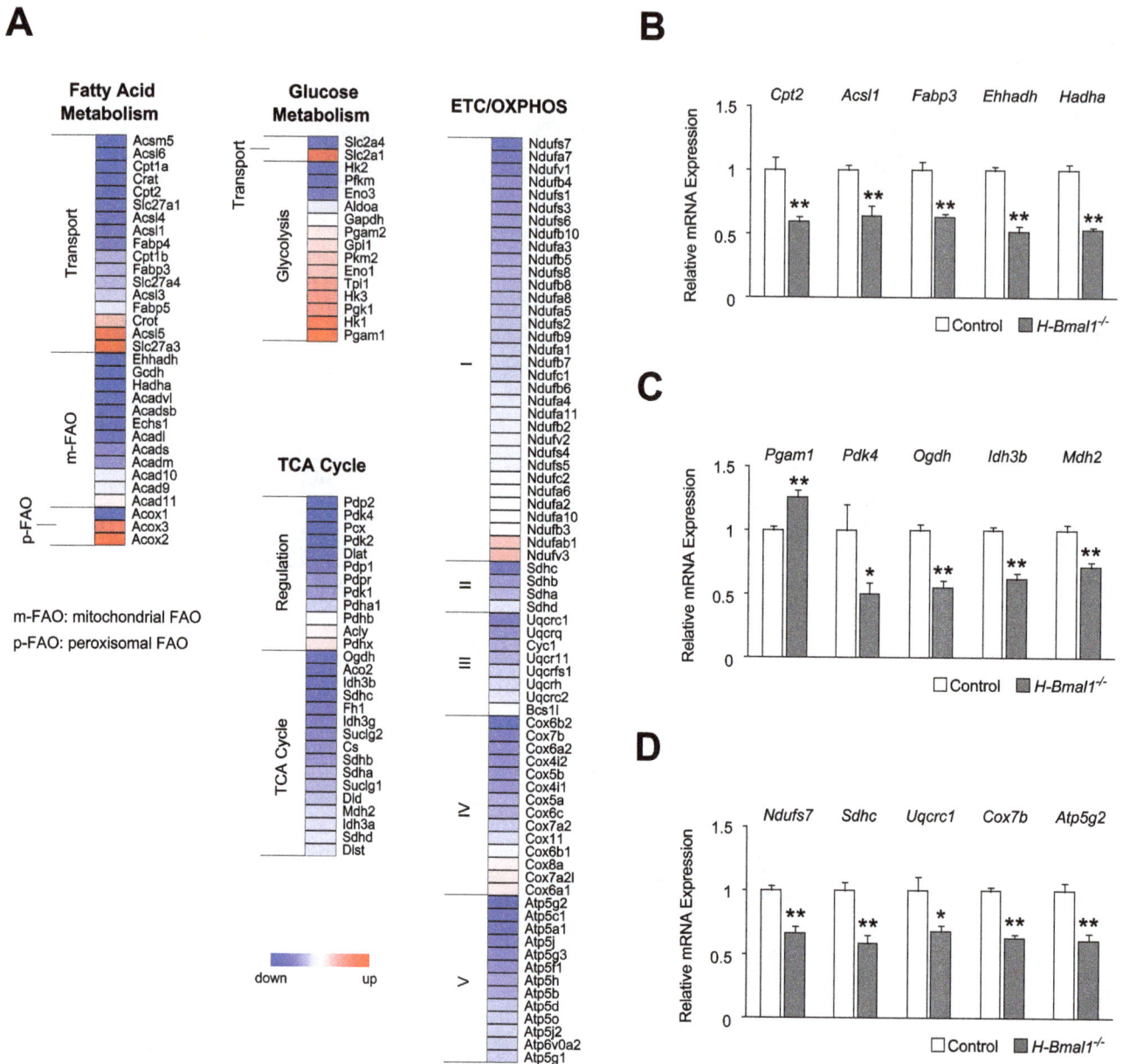

Figure 2. Gene expression profiles of cardiac energy metabolism in H-Bmal1$^{-/-}$ mice. (A) Heat-map representations of several classes of metabolic genes detected in the expression analysis of Bmal1$^{-/-}$ hearts from 12-week-old animals. (B-D) Relative expression levels of representative genes regulating (B) fatty acid metabolism, (C) glycolysis/TCA cycle, and (D) ETC/OXPHOS are shown (n = 6 per group). Data are the mean ± SEM. *P< 0.05, **P<0.01, unpaired two-tailed Student's t-test (B-D).

hearts. Indeed, in H-Bmal1$^{-/-}$ mice, the expression of the glucose transporter, which mediates basal glucose uptake into cardiomyocytes (Slc2a1, also known as Glut1), was upregulated, although the insulin-dependent glucose transporter (Slc2a4, also known as Glut4) was downregulated (Figure 2A). In addition, 9 of 14 glycolytic genes examined were upregulated in Bmal1$^{-/-}$ hearts (Figure 2A). Quantitative PCR validation analyses revealed that Pgam1 (a glycolytic enzyme) was significantly increased and that Pdk4 (a negative regulator of glucose oxidation) was decreased by approximately 50% in Bmal1$^{-/-}$ hearts (Figure 2C). Collectively, these data indicate that the energy substrate preference is altered in Bmal1$^{-/-}$ hearts.

We next examined the expression of genes controlling TCA cycle and the electron transport chain (ETC)/oxidative phosphor-

ylation (OXPHOS), common downstream pathways for both FAO and glycolysis. In contrast to the expression of FAO and glycolytic genes, which showed both upregulation and downregulation, the expression levels of the enzymatic components of TCA cycle and ETC/OXPHOS pathway were generally downregulated in Bmal1$^{-/-}$ hearts (Figure 2A). In particular, within TCA cycle, the genes encoding subunits of the enzymes that reduce NAD$^+$ to NADH (Ogdh, Idh3b, and Mdh2) were significantly decreased in Bmal1$^{-/-}$ hearts (Figure 2, A and C). Similar to TCA cycle, the majority of genes in the ETC/OXPHOS pathway were downregulated in Bmal1$^{-/-}$ hearts (Figure 2A). The significant decreases in representative genes (Ndufs7, Sdhc, Uqcrc1, Cox7b, and Atp5g2) associated with the ETC/OXPHOS pathway were also validated by quantitative PCR (Figure 2D).

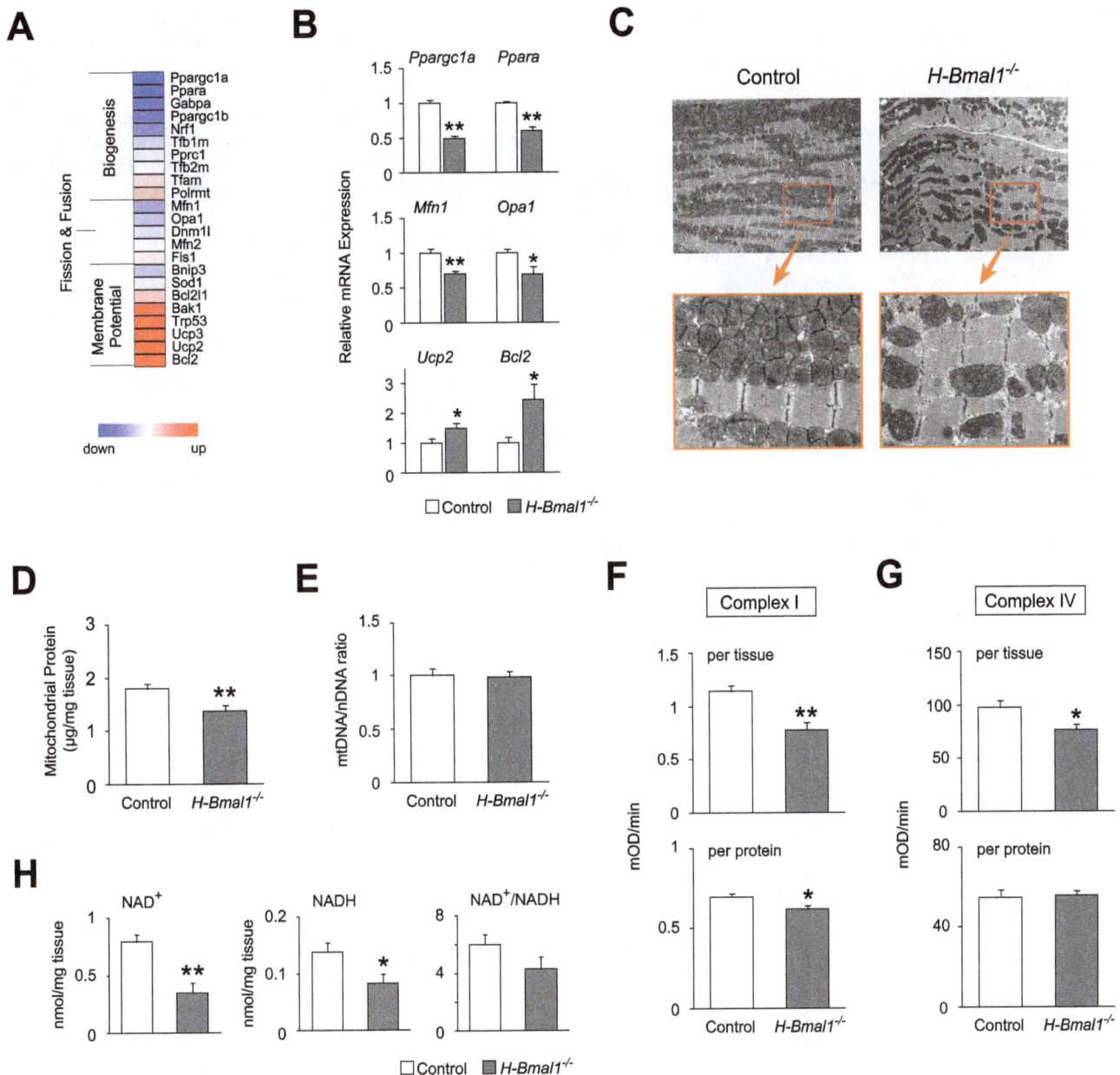

Figure 3. Mitochondrial abnormalities in the hearts of 12-week-old _H-Bmal1_$^{-/-}$ mice. (A) A heat-map representing the expression profiles of genes regulating mitochondrial structure and function in _Bmal1_$^{-/-}$ hearts. (B) Relative expression levels of genes associated with mitochondrial biogenesis, dynamics, and membrane potential in the heart tissue of control and _H-Bmal1_$^{-/-}$ animals (n = 6 per group). (C) Representative electron micrographs of sections taken from the left ventricular muscle from control and _H-Bmal1_$^{-/-}$ mice at two different magnifications. (D) Mitochondrial protein concentration in control and _Bmal1_$^{-/-}$ hearts (n = 8 per group). (E) Mitochondrial DNA to nuclear DNA ratio in control and _Bmal1_$^{-/-}$ hearts (n = 6 per group). (F-G) Enzymatic activities of (F) complex I and (G) complex IV in control and _Bmal1_$^{-/-}$ hearts (n = 8 per group). The activities of the mitochondrial respiratory enzymes are expressed either per milligram of tissue used for mitochondrial isolation (top panels) or per microgram of mitochondrial protein (bottom panels). (H) NAD$^+$ and NADH concentrations in control and _Bmal1_$^{-/-}$ hearts (n = 6 per group). Data are the mean ± SEM. *P<0.05, **P<0.01, unpaired two-tailed Student's t-test.

Defects in _Bmal1_ function alter mitochondrial dynamics in heart

Because loss of _Bmal1_ function induced a decrease in the expression of a broad range of genes controlling mitochondrial energy metabolism, we next examined the expression of transcripts that are associated with mitochondrial dynamics in _Bmal1_$^{-/-}$ hearts. Microarray analyses revealed that _Bmal1_ deficiency downregulated a variety of genes controlling mitochondrial

biogenesis as well as fission and fusion in the heart (Figure 3A). Validation analyses revealed that _Ppargc1a_ (also known as PGC-1α) and _Ppara_ (also known as PPARα), which are the genes responsible for mitochondrial biogenesis, were decreased significantly by 50% and by 40%, respectively (Figure B). In addition, a significant reduction was also found in genes associated with mitochondrial quality control (_Mfn1_ and _Opa1;_ Figure 3B). Although these mitochondrial structure-related genes were downregulated, genes associated with the loss of mitochondrial membrane potential were upregulated in $Bmal1^{-/-}$ hearts (Figure 3A). Specifically, genes controlling the permeability of the mitochondrial membrane ($Ucp2$ and $Bcl2$) were significantly increased in the heart tissue of $H\text{-}Bmal1^{-/-}$ mice (Figure 3B). These results suggest that $Bmal1^{-/-}$ hearts were defective in both the regulation of mitochondrial structure and membrane potential at the gene expression level.

Consistent with the downregulation of genes controlling mitochondrial dynamics, electron microscopic analysis showed morphological changes in the mitochondria of $Bmal1^{-/-}$ hearts. In heart tissue from 12-week-old $H\text{-}Bmal1^{-/-}$ mice, small- and large-sized mitochondria coexisted, whereas control hearts showed almost identical sizes (Figure 3C). In addition, the relative number of mitochondria was also significantly decreased in $Bmal1^{-/-}$ hearts (control 1.00 ± 0.03 vs. $Bmal1^{-/-}$ 0.70 ± 0.05; $p<0.01$). Ultrastructural examinations showed that mitochondria were sparse in $Bmal1^{-/-}$ hearts, whereas control hearts showed a dense mitochondrial population (Figure 3C). This observation was consistent with our finding that mitochondrial protein levels extracted from $Bmal1^{-/-}$ hearts were significantly reduced compared with those from control hearts (Figure 3D). Similar changes in both the size and the number of mitochondria were also observed in heart tissue from 24-week-old $H\text{-}Bmal1^{-/-}$ mice (Figure S3, A and B).

Hearts from $H\text{-}Bmal1^{-/-}$ mice showed altered mitochondrial function at the genomic and biochemical levels

In addition to the morphological examination, we also conducted functional analyses of cardiac mitochondria in $H\text{-}Bmal1^{-/-}$ mice. We first examined the ratio of mitochondrial DNA (mtDNA) to nuclear DNA (nDNA) copy number because the mtDNA copy number can vary independently of nDNA or mitochondrial number, and this ratio serves as a marker of individual mitochondrial function at the organelle level. Despite the decrease in the number of mitochondria in $Bmal1^{-/-}$ hearts (Figure 3, C and D), the ratio of mtDNA to nDNA was not different between 12-week-old control and $H\text{-}Bmal1^{-/-}$ animals (Figure 3E), which likely indicates compensation by increasing the copy number of mtDNA per mitochondrion in $Bmal1^{-/-}$ hearts. In contrast to younger animals, $Bmal1^{-/-}$ hearts from 24-week-old animals showed an approximate 30% decrease in the mtDNA/nDNA ratio compared with control hearts (Figure S3C).

We next examined biochemical activities of the ETC complexes in $Bmal1^{-/-}$ hearts. We analyzed complex I activity because the circadian clock links to redox metabolism [15,16,17], in which complex I accounts for almost all oxidative conversion from NADH to NAD$^+$ in the cell. Consistent with our microarray analyses, which showed an overall decrease in the expression of complex I genes (Figure 2A), complex I activity was significantly decreased in $Bmal1^{-/-}$ hearts (Figure 3F). Importantly, this result was not only observed when analyzed using the same weight of heart tissues (Figure 3F, top) but was also detected when normalized to mitochondrial protein concentration (Figure 3F, bottom), indicating that complex I activity in individual mitochondria from $Bmal1^{-/-}$ hearts was indeed attenuated. Complex IV activity was also significantly decreased in heart tissue from 12-week-old $H\text{-}Bmal1^{-/-}$ mice when normalized to tissue weight (Figure 3G, top). However, when normalized to mitochondrial protein concentration, complex IV activity was almost identical between control and knockout animals (Figure 3G, bottom), which indicates that the decrease in complex IV activity was not caused by reduced enzymatic activity in each mitochondrion but due to the decreased number of mitochondria in $Bmal1^{-/-}$ hearts. A

similar but more profound reduction in complex I and IV activities was also observed in heart tissue from 24-week-old $H\text{-}Bmal1^{-/-}$ animals (Figure S3, D and E).

We also examined the levels of NAD$^+$ and NADH in $Bmal1^{-/-}$ hearts because the intracellular redox state is tightly coupled to enzymatic activities of ETC complexes. In 12-week-old animals, both NAD$^+$ and NADH levels were dramatically decreased by approximately 56% and 40%, respectively, in $Bmal1^{-/-}$ hearts relative to control hearts (Figure 3H). The NAD$^+$/NADH ratio in $Bmal1^{-/-}$ hearts was also decreased, although this difference did not reach statistical significance (Figure 3H). In 24-week-old $H\text{-}Bmal1^{-/-}$ animals, we also observed similar but more severe changes in NAD$^+$ and NADH levels (approximately 78% and 83% reductions, respectively, Figure S3F). Collectively, these findings suggest that the $Bmal1$ gene is essential for mitochondrial metabolism in the heart and that loss of $Bmal1$ function causes progressive defects in cardiac mitochondrial structure and function with age.

Loss of $Bmal1$ function induces the expression of genes related to cardiac remodeling

Because cardiac remodeling processes are often observed in failing hearts, we next explored whether genes associated with cardiac remodeling were increased in $Bmal1^{-/-}$ hearts. We examined the expression levels of genes related to oxidative stress, programmed cell death, inflammation, and fibrosis, which are common cellular changes observed in remodeling hearts. Our microarray analyses revealed that the majority of oxidative stress-responsive genes (e.g., Gpx and Sod) and genes responsible for oxygen transport (e.g., $Slc38a1$ and $Cygb$) were upregulated in $Bmal1^{-/-}$ hearts (Figure 4A), indicating the existence of oxidative stress in heart tissue from $H\text{-}Bmal1^{-/-}$ animals. Increased oxidative stress can lead to programmed cell death. We observed an upregulation of genes important for the induction of apoptosis and autophagy (Figure 4B). These genes included those encoding caspases ($Casp$), proapoptotic homologs of members of the Bcl-2 family (Bax, Bak, Bad and Bid), and autophagy-associated genes (e.g., $Dram1$ and $Becn1$).

In addition, the mRNA expression levels were upregulated for a variety of inflammatory genes, such as genes encoding chemokines and their receptors (e.g., $Ccl2$, also known as $MCP\text{-}1$, and its receptor $Ccr2$) as well as interleukin-1 beta ($Il1b$), in $Bmal1^{-/-}$ hearts (Figure 4C). Furthermore, consistent with the histological analysis, which showed fibrotic changes (Figure 1H), genes responsible for pro-fibrotic changes (e.g., $Ctgf$ and $Snai1$) and remodeling enzyme genes (e.g., $Timp$ and $Serpine1$) were also increased in heart tissue from $H\text{-}Bmal1^{-/-}$ mice (Figure 4D). These results indicate that $Bmal1^{-/-}$ hearts undergo pathologic ventricular remodeling at the gene expression level.

Disruption of circadian behaviors reduces cardiac function in C57BL/6J mice with drug-induced cardiomyopathy

Having demonstrated that mitochondrial metabolism is disrupted in $Bmal1$-null hearts, which is accompanied by pathologic cardiac remodeling, we sought to examine whether cardiac mitochondrial metabolism is also impaired by the circadian disorder induced by disturbing the external LD cycle. Previous studies have shown that continuously shifting the LD and/or sleep/wake cycle is associated with increased cardiovascular morbidity or mortality in both humans and animals [37,38]. These observations, together with our findings in $H\text{-}Bmal1^{-/-}$ animals, raise the possibility that chronic desynchronization of the

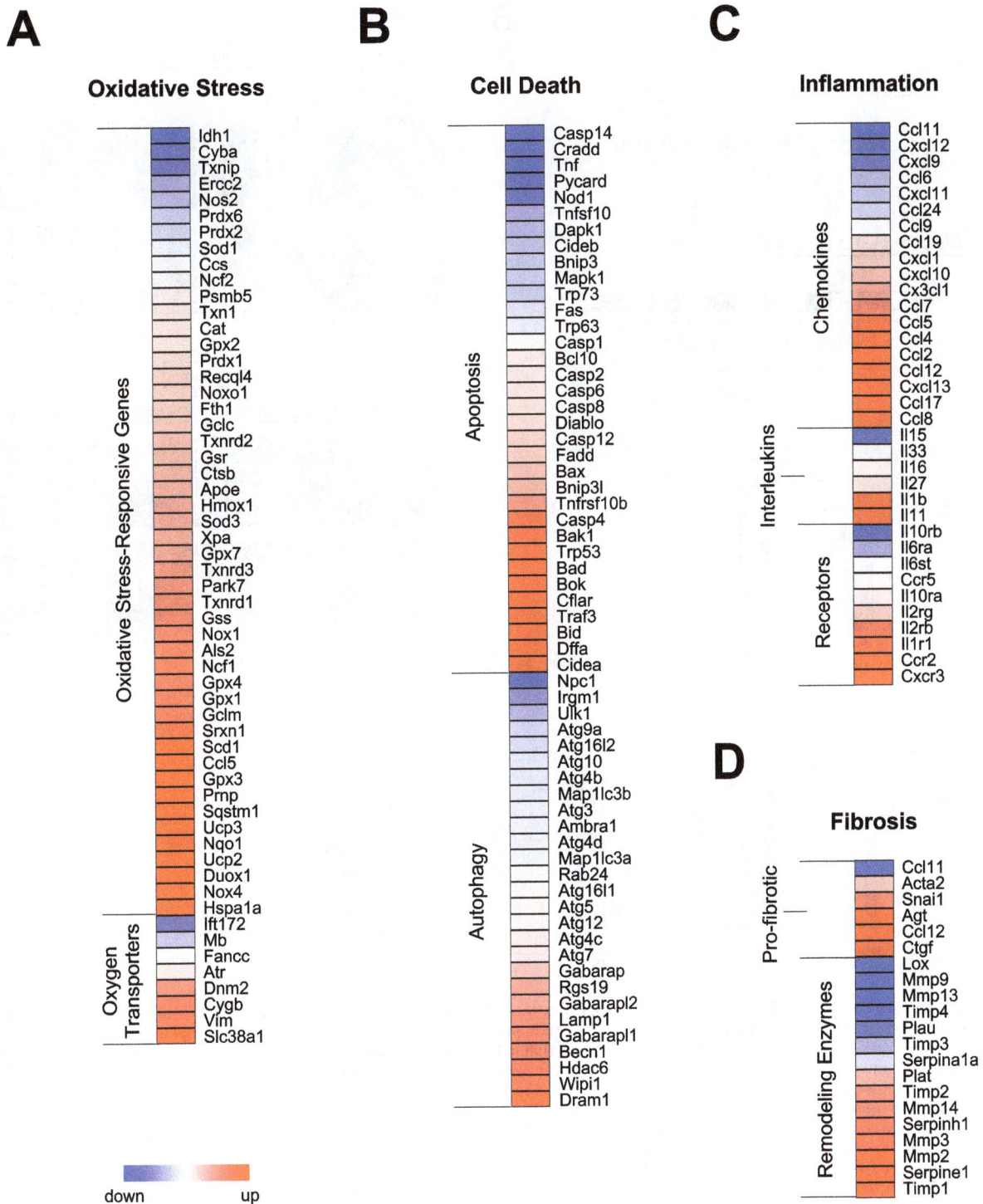

Figure 4. Expression profiles of genes associated with pathologic ventricular remodeling in *Bmal1*⁻/⁻ hearts. (A-D) Heat-map representations of genes involved in (A) the response to oxidative stress, (B) programmed cell death, (C) inflammation, and (D) fibrosis in hearts from 12-week-old *Bmal1*⁻/⁻ mice.

internal circadian clock with the external LD cycle, a condition often observed in shift work, may reduce cardiac function by altering mitochondrial metabolism in the heart. To test this hypothesis, we first examined the cardiac phenotypes of C57BL/6J mice either maintained on a fixed 12:12 h LD cycle (fixed LD cycle group) or exposed to a 12-h phase shift in the LD cycle every

3 days (disrupted LD cycle group) for 18 days (Figure 5A). To compare the effects of our disrupted LD cycle regimen between animals with healthy hearts and those with pathologic hearts, C57BL/6J mice were infused either with normal saline (NS) as a vehicle control or with phenylephrine (PE), a hypertrophic stimulus, via an osmotic pump. Infrared-based detection of

A

Fixed LD cycle group

Constant 12:12 h LD cycles

Disrupted LD cycle group

12:12 h LD for 3 days   12:12 h DL for 3 days

3 cycles

C

B

D

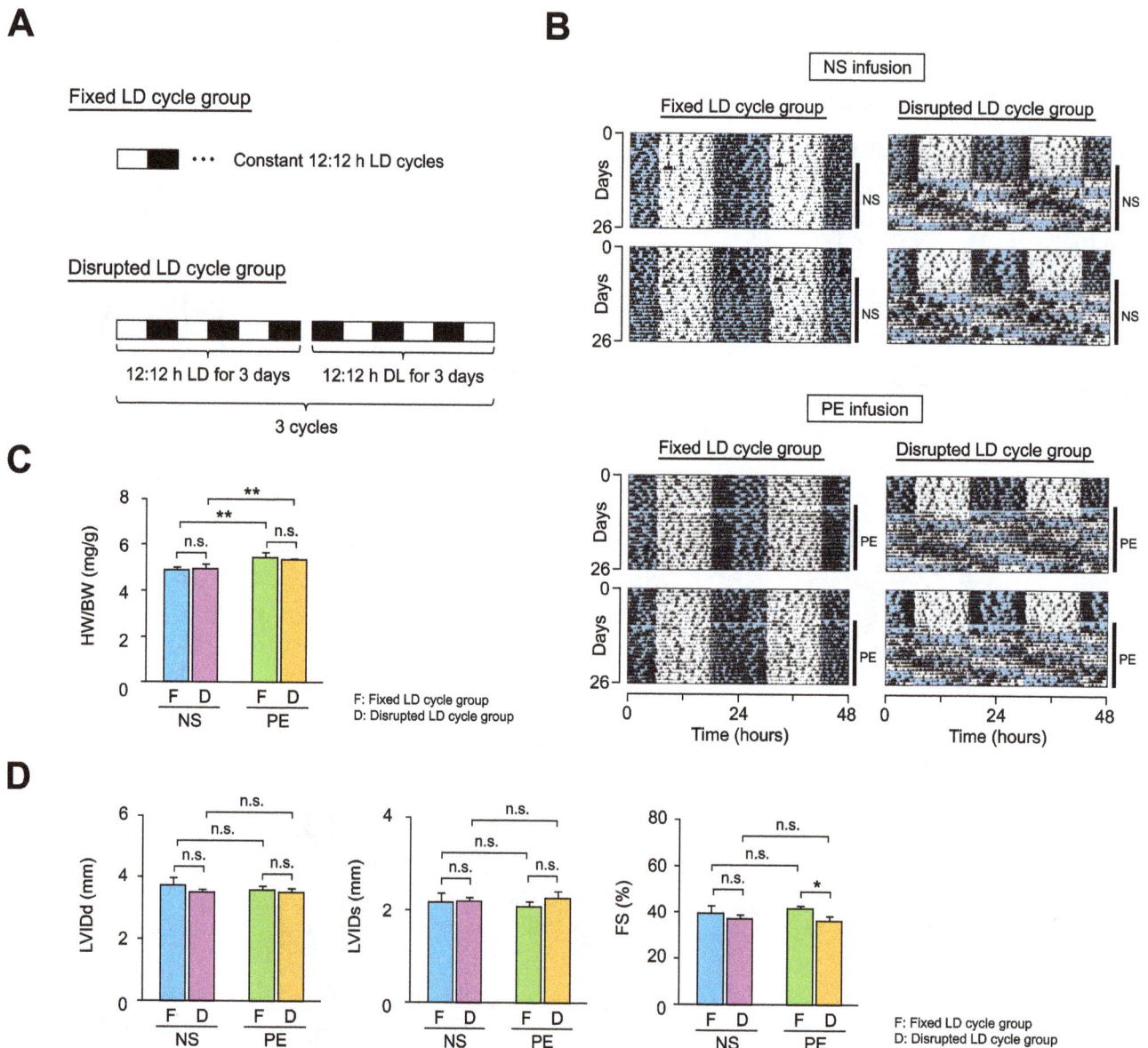

Figure 5. Circadian desynchronization reduces cardiac function in C57BL/6J mice with drug-induced cardiomyopathy. (A) A light-dark (LD) cycle regimen used to examine the effects of a variable LD schedule on cardiac function. C57BL/6J mice were either maintained on a constant LD schedule (fixed LD cycle group) or were subjected to a 12-h phase shift in LD cycle every 3 days (disrupted LD cycle group). To compare the effects of the LD schedule between animals with healthy hearts and those with cardiomyopathy, either normal saline (NS) or phenylephrine (PE) was continuously infused via an osmotic pump in each LD cycle group. (B) Locomotor activity records of NS-infused (top panel) and PE-infused (bottom panel) mice. Two representative records from animals subjected to the fixed (left column) and disrupted (right column) LD cycle are shown. Activity counts are indicated by the vertical black marks. The records are double-plotted such that 48 h are shown for each horizontal trace. The blue shaded and unshaded areas indicate the dark and light period, respectively. The duration of NS or PE infusion is indicated by the vertical line at the right margin. (C) The ratios of heart weight to body weight (HW/BW) were increased by the PE infusion but were not influenced by the disruption in LD cycle (n = 5–8 per group). (D) The effects of PE, disrupted LD cycle, or both on ventricular function were evaluated using echocardiographic measurements (n = 5–8 per group). LVIDd, LVIDs, and FS are shown in bar graph format. Data are the mean ± SEM. *$P<0.05$, **$P<0.01$, two-way ANOVA.

locomotor activity revealed that the temporal pattern of locomotor activity was sensitive to and followed the chronic shifting of the LD cycle in both NS-infused and PE-infused animals (Figure 5B), indicating that our LD reversal protocol was effective in disrupting circadian behaviors.

To examine the effects of the disrupted LD cycle on cardiac phenotypes, we first compared changes in heart morphology between the control and experimental groups. As expected, the PE

infusion increased the HW/BW ratio significantly when compared with vehicle infusion in both fixed and disrupted LD cycle groups (Figure 5C). In addition, although not statistically significant, the LV posterior diameter in diastole, as assessed by echocardiogram, was also increased in PE-infused animals (NS-infused 0.92 ± 0.11 mm vs. PE-infused 1.12 ± 0.10 mm). Although these results indicate that PE infusion induced cardiac hypertrophy, the

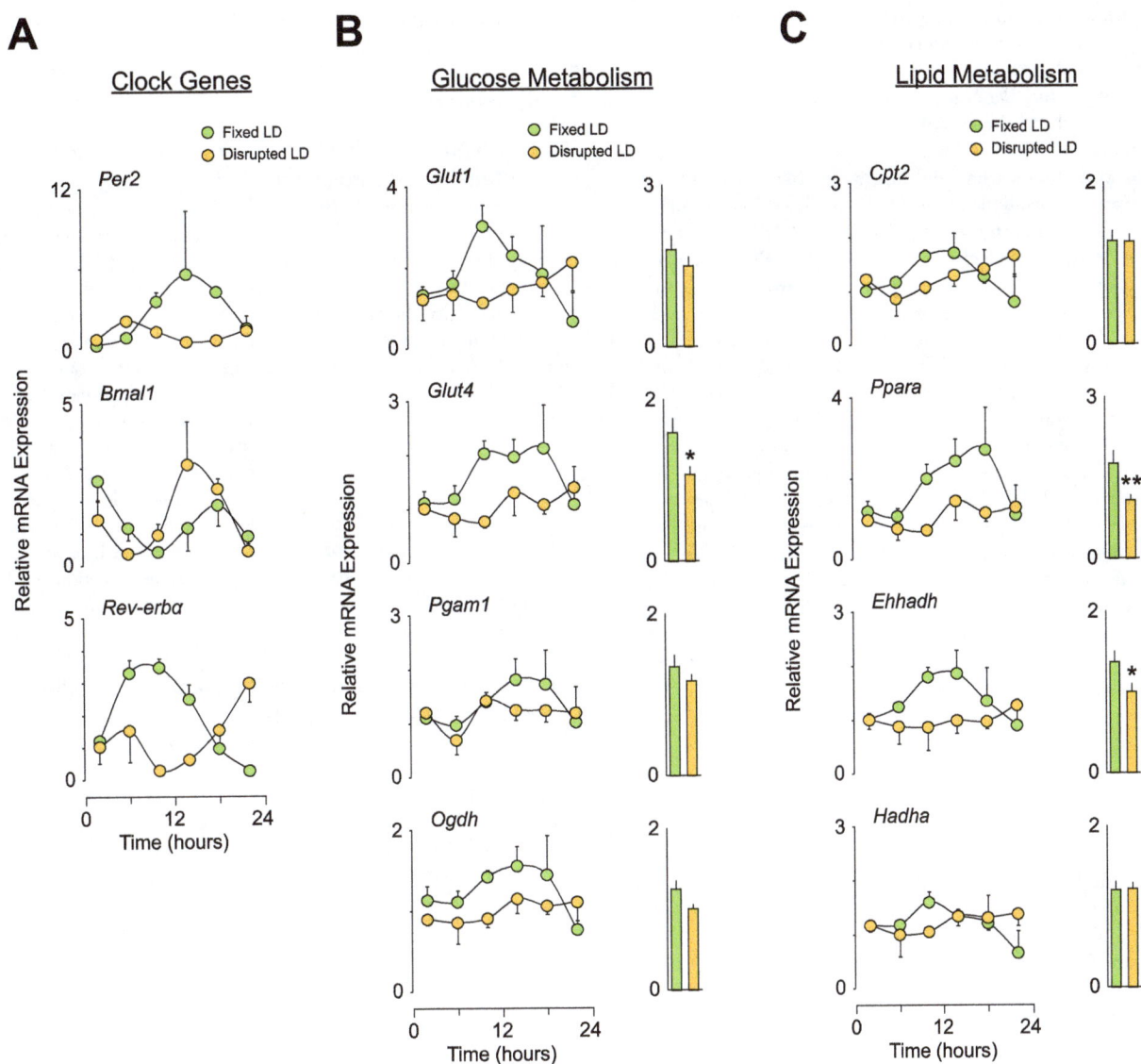

Figure 6. Circadian desynchronization not only disrupts rhythms but also reduces the expression levels of clock and metabolic genes in the heart of C57BL/6J mice with PE-induced cardiomyopathy. (A-C) Relative expression levels of genes regulating (A) clock machinery as well as (B) glucose and (C) lipid metabolism in heart. All heart tissues used were from PE-infused animals subjected to either a fixed or a disrupted LD cycle as described in Figure 5A (n = 4 per group per time point). To provide a 24-h overall mean expression level, the data over a 24-h time period in each group were also averaged and are expressed using a bar graph format. Data are the mean ± SEM. *P<0.05, **P<0.01, unpaired two-tailed Student's t-test.

disruption in LD cycle did not further affect heart weight in PE-infused animals (Figure 5C).

We next investigated whether the disrupted LD cycle altered cardiac function. Interestingly, although the disrupted LD cycle did not alter the weight of hearts (Figure 5C), heart function was significantly reduced in PE-infused but not NS-infused animals when subjected to the disrupted LD cycle (Figure 5D). Specifically, in PE-infused mice, the disrupted LD cycle significantly decreased fractional shortening, as assessed by echocardiogram, although we found no difference in LVIDd and a small, non-significant increase in LVIDs (Figure 5D). In NS-infused animals, although not statistically significant, systolic contractility (fractional shortening) tended to be lower when mice were exposed to the disrupted LD cycle (Figure 5D). These results suggest that

exposure to disrupted LD cycles adversely affects cardiac function in mice.

Disruption of circadian behaviors not only alters rhythms but also reduces levels of transcripts associated with energy metabolism in the heart

To determine the genetic basis for the reduced cardiac function caused by the disrupted LD cycle, we examined both the rhythms and expression levels of circadian and metabolic genes in heart tissue from NS-infused and PE-infused C57BL/6J mice. Under the fixed LD cycle condition, all three clock genes examined (*Per2*, *Bmal1*, and *Rev-erbα*) displayed clear diurnal rhythms in expression levels in both NS-infused and PE-infused animals (Figure S4A and Figure 6A, respectively). However, in both NS-infused and PE-infused animals, the time of peak expression of these clock genes was shifted when mice were exposed to a disrupted LD cycle (Figure 6A and Figure S4A). Furthermore, interestingly, the *Per2* gene displayed a decrease in expression levels throughout the day when under the disrupted LD cycle condition.

We also found that the disrupted LD cycle altered the expression of metabolic genes. Under the fixed LD cycle regimen, the glucose transporter genes (*Glut1* and *Glut4*), a glycolytic gene (*Pgam1*), and a gene within TCA cycle (*Ogdh*) all exhibited diurnal variations in their expression in heart tissue from both NS-infused and PE-infused mice (Figure S4B and Figure 6B, respectively). Similarly, expression levels of genes associated with fatty acid transport (*Cpt2*) and β oxidation (*Ppara*, *Ehhadh*, and *Hadha*) also displayed diurnal patterns when exposed to the fixed LD cycle condition (Figure 6C and Figure S4C). Under the disrupted LD cycle, we expected phase-shifted expression patterns of these metabolic genes, as observed with clock gene expression; however, interestingly, these metabolic genes showed changes in their expression levels (i.e., downregulation) rather than altered expression rhythms when the animals were subjected to the disrupted LD cycle (Figure 6, B and C, Figure S4, B and C). In particular, when overall 24-h expression levels were averaged, a significant decrease was observed in *Glut4*, *Ppara*, and *Ehhadh* in PE-infused animals (Figure 6, B and C) and in *Glut1* and *Ppara* in NS-infused mice (Figure S4, B and C).

Chronic circadian desynchronization disrupts mitochondrial metabolism in the heart

We next examined whether the disrupted LD cycle also altered cardiac mitochondrial metabolism. We analyzed the temporal expression levels of genes involved in mitochondrial structure (fission and fusion) and function (ETC/OXPHOS pathway). Under the fixed LD cycle, fission and fusion genes (*Mnf1*, *Mfn2*, *Drp1*, and *Opa1*) displayed diurnal variations in expression levels in heart tissues from both NS-infused and PE-infused animals (Figure S5A and Figure 7A, respectively). These time-dependent expression patterns were also observed in genes encoding components of the ETC/OXPHOS in tissue from mice exposed to the fixed LD cycle (Figure 7B and Figure S5B). However, when animals were exposed to the disrupted LD cycle, diurnal variations in the expression of genes regulating mitochondrial structure and function were dampened in both NS-infused and PE-infused animals (Figure 7, A and B, Figure S5, A and B). Importantly, the dampened expression rhythms in these mitochondrial-related genes were more profound in PE-infused animals than in NS-infused mice. Specifically, the overall 24-h expression levels of 10 of 12 examined genes were significantly decreased in PE-infused mice (Figure 7, A and B), whereas significant decreases were detected only in 7 of 12 genes in NS-infused animals (Figure S5, A and B), indicating that the regulation of gene expression associated with mitochondrial metabolism is more susceptible to disrupted LD cycles in pathologic hearts compared with healthy hearts. This idea is further supported by our observation that the disrupted LD cycle significantly reduced the enzyme activity of complex I only in PE-infused animals (Figure 7C) but not in NS-infused animals (Figure S5C). It should be noted that, although we observed dampened expression rhythms of genes regulating mitochondrial dynamics in heart tissue from animals exposed to the disrupted LD cycle (Figure 7A and Figure S5A), ultrastructural examinations did not show a difference in mitochondrial morphology between animals subjected to the fixed and disrupted LD cycles (Figure 7D and Figure S5D), indicating that the altered mitochondrial

metabolism induced by disrupted LD cycles was not due to an alteration in the number of mitochondria.

Discussion

Heart *Bmal1* is an important component for the maintenance of cardiac function

Our findings that H-$Bmal1^{-/-}$ mice develop severe, progressive heart failure with age and display a markedly shorter life span, likely due to cardiac decompensation, indicate that *Bmal1* gene expression in the heart is an important component that maintains normal cardiac function throughout life. Of note, we did not find changes in rhythms in locomotor activity and blood pressure in *H-Bmal1*$^{-/-}$ animals, suggesting that reduced function of *Bmal1*$^{-/-}$ in the heart is not secondary to alterations in behavioral and/or cardiovascular rhythms but instead is primary to cardiac *Bmal1* dysfunction. Although similar cardiac phenotypes have been shown in previous reports using whole-body *Bmal1* knockout mice [3], our findings strengthen the molecular evidence that *Bmal1* in the heart is indeed required to regulate cardiac function because whole-body *Bmal1* knockout animals exhibited various phenotypic disorders that may affect cardiac function, such as dampened diurnal rhythm in blood pressure and metabolic disorders [11,12,39].

Defects in the circadian clock system impair cellular energy metabolism in the heart

Although defects in the function of clock genes other than the *Bmal1* gene also affect the structure and/or the function of the heart [1,35], the precise molecular mechanisms that connect the circadian clock system to cardiac function have not yet been fully delineated. Recently, the regulation of cardiac electrophysiology has been reported to be controlled by clock genes [40], indicating that the primary function of cardiomyocytes is linked to the circadian clock system. Herein, we demonstrate that the expression levels of broad classes of genes that regulate fundamental cellular metabolism, such as fatty acid and glucose oxidation, are altered in *Bmal1*$^{-/-}$ hearts. In addition, we observed that heart tissue from C57BL/6J mice exposed to a disrupted LD cycle showed a significant decrease in the 24-h overall expression levels of genes that are essential for glucose and lipid metabolism, suggesting that the circadian clock system (i.e., the internal clock and the external LD cycle, which together coordinate physiological processes) is required to regulate cellular energy metabolism in the heart. These findings are consistent with previous studies demonstrating that fat metabolism in the heart is impaired in *Clock* mutant animals [2,41].

Using mouse genetic models, a considerable link between the molecular clock machinery and energy metabolism has been shown, particularly in metabolic organs including fat and liver [42,43,44,45]. For example, fat deposition is increased in the adipose tissue of mice carrying a mutant *Clock* gene [46]. Dysregulation of hepatic glucose metabolism has been demonstrated in mice without *Bmal1* function [10,11]. Although it is not yet clear how the molecular clock participates in the regulation of glucose and lipid metabolism, transcription factors and their related molecules have been proposed as key components that link the circadian and metabolic systems [42]. PPARα, a key transcriptional regulator of FAO binds to the *Bmal1* promoter [33]. The transcription of the gene encoding PPARα is in turn activated by the BMAL1 protein together with its heterodimerized partner CLOCK [34], suggesting a reciprocal regulation of gene expression between circadian and metabolic transcription factors. PGC-1α, a major regulator of mitochondrial biogenesis and

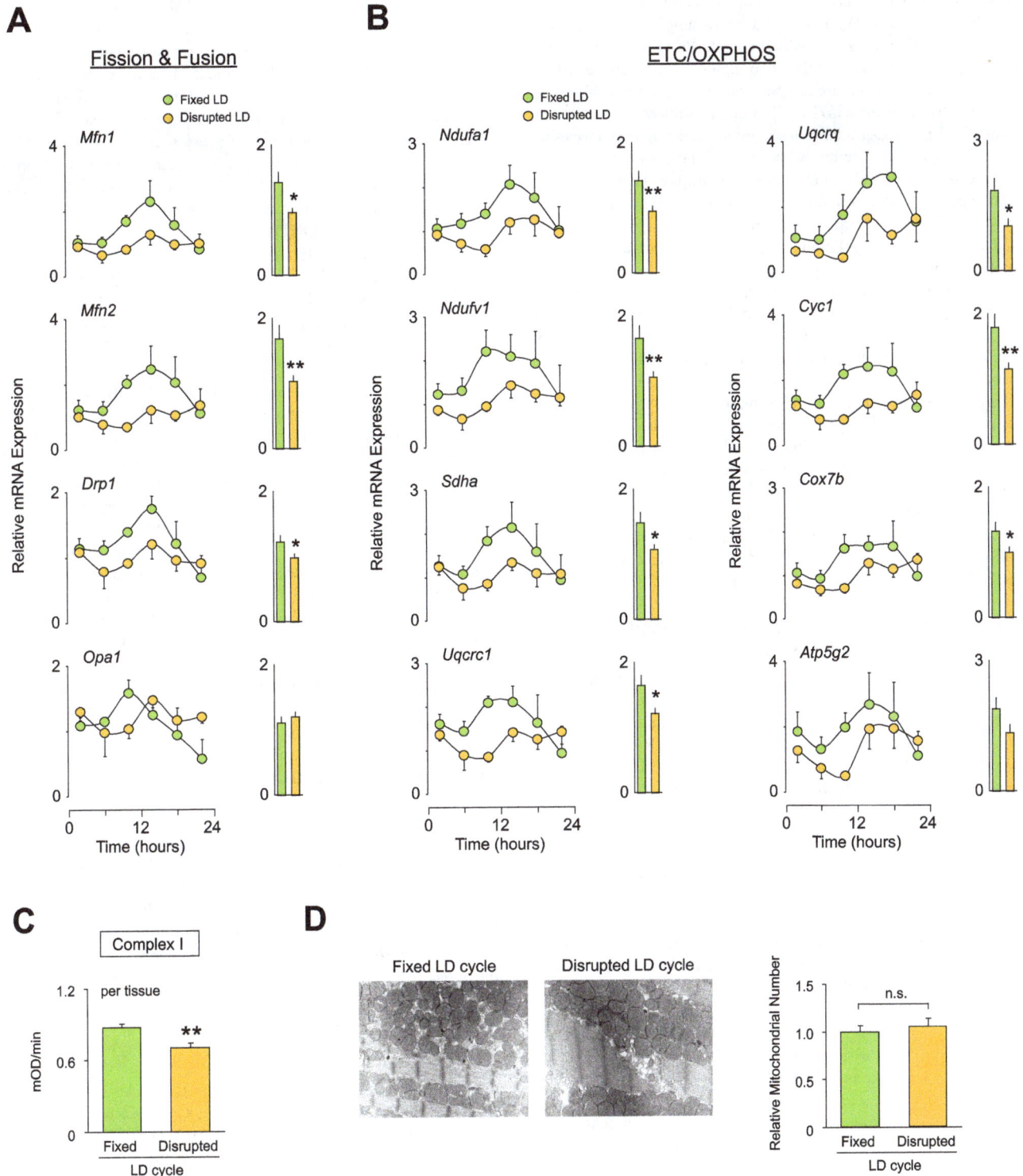

Figure 7. Circadian desynchronization impairs mitochondrial function in the hearts of C57BL/6J mice with PE-induced cardiomyopathy. (A-B) Relative expression levels of genes regulating (A) mitochondrial structure or (B) mitochondrial oxidative metabolism in heart. All heart tissues used are from PE-infused animals subjected to either a fixed or a disrupted LD cycle as described in Figure 5A (n = 4 per group per time point). To provide a 24-h overall mean expression level, the data over a 24-h time period in each group were also averaged and are expressed in bar graph format. (C) Enzymatic activity of complex I in PE-infused animals exposed to a fixed or disrupted LD cycle (n = 4–5 per group). The complex I activity is expressed per milligram of tissue used for mitochondrial isolation. (D) The relative number of mitochondria in the left ventricular muscle was counted using electron microscope images. Representative images are shown. Data are the mean ± SEM. *P<0.05, **P<0.01, unpaired two-tailed Student's t-test.

respiration, also interacts with a gene regulatory network of the circadian clock [32]. We found that expression levels of both the *Ppara* and *Ppargc1a* genes, which encode PPARα and PGC-1α, were significantly downregulated in $Bmal1^{-/-}$ hearts. Furthermore, a significant decrease of *Ppara* expression was also observed in heart tissue from C57BL/6J animals subjected to chronic reversal of the LD cycle. These results indicate that circadian disorder imposed either by the deletion of the core clock gene or by a disruption of the LD cycle may impair cellular energy metabolism in the heart due to alterations in the regulation of these metabolic transcription factors.

Because the dysregulation of transcription factors has been implicated in heart failure pathogenesis, our observation that the expression levels of metabolic transcription factors are decreased in $Bmal1^{-/-}$ hearts provides insight into the basis for reduced cardiac function in $H\text{-}Bmal1^{-/-}$ animals. Among a series of transcription factors, the dysregulation of PPARs, ERRs, and their transcriptional coactivators is involved in the progression of heart failure. Altered function of PPARα induces cardiac hypertrophy in mice [27,28]. ERRs participate in the regulation of genes associated with cardiac function not only in adult hearts but also in postnatal hearts [29,30]. In addition, deletion of the gene encoding PGC-1α alters contractile function of cardiac muscle [31]. Our findings that *Ppara* and *Ppargc1a* were downregulated in $Bmal1^{-/-}$ hearts provide additional evidence that circadian and metabolic transcription factors coordinate a gene regulatory network that affects cardiac performance.

The circadian clock is required for mitochondrial dynamics and bioenergetics in the heart

Increasing evidence suggests that dysregulation of metabolic transcription factors, such as PPARα and PGC-1α, underlie the defects in mitochondrial structure and function in the failing heart. An alteration in transcriptional control by PPARα induces the dysregulation of a series of genes encoding mitochondrial FAO enzymes in the heart [47]. In addition, dysfunction of PGC-1α also causes abnormalities in both mitochondrial structure and function in the failing heart [48]. Consistent with these previous studies, we observed that a majority of genes associated with mitochondrial FAO, fission and fusion, and the ETC/OXPHOS are significantly decreased in $Bmal1^{-/-}$ hearts. These data support the conclusion that the reduced cardiac function observed in $H\text{-}Bmal1^{-/-}$ mice is caused by defects in the generation of mitochondrial bioenergy due to the abnormal regulation of metabolic transcription factors in the heart. In addition, we also observed that a circadian disorder imposed by the disrupted LD cycle also reduces the expression of genes involved in the ETC/OXPHOS pathway, suggesting that perturbation of overall function of the circadian clock may also induce mitochondrial defects in the heart. Of note, these changes were enhanced in PE-infused pathologic hearts compared with healthy hearts, indicating that mitochondrial function in the diseased heart is more susceptible to disorders of the circadian system.

A close association between the circadian clock and mitochondria-based cellular metabolism has been indicated in a series of studies focusing on the regulation of cellular redox. For example, the DNA binding activity of CLOCK:BMAL1 is influenced by the ratio of NAD^+ and NADH [15] and PARP-1 (poly(ADP-ribose) polymerase 1) [49], an NAD^+-dependent ADP-ribosyltransferase. Furthermore, CLOCK:BMAL1 is, in turn, involved in NAD^+ synthesis by directly regulating the expression of NAMPT (nicotinamide phosphoribosyltransferase) [50,51], the rate-limiting enzyme of the NAD^+ salvage pathway. Consistent with these previous studies, we noted a significant decrease in NAD^+ levels in

$Bmal1^{-/-}$ hearts. The decrease in NAD^+ levels in $Bmal1^{-/-}$ hearts may also be caused by reduced activity of mitochondrial complex I because a large portion of the oxidative conversion from NADH to NAD^+ in the cell is undertaken by complex I. Of interest, $Bmal1^{-/-}$ hearts also showed reduced NADH levels. This observation is consistent with our gene expression analysis, which showed that a wide range of genes that are associated with TCA cycle and FAO, two major pathways that convert NAD^+ into NADH, are downregulated in $Bmal1^{-/-}$ hearts. The reduction of both NAD^+ and NADH levels indicates the uncoupling of the TCA cycle and ETC/OXPHOS pathway in $Bmal1^{-/-}$ hearts because the TCA cycle functions as a direct source of intramitochondrial NADH for the ETC, which, in turn, supplies NAD^+ to the TCA cycle. Although the precise mechanisms that cause the decrease in expression of genes involved in FAO, TCA cycle, and ETC/OXPHOS in $Bmal1^{-/-}$ hearts remains to be elucidated, our data suggest that clock machinery is required to maintain cellular redox and mitochondrial function in the heart.

In conclusion, our findings that $H\text{-}Bmal1^{-/-}$ mice show mitochondrial defects in the heart, together with pathologic cardiac remodeling, indicate that the molecular clock machinery plays an important role in maintaining cardiac function by regulating mitochondrial dynamics and bioenergetics. In addition to the genetic model, we found that a circadian disorder imposed by disrupting the LD cycle also induced a striking effect on the expression levels of genes regulating mitochondrial energy metabolism, suggesting that the overall function of the internal clock, which is coupled to the external LD cycle, is essential for the maintenance of mitochondrial energy metabolism in the heart. Gaining further insight on how the molecular clock balances the capacity of mitochondrial metabolism in the heart based on the external cycle may pave new avenues to understand the pathogenesis of heart failure.

Materials and Methods

Mice

C57BL/6J mice bearing the modified *Bmal1* gene containing *loxP* sites [B6.129S4(Cg)-*Arntl*^{tm1Weit}/J; stock number 007668] and transgenic mice expressing Cre recombinase driven by the αMHC promoter [B6.FVB-Tg(Myh6-cre)2182Mds/J; stock number 011038] were purchased from Jackson Laboratory and crossed to generate $H\text{-}Bmal1^{-/-}$ mice. We used littermate animals that harbored the floxed *Bmal1* gene but not the *Cre* transgene as controls. Both the strategy used and confirmation of the heart-specific deletion of the *Bmal1* gene are delineated in Figure S6. During the experiments, these mice were maintained on a 12:12 LD cycle unless otherwise noted.

In all experiments in which disrupted LD cycles were applied, we used C57BL/6J mice obtained from Charles River Laboratories Japan, Inc. (Yokohama, Japan). All mice in the disrupted LD cycle group were subjected to reversals of the LD cycle every 3 days for a total 18 days. The LD reversal regimen began when the animals were 9 weeks of age. The animals in the fixed LD cycle group were exposed to a constant 12:12 LD cycle until the end of the experiment. Osmotic pumps (ALZET model 1004, DURECT, Cupertino, CA) were used for long-term treatment with PE (30 mg/kg/day) or NS (control). Pumps were implanted subcutaneously under light anesthesia on the day when the disrupted (or fixed) LD regimen began.

In all experiments, male animals that had ad libitum access to food and water were used.

Behavioral analysis

The locomotor activity of the animals was monitored using a Supermex system (Muromachi Kikai, Tokyo, Japan). In this system, a sensor counts the movements of the mouse, which is individually housed in a home cage, by detecting the radiated body heat. Data were recorded continuously in 1-min bins using a data collection program (CompACT AMS, Muromachi Kikai).

Echocardiographic analysis

Transthoracic echocardiography was performed using a 15-MHz linear-array probe. The images were obtained in M-mode (left parasternal short-axis). All echocardiograms were performed blinded to the mouse genotype or condition used. LV fractional shortening was calculated using the formula (LVIDd-LVIDs)/LVIDd x 100.

Life-span analysis

Animals for the longevity study were not used for any other physiological, biochemical, or molecular experiments. All mice for the longevity study were carefully inspected every day. The endpoint of life was when the animal was found dead during daily inspections. Moribund animals were humanely euthanized under isoflurane anesthesia by cervical dislocation upon presentation of defined criteria (diminished response to stimuli, lethargy, and failure to thrive), and the time of euthanasia was used as the endpoint. Any animals which did not meet these criteria were allowed to proceed to a natural death; however, alternative endpoints were considered when any of the following symptoms was found: 1) inability to reach food and water, 2) inability to remain upright, 3) weight loss (more than 15%). All efforts (ex. having food and water easily accessible) were made to minimize suffering of the animals. The survival data for each genotype were analyzed by plotting the Kaplan-Meier curves and performing log-rank tests.

Tissue collection and RNA extraction

The left ventricles of hearts were used to extract total RNA. Tissues from $H\text{-}Bmal1^{-/-}$ (n = 6) and their controls (n = 6) were collected at ZT2 of the LD cycle. In the experiments employing the LD reversal regimen, heart tissues from C57BL/6J mice were obtained every 4 hours beginning at ZT2 on day 19 of the LD reversal regimen (n = 4 per group per time point). Total RNA was extracted from frozen tissue with TRIzol reagent (Invitrogen, Carlsbad, CA). The specific procedures for microarray analysis and quantitative PCR are described below.

Microarray analysis

Heart total RNA extracted from six animals per genotype (control and $H\text{-}Bmal1^{-/-}$) was pooled and then used for a microarray analysis. We used a commercially available DNA microarray service (Takara Bio, Otsu, Japan). For this service, cyanine-3 (Cy3)-labeled cRNA was prepared from 0.5 µg total RNA. Cy3-labeled cRNA (1.65 µg) was fragmented and hybridized to the SurePrint G3 Mouse (8x60K) Microarray (Agilent Technologies, Santa Clara, CA) using standard procedures. After hybridization, the microarrays were washed with GE Wash Buffer 1 (Agilent) for 1 min at room temperature and for another 1 min with GE Wash Buffer 2 (Agilent). The microarrays were then dried by centrifugation and scanned by an Agilent DNA Microarray Scanner (G2565CA). For data processing, the scanned images were analyzed using Feature Extraction Software 10.5.1.1 (Agilent) to obtain (background-corrected and normalized) signal intensities. Gene expression levels were graphically represented in heat maps, which were created by the freely available software MultiExperiment Viewer (MeV; http://www.tm4.org/mev.html) [52].

Quantitative RT-PCR

First-strand cDNA was synthesized using 0.25 µg of total RNA and the High Capacity cDNA Reverse Transcription Kit (Applied Biosystems, Foster City, CA). Quantitative PCR was performed and analyzed using a TP850 Thermal Cycler Dice Real-time System (Takara Bio). Samples contained 1 X SYBR Premix Ex Taq II (Takara Bio), 1000 nM of each primer, and cDNA in a 10 µl volume. The PCR conditions were as follows: 30 sec at 95°C, then 35 cycles of 5 sec at 95°C and 30 sec at 60°C. Expression levels relative to $Gapdh$ were calculated using the comparative C_T method. The sequences of primers used for the quantitative RT-PCR are shown in Table S1.

Histology and electron microscopy

Tissues were fixed in 4% paraformaldehyde, processed and embedded in paraffin prior to sectioning (4 microns), and stained with H&E for overall morphology and with Masson's trichrome to detect fibrosis. Heart tissues for transmission electron microscopy were cut into 1-mm pieces that were fixed immediately after collection in 2.5% glutaraldehyde in 0.1 M phosphate buffer (pH 7.4), and stored at 4°C. Post-fixation was performed in 2% OsO4 (4°C). Subsequently, samples were dehydrated and embedded in epon. Ultrathin sections were examined using a H-7100 electron microscope (Hitachi High-Technologies, Tokyo, Japan). The number of cardiac mitochondria was determined from electron microscope images. For each mouse genotype (i.e., control vs. $H\text{-}Bmal1^{-/-}$) and condition used (i.e., fixed vs. disrupted LD cycle and NS vs. PE infusion), at least 5 different images were examined in a blinded fashion.

mtDNA quantification

To extract DNA, a small piece of heart tissue was incubated in 50 mM NaOH at 95°C and then neutralized with Tris buffer (pH 5.5). Quantitative PCR was performed using 200-fold diluted DNA, 1 X SYBR Premix Ex Taq II (Takara Bio), and 1000 nM of each primer [mtDNA-specific primers (16S rRNA): forward 5'-CCGCAAGGGAAAGATGAAAGAC-3', reverse 5'-TCGTTT-GGTTTCGGGGTTTC-3'; nDNA specific primers (hexokinase 2): forward 5'-GCCAGCCTCTCCTGATTTTAGTGT-3', reverse 5'-GGGAACACAAAAGACCTCTTCTGG-3'] in a 10 µl volume. The PCR conditions were as follows: 30 sec at 95°C and then 40 cycles of 5 sec at 95°C and 30 sec at 56°C. Results were calculated based on differences in threshold cycle values for mtDNA and nDNA. The data are expressed as the ratio of mtDNA to nDNA copy number.

Mitochondrial isolation and protein assay

Mitochondria were isolated from heart tissues using a differential centrifugation method following the manufacturer's protocol (BioChain Institute, Newark, CA). Briefly, 50-mg heart tissue samples were homogenized in isolation buffer. The homogenates were transferred to tubes and centrifuged at 600 g for 10 min at 4°C. The supernatants were again centrifuged at 12 000 g for 15 min at 4°C to isolate the mitochondrial pellet. The pellet was resuspended in mitochondrial isolation buffer. These centrifugation processes were then repeated once more to obtain the pure mitochondrial sample. The protein concentrations of the mitochondrial samples were determined using a Bradford protein assay kit (Nacalai Tesque, Kyoto, Japan).

Biochemical assays

The enzymatic activities of mitochondrial complex I and IV were determined using the pure mitochondria isolated from the heart tissue. Complex I activity was analyzed using a microplate assay kit (EMD Millipore, Darmstadt, Germany). For this assay, complex I activity is measured based on the oxidation of NADH to NAD^+, which leads to an increase in absorbance at 450 nm. The activity of complex IV was measured using the Cytochrome C Oxidase Activity Assay Kit (BioChain Institute), for which the oxidation of reduced cytochrome c at 550 nm was monitored.

NAD^+ and NADH measurements

The NAD^+ and NADH levels in heart tissue were determined using the EnzyChrom NAD^+/NADH Assay Kit (BioAssay Systems, Hayward, CA) according to the manufacturer's instructions.

Statistical analyses

All results are presented as the means ± SEM. Survival analysis was performed using the Kaplan-Meier method, and significance was calculated based on log-rank tests. Two-way ANOVAs followed by Scheffe's post-hoc tests were used when data contained two variable factors (i.e., fixed vs. disrupted LD cycle and NS vs. PE infusion). All other comparisons were performed using unpaired two-tailed Student's t-tests to determine significance. In all cases, a P value of less than 0.05 was considered significant.

Study approval

All animal care and use procedures were approved by the Wakayama Medical University Institutional Animal Care and Use Committee (Wakayama Medical University Permit Number: 497).

Supporting Information

Figure S1 Behavioral and cardiovascular rhythms are unaltered in H-$Bmal1^{-/-}$ mice. (A) Two representative actograms from control (left column) and H-$Bmal1^{-/-}$ (right column) animals. Activity counts are indicated by the vertical black marks. The records are double plotted such that each day's record is presented both to the right of and beneath that of the previous day. For the first 10 days, animals were maintained on a 12:12 h light-dark (LD) cycle, denoted by the bar above the record. The animals were then transferred to constant darkness (DD) on the day indicated by the horizontal line at the right margin. The free-running period and the amplitude of the circadian rhythm in control and H-$Bmal1^{-/-}$ mice are shown in bar graphs (n = 8 per genotype). The free-running period was calculated as the duration of time between the major activity periods on consecutive days. The amplitude of the locomotor activity rhythm was determined using fast Fourier transformation (FFT), which estimates the relative power of the approximately 24-h periodic rhythm compared with all other periodicities. (B) Profiles of 24-h systolic and diastolic blood pressures (BPs) and heart rates (HRs) in control (black lines, n = 4) and H-$Bmal1^{-/-}$ (gray lines, n = 4) mice. Animals were maintained on a 12:12-h LD cycle (indicated by the bar at the bottom). Each cardiovascular parameter was averaged over the 12-h light and 12-h dark periods and is expressed using a bar graph. Data are the mean ± SEM. Data were compared using unpaired two-tailed Student's t-test.

Figure S2 Severely impaired cardiac function in 24-week-old H-$Bmal1^{-/-}$ mice. (A) Ratios of heart weight to

body weight (HW/BW) at 24 weeks of age (n = 6 per group). (B) Echocardiographic analysis in 24-week-old control and H-$Bmal1^{-/-}$ animals (n = 9–12 per group). LV internal diameter at diastole (LVIDd) and at systole (LVIDs) and fractional shortening (FS) are shown in bar graph format. Data are the mean ± SEM. *$P<0.05$, **$P<0.01$, unpaired two-tailed Student's t-test.

Figure S3 Mitochondrial abnormalities in the hearts of 24-week-old H-$Bmal1^{-/-}$ mice. (A) Representative electron micrographs of sections taken from the left ventricular muscle from control and H-$Bmal1^{-/-}$ mice at two different magnifications. (B) The mitochondrial protein concentration of control and $Bmal1^{-/-}$ heart preparations (n = 8 per group). (C) The mitochondrial DNA to nuclear DNA ratio in control and $Bmal1^{-/-}$ hearts (n = 6 per group). (D-E) Enzymatic activities of (D) complex I and (E) complex IV in control and $Bmal1^{-/-}$ hearts (n = 8 per group). The activities of these mitochondrial respiratory enzymes are expressed either per milligram of tissue used for mitochondrial isolation (top panels) or per microgram of mitochondrial protein (bottom panels). (F) NAD^+ and NADH concentrations in control and $Bmal1^{-/-}$ hearts (n = 6 per group). Data are the mean ± SEM. *$P<0.05$, **$P<0.01$, unpaired 2-tailed Student's t-test.

Figure S4 Circadian desynchronization not only disrupts rhythms but also reduces the expression levels of clock and metabolic genes in the hearts of NS-infused C57BL/6J mice. (A-C) Relative expression levels of genes regulating (A) clock machinery as well as (B) glucose and (C) lipid metabolism in the heart. All heart tissues used were from the NS-infused animal group subjected to either a fixed or a disrupted LD cycle as described in Figure 5A (n = 4 per group per time point). To provide a 24-h overall mean expression level, the 24-h data were also averaged and are expressed using a bar graph format. Data are the mean ± SEM. **$P<0.01$, unpaired 2-tailed Student's t-test.

Figure S5 Circadian desynchronization impairs mitochondrial function in the hearts of NS-infused C57BL/6J mice. (A-B) Relative expression levels of genes regulating (A) mitochondrial structure and (B) mitochondrial oxidative metabolism in heart. All heart tissues used were from the NS-infused animal group subjected to either a fixed or a disrupted LD cycle as described in Figure 5A (n = 4 per group per time point). To provide overall 24-h mean expression levels, data from over a 24-h time period in each group were also averaged and are expressed using a bar graph format. (C) Enzymatic activity of complex I in NS-infused animals exposed to a fixed or disrupted LD cycle (n = 8 per group). The complex I activity is expressed per milligram of tissue used for mitochondrial isolation. (D) The relative number of mitochondria in the left ventricular muscle was counted using electron microscope images. Representative images are shown. Data are the mean ± SEM. *$P<0.05$, unpaired two-tailed Student's t-test.

Figure S6 Heart-specific disruption of the $Bmal1$ conditional allele. (A) Scheme showing the conditional knockout of exon 8, which encodes the basic helix-loop-helix domain of the BMAL1 protein. Triangles, $loxP$ sites; bars with base pair (bp) markers, sites and sizes of PCR products diagnostic of the heart-specific disruption of the $Bmal1$ gene. (B) Confirmation of the heart-specific deletion of exon 8 in the *Bmal1* gene of *H-Bmal1*$^{-/-}$ mice. PCR products amplified from genomic DNA, which were extracted from the heart tissue of a control mouse (homozygous *Bmal1* conditional, no *Cre*) and from the heart, lung, and kidney tissues of an *H-Bmal1*$^{-/-}$ mouse (homozygous *Bmal1* conditional, *Myh6Cre*).

Table S1 Primer sequences used for quantitative RT-PCR.

Materials S1 Materials and methods for analysis of cardiovascular parameters.

Acknowledgments

We thank A. Hatada, S. Yamanaka, J. Yokoi, T. Hyo, K. Miyazaki, and H. Yokoyama for technical assistance.

Author Contributions

Conceived and designed the experiments: AK. Performed the experiments: AK PD IH TN YD SSG HW YM. Analyzed the data: AK PD IH TN YD SSG HW YM. Contributed reagents/materials/analysis tools: MM. Wrote the paper: AK.

References

1. Bray MS, Shaw CA, Moore MW, Garcia RA, Zanquetta MM, et al. (2008) Disruption of the circadian clock within the cardiomyocyte influences myocardial contractile function, metabolism, and gene expression. Am J Physiol Heart Circ Physiol 294: H1036–1047.
2. Tsai JY, Kiensesberger PC, Pulinilkunnil T, Sailors MH, Durgan DJ, et al. (2010) Direct regulation of myocardial triglyceride metabolism by the cardiomyocyte circadian clock. J Biol Chem 285: 2918–2929.
3. Lefta M, Campbell KS, Feng HZ, Jin JP, Esser KA (2012) Development of dilated cardiomyopathy in Bmal1-deficient mice. Am J Physiol Heart Circ Physiol 303: H475–485.
4. Arjona A, Silver AC, Walker WE, Fikrig E (2012) Immunity's fourth dimension: approaching the circadian-immune connection. Trends Immunol 33: 607–612.
5. Mavroudis PD, Scheff JD, Calvano SE, Androulakis IP (2013) Systems biology of circadian-immune interactions. J Innate Immun 5: 153–162.
6. Pacha J, Sumova A (2013) Circadian regulation of epithelial functions in the intestine. Acta Physiol (Oxf) 208: 11–24.
7. Hussain MM, Pan X (2009) Clock genes, intestinal transport and plasma lipid homeostasis. Trends Endocrinol Metab 20: 177–185.
8. Firsov D, Bonny O (2010) Circadian regulation of renal function. Kidney Int 78: 640–645.
9. Stow LR, Gumz ML (2011) The circadian clock in the kidney. J Am Soc Nephrol 22: 598–604.
10. Lamia KA, Storch KF, Weitz CJ (2008) Physiological significance of a peripheral tissue circadian clock. Proc Natl Acad Sci U S A 105: 15172–15177.
11. Rudic RD, McNamara P, Curtis AM, Boston RC, Panda S, et al. (2004) BMAL1 and CLOCK, two essential components of the circadian clock, are involved in glucose homeostasis. PLoS Biol 2: e377.
12. Marcheva B, Ramsey KM, Buhr ED, Kobayashi Y, Su H, et al. (2010) Disruption of the clock components CLOCK and BMAL1 leads to hypoinsulinaemia and diabetes. Nature 466: 627–631.
13. Shimba S, Ishii N, Ohta Y, Ohno T, Watabe Y, et al. (2005) Brain and muscle Arnt-like protein-1 (BMAL1), a component of the molecular clock, regulates adipogenesis. Proc Natl Acad Sci U S A 102: 12071–12076.
14. Grimaldi B, Bellet MM, Katada S, Astarita G, Hirayama J, et al. (2010) PER2 controls lipid metabolism by direct regulation of PPARgamma. Cell Metab 12: 509–520.
15. Rutter J, Reick M, Wu LC, McKnight SL (2001) Regulation of clock and NPAS2 DNA binding by the redox state of NAD cofactors. Science 293: 510–514.
16. Musiek ES, Lim MM, Yang G, Bauer AQ, Qi L, et al. (2013) Circadian clock proteins regulate neuronal redox homeostasis and neurodegeneration. J Clin Invest 123: 5389–5400.
17. Wang TA, Yu YV, Govindaiah G, Ye X, Artinian L, et al. (2012) Circadian rhythm of redox state regulates excitability in suprachiasmatic nucleus neurons. Science 337: 839–842.
18. Woldt E, Sebti Y, Solt LA, Duhem C, Lancel S, et al. (2013) Rev-erb-alpha modulates skeletal muscle oxidative capacity by regulating mitochondrial biogenesis and autophagy. Nat Med 19: 1039–1046.
19. Peek CB, Affinati AH, Ramsey KM, Kuo HY, Yu W, et al. (2013) Circadian clock NAD+ cycle drives mitochondrial oxidative metabolism in mice. Science 342: 1243417.
20. Mohawk JA, Green CB, Takahashi JS (2012) Central and peripheral circadian clocks in mammals. Annu Rev Neurosci 35: 445–462.
21. Dibner C, Schibler U, Albrecht U (2010) The mammalian circadian timing system: organization and coordination of central and peripheral clocks. Annu Rev Physiol 72: 517–549.
22. Marin-Garcia J, Akhmedov AT, Moe GW (2013) Mitochondria in heart failure: the emerging role of mitochondrial dynamics. Heart Fail Rev 18: 439–456.
23. Marin-Garcia J (2005) Mitochondria and the Heart. New York: Springer Science+Business Media, Inc.
24. Leone TC, Kelly DP (2011) Transcriptional control of cardiac fuel metabolism and mitochondrial function. Cold Spring Harb Symp Quant Biol 76: 175–182.
25. Scarpulla RC, Vega RB, Kelly DP (2012) Transcriptional integration of mitochondrial biogenesis. Trends Endocrinol Metab 23: 459–466.
26. Scarpulla RC (2008) Transcriptional paradigms in mammalian mitochondrial biogenesis and function. Physiol Rev 88: 611–638.
27. Finck BN, Lehman JJ, Leone TC, Welch MJ, Bennett MJ, et al. (2002) The cardiac phenotype induced by PPARalpha overexpression mimics that caused by diabetes mellitus. J Clin Invest 109: 121–130.
28. Smeets PJ, Teunissen BE, Willemsen PH, van Nieuwenhoven FA, Brouns AE, et al. (2008) Cardiac hypertrophy is enhanced in PPAR alpha-/- mice in response to chronic pressure overload. Cardiovasc Res 78: 79–89.
29. Huss JM, Imahashi K, Dufour CR, Weinheimer CJ, Courtois M, et al. (2007) The nuclear receptor ERRalpha is required for the bioenergetic and functional adaptation to cardiac pressure overload. Cell Metab 6: 25–37.
30. Alaynick WA, Kondo RP, Xie W, He W, Dufour CR, et al. (2007) ERRgamma directs and maintains the transition to oxidative metabolism in the postnatal heart. Cell Metab 6: 13–24.
31. Arany Z, He H, Lin J, Hoyer K, Handschin C, et al. (2005) Transcriptional coactivator PGC-1 alpha controls the energy state and contractile function of cardiac muscle. Cell Metab 1: 259–271.
32. Liu C, Li S, Liu T, Borjigin J, Lin JD (2007) Transcriptional coactivator PGC-1alpha integrates the mammalian clock and energy metabolism. Nature 447: 477–481.
33. Canaple L, Rambaud J, Dkhissi-Benyahya O, Rayet B, Tan NS, et al. (2006) Reciprocal regulation of brain and muscle Arnt-like protein 1 and peroxisome proliferator-activated receptor alpha defines a novel positive feedback loop in the rodent liver circadian clock. Mol Endocrinol 20: 1715–1727.
34. Oishi K, Shirai H, Ishida N (2005) CLOCK is involved in the circadian transactivation of peroxisome-proliferator-activated receptor alpha (PPARalpha) in mice. Biochem J 386: 575–581.
35. Durgan DJ, Tsai JY, Grenett MH, Pat BM, Ratcliffe WF, et al. (2011) Evidence suggesting that the cardiomyocyte circadian clock modulates responsiveness of the heart to hypertrophic stimuli in mice. Chronobiol Int 28: 187–203.
36. Storch KF, Lipan O, Leykin I, Viswanathan N, Davis FC, et al. (2002) Extensive and divergent circadian gene expression in liver and heart. Nature 417: 78–83.
37. Penev PD, Kolker DE, Zee PC, Turek FW (1998) Chronic circadian desynchronization decreases the survival of animals with cardiomyopathic heart disease. Am J Physiol 275: H2334–2337.
38. Vyas MV, Garg AX, Iansavichus AV, Costella J, Donner A, et al. (2012) Shift work and vascular events: systematic review and meta-analysis. BMJ 345: e4800.
39. Curtis AM, Cheng Y, Kapoor S, Reilly D, Price TS, et al. (2007) Circadian variation of blood pressure and the vascular response to asynchronous stress. Proc Natl Acad Sci U S A 104: 3450–3455.
40. Jeyaraj D, Haldar SM, Wan X, McCauley MD, Ripperger JA, et al. (2012) Circadian rhythms govern cardiac repolarization and arrhythmogenesis. Nature 483: 96–99.
41. Durgan DJ, Trexler NA, Egbejimi O, McElfresh TA, Suk HY, et al. (2006) The circadian clock within the cardiomyocyte is essential for responsiveness of the heart to fatty acids. J Biol Chem 281: 24254–24269.
42. Asher G, Schibler U (2011) Crosstalk between components of circadian and metabolic cycles in mammals. Cell Metab 13: 125–137.
43. Bass J, Takahashi JS (2010) Circadian integration of metabolism and energetics. Science 330: 1349–1354.
44. Sahar S, Sassone-Corsi P (2012) Regulation of metabolism: the circadian clock dictates the time. Trends Endocrinol Metab 23: 1–8.
45. Peek CB, Ramsey KM, Marcheva B, Bass J (2012) Nutrient sensing and the circadian clock. Trends Endocrinol Metab 23: 312–318.
46. Turek FW, Joshu C, Kohsaka A, Lin E, Ivanova G, et al. (2005) Obesity and metabolic syndrome in circadian Clock mutant mice. Science 308: 1043–1045.
47. Madrazo JA, Kelly DP (2008) The PPAR trio: regulators of myocardial energy metabolism in health and disease. J Mol Cell Cardiol 44: 968–975.
48. Lehman JJ, Barger PM, Kovacs A, Saffitz JE, Medeiros DM, et al. (2000) Peroxisome proliferator-activated receptor gamma coactivator-1 promotes cardiac mitochondrial biogenesis. J Clin Invest 106: 847–856.
49. Asher G, Reinke H, Altmeyer M, Gutierrez-Arcelus M, Hottiger MO, et al. (2010) Poly(ADP-ribose) polymerase 1 participates in the phase entrainment of circadian clocks to feeding. Cell 142: 943–953.

50. Ramsey KM, Yoshino J, Brace CS, Abrassart D, Kobayashi Y, et al. (2009) Circadian clock feedback cycle through NAMPT-mediated NAD+ biosynthesis. Science 324: 651–654.

51. Nakahata Y, Sahar S, Astarita G, Kaluzova M, Sassone-Corsi P (2009) Circadian control of the NAD+ salvage pathway by CLOCK-SIRT1. Science 324: 654–657.

52. Saeed AI, Sharov V, White J, Li J, Liang W, et al. (2003) TM4: a free, open-source system for microarray data management and analysis. Biotechniques 34: 374–378.

6

Preventive Effect of Daiokanzoto (TJ-84) on 5-Fluorouracil-Induced Human Gingival Cell Death through the Inhibition of Reactive Oxygen Species Production

Kaya Yoshida[1]*, Masami Yoshioka[2], Hirohiko Okamura[3], Satomi Moriyama[4], Kazuyoshi Kawazoe[5], Daniel Grenier[6], Daisuke Hinode[4]

1 Department of Oral Healthcare Education, Institute of Health Biosciences, University of Tokushima Graduate School, Tokushima, Japan, 2 Department of Oral Health Science and Social Welfare, Institute of Health Biosciences, University of Tokushima Graduate School, Tokushima, Japan, 3 Department of Histology and Oral Histology, Institute of Health Biosciences, University of Tokushima Graduate School, Tokushima, Japan, 4 Department of Hygiene and Oral Health Science, Institute of Health Biosciences, University of Tokushima Graduate School, Tokushima, Japan, 5 Department of Clinical Pharmacy, Institute of Health Biosciences, University of Tokushima Graduate School, Tokushima, Japan, 6 Oral Ecology Research Group, Faculty of Dentistry, Laval University, Quebec City, QC, Canada

Abstract

Daiokanzoto (TJ-84) is a traditional Japanese herbal medicine (Kampo formulation). While many Kampo formulations have been reported to regulate inflammation and immune responses in oral mucosa, there is no evidence to show that TJ-84 has beneficial effects on oral mucositis, a disease resulting from increased cell death induced by chemotherapeutic agents such as 5-fluorouracil (5-FU). In order to develop effective new therapeutic strategies for treating oral mucositis, we investigated (i) the mechanisms by which 5-FU induces the death of human gingival cells and (ii) the effects of TJ-84 on biological events induced by 5-FU. 5-FU-induced lactate dehydrogenase (LDH) release and pore formation in gingival cells (Sa3 cell line) resulted in cell death. Incubating the cells with 5-FU increased the expression of nucleotide-binding domain and leucine-rich repeat containing PYD-3 (NLRP3) and caspase-1. The cleavage of caspase-1 was observed in 5-FU-treated cells, which was followed by an increased secretion of interleukin (IL)-1β. The inhibition of the NLRP3 pathway slightly decreased the effects of 5-FU on cell viability and LDH release, suggesting that NLRP3 may be in part involved in 5-FU-induced cell death. TJ-84 decreased 5-FU-induced LDH release and cell death and also significantly inhibited the depolarization of mitochondria and the up-regulation of 5-FU-induced reactive oxygen species (ROS) and nitric oxide (NO) production. The transcriptional factor, nuclear factor-κB (NF-κB) was not involved in the 5-FU-induced cell death in Sa3 cells. In conclusion, we provide evidence suggesting that the increase of ROS production in mitochondria, rather than NLRP3 activation, was considered to be associated with the cell death induced by 5-FU. The results also suggested that TJ-84 may attenuate 5-FU-induced cell death through the inhibition of mitochondrial ROS production.

Editor: David M. Ojcius, University of California Merced, United States of America

Funding: This work was funded by a Grant-in-Aid for Scientific Research (B), Ministry of Education, Culture, Sports, Science & Technology (MEXT) (DH, 24390471). The funders had no role in study design, data collection and analysis, decision to publish, or preparation of the manuscript.

Competing Interests: The authors have declared that no competing interests exist.

* Email: kaya@tokushima-u.ac.jp

Introduction

Kampo formulations, which are traditional Japanese herbal medicines composed of crude herb extracts, have been prescribed in Japan for a wide variety of diseases for over 1500 years [1]. However, little research has been conducted on their potential beneficial effects on oral health. In a previous study, we investigated the effects of 27 Kampo formulations on the growth and virulence properties of *Porphyromonas gingivalis* (*P. gingivalis*), which is a major pathogen of chronic periodontitis and showed that Kampo formulations containing Rhubarb Rhizome (Daio), including Daiokanzoto (TJ-84), can decrease the growth of *P. gingivalis* and its adherence to oral epithelial cells, suggesting that they may have potential for preventing periodontal diseases [2]. Moreover, *in vitro* evidence has shown that some Kampo formulations can decrease inflammation and bacterial infections of oral mucosa. For example, Shosaikoto and Orento decrease the production of the inflammatory mediator prostaglandin E_2 by lipopolysaccharide (LPS)-treated human gingival fibroblasts [3,4]. Shosaikoto also increases the gene expression of antimicrobial peptides such as calprotectin by human oral epithelial cells [5]. Lastly, Rokumigan has been reported to reduce IL-6 secretion by LPS-stimulated gingival epithelial cells and fibroblasts and to promote wound healing in a fibroblast model [6]. These results indicate that Kampo formulations may be promising new drugs for the prevention and treatment of oral mucosal diseases in which an inflammatory host response is involved.

5-fluorouracil (5-FU) is a widely used chemotherapeutic agent in the treatment of cancers. While 5-FU displays beneficial antitumor effects by inhibiting DNA synthesis [7], it also induces a high rate of oral mucositis (20–50%) in patients receiving multicycle chemotherapy [8]. Oral mucositis results from increased inflammation and the death of oral mucosal cells (epithelial cells and fibroblasts), and has specific symptoms such as erythema, bleeding, ulcer formation, and localized oral superinfections. The development of oral mucositis causes severe pain, which in turn makes it difficult to eat and drink, leading to malnutrition. Furthermore, the loss of the integrity of the oral mucosal epithelium favors the destruction of the mucosal barrier and increases the risk of local infections by oral pathogenic microorganisms such as Candida albicans, herpes simplex virus (HSV), and Gram-negative bacilli [9]. It has also been reported that the high prevalence of local infections associated with oral mucositis may increase the risk of systemic bacterial infections [10]. The prevention or treatment of oral mucositis may thus play a significant role in improving the quality of life and clinical outcomes of patients with cancer. While various strategies to prevent or treat oral mucositis have been evaluated, there is currently no effective therapeutic modality for this disease [11].

The reactive oxygen species (ROS) are involved in multiple biological processes leading to oral mucositis by both direct and indirect mechanisms [12,13]. More specifically, 5-FU-induced ROS cause oxidative stress that damages DNA and proteins in epithelial cells, leading to cell death and ulcer formation, a characteristic of oral mucositis. The factors which are released from injured tissues affect the initiation and development of chemotherapy-induced mucositis [14]. 5-FU-induced ROS production also causes indirect effects through the activation of a number of signal transduction pathways that regulate transcriptional factors such as nuclear factor-κB (NF-κB). NF-κB modulates the expression of many genes that play critical roles in inflammatory cytokine secretion. The inflammatory response induced by these cytokines contributes to a loss of mucosal integrity and the progression of oral mucositis.

5-FU also induces and activates inflammasomes, multi-protein complexes formed by the intracellular nucleotide-binding domain and leucine-rich repeat containing PYD (NLRP) family, as well as apoptosis-associated speck-like protein containing a CARD (ASC) [15,16]. NLRP3 inflammasomes have been extensively studied as they have been associated with many diseases, including type 2 diabetes mellitus [17,18], cancer [19], Alzheimer's disease [20], and atherosclerosis [21]. When NLRP3 inflammasomes recognize pathogenic microorganisms and danger signals they are activated and cleave pro-caspase-1. Caspase-1 possesses enzymatic activity and can induce inflammatory cell death called pyroptosis. Activated caspase-1 also leads to the cleavage and secretion of the biologically active form of interleukin (IL)-1β, an inflammatory cytokine. This regulation of cell death and cytokine production by NLRP3 inflammasomes may play important roles in immune and inflammatory responses [22]. It has recently been suggested that ROS, which are produced in mitochondria in response to various stimuli, trigger the activation of inflammasomes [23]. For example, ATP-mediated ROS increases the activation of caspase-1 and IL-1β and IL-18 secretion through by phosphatidylinositol 3-kinase (PI3 K) pathway in macrophages [24]. Asbestos and silica can induce ROS generation by NADPH oxidase, which leads to NLRP3 inflammasome activation [25]. It has also been proposed that ROS generation resulted from mitochondria dysfunction are required to activate NLRP3 inflammasomes [26–28]. Indeed, an NLRP3 gene mutation has been shown to induce autoinflammatory diseases such as cryopyrin-associated periodic syndrome

(CAPS) and to alter the basal redox state of monocytes of patients with CAPS [29].

The findings described above suggest that NLRP3 inflammasomes are involved in the pathogenesis of 5-FU-associated oral mucositis through ROS production, although its role in oral mucositis has not yet been examined. In the present study, we looked at whether the NLRP3 inflammasome pathway is involved in 5-FU-induced Sa3 cell death with the ultimate goal of developing effective strategies to prevent or treat oral mucositis. We also looked at whether Kampo formulation Daiokanzoto (TJ-84) has a beneficial effect on oral mucositis by affecting the biological processes induced by 5-FU such as cell death, mitochondrial dysfunction, ROS generation, and NLRP3 inflammasome activation.

Results

5-FU-induced Sa3 cell death

To investigate the involvement of 5-FU in cell death, Sa3 cells were incubated with different concentrations of 5-FU for 24 h prior to measuring cell viability. The incubation of the cells with increasing concentrations of 5-FU (1.25–5 mg/mL) resulted in decrease in cell viability over a 24-h period (Fig. 1A). A time-course study showed that there was a significant decrease in cell viability from 1 to 24 h following the exposure of the Sa3 cells to 5 mg/mL of 5-FU (Fig. 1B). In order to determine whether 5-FU-induced cell death was related to cell lysis, the release of cytosolic LDH into the extracellular environment was quantified. A significant release of LDH into the supernatant was observed within 3 h and increased up to 24 h following the incubation with 5-FU (Fig. 1C). Given that 5-FU-induced LDH release suggested that 5-FU led to cell lysis, we investigated pore formation in response to 5-FU by assessing the uptake of propidium iodide (PI) and Hoechst 33342. As shown in Figure 1D, all the cells were stained with the membrane-permeant dye Hoechst 33342 (Fig. 1D a, d, g) whereas only cells with membrane pores allowed the membrane-impermeant dye PI to diffuse into the cells (Fig. 1D e, h). The influx of PI was observed 24 h after the incubation with 5-FU (Fig. 1D e, h) whereas PI did not diffuse into cells that had not been incubated with 5-FU (Fig. 1D b). These results indicated that 5-FU induced pore formation in Sa3 cells.

Involvement of NLRP3 inflammasomes in 5-FU-induced cell death

We hypothesized that if NLRP3 inflammasomes are activated in response to 5-FU, caspase-1 would be cleaved to the p20 subunit, which in turn would produce and release the mature form of IL-1β. To verify this hypothesis, we used Western blot analyses to determine whether 5-FU affects the expression of NLRP3 and the caspase-1 p20 subunit in Sa3 cells. The incubation with 5-FU increased NLRP3 protein expression between 3 and 12 h after the initiation of the incubation. Pro-caspase-1 levels increased at 6 h after the initiation of the incubation with 5-FU (Fig. 2A). Caspase-1 was cleaved and secreted into the supernatant at 24 h after the initiation of the incubation (Fig. 2B). IL-1β secretion into the supernatant was quantified by ELISA following a 6-h or 24-h incubation of the cells with 5 mg/mL 5-FU. Increased secretion of IL-1β at 24 h post-5-FU incubation was observed compared to cells that had not been incubated with 5-FU (Fig. 2C). Given that 5-FU activated the inflammasome pathway, we then investigated whether the NLRP3 inflammasome pathway regulates 5-FU-induced Sa3 cell death. Sa3 cells were pre-incubated for 30 min with 0 to 100 μM carbobenzoxy-valyl-alanyl-aspartyl-[O-methyl]-fluoromethylketone (zVAD-FMK), a caspase inhibitor that binds

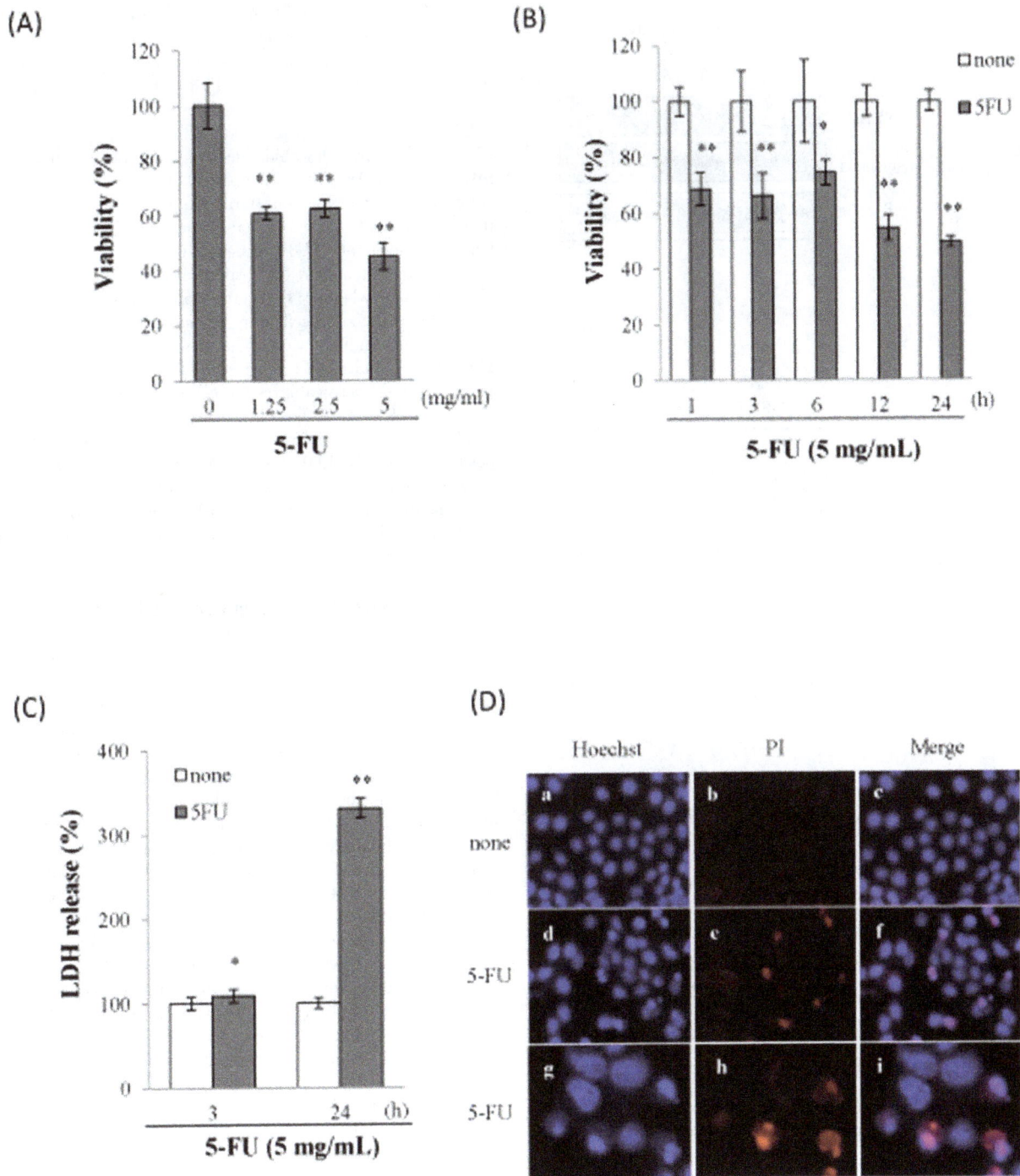

Figure 1. 5-FU induced Sa3 cell death. (A), Viability of Sa3 cells incubated with various concentrations of 5-FU for 24 h. Values are means ± S.E.M. (n = 8). **$p<0.01$ compared to untreated cells. (B), Time course of cell viability of Sa3 cells incubated with 5 mg/mL of 5-FU. Values are means ± S.E.M. (n = 8). *$p<0.05$, **$p<0.01$ compared to control cells. (C), LDH levels in culture media after 3-h and 24-h incubations with 5-FU. Data shown are percentages with respect to control cells at each time point. Values are means ± S.E.M. (n = 8). *$p<0.05$, **$p<0.01$ compared to Sa3 cells incubated without 5-FU. (D), Micrographs of Sa3 cells incubated without (a–c) or with (d–i) 5-FU for 24 h. Hoechst 33342- (a, d, g) and PI-stained cells (b, e, h) and merged images (c, f, i) are shown. Micrographs of cells incubated with 5-FU for 24 h at high magnification (×1,000) (g–i).

to the catalytic site of the enzyme. They were then incubated with 5 mg/mL of 5-FU for 24 h after which cell viability was measured. Figure 3A shows that pre-incubating the cells with zVAD-FMK attenuated the decrease in cell viability induced by 5-FU. We then knocked down NLRP3 using siRNA and determined whether the reduction in NLRP3 expression affects cell viability and LDH release in response to 5-FU. The expression of NLRP3 mRNA was suppressed in siRNA-treated cells but not in control cells. The scrambled oligonucleotide did not affect NLRP3 mRNA expression (Fig. 3B). Cell viability was not altered by the NLRP3 knock-

(A)

5-FU (5 mg/ml)

0 1 3 6 12 24 (h)

NLRP3

pro casp-1

β-actin

Lys

(B)

5-FU (5 mg/ml)

0 1 3 6 12 24 (h)

casp-1
(p20)

— 25
— 20

(kDa)

SN

(C)

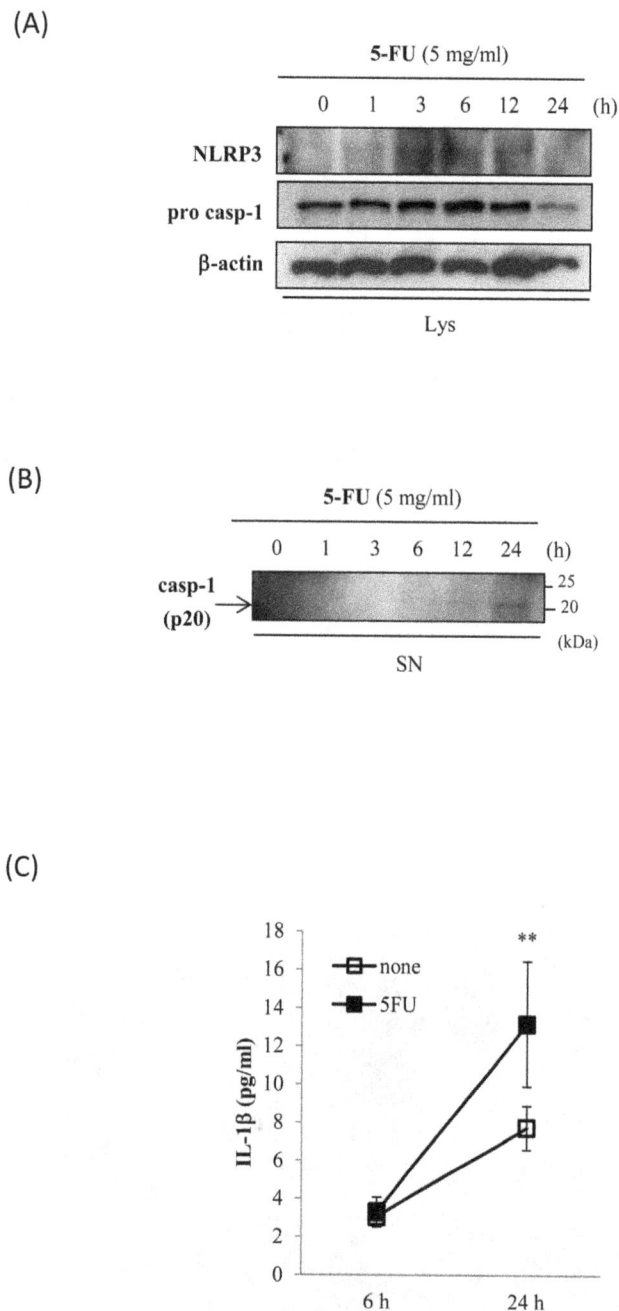

Figure 2. 5-FU-activated inflammasome pathway. Sa3 cells were incubated with 5 mg/mL of 5-FU for 0 to 24 h. (A), Western blot analysis of the expression of NLRP3 and the precursor of caspase-1 (pro-casp-1) in cell lysates. (B), Western blot analysis of cleaved caspase-1 (p20) in supernatants. Arrow indicates p20-specific bands. (C), ELISA assay of IL-1β in supernatants of Sa3 cells incubated without (open box) or with (closed box) 5 mg/mL of 5-FU for 6 h and 24 h. Values are means ± S.E.M. (n = 4). **$p < 0.01$ compared to Sa3 cells incubated without 5-FU.

down itself, while 5-FU-suppressed cell viability (68.08±4.62%) was slightly higher in NLRP3 knock-down cells (74.26±6.28%, p = 0.042) (Fig. 3C). The siRNA knock-down of NLRP3 decreased LDH release (123.18±11.87%, p = 0.045), the scrambled oligonucleotide had no effect (124.99±32.90%, p = 0.945), while 5-FU increased LDH release (136.87±12.99%) (Fig. 3D).

Preventive effects of TJ-84 on 5-FU-induced cell death

To determine whether TJ-84 can prevent 5-FU-induced cell death, we first evaluated the cytotoxic effect of TJ-84 on Sa3 cells. Sa3 cells were incubated with TJ-84 at concentrations up to 5000 µg/mL for 24 h, and cell viability was then assessed using a WST-8 assay. While up to 2500 µg/mL of TJ-84 had no toxic effect on Sa3 cells, cell viability decreased significantly at 5000 µg/mL (Fig. 4A). We then incubated Sa3 cells with TJ-84 at concentrations ranging from 0 to 1000 µg/mL for 1 h, incubated them with 5-FU for 24 h, and then assessed cell viability. Concentrations of TJ-84 ranging from 250 µg/mL up to 1000 µg/mL attenuated 5-FU-suppressed cell viability (Fig. 4B). To determine whether TJ-84 attenuates the secretion of LDH induced by 5-FU, Sa3 cells were pre-incubated with 500 mg/mL of TJ-84 for 1 h. They were then incubated with 5 mg/mL of 5-FU for 24 h, and LDH levels in the supernatant were measured. Since Triton-X permeabilizes the cell membrane, which leads to the release of the cytosolic contents into the medium, we used cells incubated with 0.1% Triton-X for 5 min at room temperature as a positive control for LDH release. As expected, the incubation of the cells with 0.1% Triton-X increased the release of LDH. The 24-h incubation with 5 mg/mL 5-FU increased LDH release, while the pre-incubation with TJ-84 significantly attenuated LDH release (Fig. 4C).

TJ-84 reduces mitochondrial ROS production

We examined the effects of 5-FU and TJ-84 on ROS production by mitochondria to investigate the molecular mechanisms by which TJ-84 attenuates the death of 5-FU-incubated Sa3 cells. Since mitochondrial depolarization occurs in the early stages of cell death, we first determined whether 5-FU and TJ-84 modify the membrane potential of mitochondria using JC-1. Monomer JC-1 is excited by green fluorescence (488 nm) and selectively accumulates in the mitochondrial matrix where it forms red fluorescence (568) JC-1 aggregates. Mitochondrial depolarization can thus be visualized as a shift in fluorescence from red to green. Cells were incubated with 5 mg/mL of 5-FU for 3 h followed by 500 µg/mL of TJ-84 for 1 h and then with 1 µg/mL of JC-1 for 30 min. The cells were observed by fluorescence microscopy. The accumulation of JC-1 aggregates (red fluorescence) decreased in 5-FU-incubated cells, whereas the accumulation of JC-1 monomers (green fluorescence) increased compared to cells that had not been incubated with 5-FU (Fig. 5A). These results suggested that 5-FU may decrease the accumulation of JC-1 in the mitochondrial membrane by inducing its depolarization. The 5-FU-induced reduction in membrane depolarization was recovered by a pre-incubation with TJ-84 (Fig. 5A). Fluorescence intensity was quantified in each group using NIH ImageJ analysis software, and the red/green fluorescence intensity ratio was calculated. Consistent with the results shown in Figure 5A, the red/green fluorescence intensity ratio was lower in Sa3 cells incubated with 5-FU for 3 h (2.948±0.876, p = 0.002) than in cells that had not been incubated with 5-FU (3.970±0.947). The red/green ratio suppressed by 5-FU wacorrecteds recovered by a pre-incubation with TJ-84 (3.682±0.826, p = 0.026) (Fig. 5B). These results suggested that TJ-84 inhibits the 5-FU-induced depolarization of the mitochondrial membrane of Sa3 cells.

Mitochondrial impairment results in the production of ROS. We thus examined the effects of 5-FU and TJ-84 on the generation of mitochondria-specific ROS. We assessed mitochondria-specific ROS levels using MitoSOX Red, which selectively detects mitochondria-derived $O \cdot_2$ but not other ROS such as hydrogen peroxide (H_2O_2), hydroxyl radicals (OH·), and reactive nitrogen species. The localization of mitochondria was also assessed using

(A)

(B)

(C)

(D)

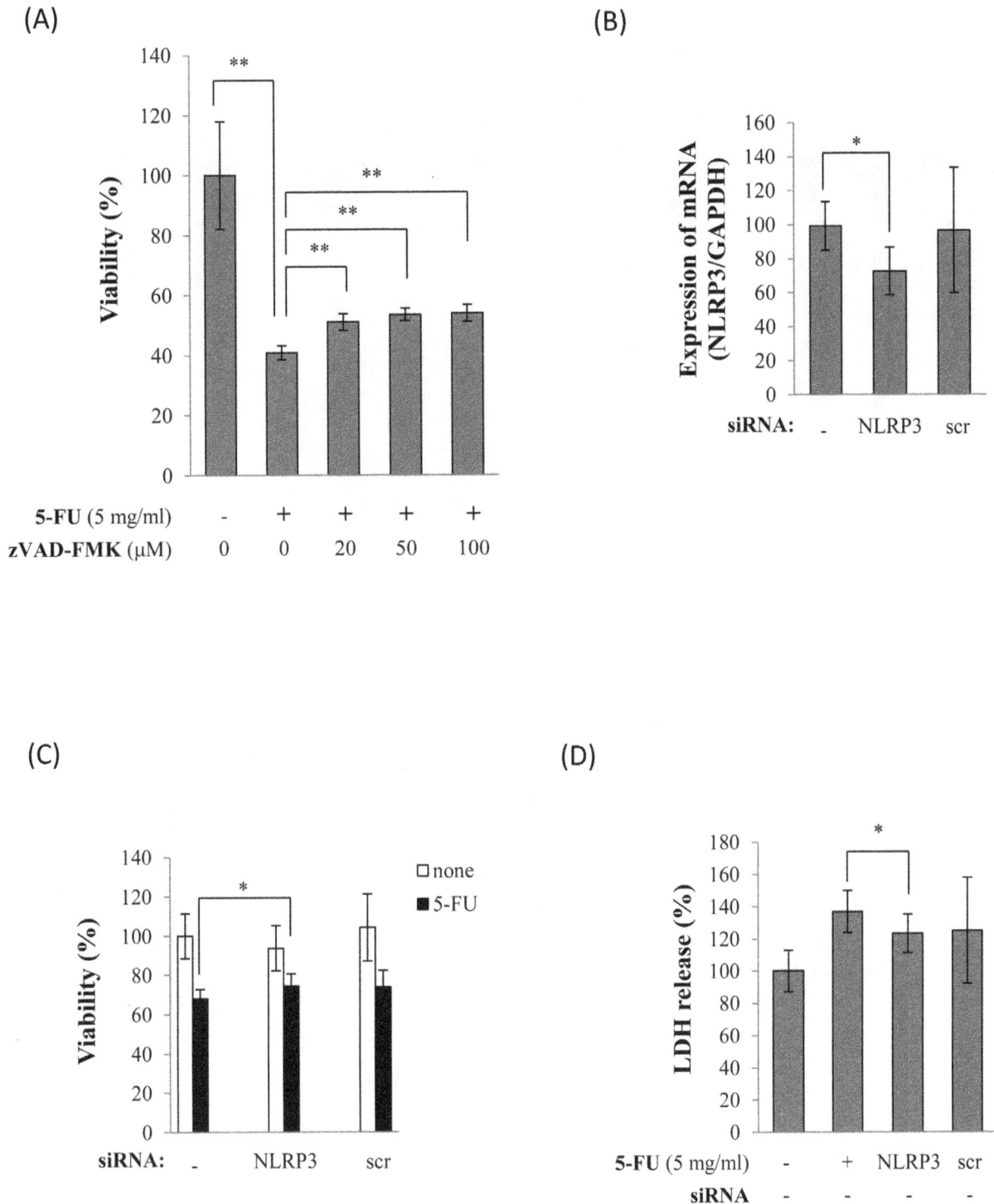

Figure 3. Inhibition of NLRP3 inflammasomes decreased 5-FU-induced cell death. (A), Cell viability of Sa3 cells incubated with 5 mg/mL of 5-FU for 24 h after a 30-min pre-incubation with caspase inhibitor at each concentration. Values are means ± S.E.M. (n = 4). **$p < 0.01$ compared to the control group. (B), Expression of NLRP3 mRNA in Sa3 cells treated with NLRP3 siRNA (NLRP3) or scrambled oligo (scr). Values are means ± S.E.M. (n = 4). *$p < 0.05$ compared to cells without oligo (–). (C), Cell viability of cells transduced without (–) or with NLRP3 siRNA (NLRP3) or control oligo (scr) following an incubation with (closed bar) or without (open bar) 5 mg/mL of 5-FU for 3 h. Data are given as percentages compared to the group that was not incubated with 5-FU. Values are means ± S.E.M. (n = 6). *$p < 0.05$ compared to the control group. (D), Effect of transfection with NLRP3 siRNA (NLRP3) or scrambled oligo (scr) on 5-FU-induced LDH release. The LDH levels in the supernatants are given as percentages of cells not incubated with 5-FU and not transduced with siRNA. Values are means ± S.E.M. (n = 8). *$p < 0.05$ compared to the control cells.

(A)

(B)

(C)

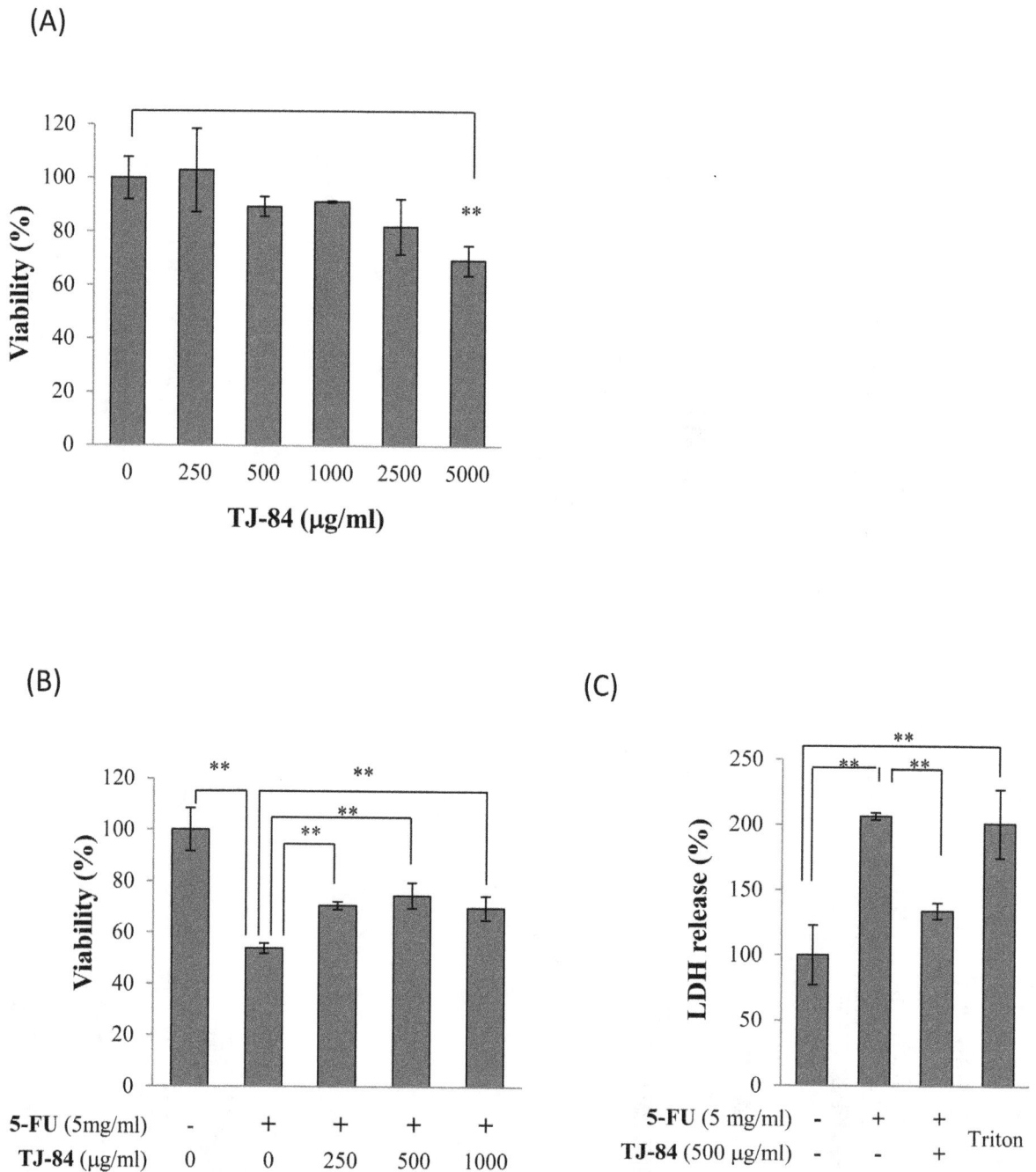

Figure 4. TJ-84 reduced 5-FU-induced cell death. (A), Cytotoxicity of TJ-84. Sa3 cells were incubated with various concentrations of TJ-84 for 24 h, and cell viability was then measured using WST-8 kits. Values are means ± S.E.M. (n = 3). **$p<0.01$ compared to control cells that had not been incubated with TJ-84. (B), Viability of cells incubated with various concentrations of TJ-84 for 1 h and then incubated with 5-FU for 24 h. Results are expressed as percentages with respect to control cells that had not been incubated with TJ-84 and 5-FU. Values are means ± S.E.M. (n = 4). **$p<0.01$ compared to control cells. (C), The effect of TJ-84 on LDH release from cells incubated with 5-FU for 24 h was assessed using WST-8 kits. The supernatant of Sa3 cells incubated with 0.1% Triton-X for 5 min was used as a positive control (Triton). The LDH levels in the supernatants are expressed as percentages with respect to cells that had not been incubated with 5-FU. Values are means ± S.E.M. (n = 4). *$p<0.05$ compared to the control cells.

Mitotracker Green. Red fluorescence detected by MitoSOX Red was higher in Sa3 cells incubated with 5 mg/ml 5-FU for 6 h than in cells that had not been incubated with 5-FU (Fig. 6A). The 5-FU-induced increase in red fluorescence was suppressed by a 1-h pre-incubation with TJ-84 (Fig. 6A). These results were quantified using NIH ImageJ analysis software. The incubation with 5-FU

increased the intensity of red fluorescence after 3 h (14.435 ± 2.852, $p<0.01$) compared to cells that had not been incubated with 5-FU (10.901 ± 1.429), while TJ-84 significantly inhibited the effect of 5-FU (11.058 ± 1.284, $p<0.01$) (Fig. 6B). These results indicated that TJ-84 decreases the generation of mitochondria-derived-O_2·that is up-regulated by 5-FU.

(A)

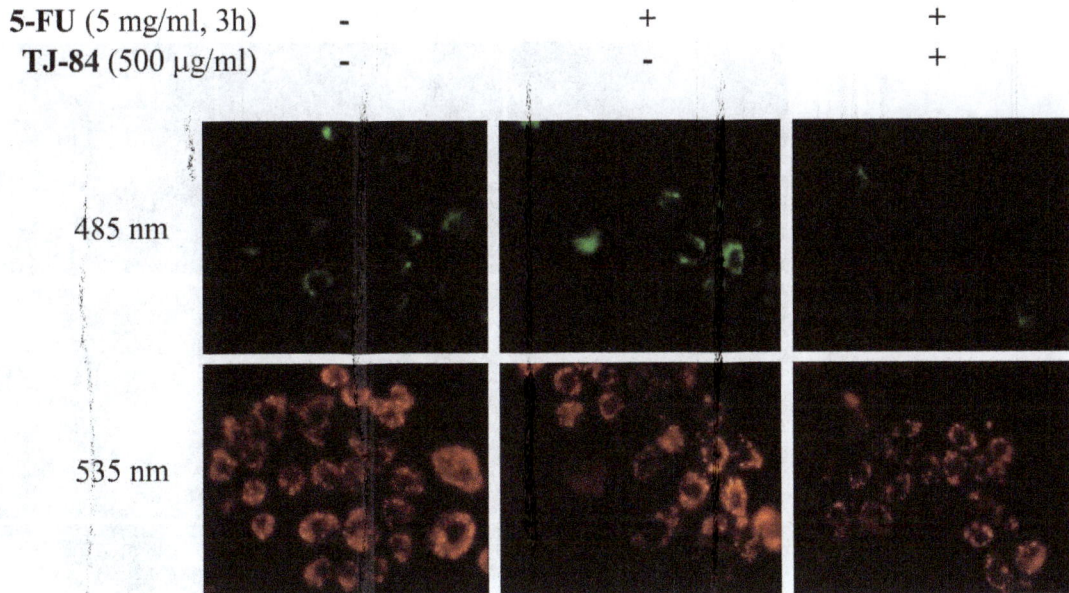

(B)

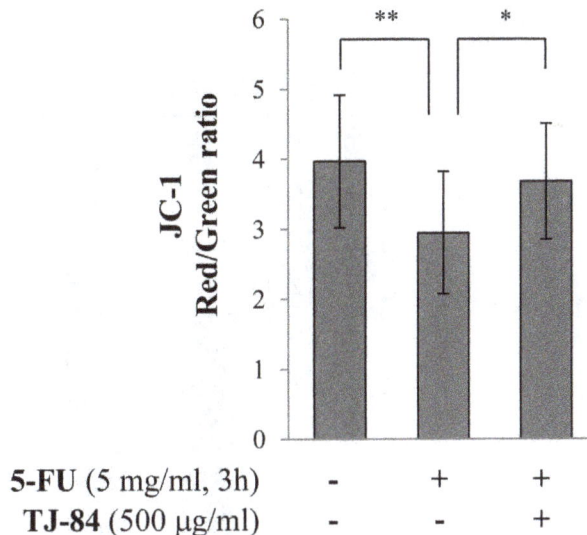

Figure 5. TJ-84 attenuated 5-FU-induced mitochondrial depolarization. (A), Sa3 cells were incubated with or without 5 mg/mL of 5-FU for 3 h following a 1-h pre-incubation with 500 μg/mL of TJ-84. JC-1 (1 μg/mL) was then loaded for 30 min. JC-1 aggregates (red) and monomers (green) were detected by fluorescence microscopy. (B), The fluorescence intensity per cell was calculated using ImageJ. The calculation of the red/green ratio is shown on the graph. Values are means ± S.E.M. (n = 20, 16, 14). **$p<0.01$, *$p<0.05$ compared to the control cells.

The effects of NF-κB and nitric oxide on 5-FU-induced cell death

To investigate the further mechanisms of 5-FU-induced oral mucositis, we examined the effects of 5-FU on the transcriptional factor, NF-κB in Sa3 cells. Activated NF-κB translocates from the cytosol to the nucleus where it regulates gene expression. We thus determined whether 5-FU affects the localization of NF-κB in Sa3 cells by immunocytochemistry using an antibody directed against p65, a subunit of NF-κB. NF-κB translocated from the cytoplasm to the nucleus 3 h after the initiation of the 5-FU incubation (Fig. 7A), suggesting that 5-FU increased the active form of NF-κB. We next investigated the effect of NF-κB on 5-FU-induced cell

| 5-FU (5 mg/ml, 6h) | - | + | + |
| TJ-84 (500 µg/ml) | - | - | + |

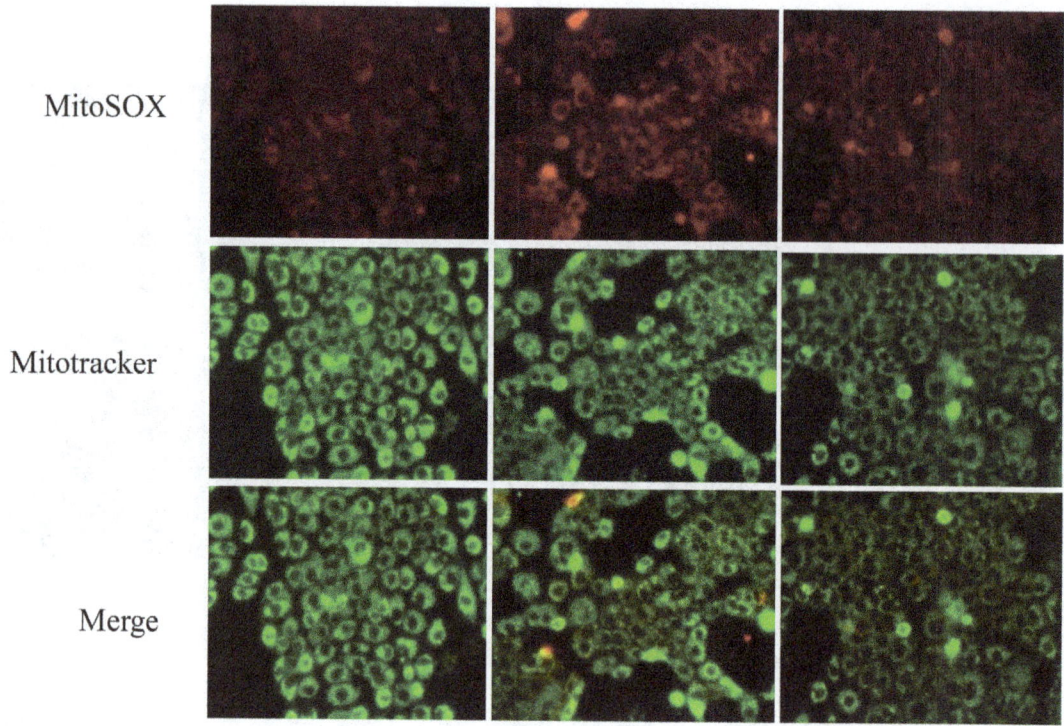

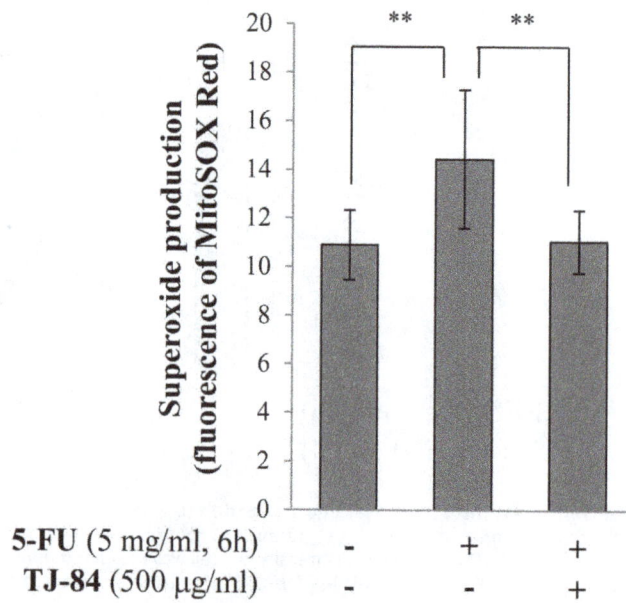

MitoSOX

Mitotracker

Merge

Superoxide production (fluorescence of MitoSOX Red)

| 5-FU (5 mg/ml, 6h) | - | + | + |
| TJ-84 (500 µg/ml) | - | - | + |

Figure 6. TJ-84 suppressed 5-FU-induced ROS production in mitochondria. (A), Sa3 cells were incubated with or without 5 mg/mL of 5-FU for 6 h following a 1-h pre-incubation with 500 µg/mL of TJ-84. MitoSOX Red (5 µM) and 100 µM Mitotracker Green were then loaded for 30 min. ROS (red) and mitochondria (green) were detected by fluorescence microscopy. (B), The red fluorescence intensity per cell was calculated using ImageJ and is shown on the graph. Values are means ± S.E.M. (n = 63, 71, 75). **$p < 0.01$ compared to the control cells.

(A)

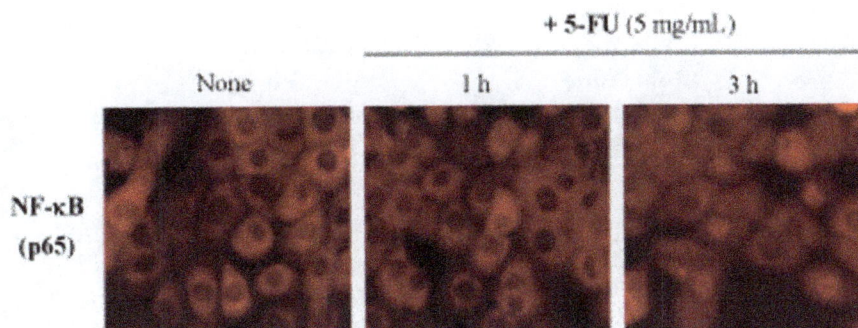

(B)

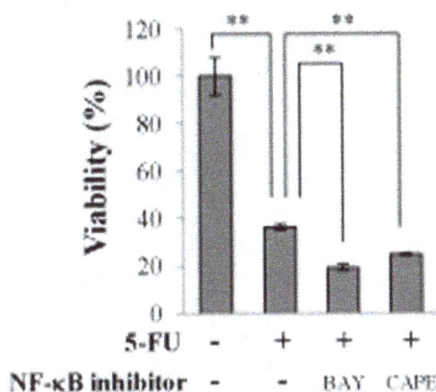

Figure 7. NF-κB did not attenuate 5-FU-induced cell death. (A), Sa3 cells were incubated with or without 5 mg/mL of 5-FU for 1 h and 3 h. The cells then stained with anti-p65 antibody and the localization of p65 was detected by florescence microscopy. (B), Cell viability of Sa3 cells incubated with 5 mg/mL of 5-FU for 24 h after a 30-min pre-incubation with NF-κB inhibitors, 5 μM BAY or 200 μM CAPE. Values are means ± S.E.M. (n = 4). **$p < 0.01$ compared to the control group.

death by a pre-incubation with BAY 11-7085 and caffeic acid phenethyl ester (CAPE), two inhibitors of NF-κB activation. The inhibition of NF-κB by BAY 11-7085 or CAPE did not allow the 5-FU-induced loss of Sa3 cell viability to be recovered (Fig. 7B).

We next examined the effect of 5-FU on the production of nitric oxide (NO) by Sa3 cells. The production of NO was determined by DAF-2DA, a cell-permeable sensitive fluorescent indicator. After the incubation with DAF-2DA, cells were analyzed in fluorescence microscope. Green fluorescence detected by DAF-2DA was higher in Sa3 cells incubated with 5 mg/mL 5-FU for 3h (Fig. 8A, b) and 6 h (Fig. 8A, c) than in cells that had not been incubated with 5-FU (Fig. 8A, a). The 5-FU-induced increase in green fluorescence was suppressed by a 1-h pre-incubation with TJ-84 (Fig. 8A, d, e). These results were quantified using NIH ImageJ software. The incubation with 5-FU increased the intensity of green fluorescence after 3 h (71.69 ± 14.11, $p < 0.01$) and 6 h (48.53 ± 7.40, $p < 0.01$) compared to cells that incubated without 5-FU (24.47 ± 3.05), while TJ-84 significantly inhibited the effect of

5-FU at both 3 h (36.46 ± 5.95, $p < 0.01$) and 6 h (35.46 ± 6.64, $p < 0.01$) (Fig. 8B).

Discussion

We report two main findings: (i) 5-FU-activated NLRP3 inflammasome induces gingival cell death at low NLRP3 levels, and (ii) TJ-84 suppresses 5-FU-induced mitochondrial ROS production and, as a result, 5-FU-induced cell death.

Inflammasomes, including NLRP3 and NLRP1, are expressed at high levels in hematopoietic cells such as granulocytes, dendritic cells, and B and T cells. However, Kummer et al. reported that NLRP3 inflammasomes, but not NLRP1 inflammasomes, are also expressed in the epithelia of the oropharynx and esophagus, suggesting that NLRP3 inflammasomes in the digestive tract may allow the sensing of invading pathogens [30]. It has also been reported that NLRP3 inflammasomes are expressed in fibroblasts and epithelial cells of the oral mucosa and that their expression is

(A)

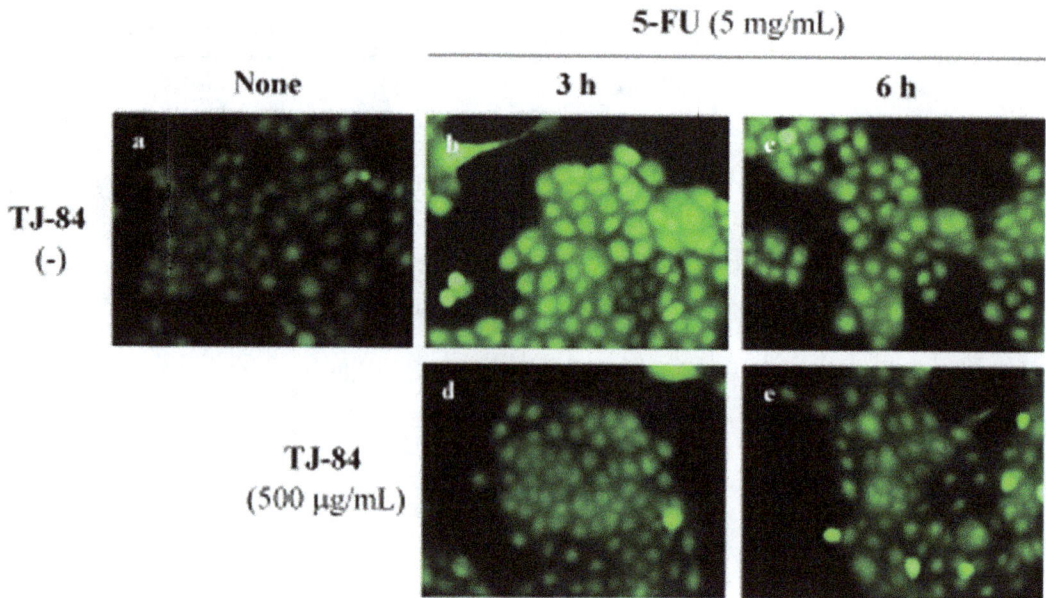

(B)

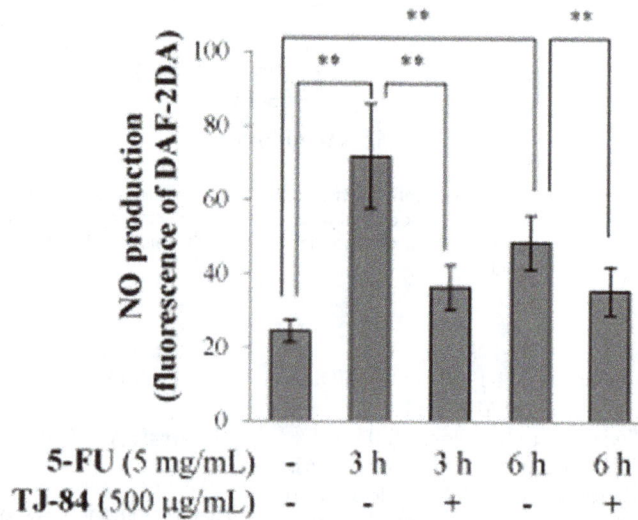

Figure 8. TJ-84 attenuated 5-FU-induced NO production. (A), Sa3 cells were incubated with or without 5 mg/mL of 5-FU for 3 h following a 1-h pre-incubation with 500 µg/mL of TJ-84. DAF-2DA was then loaded for 30 min. The production of NO (green) were detected by fluorescence microscopy. (B), The fluorescence intensity per cell was calculated using ImageJ. The calculation of the red/green ratio is shown on the graph. Values are means ± S.E.M. (n = 50). **$p < 0.01$ compared to the control cells.

regulated by oral bacterial infections [31–33]. Consistent with these reports, we detected NLRP3 in Sa3 cells, which are derived from human gingiva epithelial cells, and observed that NLRP3 was up-regulated by 5-FU (Fig. 2A).

Pyroptosis triggered by caspase-1 activation through the inflammasome pathway is programmed cell death associated with inflammation and is different from apoptosis. In our experimental models, 5-FU increased the cleavage of caspase-1, which led to the release of IL-1β by Sa3 cells (Fig. 2). Based on these results, 5-FU-induced cell death appeared to be related to pyroptosis, while, in the past, 5-FU has been reported to induce apoptosis, resulting in the progression of oral mucositis. To confirm that 5-FU-induced cell death was indeed pyroptosis, we analyzed the features of pyroptotic cell death, including cell lysis and pore formation, in Sa3 cells. Pyroptosis results in cell lysis and the release of cytosolic contents such as LDH [34]. In contrast, cytosolic contents are not released during apoptosis because they are contained in vesicles called apoptotic bodies, which are shed by blebbing [35]. As shown in Figure 1C, the release of LDH into the supernatant increased significantly between 3 and 24 h post-5-FU incubation. Pore formation was assessed using the membrane impermeable dye PI since active caspase-1 induces ion-permeable pores in the plasma membrane in cells dying by pyroptosis [36]. PI was incorporated into Sa3 cells within 24 h of the 5-FU incubation (Fig. 1D). These results suggested that Sa3 cell death induced by 5-FU is related to pyroptosis. Moreover, 5-FU-reduced cell viability was slightly increased by inhibiting the NLRP3 inflammasome pathway using a caspase inhibitor or NLRP3 siRNA (Fig. 3A, C). 5-FU-induced release of LDH was also suppressed by NLRP3 siRNA (Figure 3D). These results suggested that the pyroptotic cell death induced by 5-FU is regulated, at least in part, by the caspase-1/NLRP3 inflammasome pathway.

It was recently reported that ROS-activated inflammasome increases intestinal mucositis in mice treated with chemotherapeutic agent, irinotecan [37]. Administration of IL-1 receptor antagonist to the mouse model also reduces 5-FU-induced intestinal mucositis [38]. These observations indicate that inflammasome plays critical roles in chemotherapy-induced intestinal mucositis. In present study, 5-FU slightly increased the secretion of IL-1β (Fig. 2C), while inhibiting NLRP3 did not entirely recover the 5-FU-induced decrease in cell viability (Fig. 3C), suggesting that other mechanisms may be involved in 5-FU-induced cell death in addition to the NLRP3/caspase-1 pathway. 5-FU also significantly increased mitochondrial ROS production (Figs. 5 and 6). It is thus possible that other factors induced by ROS are involved in 5-FU-induced cell death. We verified two possible candidates: the NF-κB-regulated apoptosis pathway and NO production.

ROS can act as a modulator of signal transduction following the activation of transcriptional factors such as NF-κB, AP-1, and p53. In chemotherapy-induced oral mucositis, NF-κB is the most important transcriptional factor [39] and can cause apoptosis by increasing the expression of BCL-2 family genes. As shown in Figure 7, NF-κB was translocated to nucleus by 5-FU treatment, however, the treatment with NF-κB inhibitors did not recover the 5-FU-induced loss of Sa3 cell viability. These results indicated that NF-κB is activated by 5-FU but is not involved in 5-FU-induced cell death in our experimental model.

In contrast, our results indicated that NO, which was regulated by 5-FU, may be involved in 5-FU-induced Sa3 cell death (Figure 8). However, the role of NO is controversial since it can contribute both positively and negatively to cell death [40,41]. The unpaired electron (NO·) can react with a superoxide radical (O·$_2$) to form the powerful oxidant peroxynitrite (ONOO·$^-$), which is thought to induce apoptosis via multiple mechanisms, including the induction of p53 and ER stresses, the release of cytochrome c by mitochondrial transition, and the activation of p38 or other MAP kinases [42]. Our results with MitoSOX Red showed that 5-FU increased the production of mitochondria-derived superoxide radicals and that TJ-84 inhibited this production (Fig. 6),

suggesting that peroxynitrite, which is produced by a reaction between the superoxide radical and ROS, may contribute to 5-FU-induced Sa3 cell death. On the other hand, low levels of NO are thought to inhibit cell death [43,44], while NO negatively regulates NLRP3 inflammasomes via the S-nitrosylation of NLRP3 [45]. It remains unclear whether 5-FU-induced NO can form peroxynitrite and mediate the activation of pro-apoptotic pathways. Further studies will be required to examine the effect of NO on 5-FU-induced cell death and its relationship with ROS.

Our results clearly showed that TJ-84 attenuates the 5-FU-induced decrease in Sa3 cell viability, indicating that Kampo formulation TJ−84 shows potential as a therapeutic agent for the treatment of 5-FU-induced oral mucositis. A number of naturally occurring compounds in plants, including Kampo formulations, have been investigated for their ability to reduce the severity of 5-FU-induced mucositis. For example, Iberogast, a herbal formula composed of nine extracts, possesses anti-inflammatory properties and has been shown to partially improve the histopathological features of mucositis in the small intestines of rats injected intraperitoneally with 5-FU [46]. Moreover, topically applying Kampo formulation Hangeshashinto to the oral mucosa decreases the symptoms of oral mucositis in patients with advanced colorectal cancer undergoing chemotherapy [47].

Additional properties associated to TJ-84 or its ingredients may also contribute to maintaining healthy oral mucosa. Unpublished data obtained in our laboratory showed that TJ-84 possesses an anti-inflammatory activity resulting in a decreased secretion of inflammatory cytokines by lipopolysaccharide-stimulated gingival epithelial cells and fibroblasts. Moreover, both a licorice extract and emodin, an anthraquinone derivative from rhubarb, have been shown to possess wound healing properties. In a preliminary study, Das et al. reported that the use of a mouthwash containing a deglycerinized licorice extract for two weeks tends to provide pain relief and accelerate the healing of aphthous ulcers [48]. More recently, Tang et al. showed in a rat model that emodin promotes wound healing through transforming growth factor-β1 (TGF-β1)/Smad signaling pathway [49]. These reports support our results and suggest that Kampo formulations can improve oral mucositis.

Kampo formulations have been used to treat a number of diseases, and their beneficial effects have been widely acknowledged. However, the mechanisms by which Kampo formulations produce their effects are not well understood. In the present study, the treatment of TJ-84 suppressed 5-FU-induced mitochondrial ROS production (Fig. 6) and NO production (Fig. 8). It has been reported that Kampo formulation Inchinkoto possesses antioxidant properties that act via a nuclear factor-E2 (Nrf2)-dependent mechanism [50], and that Inchinkoto suppresses Fas-mediated apoptosis in the liver [51]. Based on our results and these reports, it is possible that TJ-84 may decrease 5-FU-induced cell death by decreasing mitochondrial-associated oxidative stresses in Sa3 cells.

In conclusion, we showed that 5-FU-induced Sa3 cell death involves ROS and the NLRP3 inflammasome pathway (Fig. 9). 5-FU caused mitochondrial depolarization and an up-regulation of ROS production, which triggered the activation of NLRP3 inflammasomes and caspase-1, resulting in an increase in cell death. In addition to the NLRP3 inflammasome pathway, another unknown mechanism appears to participate in 5-FU-induced cell death. Kampo formulation TJ-84 may prevent the loss of cell viability by inhibiting the effect of 5-FU on mitochondrial ROS production. Our findings point to a new mechanism by which 5-

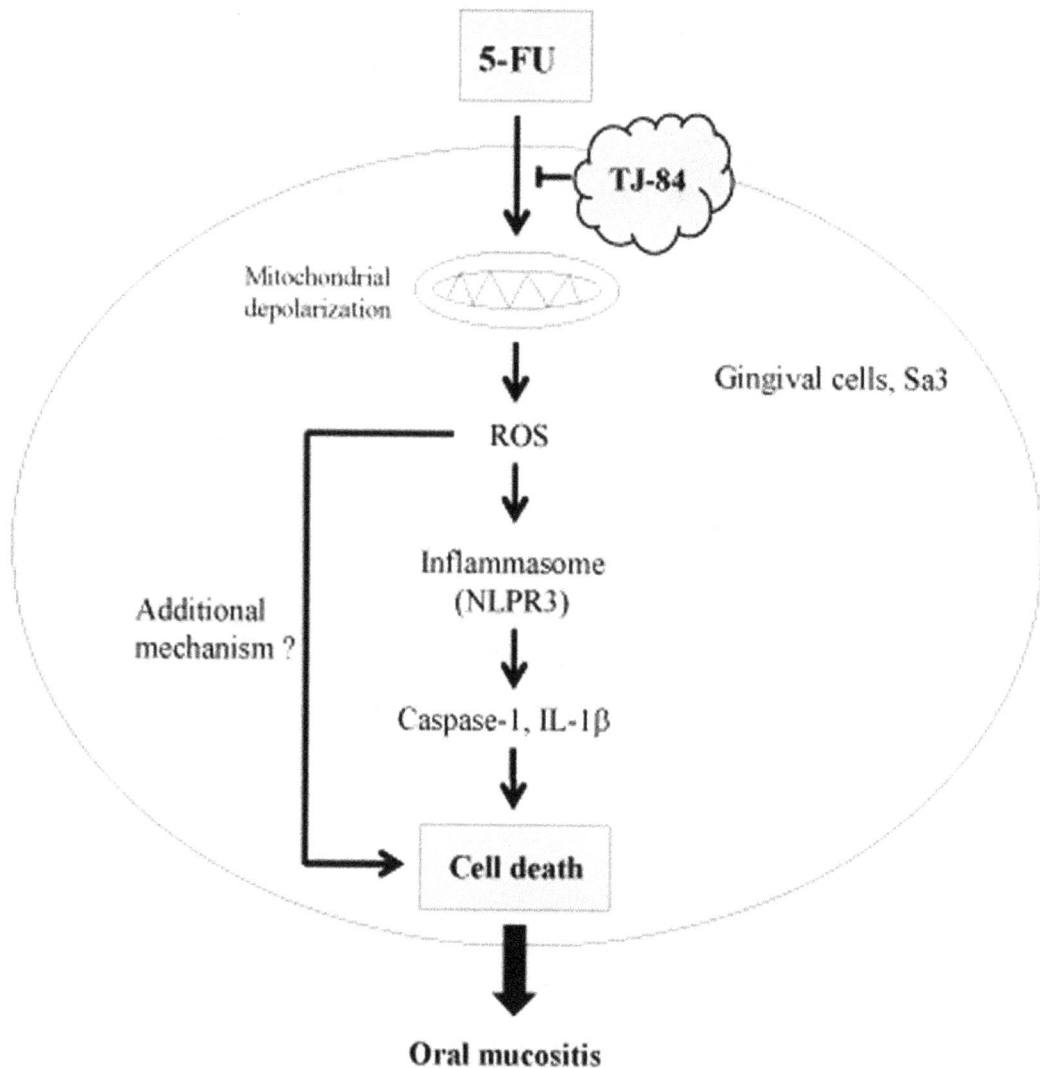

Figure 9. Hypothetical model of how TJ-84 decreases 5-FU-induced Sa3 cell death. TJ-84 decreased 5-FU-induced cell death by inhibiting ROS production. See text for details.

FU induces cell death in oral mucositis and suggest that TJ-84 may be useful in treating 5-FU-induced oral mucositis.

Materials and Methods

Drugs

Daiokanzoto (TJ-84) was obtained from Tsumura & Co (Tokyo, Japan). TJ-84 is manufactured as a powdered extract obtained from spray drying a decoction of 2 medicinal plants: 4.0 g of Rhubarb Rhizome (Daio) and 2.0 g of Glycyrrhiza Root (Kanzo). TJ-84 was dissolved in hot water homogeneously at a concentration of 500 mg/mL and used for experiments. 5-fluorouracil (5-FU injection 250 Kyowa) was purchased from Kyowa Hakko Kirin (Tokyo, Japan).

Cell cultures

Sa3 OSCC cells were kindly provided by the RIKEN BioResource Center through the National BioResource Project of MEXT (Ministry of Education, Culture, Sports, Science & Technology, Tokyo, Japan). The Sa3 cells were plated in plastic dishes at a density of 10×10^4 cells/mL and were cultured in

DMEM supplemented with 10% fetal bovine serum (FBS) at 37°C in a humidified 5% CO_2/95% air atmosphere. After reaching 70−80% confluence, the cells were used for the experiments.

Cell viability

Cell viability was assessed using WST-8 Cell Counting Kit-8 assays (Dojindo Laboratories, Kumamoto, Japan, cat. no 347-07621). Briefly, Sa3 cells were plated at a density of 1×10^4 cells per well in 96-well plates. After the cells had been incubated with 5-FU for the indicated periods, 10 μL of kit reagent was added to the wells. Following a 30-min incubation, cell viability was assessed using a ELISA plate reader.

LDH release

Sa3 cells were incubated with 5-FU for the indicated periods, and aliquots of culture medium were then collected to measure extracellular LDH activity. As a positive control for LDH release, 0.1% Triton-X was added into the medium and incubated for 5 min at room temperature.LDH activity was monitored using LDH cytotoxicity assay kits (Cayman Chemical Company, Ann

Arbor, MI, USA, cat no 10008882) according to the manufacturer's protocol.

Propidium iodide (PI) and Hoechst 33342 staining

Sa3 cells were seeded onto coverslips and were incubated with or without 5-FU for the indicated periods. Following the 5-FU incubation, the cells were incubated with 1 μg/mL of PI or 1 μg/mL of Hoechst 33342 for 15 min. Adhered cells were fixed, mounted, and examined under a microscope. Images were acquired using an ECLIPSE Ti-U microscope and NIS-Elements software (Nikon, Tokyo, Japan).

IL-1β release

Sa3 cells were incubated with 5-FU for the indicated periods, and aliquots of culture medium were then collected. IL-1β was measured with a Quantikine ELISA (R&D systems, Minneapolis, MN, USA) according to manufacture instructions (cat no. DLB50).

siRNA

An RNA duplex targeting the 5′-GUUGCAAGAUCUCU-CAGCA-3′ sequence of human NLRP3 (NM_004894.5) was synthesized and transfected into Sa3 cells using Lipofectamine 2000 (Invitrogen, Carlsbad, CA, USA, cat no 11668027). A scrambled oligonucleotide, which was designed to have no homology to known gene sequences, was transfected into Sa3 cells as a negative control. After 24 h, the transfected cells were incubated with or without 5-FU and were used for the experiments.

Real-time PCR

Total RNA was isolated from Sa3 cells using ISOGEN (Nippon Gene, Tokyo, Japan, cat no 347-07621), followed by phenol extraction and ethanol precipitation. The cDNA was synthesized using Prime Script RT reagent kits (Takara Bio, Kyoto, Japan, cat no RR037A). Real-time PCR was performed with a 7300 Real-Time PCR system (Applied Biosystems, Carlsbad, CA, USA) using SYBR Premix Ex Taq (Takara Bio, cat no RR820A). The primer sequences were as follows: human GAPDH (NM_002046): forward, 5′-GCACCGTCAAGGCTGAGAAC-3′, reverse, 5′-TGGTGAAG-ACGCCAGTGGA-3′; human NLRP3 (NM_004894.5): forward, 5′-AAGCACCTGTTGTGCAATCTGAAG-3′, reverse, 5′-GGG-AATGGCTGGTGCTCAATAC-3′.

SDS-PAGE and Western blot analyses

Sa3 cells were washed with PBS and were scraped into TN lysis buffer (50 mM Tris [pH 8.0], 150 mM NaCl, 0.1% NP-40) supplemented with protease inhibitors (4 μg/mL of aprotinin, 1 μg/mL of leupeptin, 0.2 mM phenylmethylsulfonyl fluoride (PMSF)). The surpernatants were collected and aliquots of them (3 mL) were concentrated to 250 μL using 10 kDa Amicon Ultra centrifuge tubes (Merck KGaA, Darmstadt, Gemany, cat no UFC801024). The protein extracts were immunoblotted using a previously published protocol [52]. Anti-caspase-1 p20 antibody was from Enzo Life Science (Farmingdale, NY, USA, cat no ALX-210-804). Anti-pro-caspase-1 and anti-β-actin antibodies were from Santa Cruz Biotechnology (Santa Cruz, CA, USA, cat no sc-515, sc-47778). Anti-NLRP3 antibody (Cryo-2) was from Adipogen (Liestal, Switzerland, cat no AG-20B-0014).

Mitochondrial depolarization

Sa3 cells were incubated with 1 μg/mL of 5, 5′, 6, 6′-tetrachloro-1, 1′, 3, 3′-tetraethyl-benzimidazol-carbocyanin iodide

(JC-1) (Molecular Probe, Invitrogen, Milan, Italy, cat no) for 30 min and were detected at 590/610 nm (excitation/emission) for JC-1 aggregates and 485/535 nm (excitation/emission) for JC-1 monomers by fluorescence microscopy. The fluorescence intensity per cell was quantified using NIH ImageJ analysis software, and the ratio of red/green was calculated.

Mitochondrial ROS production

To detect surperoxide in mitochondria, MitoSOX Red (Molecular Probe, Invitrogen, cat no M36008) and MitoTracker Green FM (Molecular Probe, Invitrogen, cat no M-7514) were used according to the manufacturers' instructions. Briefly, cells were incubated with 5-FU or/and TJ-84 for the indicated periods and were then loaded with MitoSOX Red (5 μM) and MitoTracker Green FM (100 μM) in balanced salt solution [BSS; 135 mM NaCl, 5.6 mM KCl, 1.2 mM MgSO$_4$, 2.2 mM CaCl$_2$, 10 mM glucose, and 20 mM [4-(2-hydroxyethyl)-1-piperazine ethanesulfonic acid (HEPES)/NaOH, pH 7.4] for 20 min. The cells were rinsed with BSS, and the locations of ROS and mitochondria were observed using a fluorescence microscope (ECLIPSE Ti-U, Nikon). The red fluorescence intensity per cell from images obtained by fluorescence microscopy was quantified by NIH ImageJ software.

Immunocytochemistry and NF-κB inhibitors

Sa3 cells were cultured on sterile 18-mm round coverslips. After reaching 70−80% confluence, the cells were treated with 5 mg/mL 5-FU for 1 or 3 h. For immunocytochemistry, the cells were fixed with 3.0% formalin for 30 min and then permeabilized with 0.1% Triton X-100 in PBS for 2 min at 4°C. After blocking of nonspecific binding sites, cells were incubated with anti-NF-κB antibody (sc-372) (Santa cruz) diluted 1:200 in 4% BSA in PBS overnight at 4°C, followed by Alexa 546-conjugated anti-rabbit IgG (Molecular Probes, Eugene, OR, cat no A11010), diluted 1:500 in 4% BSA in PBS for 60 min at ambient temperature. The samples were mounted and examined under a microscope equipped with epifluorescence illumination (ECLIPSE Ti-U, Nikon). To exmamine the effect of NF-κB on 5-FU-suppressed cellular viability, Sa3 cells incubated with NF-κB inhibitors, 5 μM BAY 11-7085 (Enzo Life Sciences, cat no EI-279) and 200 μM Caffeic acid phenethyl ester (CAPE) (Calbiochem, San Diego, CA, cat no 211200) for 1 h. Then cells were treated with 5 mg/mL 5-FU for 24 h and cellular viability was mesured by WST-8 assay.

NO production

Sa3 cells were pre-treated with 500 mg/mL TJ-84 for 1 h before treatment with 5 mg/mL 5-FU for 3 or 6 h. Then the cells were incubated with the 10% FBS DMEM containing the mixture of 10 μM diaminofluorescein-2 diacetate (DAF-2 DA, Sekisui medical, Kyoto, Japan, cat no 423727) and pluronic F-127 (50/50%, v/v) for 1 h, and then fixed with 10% formalin. The levels of NO were assessed by fluorescence microscopy at 488 nm (ECLIPSE Ti-U, Nikon). The fluorescence intensity was quantified by Image J software.

Statistical analysis

All data are expressed as means ± S.E.M. A minimum of three independent experiments were performed for each assay. The Student's t-test was used for the statistical analyses.

Acknowledgments

We thank Dr. Masayuki Syono (Institute of Health Biosciences, University of Tokushima Graduate School) for help with fluorescence microscopy and for constructive discussions. We also thank the Support Center for Advanced Medical Sciences (Institute of Health Biosciences, University of Tokushima Graduate School) for technical support.

Author Contributions

Conceived and designed the experiments: KY MY DH. Performed the experiments: KY SM HO. Analyzed the data: KY DH. Contributed reagents/materials/analysis tools: KK. Contributed to the writing of the manuscript: KY DG.

References

1. Watanabe K, Matsuura K, Gao P, Hottenbacher L, Tokunaga H, et al. (2011) Traditional Japanese Kampo medicine: Clinical research between modernity and traditional medicine–The state of research and methodological suggestions for the future. Evid Based Complement Alternat Med 2011, doi: 10.1093/ecam/neq067.
2. Liao J, Zhao L, Yoshioka M, Hinode D, Grenier D (2013) Effects of Japanese traditional herbal medicines (Kampo) on growth and virulence properties of *Porphyromonas gingivalis* and viability of oral epithelial cells. Pharm Biol 51: 1538–1544.
3. Ara T, Maeda Y, Fujinami Y, Imamura Y, Hattori T, et al. (2008) Preventive effects of a Kampo medicine, Shosaikoto, on inflammatory responses in LPS-treated human gingival fibroblasts. Biol Pharm Bull 3:, 1141–1144.
4. Ara T, Honjo K, Fujinami Y, Hattori T, Imamura Y, et al. (2010) Preventive effects of a Kampo medicine, Orento on inflammatory responses in lipopolysaccharide treated human gingival fibroblasts. Biol Pharm Bull 33: 611–616.
5. Hiroshima Y, Bando M, Kataoka M, Shinohara Y, Herzberg MC, et al. (2009) Shosaikoto increases calprotectin expression in human oral epithelial cells. J Periodont Res 2010 45: 79–86.
6. Liao J, Jabrane A, Zhao L, Yoshioka M, Hinode D, et al. (2014) The Kampo medicine Rokumigan possesses antibiofilm, anti-inflammatory, and wound healing properties. Biomed Res Int 2014, Article ID 436206.
7. Fata F, Ron IG, Kemeny N, O'Reilly E, Klimstra D, et al. (1999) 5-fluorouracil-induced small bowel toxicity in patients with colorectal carcinoma. Cancer 86: 1129–1134.
8. Peterson DE, Bensadoun RJ, Roila F (2011) Management of oral and gastrointestinal mucositis: ESMO Clinical Practice Guidelines. Ann Oncol 22: vi78-vi84.
9. Dreizen S, Bodey GP, Valdivieso M (1983) Chemotherapy-associated oral infections in adults with solid tumors. Oral Surg Oral Med Oral Pathol 55: 113–120.
10. Elting LS, Rubenstein EB, Rolston KV, Bodey GP (1997) Outcomes of bacteremia in patients with cancer and neutropenia: observations from two decades of epidemiological and clinical trials. Clin Infect Dis 25: 247–259.
11. Saadeh CE, Pharm D (2005) Chemotherapy-and radiotherapy-induced oral mucositis: review of preventive strategies and treatment. Pharmacotherapy 25: 540–554.
12. Sonis ST (2004) The pathobiology of mucositis. Nat Rev 4: 277–284.
13. Yoshino F, Yoshida A, Nakajima A, Wada-Takahashi S, Takahashi S, et al. (2013) Alteration of the redox state with reactive oxygen species for 5-fluorouracil-induced oral mucositis in hamsters. PLoS ONE 8: e82834.
14. Sonis ST (2010)New thoughts on the initiation of mucositis. Oral Dis 16: 597–600.
15. Masters SL, Gerlic M, Metcalf D, Preston S, Pellegrini M, et al. (2012) NLRP1 inflammasome activation induces pyroptosis of hematopoietic progenitor cells. Immunity 37: 1009–1023.
16. Bruchard M, Mignot G, Derangère V, Chalmin F, Chevriaux A, et al. (2013) Chemotherapy-triggered cathepsin B release in myeloid-derived suppressor cells activates the Nlrp3 inflammasome and promotes tumor growth. Nat Med 19: 57–64.
17. Schroder K, Zhou R, Tschopp J (2010) The NLRP3 inflammasome: A sensor for metabolic danger? Science 327: 296–300.
18. Lee BC, Lee J (2014) Cellular and molecular players in adipose tissue inflammation in the development of obesity-induced insulin resistance. Biochim Biophys Acta 1842: 446–462.
19. Okamoto M, Liu W, Luo Y, Tanaka A, Cai X, et al. (2010) Constitutively active inflammasome in human melanoma cells mediating autoinflammation via caspase-1 processing and secretion of interleukin-1β. J Biol Chem 285: 6477–6488.
20. Halle A, Hornung V, Petzold GC, Stewart CR, Monks BG, et al. (2008) The NALP3 inflammasome is involved in the innate immune response to amyloid-β. Nat immunol 9: 857–865.
21. Duewell P, Kono H, Rayner KJ, Sirois CM, Vladimer G, et al. (2010) NLRP3 inflammasomes are required for atherogenesis and activated by cholesterol crystals. Nature 464: 1357–1362.
22. Ouyang X, Ghani A, Mehal WZ (2013) Inflammasome biology in fibrogenesis. Biochimica et Biophysica Acta 183: 979–988.
23. Lawlor KE, Vince JE (2014) Ambiguities in NLRP3 inflammasome regulation: Is there a role for mitochondria? Biochim Biophys Acta 1840: 1433–1440.
24. Cruz CM, Rinna A, Forman HJ, Ventura AL, Persechini PM, et al. (2007) ATP activates a reactive oxygen species-dependent oxidative stress response and secretion of proinflammatory cytokines in macrophages. J Biol Chem 282: 2871–2879.
25. Dostert C, Pétrilli V, Bruggen RV, Steele C, Mossman BT, et al. (2008) Innate immune activation through Nalp3 inflammasome sensing of asbestos and silica. Science 320: 674–677.
26. Nakahira K, Haspel JA, Rathinam VAK, Lee SJ, Dolinay T, et al. (2011) Autophagy proteins regulate innate immune responses by inhibiting the release of mitochondrial DNA mediated by the NALP3 inflammasome. Nat immunol 8: 222–231.
27. Zhou R, Yazdi AS, Menu P, Tschopp J (2011) A role for mitochondria in NLRP3 inflammasome activation. Nature 469: 221–225.
28. Jabaut J, Ather JL, Taracanova A, Poynter ME, Ckless K (2013) Mitochondria-targeted drugs enhance Nlrp3 inflammasome-dependent IL-1β secretion in association with alterations in cellular redox and energy status. Free Rad Biol Med 60: 233–245.
29. Tassi S, Carta S, Delfino L, Caorsi R, Martini A, et al. (2010) Altered redox state of monocytes from cryopyrin-associated periodic syndromes causes accelerated IL-1beta secretion. Proc Natl Acad Sci U S A 107: 9789–9794.
30. Kummer JA, Broekhuizen R, Everett H, Agostini L, Kuijk L, et al. (2007) Inflammasome components NALP 1 and 3 show distinct but separate expression profiles in human tissue suggesting a site-specific role in the inflammatory response. J Histochem Cytochem 55: 443–452.
31. Belibasakis GN, Guggenheim B, Bostanci N (2013) Down-regulation of NLRP3 inflammasome in gingival fibroblasts by subgingival biofilms: involvement of *Porphyromonas gingivalis*. Innate Immun 19: 3–9.
32. Bostanci N, Meier A, Guggenheim B, Belibasakis GN (2011) Regulation of NLRP3 and AIM2 inflammasome gene expression levels in gingival fibroblasts by oral biofilms. Cell Immunol 270: 88–93.
33. Yilmaz O, Sater AA, Yao L, Koutouzis T, Pettengill M, et al. (2010) ATP-dependent activation of an inflammasome in primary gingival epithelial cells infected by *Porphyromonas gingivalis*. Cell Microbiol 12: 188–198.
34. Duprez L, Wirawan E, Berghe TV, Vandenabeele P (2009) Major cell death pathways at a glance. Microbes Infect 11: 1050–1062.
35. Majno G, Joris I (1995) Apoptosis, oncosis, and necrosis. An overview of cell death. Am J Pathol 146: 3–15.
36. Fink SL, Cookson BT (2006) Caspase-1-dependent pore formation during pyroptosis leads to osmotic lysis of infected host macrophages. Cell Microbiol 8: 1812–1825.
37. Arifa RD, Madeira MF, de Paula TP, Lima RT, Tavares LD, et al. (2014) Inflammasome activation is reactive oxygen species dependent and mediates irinotecan-induced mucositis through IL-1β and IL-18 in mice. Am J Pathol 184: 2023–2034.
38. Wu Z, Han X, Qin S, Zheng Q, Wang Z, et al. (2010) Interleukin 1 receptor atagonist reduces lethality and intestinal toxicity of 5-fluorouracil in a mouse mucositis model. Biomed Pharmacother 64: 589–593.
39. Sonis ST (2002) The biologic role for nuclear factor-kappaB in disease and its potential involvement in mucosal injury associated with anti-neoplastic therapy. Crit Rev Oral Biol Med 13: 380–389.
40. Brüne B, von Knethen A, Sandau KB (1998) Nitric oxide and its role in apoptosis. Eur J Pharmacol 351: 261–272.
41. Brown GC (2010) Nitric oxide and neuronal death. Nitric Oxide 23: 153–165.
42. Brown GC, Borutaite V (2002) Nitric oxide inhibition of mitochondrial respiration and its role in cell death. Free Radic Biol Med 33: 1440–1450.
43. Takuma K, Phuagphong P, Lee E, Mori K, Baba A, et al. (2001) Anti-apoptotic effect of cGMP in cultured astrocytes: inhibition by cGMP-dependent protein kinase of mitochondrial permeable transition pore. J Biol Chem 276: 48093–48099.
44. Thomas DD, Ridnour LA, Isenberg JS, Flores-Santana W, Switzer CH, et al. (2008) The chemical biology of nitric oxide: implications in cellular signaling. Free Rad Biol Med 45: 18–31.
45. Hernandez-Cuellar E, Tsuchiya K, Hara H, Fang R, Sakai S, et al. (2012) Cutting edge: nitric oxide inhibits the NLRP3 inflammasome. J Immunol 189: 5113–5117.
46. Wright TH, Yazbeck R, Lymn KA, Whitford EJ, Cheah KY, et al. (2009) The herbal extract, Iberogast, improves jejunal integrity in rats with 5-fluorouracil (5-FU)-induced mucositis. Cancer Biol Ther 8: 923–929.
47. Kono T, Satomi M, Chisato N, Ebisawa Y, Suno M, et al. (2010) Topical application of Hangeshashinto (TJ-14) in the treatment of chemotherapy-induced oral mucositis. World J Oncol 1: 232–235.
48. Das SK, Das V, Guati AK, Singh VP (1989) Deglycyrrhizinated liquorice in aphthous ulcers. J Assoc Physicians India 37: 647.
49. Tang T, Yin Longwu, Yang Jing, Shan G (2007) Emodin, an antharquinone derivative from *Rheum officinale* Baill, enhances cutaneous wound healing in rats. Eur J Pharmacol 567: 177–185.
50. Okada K, Shoda J, Kano M, Suzuki S, Ohtake N, et al. (2007) Inchinkoto, a herbal medicine, and its ingredients dually exert Mrp2/MRP2-mediated

choleresis and Nrf2-mediated antioxidative action in rat livers. Am J Gastrointest Liver Physiol 292: G1450–G1463.

51. Yamamoto M, Miura N, Ohtake N, Amagaya S, Ishige A, et al. (2000) Genipin, a metabolite derived from the herbal medicine Inchin-ko-to, and suppression of Fas-induced lethal liver apoptosis in mice. Gastroenterology 2000 118: 380–389.

52. Okamura H, Yang D, Yoshida K, Haneji T (2013) Protein phosphatase 2A Cα is involved in osteoclastogenesis by regulating RANKL and OPG expression in osteoblasts. FEBS Lett 587: 48–53.

7

Accelerated Recovery of Mitochondrial Membrane Potential by GSK-3β Inactivation Affords Cardiomyocytes Protection from Oxidant-Induced Necrosis

Daisuke Sunaga, Masaya Tanno, Atsushi Kuno, Satoko Ishikawa, Makoto Ogasawara, Toshiyuki Yano, Takayuki Miki, Tetsuji Miura*

Department of Cardiovascular, Renal and Metabolic Medicine, Sapporo Medical University School of Medicine, Sapporo, Japan

Abstract

Loss of mitochondrial membrane potential ($\Delta\Psi_m$) is known to be closely linked to cell death by various insults. However, whether acceleration of the $\Delta\Psi_m$ recovery process prevents cell necrosis remains unclear. Here we examined the hypothesis that facilitated recovery of $\Delta\Psi_m$ contributes to cytoprotection afforded by activation of the mitochondrial ATP-sensitive K^+ (mK_{ATP}) channel or inactivation of glycogen synthase kinase-3β (GSK-3β). $\Delta\Psi_m$ of H9c2 cells was determined by tetramethylrhodamine ethyl ester (TMRE) before or after 1-h exposure to antimycin A (AA), an inducer of reactive oxygen species (ROS) production at complex III. Opening of the mitochondrial permeability transition pore (mPTP) was determined by mitochondrial loading of calcein. AA reduced $\Delta\Psi_m$ to $15\pm1\%$ of the baseline and induced calcein leak from mitochondria. $\Delta\Psi_m$ was recovered to $51\pm3\%$ of the baseline and calcein-loadable mitochondria was $6\pm1\%$ of the control at 1 h after washout of AA. mK_{ATP} channel openers improved the $\Delta\Psi_m$ recovery and mitochondrial calcein to $73\pm2\%$ and $30\pm7\%$, respectively, without change in $\Delta\Psi_m$ during AA treatment. Activation of the mK_{ATP} channel induced inhibitory phosphorylation of GSK-3β and suppressed ROS production, LDH release and apoptosis after AA washout. Knockdown of GSK-3β and pharmacological inhibition of GSK-3β mimicked the effects of mK_{ATP} channel activation. ROS scavengers administered at the time of AA removal also improved recovery of $\Delta\Psi_m$. These results indicate that inactivation of GSK-3β directly or indirectly by mK_{ATP} channel activation facilitates recovery of $\Delta\Psi_m$ by suppressing ROS production and mPTP opening, leading to cytoprotection from oxidant stress-induced cell death.

Editor: Hossein Ardehali, Northwestern University, United States of America

Funding: This work was supported by the Japanese Society for the Promotion of Science Grants-in-Aid for Scientific Research #23501086 (to T. Miura) and a grant from Chugai Pharmaceutical Co. Ltd. Neither funder played any role in the study design, data collection and analysis, decision to publish or preparation of the manuscript.

Competing Interests: Tetsuji Miura received research grant support from Chugai Pharmaceutical Co. Ltd.

* Email: miura@sapmed.ac.jp

Introduction

Mitochondrial membrane potential ($\Delta\Psi_m$) is crucial for cell viability. Loss of $\Delta\Psi_m$ by opening of the mitochondrial permeability transition pore (mPTP) is a major mechanism of myocardial infarction [1,2] and cerebral infarction [3] after ischemia/reperfusion. Dissipation of $\Delta\Psi_m$ has also been shown to precede shrinkage and fragmentation of cells, contributing to programed cell death [4,5]. Loss of $\Delta\Psi_m$ by irreversible opening of the mPTP leads to arrest of mitochondrial ATP synthesis, mitochondrial swelling and outer membrane permeabilization. During ischemia, $\Delta\Psi_m$ is temporarily maintained by consumption of ATP by mitochondrial ATPase, and its recovery after reperfusion depends on ischemia-induced injury of the mitochondrial machinery and on the level of stimuli for mPTP opening upon reperfusion [6,7]. Involvement of mPTP opening and $\Delta\Psi_m$ loss in reperfusion-induced cell death has been supported by results of studies showing that protection against reperfusion injury afforded by ischemic preconditioning (IPC) and IPC mimetics was

associated with inhibition of mPTP opening [1,8–11]. On the other hand, few studies have examined if manipulation of $\Delta\Psi_m$ recovery protects the heart.

A rationale for the hypothesis that acceleration of $\Delta\Psi_m$ recovery protects cells from necrosis has been provided by several lines of evidence. First, preserved $\Delta\Psi_m$ is necessary for mitochondrial ATP synthesis, and recovery of ATP synthesis is crucial for restoration of intracellular Na^+ and Ca^{2+} homeostasis [5,7]. Second, there are differences between mitochondria within a cell in susceptibility to Ca^{2+}-induced mPTP opening, production of reactive oxygen species (ROS) and structural changes after ischemia/reperfusion [9,10]. ROS-induced ROS release and Ca^{2+}-induced Ca^{2+} release from mitochondria have been reported as mechanisms of accelerated ROS production and Ca^{2+} overload [12–14]. Hence, the recovery of $\Delta\Psi_m$ after withdrawal of insults is also likely to be heterogeneous in mitochondria within a cell. Third, mPTP opening is not always irreversible and re-closure of mPTPs after reperfusion has been shown in rat hearts by use of D-^{3}H-2-deoxyglucose as a tracer of opened mPTPs [15,16], suggesting that recovery of $\Delta\Psi_m$ can be achieved by re-closure of mPTPs. Collectively, these findings indicate the possibility that the percentage of mitochondria with unrecoverable $\Delta\Psi_m$ within a cell upon removal of an insult (for example, ischemia/reperfusion) determines mortality of the cell.

We hypothesized that acceleration of $\Delta\Psi_m$ recovery by pro-survival signaling protects myocytes from necrosis. To test this hypothesis, we examined the time course of $\Delta\Psi_m$ recovery in response to a period of exposure to ROS in H9c2 and C2C12 cells and possible modification of the time course by activation of the mitochondrial ATP-sensitive K^+ (mK_{ATP}) channel and by inhibition of glycogen synthase kinase-3β (GSK-3β). The mK_{ATP} channel and GSK-3β were selected for testing effects on $\Delta\Psi_m$ as these two localize within mitochondria and are known to play roles in regulation of the threshold for mPTP opening [2,17,18].

Materials and Methods

Cell culture and experimental protocols

H9c2 cells, C2C12 cells and human embryonic kidney cells (HEK-293 cells) were obtained from ATCC (American Type Culture Collection). The cells were cultured in DMEM (4.5 g/L glucose) supplemented with 10% fetal bovine serum and antibiotics. All experiments using H9c2 and C2C12 cells were started after serum deprivation for 24 h. An inhibitor of complex III, antimycin A (AA, 40 μM), was used to induce mitochondrial ROS generation. After 60-min treatment with AA, AA was washed out from the culture medium by replacing the medium with pre-warmed fresh medium without AA. Hypoxia/reoxygenation was not employed to induce oxidant stress in this study, since hypoxia alone reduces $\Delta\Psi_m$ and activates multiple pathways of intracellular signaling, which complicates analysis of mPTP-relevant signaling.

To examine the effects of activation of the mK_{ATP} channel, an NO donor and inhibition of GSK-3β, H9c2 cells were incubated with a vehicle, nicorandil (300 μM), diazoxide (300 μM), S-nitroso-N-acetyl-DL-penicillamine (SNAP, 1 μM), or LiCl (30 mM), for 1 h before addition of AA to the medium, and the treatment was continued until the end of the experiment. Treatment with 5-hydroxydecanote (5-HD, 100 μM) to inhibit opening of the mK_{ATP} channel and treatment with mercaptopro-pionyl glycine (MPG, 30 μM) to suppress ROS during nicorandil treatment were commenced 30 min before treatment with a K_{ATP} channel opener and discontinued with the onset of AA treatment. To examine the possible effect of ROS on recovery of H9c2 cells, MPG (30 μM) or N-acetyl-cysteine (NAC, 1 mM) was added to the medium during AA treatment or to the medium after AA treatment (i.e., a non-AA-containing medium). The effect of inhibition of glycolysis was examined by use of iodoacetate (IAA, 30 μM).

Monitoring of mitochondrial membrane potential

Mitochondrial membrane potential was monitored by tetra-methylrhodamine ethyl ester (TMRE) fluorescence as previously reported [19,20]. H9c2 cells or C2C12 cells were loaded with TMRE (100 nM) 1 h before AA treatment for assessing the effects of AA on $\Delta\Psi_m$ or at the time of AA washout for assessing recovery of $\Delta\Psi_m$ from oxidant stress by AA. Level of TMRE fluorescence at each time point was expressed as percentage of values in time controls without AA. Fluorescence was recorded by fluorescence microscopy (Olympus IX-70), and images of TMRE fluorescence taken at a magnification of 400× were quantified by pixel counts after cutting off background fluorescence using a threshold value. Data from three regions of interest (ROI) were averaged for each

well in the culture plate, and the numbers of cells within the ROI were made comparable between treatment groups. In cells loaded with TMRE after washout of AA, TMRE fluorescence level was normalized by values of time controls in the same culture plate since we had confirmed that TMRE fluorescence level did not significantly change for 3 h of the control period.

Monitoring of mitochondrial permeability by calcein

In addition to TMRE, a membrane potential-independent tracer, calcein, was used for detection of opening of the mPTP as previously reported [21,22]. Briefly, in experiments in which the effect of AA on the mPTP was examined, cells were incubated with 0.25 μM calcein for 15 min, and the medium was changed to calcein-free medium containing 4 mM of cobalt chloride ($CoCl_2$) before treatment with AA. In experiments in which mPTP opening status after AA washout was examined, calcein was loaded for 15 min immediately after washout of AA. The medium was then replaced with a calcein-free medium containing 4 mM $CoCl_2$. Cells were pretreated with 1 μM MitoTracker red for 15 min to stain mitochondria before AA treatment. The calcein-stained area overlapped with the MitoTraker red-stained area was used as an index of mitochondria with closed mPTP and quantified as the level of TMRE described above. Data were normalized by time control cell data. In pilot experiments (n = 6), MitoTracker red was found to be lost from some mitochondria after AA treatment, and 68.3±6.2% of mitochondria retained MitoTracker red at 60 min after washout of AA. Thus, the extent of mPTP opening was somewhat underestimated by the present method.

Isolation of mitochondria and cytosol fractions, Western blotting, and immunoprecipitation

Mitochondrial and cytosolic fractions of H9c2 cells were prepared by using a mitochondrial isolation kit (Pierce Biotech-nology, Rockford, IL) according to the manufacturer's protocol. The samples were subjected to SDS-PAGE, followed by transfer to a polyvinylidene difluoride membrane. After blocking with TBS-T with 5% skim milk or 5% BSA, the membrane was incubated with the primary antibody at 4°C overnight. After incubation with the secondary antibody, the bands were visualized by a standard ECL technique. Interaction of GSK-3β and Reiske was analyzed by immunoprecipitation experiments. Precleared cell lysates (500 μg) were incubated with 2 μg of anti-Rieske antibody in IP buffer (20 mM Tris–HCl [pH 7.4], 1 mM EGTA, 5 mM NaN_3, 50 mM NaCl, 1 mM PMSF, 50 mM Na_3VO_4, 1% Triton X-100, 0.5% NP-40 and a protease inhibitor cocktail) at 4°C overnight with rotation. The antibody-Rieske complex was collected with magnet beads and washed with IP buffer. The immunoprecipitates were subjected to Western blotting as described above. Antibodies used were rabbit monoclonal anti-GSK-3β (#9315, Cell Signaling), rabbit polyclonal anti-phospho-(Ser9) GSK-3β (#9336, Cell Signaling), rabbit polyclonal anti-glycogen synthase (#3893, Cell Signaling), rabbit polyclonal anti-phospho-(Ser641/645) glycogen synthase (44-1092G, Invitrogen), mouse monoclonal anti-Rieske (ab14746, Abcam), goat polyclonal anti-ANT (sc-9299, Santa Cruz), mouse monoclonal anti-VDAC1 (ab14734, Abcam), mouse monoclonal anti-cyclophilin D (AP1035, Calbiochem), mouse monoclonal anti-prohibitin (sc-56346, Santa Cruz), mouse mono-clonal anti-β-actin (A5316, SIGMA) and rabbit polyclonal anti-inorganic phosphate carrier (custom-made antibody [23]).

Transfection of siRNA

Knockdown of GSK-3β was performed by transfection of siRNA against rat GSK-3β (Mission siRNA, SA-SI_Rn01_00035806, Sigma-Aldrich) using Nucleofection (Lonza Walkersville, MD) according to the manufacturer's protocol. Experiments were completed 48 h after transfection.

Determination of cell necrosis and apoptosis

Cell necrosis was analyzed by determination of lactate dehydrogenase (LDH) released into the incubation medium. LDH activity in the culture medium and LDH activity in the medium after freeze-thawing of the cells (total cellular LDH activity) were measured by using a CytoTox 96 Non-Radioactive Cytotoxicity assay kit (Promega, Madison, WI) according to the manufacturer's protocol. LDH activity in the medium as a percentage of the total cellular LDH activity was used as an index of cell necrosis. To quantify apoptosis, cells were stained with Hoechst33342 as previously reported [24]. Apoptosis of cells was defined as nuclear condensation revealed by Hoechst33342.

Determination of ROS production

Intracellular ROS levels were monitored by 2′-7′-dichlorofluorescein (DCF) fluorescence. H9c2 cells were loaded with DCF according to the manufacturer's protocol. DCF fluorescence was recorded by FLoid Cell Imaging Station (Life Technologies, CA) at 30 and 60 min after the start of incubation in an AA-containing medium and at 15 and 60 min after washout of AA.

Determination of ATP level

H9c2 cells were subjected to 60-min treatment with AA (40 μM), 120-min treatment with IAA (30 μM), combination of AA and IAA treatments or 120-min treatment with a vehicle, and their ATP levels were determined by an assay kit, CellTiter-Glo Luminescent Cell Viability Assay G7570 (Promega, Madison, USA).

Succinate dehydrogenase activity assay

Cells were subjected to 24-hour serum-deprived culture and then incubated in AA (40 μM)-containing medium or normal medium for 60 min with or without 300 μM nicorandil. For cells treated with both AA and nicorandil, nicorandil was added to the medium 60 min before the onset of incubation with AA. Cellular succinate dehydrogenase (complex II) activity was measured by using a Complex II Enzyme Activity Microplate Assay Kit (Abcam, Cambridge, UK) according to the manufacturer's protocol.

Chemical compounds

AA, diazoxide, 5-HD, LiCl, MPG, and cyclosporine A were purchased from Sigma Aldrich (St. Louis, MO). NAC was from Wako Pure Chemical Industries (Osaka, Japan). Nicorandil was provided by Chugai Pharmaceutical Co. Ltd. (Tokyo, Japan). TMRE, calcein and DCF were purchased from Invitrogen (Carlsbad, CA).

Statistical analysis

Data are presented as means ± standard error of the mean. One-way or two-way analysis of variance (ANOVA) was used to detect significant differences between group means in the treatment groups. When ANOVA indicated a significant overall difference, multiple comparisons of the groups were performed by the Student-Newman-Keuls *post-hoc* test. A difference was considered to be statistically significant if p was less than 0.05.

Results

Activation of the mK$_{ATP}$ channel promotes recovery of ΔΨ$_m$ dissipated by AA

During treatment with AA, TMRE fluorescence was reduced to 15% of the control and the rod-shaped structure of mitochondria became blurred, indicating reduction in ΔΨ$_m$ and mPTP opening (Figure 1). At 60 min after washout of AA, the level of TMRE fluorescence was 51% of the baseline (Figure 1B and C). Since AA not only induces ROS generation but also inhibits oxidative phosphorylation, both effects were possibly responsible for the change in ΔΨ$_m$. To assess the impact of ATP deletion on the time course of ΔΨ$_m$ after AA treatment, we compared its effect with those of IAA and AA+IAA. AA reduced ATP level to $23.9 \pm 1.2\%$ of vehicle controls, and a significantly greater reduction in ATP level was achieved by IAA ($13.0 \pm 0.4\%$). The combination of AA and IAA almost completely depleted ATP ($0.6 \pm 0.2\%$). Although depletion of ATP was significantly less in the AA-treated group, time courses of TMRE fluorescence were similar in the AA-treated and IAA-treated groups (Figure 1D). Together with the effects of ROS scavengers on TMRE fluorescence (see "Relationship between ROS and recovery of ΔΨ$_m$" below), these results indicate that ROS, in addition to ATP depletion, was responsible for loss of ΔΨ$_m$ by AA treatment.

Opening of the mPTP by AA was indicated by the finding that calcein loaded in mitochondria leaked into the cytosol after AA treatment (Figure 2A). The ratio of mitochondria positive for calcein was 6% of the baseline at 60 min after AA treatment, and the ratios were 10% and 18% of time control value at 60 and 120 min after AA washout, respectively (Figure 2B and C). In the vehicle-treated controls, the ratio of mitochondria positive for calcein was not 100% (Figure 2C) since the threshold for calcein fluorescence was set at a relatively high level in order to include clearly discrete mitochondrial calcein.

Contribution of mPTP opening to reduction in TMRE fluorescence and cell necrosis after washout of AA was supported by results showing that cyclosporine A, a direct mPTP inhibitor, attenuated the effect of AA treatment on TMRE fluorescence and LDH release after AA washout by 24% (Figure 3). Protein levels of putative regulatory subunits of the mPTP (adenine nucleotide translocase, voltage-dependent anion channel, inorganic phosphate carrier and cyclophilin D) were not changed by AA (Figure S1).

Pretreatment with an mK$_{ATP}$ channel activator (nicorandil or diazoxide) did not affect reduction of TMRE fluorescence during AA treatment but significantly enhanced its recovery after washout of AA (Figure 1B and C). This effect of mK$_{ATP}$ channel openers was abolished by 5-HD, an mK$_{ATP}$ channel blocker (Figure 1B). SNAP, an NO donor, did not mimic the effect of nicorandil on TMRE fluorescence after AA treatment, indicating that the nitrate property of nicorandil was not involved in its effect on ΔΨ$_m$ (Figure S2). Although SNAP at relatively high doses (0.1–1 mM) has been shown to activate the mK$_{ATP}$ channel [25,26], we selected a low dose (1 μM) to avoid the effect on the mK$_{ATP}$ channel in this study. Improvement by diazoxide in recovery of TMRE fluorescence after washout of AA was also confirmed in C2C12 cells (Figure 4), indicating that role of the mK$_{ATP}$ channel in ΔΨ$_m$ regulation is not unique to H9c2 cells.

Inhibition of GSK-3β by activation of the mK$_{ATP}$ channel

Like diazoxide in our previous study [27], nicorandil increased the levels of Ser9-phospho-GSK-3β in the mitochondria and cytosol by 24% and 37%, respectively (Figure 5AB). Phosphorylation of GSK-3β in mitochondria by nicorandil was inhibited by

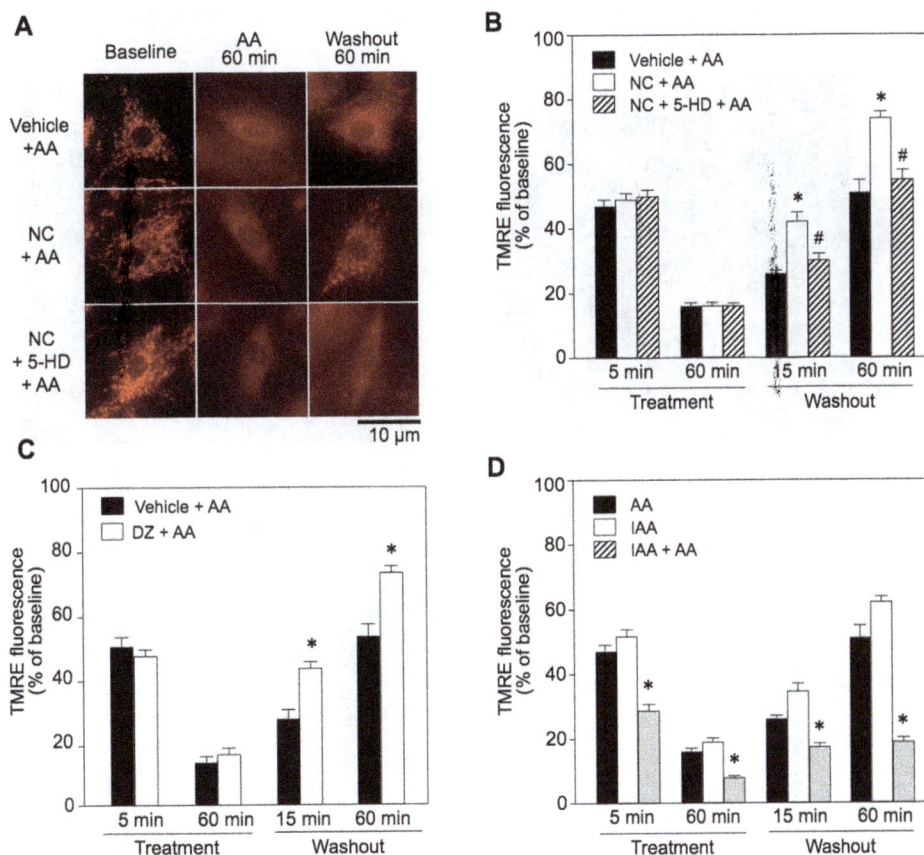

Figure 1. Effects of mK$_{ATP}$ channel openers and a glycolysis inhibitor on antimycin A-induced loss of $\Delta\Psi_m$ and its recovery in H9c2 cells. A: TMRE images before and after antimycin A (AA) treatment (Images were from different cells). B and C: TMRE fluorescence in NC- (B) and DZ-pretreated cells (C). D: TMRE fluorescence in IAA-treated cells. TMRE fluorescence in NC = nicorandil, DZ = diazoxide, 5-HD = 5-hydoxydecanote, IAA = iodoacetate, Treatment = time after onset of treatment with AA, Washout = time after washout of AA. *p<0.05 vs. Vehicle+AA or AA, #p<0.05 vs. NC+AA. N = 8.

MPG (Figure 5C), indicating that ROS generated by mK$_{ATP}$ channel activation [28,29] mediated GSK-3β phosphorylation. Elevation of mitochondrial phospho-GSK-3β level by nicorandil was maintained at 5 min after AA treatment, though the effect was not significant afterwards (Figure 6A and B). If inactivation of GSK-3β by phosphorylation at Ser9 contributes to facilitated recovery of $\Delta\Psi_m$ after AA treatment by nicorandil, the effect of nicorandil should be mimicked by an inhibitor of GSK-3β, LiCl, and by knockdown of GSK-3β. That was indeed the case as shown in Figure 6C and D. Inhibition of GSK-3β activity by LiCl was confirmed by the results that LiCl increased Ser9-phospho-GSK-3β level, reflecting suppression of a GSK-3β activity-dependent phosphatase, and reduced phosphorylation of glycogen synthase (GS), a downstream target of GSK-3β (Figure 6C, lower panel).

Relationship between ROS and recovery of $\Delta\Psi_m$

The level of ROS determined by DCF was significantly elevated during AA treatment and then decreased time-dependently after washout of AA (Figure 7A and B). Treatment with nicorandil or LiCl reduced the level of ROS during AA treatment and after AA washout (Figure 7B). Although inhibition of the activity of succinate dehydrogenase was reported as a mechanism by which mK$_{ATP}$ channel activation reduces ROS production in the heart [30], succinate dehydrogenase activity in H9c2 cells was not significantly changed by nicorandil or AA: 0.53±0.11 (mOD/min) in controls (vehicle-treated cells), 0.61±0.07 in AA-treated cells,

0.67±0.20 in nicorandil-treated cells and 0.69±0.08 in nicorandil plus AA-treated cells (n = 5 in each treatment).

To determine whether ROS during AA treatment or residual ROS being produced after washout of AA inhibit recovery of $\Delta\Psi_m$, we tested the effects of an ROS scavenger during AA treatment or after washout of AA. MPG (30 μM) added to the medium only during AA treatment period significantly reduced ROS both during and after AA treatment (Figure 7C). The effect of MPG on ROS was associated with partial preservation of TMRE fluorescence during AA treatment and improved recovery of TMRE fluorescence (Figure 8A). Treatment with MPG or NAC (1 mM) commenced at the time of washout of AA, which reduced ROS level (Figure 7D), also significantly improved recovery of TMRE fluorescence (Figure 8B), indicating contribution of persistent ROS production after washout of AA to continual mPTP opening.

Effect of facilitated $\Delta\Psi_m$ recovery on cell necrosis and apoptosis

LDH released into the medium was determined at the end of AA treatment and at 2 h after AA removal with or without mK$_{ATP}$ channel opener pretreatment. As shown in Figure 9A, LDH release at the end of AA treatment was not reduced by nicorandil or diazoxide. However, LDH release after removal of AA was significantly reduced by nicorandil, diazoxide and LiCl (Figure 9B

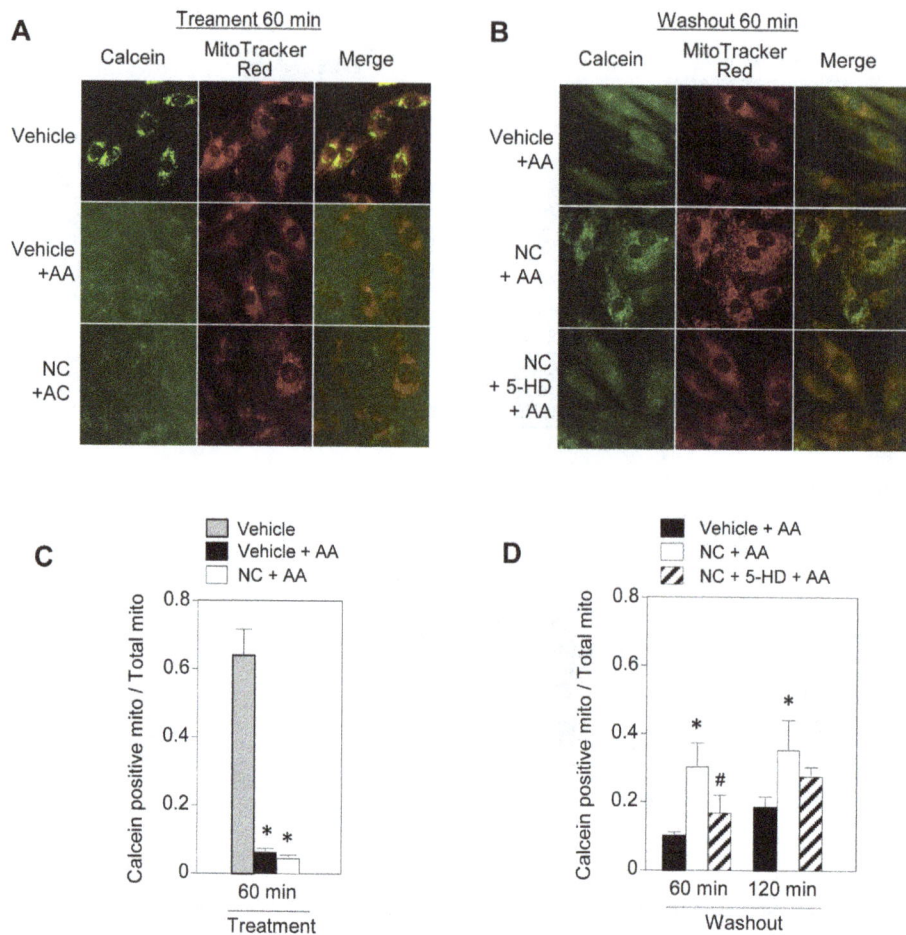

Figure 2. Effects of mK$_{ATP}$ channel openers on antimycin A-induced mPTP opening and its recovery in H9c2 cells. A and B: Calcein and MitoTracker images before (A) and after (B) antimycin A (AA) treatment. C and D: Levels of calcein-positive mitochondria 60 min after AA treatment (C) and 60 and 120 min after washout of AA (D). Level of calcein-positive mitochondria is expressed as the ratio of calcein-positive area to MitoTracker-positive area. *p<0.05 vs. Vehicle+AA, #p<0.05 vs. NC+AA.

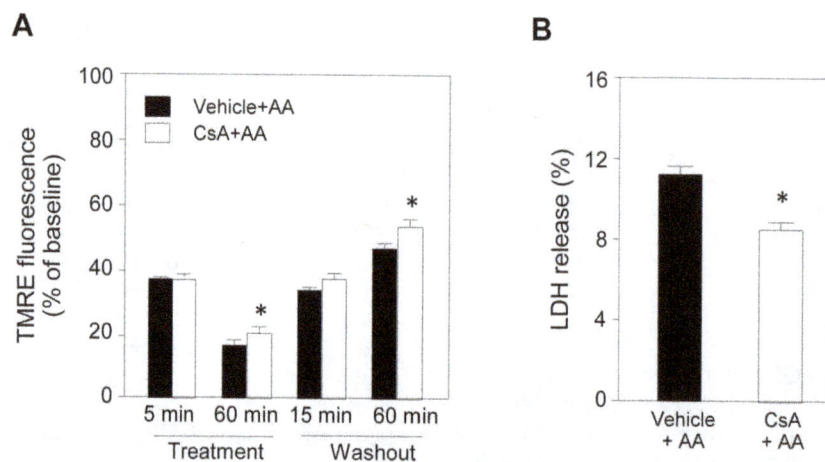

Figure 3. Effects of cyclosporine A on antimycin A-induced changes in $\Delta\Psi_m$ and LDH release. A: Level of TMRE fluorescence was determined as an index of $\Delta\Psi_m$. Cyclosporine A (CsA, 0.5 μM) was added to the medium 60 min before antimycin A (AA) treatment. AA-induced reduction of TMRE fluorescence at 60 min after the onset of AA treatment was slightly attenuated in the CsA-treated group compared to that in the AA group, indicating partial suppression of ROS-induced mPTP opening. Recovery of TMRE fluorescence at 60 min after washout of AA was also slightly improved by CsA. N = 5 per group. B: LDH released after washout of AA was significantly reduced by CsA. *p<0.05 vs. Vehicle+AA. N = 8 per group.

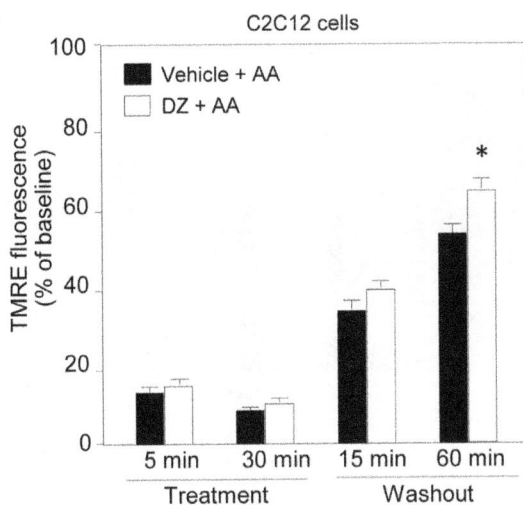

Figure 4. Effects of an mK$_{ATP}$ channel opener on antimycin A-induced loss of ΔΨ$_m$ and its recovery in C2C12 cells. TMRE fluorescence in untreated and diazoxide-pretreated cells. AA = antimycin A, DZ = diazoxide, Treatment = time after onset of treatment with AA, Washout = time after washout of AA. *p<0.05 vs. Vehicle+AA. N = 8.

and C), and the protective effects of mK$_{ATP}$ channel openers were inhibited by 5-HD. Apoptosis after washout of AA was also suppressed by nicorandil in a 5-HD-sensitive manner (Figure 9D and E). Collectively, these results indicate that accelerated recovery of ΔΨ$_m$ after oxidant stress by suppression of ROS production via GSK-3β inactivation prevents cell necrosis and apoptosis.

Modification of mitochondrial complex III by activation of the mK$_{ATP}$ channel

Since AA-induced ROS production was suppressed by LiCl or nicorandil (Figure 7B), we examined the interaction of complex III, a target for AA to generate ROS, and GSK-3β. Interaction of a complex III subunit, Rieske protein (Rieske), and GSK-3β was significantly increased by AA. The Rieske-GSK-3β interaction was attenuated by mK$_{ATP}$ channel openers in H9c2 cells and also in HEK293 cells (Figure 10).

Discussion

The time course of ΔΨ$_m$ recovery or mPTP re-closure and their relationships with development of tissue injury have not been characterized. The present study showed that activation of the mK$_{ATP}$ channel (Figures 1BC and 4), reduction of GSK-3β activity (Figure 6C and D) or suppression of persistent ROS production (Figure 8B) significantly improved recovery of ΔΨ$_m$ from ROS-induced dissipation. Furthermore, the improved ΔΨ$_m$ recovery was associated with suppressed LDH release during the recovery process and apoptosis (Figure 9). These results support the notion that facilitation of ΔΨ$_m$ recovery protects cell from cell death.

ROS production induced by AA, an inhibitor of complex III, was suppressed by inactivation of GSK-3β (Figure 6), indicating that this kinase enhanced ROS production. Involvement of GSK-3β in mitochondrial ROS production is not specific to AA-induced ROS production. Our recent study has shown that mitochondrial translocation of GSK-3β triggered by exogenous hydrogen peroxide induced enhanced ROS production and that both

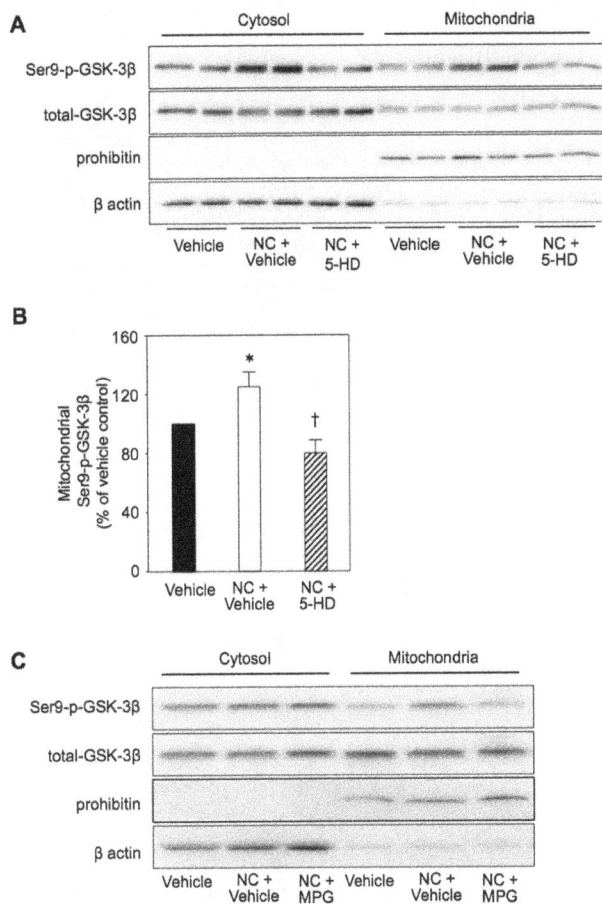

Figure 5. Effects of nicorandil and LiCl on GSK-3β phosphorylation. A: Representative Western blotting for Ser9-phospho-GSK-3β and total GSK-3β in cytosolic and mitochondrial fractions of vehicle-treated, nicorandil-treated (NC) and 5-hydroxydecanoate (5-HD) plus NC-treated cells. B: Group means of mitochondrial Ser9-phospho-GSK-3β levels. NC and 5-HD were added to the culture medium 60 min and 90 min before collection of cells for Western boltting, respectively. *p<0.05 vs. Vehicle, †p<0.05 vs. NC+Vehicle. N = 8 per group. C: Effects of MPG (mercaptopropionyl glycine) on the effect of NC-induced phosphorylation of GSK-3β.

mitochonrial translocation of GSK-3β and ROS production were dependent on GSK-3β kinase activity [20].

Since opening of the mPTP is a major mechanism of failure of ΔΨ$_m$ recovery after ischemia/reperfusion [2,3,5,10], we assessed the impact of mK$_{ATP}$ channel activation on mPTP opening and re-closure. As expected from massive ROS production by AA, the level of mPTP opening was unaffected by activation of the mK$_{ATP}$ channel. While level of ROS production during AA treatment was different between vehicle-treated and nicorandil-treated cells (Figure 7B), TMRE fluorescence was similarly suppressed by AA to 20% of baseline level in both treatment groups (Figure 1B), suggesting presence of ROS threshold level for collapsing ΔΨ$_m$. However, level of the calcein-loadable mitochondria was higher in the nicorandil-pretreated group at 60 and 120 min after washout of AA (Figure 2D). Since MitoTracker red is a ΔΨ$_m$-dependent probe, use of this probe for identifying calcein localized in mitochondria probably underestimated mitochondria with closed mPTPs after washout of AA. In fact, MitoTracker positive area was reduced to 70~80% of time control after AA treatment. Nevertheless, there was a significant difference between levels of

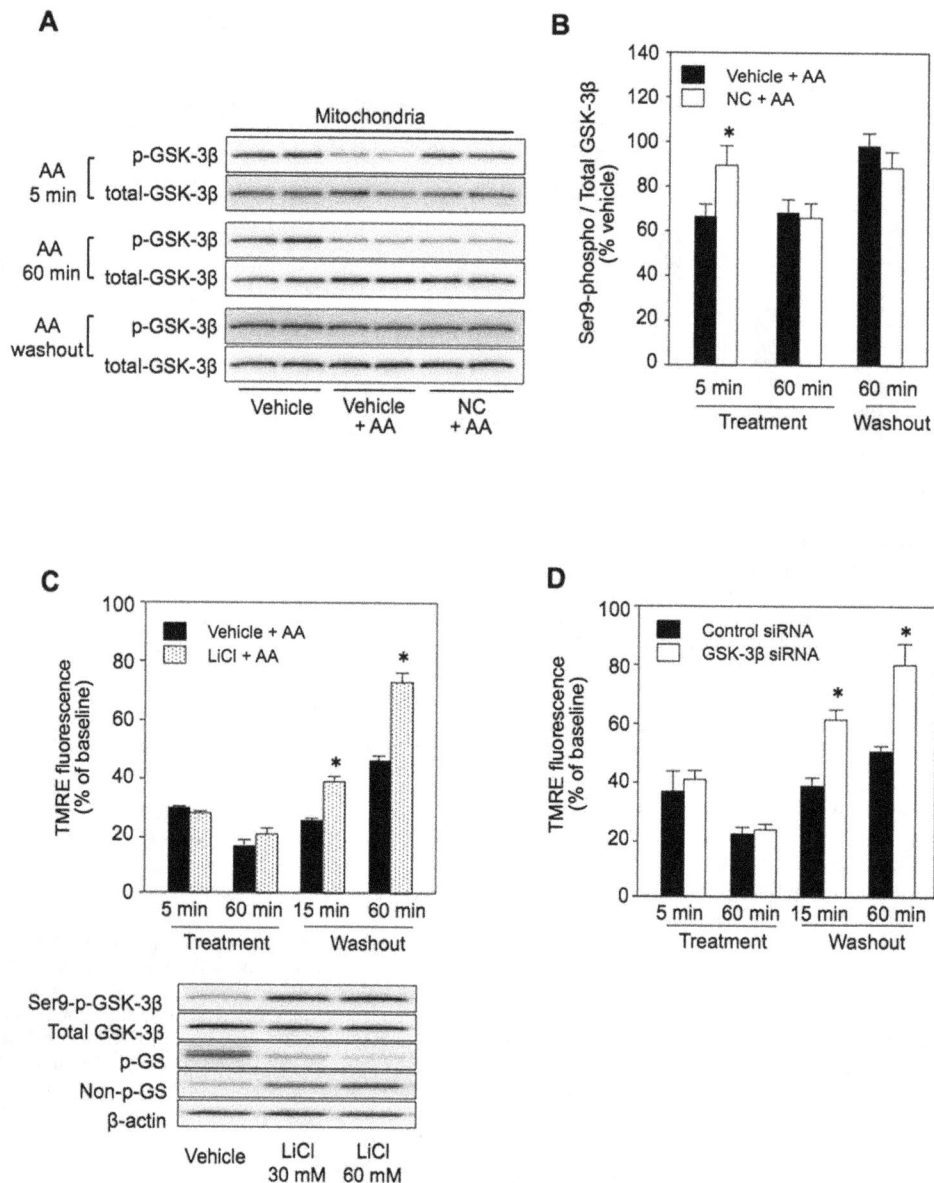

Figure 6. Effects of inhibition of GSK-3β on antimycin A-induced changes in ΔΨ$_m$ and its recovery. A: Western blotting for Ser9-phospho- and total GSK-3β in mitochondria, B: Effects of antimycin A (AA) on phospho-GSK-3β level. C: TMRE fluorescence after AA treatment in vehicle- and LiCl-pretreated cells. Western blotting for Ser9-phospho-GSK-3β, total GSK-3β, Ser641/645-phospho-glycogen synthase (GS), non-phospho-GS and β-actin (loading control) in total lysates of vehicle-treated and LiCl-treated cells. Treatments with 30 mM and 60 mM LiCl for 60 min induced phosphorylation of GSK-3β and dephosphorylation of GS. Increased phosphorylation of GSK-3β by LiCl reflects reduced activity of protein phosphatase 1, which is positively regulated by GSK-3β activity. D: TMRE fluorescence after AA treatment in control siRNA- and GSK-3β-siRNA-pretreated cells. NC = nicorandil. Treatment = time after onset of treatment with AA, Washout = time after washout of AA. *$p < 0.05$ vs. Vehicle or Control siRNA. N = 8.

calcein-positive mitochondria in nicorandil-treated and untreated cells. These findings are consistent with the notion that facilitated re-closure of mPTPs by mK$_{ATP}$ channel activation contributed to improved recovery of ΔΨ$_m$ from ROS-induced collapse. However, the possibility that improved preservation of respiratory chain complexes was involved in better recovery of ΔΨ$_m$, leading to restoration of mPTP status, cannot be excluded.

As a possible mechanism by which nicorandil promoted re-closure of the mPTP, we postulated that withdrawal of mPTP-opening stimuli after washout of AA was facilitated by nicorandil. Of the known mPTP opening factors [2], we focused on ROS and found that nicorandil and a GSK-3β inhibitor, LiCl, suppressed

ROS production during AA treatment and after washout of AA. The effects of these agents on ROS level are consistent with results of previous studies showing that mK$_{ATP}$ channel openers suppress burst production of ROS upon reperfusion in isolated perfused hearts [30,31] and that a mitochondria-targeting GSK-3β mutant increased ROS production in SH-SY5Y cells [32]. Interestingly, treatment with ROS scavengers only after removal of AA was sufficient to reproduce the effects of pretreatment with nicorandil or LiCl on ΔΨ$_m$ (Figure 8B). In contrast, suppression of ROS during AA treatment resulted in parallel up-ward shift of TMRE fluorescence throughout experiments (Figure 8A). These results indicate that suppression of ROS production by inactivation of

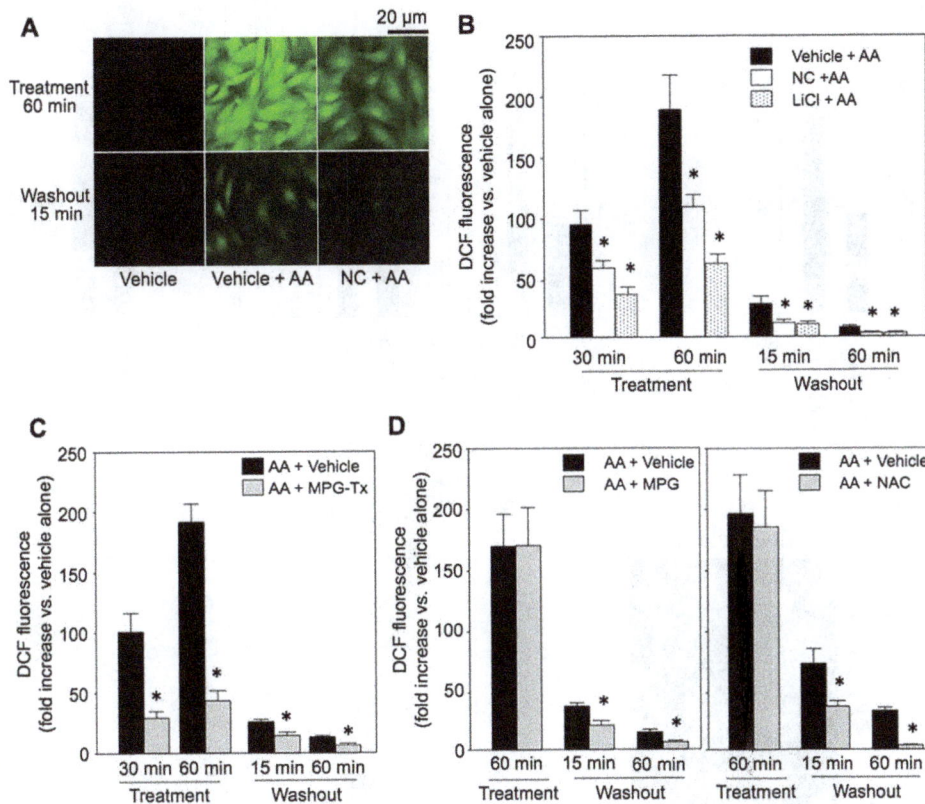

Figure 7. ROS generated by antimycin A. A: Representative DCF images. B: DCF fluorescence during and after antimycin A (AA) treatment in vehicle-, NC- and LiCl-pretreated cells. C: Effects of MPG administered during AA treatment on DCF. D: Effects of MPG or NAC treatment commenced at the time of AA washout on DCF. MPG-Tx = treatment with mercaptopropionyl glycine (MPG) during AA treatment, MPG = MPG treatment commenced at the time of AA washout, NAC = N-acetylcysteine treatment commenced at the time of AA washout. Treatment = time after onset of treatment with AA, Washout = time after washout of AA. *p<0.05 vs. Vehicle+AA or AA+Vehicle. N = 8.

GSK-3β mediates facilitation of $\Delta\Psi_m$ recovery, possibly via mPTP re-closure, by mK_{ATP} channel activation.

Nicorandil induced phosphorylation of GSK-3β at Ser9 in H9c2 cells (Figures 5 and 6) as did diazoxide in the rat myocardium *in vivo* [27], though its level declined during AA treatment. Interestingly, despite its transient effect on GSK-3β

during the early phase of AA treatment, nicorandil improved recovery of $\Delta\Psi_m$ similarly to inactivation of GSK-3β by LiCl throughout the AA treatment period (Figure 1). Hence, a signal mechanism downstrem of this kinase needs to be postulated for suppression of ROS production. However, relationships between mitochondrial GSK-3β and molecules regulating ROS production

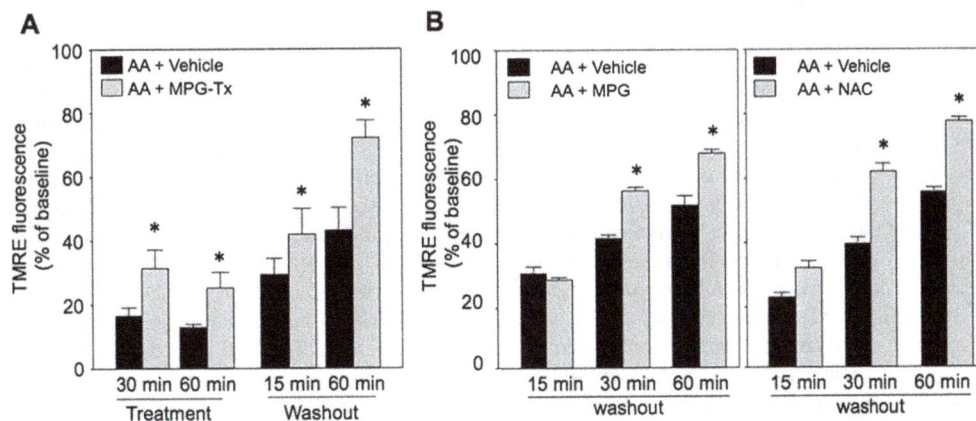

Figure 8. Effects of ROS scavengers on time course of $\Delta\Psi_m$. A and B: Effects of MPG administered during AA treatment (A) and effects of treatment with MPG or NAC commenced at the time of AA washout (B) on TMRE fluoresence. MPG-Tx = treatment with mercaptopropionyl glycine (MPG) during AA treatment, MPG = MPG treatment commenced at the time of AA washout, NAC = N-acetylcysteine treatment commenced at the time of AA washout. Treatment = time after onset of treatment with AA, Washout = time after washout of AA. *p<0.05 vs. AA+Vehicle. N = 8.

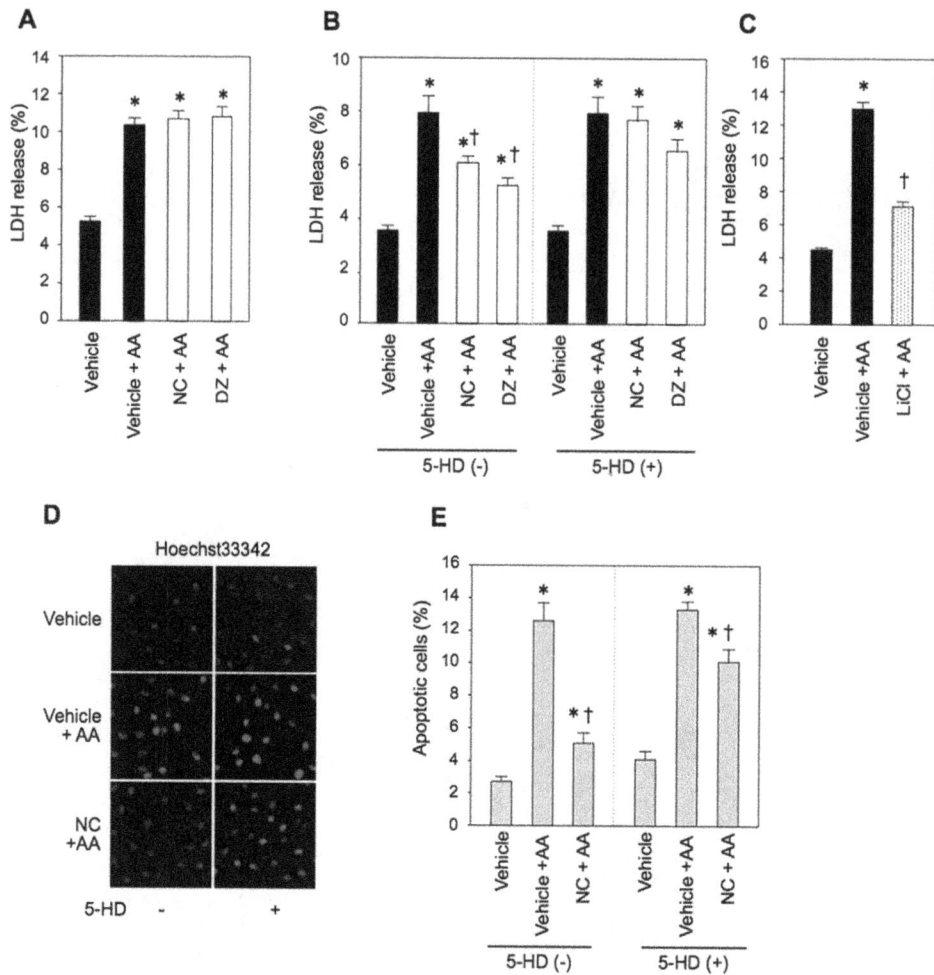

Figure 9. Necrosis and apoptosis after antimycin A treatment. A–C: Cell necrosis indicated by LDH release. LDH release at the end of AA treatment (A) and during a 2-h period after washout of AA (B, C) are shown. D and E: representative images of nuclear staining with Hoechst33342 (D) and apoptosis at 2 h after washout of AA (E). AA = antimycin A, NC = nicorandil, DZ = diazoxide, 5-HD = 5-hydroxydecanoate. *p<0.05 vs. Vehicle, †p<0.05 vs. Vehicle+AA. N=8~12.

or ROS elimination remain unclear. A possible explanation is involvement of GSK-3β in persistent ROS production triggered by inhibitionn of comlex III at the Qi site. Recent studies by Viola et al. [33,34] indicated that transient exposure of cardiomyocytes to hydrogen peroxide induced persistent ROS production by modification of the Qo site of complex III. In addition, ROS-induced ROS release, which potentially leads to chain reactions of mitochondrial ROS production, has been reported [14]. Hence, it is possible that inactivation of GSK-3β during the early phase of AA treatment has some impact on the level of persistent ROS production afterwards. Although inhibition of succinate dehydrogenase activity has been reported as a mechanism of ROS suppression by mK$_{ATP}$ channel openers in rat mitochondria [30], we could not detect a significant change in succinate dehydrogenase activity by nicorandil in the present H9c2 cell preparation.

The mechanism by which inactivation of GSK-3β suppresses ROS production remains unclear. Interestingly, we found that interaction of GSK-3β with complex III was induced by AA in association with ROS production and that mK$_{ATP}$ channel openers suppressed both GSK-3β-Rieske interaction and ROS production by AA. AA induces ROS production by interaction with the Qi site of the cytochrome bc$_1$ complex, and Rieske is

involved in ROS production in complex III [35–37]. Reiske is a major subunit of complex III, and its deletion does not prevent assembly of other submits but abolishes enzymatic activity [37]. Deletion of Rieske also leads to reduction in protein levels of complexes I and IV. ROS production in mitochondria are reportedly increased by Rieske knockout fibroblasts [37], but ROS production during hypoxia in pulmonary artery smooth muscle cells has been shown to be reduced by deletion of Rieske [36]. Modification of the expression levels of complexes I and IV may be involved in the apparently opposite effects of deletion of Rieske on ROS production.

GSK-3β translocates from the cytosol to mitochondria and interacts with adenine nucleotide translocase, a protein in the mitochondrial inner membrane, after ischemia/reperfusion [20,38]. A role of GSK-3β translocated to mitochondria in ROS production is supported by the finding that selective expression of unregulated GSK-3β in mitochondria significantly increased ROS production in SH-SY5Y cells [32]. However, whether GSK-3β-Rieske interaction is indeed causally related to ROS production at complex III remains to be tested in future projects.

There are limitations in the present study. First, cell necrosis induced by the present dose of AA was modest, and the impact of

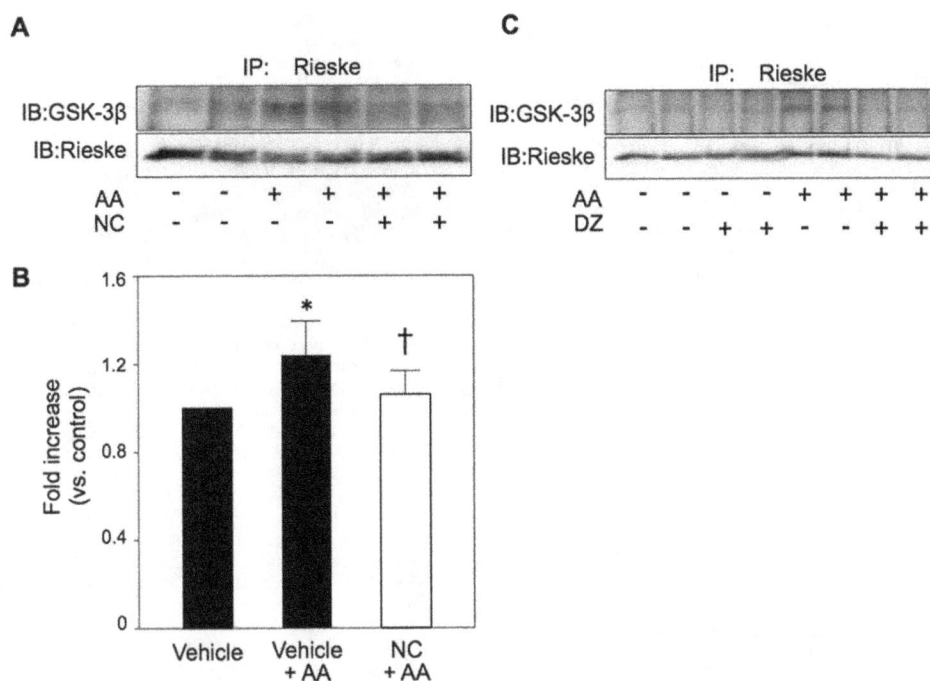

Figure 10. Interaction of Rieske protein and GSK-3β. Cell lysates after 5 min of AA treatment were immunoprecipitated with anti-Rieske antibodies and used for Western blotting for GSK-3β. A and B: Interaction of Rieske with GSK-3β was increased by antimycin A (AA), and the AA-induced changes were attenuated by nicorandil (NC) in H9c2 cells. N = 10~11 per group. *$p < 0.05$ vs. Vehicle, †$p < 0.05$ vs. Vehicle+AA. C: Similar results were obtained in HEK293 cells by use of diazoxide (DZ).

improvement of $\Delta\Psi_m$ recovery on cell survival has not been characterized fully. However, this is a technical limitation and larger doses of AA used in preliminary experiments increased massive cell necrosis during AA treatment, making it difficult to analyze recovery of $\Delta\Psi_m$ in cells that survived after washout of AA. Second, levels of mPTP opening after washout of AA (Figure 2D) determined by the present method are presumably underestimated since MitoTracker red, a marker of mitochondria, was lost from some mitochondria during AA treatment. However, a significant increase in the percentage of calcein-positive mitochondria after washout of AA was shown, and the results still support the notion that mPTPs re-close after withdrawal of oxidant stress. Third, we mainly used H9c2 cells, a rat cardiomyoblast cell line, and it is unclear whether the present results can be extrapolated to adult cardiomyocytes. However, the effects of mK$_{ATP}$ channel activation on $\Delta\Psi_m$ recovery and on Rieske-GSK-3β interaction were also observed in C2C12 cells and HEK293 cells, respectively, excluding the possibility that response of $\Delta\Psi_m$ to ROS and its modification by activation of the mK$_{ATP}$ channel is unique to H9c2 cells.

In conclusion, the results indicate that facilitated recovery of $\Delta\Psi_m$ from ROS-induced injury can be achieved by inactivating

GSK-3β directly or indirectly by activation of the mK$_{ATP}$ channel in isolated cardiomyocytes. The improvement of $\Delta\Psi_m$ recovery protects cardiomyocytes from necrosis by burst production of ROS. Significance of this mechanism in the myocardium *in vivo* remains to be investigated.

Supporting Information

Figure S1 Effects of antimycin A on protein levels of putative subunits of the mPTP.

Figure S2 Effects of S-nitroso-N-acetyl-DL-penicilla-mine on antimycin A-induced changes in $\Delta\Psi$m.

Author Contributions

Conceived and designed the experiments: T. Miura T. Miki. Performed the experiments: DS AK SI MO. Analyzed the data: MO TY. Wrote the paper: DS MT T. Miura.

References

1. Hausenloy DJ, Maddock HL, Baxter GF, Yellon DM (2002) Inhibiting mitochondrial permeability transition pore opening: a new paradigm for myocardial preconditioning? Cardiovasc Res 55: 534–543.
2. Miura T, Tanno M (2012) The mPTP and its regulatory proteins: final common targets of signalling pathways for protection against necrosis. Cardiovasc Res 94: 181–189.
3. Sims NR, Muyderman H (2010) Mitochondria, oxidative metabolism and cell death in stroke. Biochim Biophys Acta 1802: 80–91.
4. Zamzami N, Marchetti P, Castedo M, Decaudin D, Macho A, et al. (1995) Sequential reduction of mitochondrial transmembrane potential and generation of reactive oxygen species in early programmed cell death. J Exp Med 182: 367–377.
5. Kroemer G, Galluzzi L, Brenner C (2007) Mitochondrial membrane permeabilization in cell death. Physiol Rev 87: 99–163.
6. Di Lisa F, Menabò R, Canton M, Petronilli V (1998) The role of mitochondria in the salvage and the injury of the ischemic myocardium. Biochim Biophys Acta 1366: 69–78.
7. Garcia-Dorado D, Ruiz-Meana M, Inserte J, Rodriguez-Sinovas A, Piper HM (2012) Calcium-mediated cell death during myocardial reperfusion. Cardiovasc Res 94: 168–180.

8. Baines CP, Kaiser RA, Purcell NH, Blair NS, Osinska H, et al. (2005) Loss of cyclophilin D reveals a critical role for mitochondrial permeability transition in cell death. Nature 434: 658–662.

9. Javadov SA, Clarke S, Das M, Griffiths EJ, Lim KH, et al. (2003) Ischaemic preconditioning inhibits opening of mitochondrial permeability transition pores in the reperfused rat heart. J Physiol 549: 513–524.

10. Juhaszova M, Zorov DB, Kim SH, Pepe S, Fu Q, et al. (2004) Glycogen synthase kinase-3beta mediates convergence of protection signaling to inhibit the mitochondrial permeability transition pore. J Clin Invest. 113: 1535–1549.

11. Yano T, Miki T, Tanno M, Kuno A, Itoh T, et al. (2011) Hypertensive hypertrophied myocardium is vulnerable to infarction and refractory to erythropoietin-induced protection. Hypertension 57: 110–115.

12. Ichas F, Jouaville LS, Mazat JP (1997) Mitochondria are excitable organelles capable of generating and conveying electrical and calcium signals. Cell 89: 1145–1153.

13. Ichas F, Jouaville LS, Sidash SS, Mazat JP, Holmuhamedov EL (1994) Mitochondrial calcium spiking: a transduction mechanism based on calcium-induced permeability transition involved in cell calcium signalling. FEBS Lett 348: 211–215.

14. Zorov DB, Juhaszova M, Sollott SJ (2006) Mitochondrial ROS-induced ROS release: an update and review. Biochim Biophys Acta 1757: 509–517.

15. Griffiths EJ, Halestrap AP (1995) Mitochondrial non-specific pores remain closed during cardiac ischaemia, but open upon reperfusion. Biochem J 307: 93–98.

16. Kerr PM, Suleiman MS, Halestrap AP (1999) Reversal of permeability transition during recovery of hearts from ischemia and its enhancement by pyruvate. Am J Physiol 276: H496–H502.

17. Garlid KD, Costa AD, Quinlan CL, Pierre SV, Dos Santos P (2009) Cardioprotective signaling to mitochondria. J Mol Cell Cardiol 46: 858–866.

18. O'Rourke B (2004) Evidence for mitochondrial K$^+$ channels and their role in cardioprotection. Circ Res 94: 420–432.

19. Chanoit G, Zhou J, Lee S, McIntosh R, Shen X, et al. (2011) Inhibition of phosphodiesterases leads to prevention of the mitochondrial permeability transition pore opening and reperfusion injury in cardiac H9c2 cells. Cardiovasc Drugs Ther 25: 299–306.

20. Tanno M, Ishikawa S, Kuno A, Miki T, Kouzu H, et al. (2014) Translocation of GSK-3β, a trigger of permeability transition, is kinase activity-dependent and mediated by interaction with VDAC2. J Biol Chem Sep 3. pii: jbc.M114.563924. [Epub ahead of print].

21. Petronilli V, Miotto G, Canton M, Brini M, Colonna R, et al. (1999) Transient and long-lasting openings of the mitochondrial permeability transition pore can be monitored directly in intact cells by changes in mitochondrial calcein fluorescence. Biophys J 76: 725–734.

22. Tominaga H, Katoh H, Odagiri K, Takeuchi Y, Kawashima H, et al. (2008) Different effects of palmitoyl-L-carnitine and palmitoyl-CoA on mitochondrial function in rat ventricular myocytes. Am J Physiol Heart Circ Physiol. 295: H105–112.

23. Itoh T, Kouzu H, Miki T, Tanno M, Kuno A, et al. (2012) Cytoprotective regulation of the mitochondrial permeability transition pore is impaired in type 2 diabetic Goto-Kakizaki rat hearts. J Mol Cell Cardiol 53: 870–879.

24. Ohori K, Miura T, Tanno M, Miki T, Sato T, et al. (2008) Ser9 phosphorylation of mitochondrial GSK-3beta is a primary mechanism of cardiomyocyte protection by erythropoietin against oxidant-induced apoptosis. Am J Physiol Heart Circ Physiol. 295: H2079–H2086.

25. Sasaki N, Sato T, Ohler A, O'Rourke B, Marbán E (2000) Activation of mitochondrial ATP-dependent potassium channels by nitric oxide. Circulation. 101: 439–445.

26. Cuong DV, Kim N, Youm JB, Joo H, Warda M, et al. (2006) Nitric oxide-cGMP-protein kinase G signaling pathway induces anoxic preconditioning through activation of ATP-sensitive K+ channels in rat hearts. Am J Physiol Heart Circ Physiol. 290: H1808–H1817.

27. Terashima Y, Sato T, Yano T, Maas O, Itoh T, et al. (2010) Roles of phospho-GSK-3β in myocardial protection afforded by activation of the mitochondrial K_{ATP} channel. J Mol Cell Cardiol 49: 762–770.

28. Ozcan C, Bienengraeber M, Dzeja PP, Terzic A (2002) Potassium channel openers protect cardiac mitochondria by attenuating oxidant stress at reoxygenation. Am J Physiol Heart Circ Physiol 282: H531–H539.

29. Downey JM, Davis AM, Cohen MV (2007) Signaling pathways in ischemic preconditioning. Heart Fail Rev 12: 181–188.

30. Dost T, Cohen MV, Downey JM (2008) Redox signaling triggers protection during the reperfusion rather than the ischemic phase of preconditioning. Basic Res Cardiol 103: 378–84.

31. Pasdois P, Beauvoit B, Tariosse L, Vinassa B, Bonoron-Adèle S, et al. (2008) Effect of diazoxide on flavoprotein oxidation and reactive oxygen species generation during ischemia-reperfusion: a study on Langendorff-perfused rat hearts using optic fibers. Am J Physiol Heart Circ Physiol 294: H2088–2097.

32. King TD, Clodfelder-Miller B, Barksdale KA, Bijur GN (2008) Unregulated mitochondrial GSK3beta activity results in NADH: ubiquinone oxidoreductase deficiency. Neurotox Res 14: 367–382.

33. Viola HM, Arthur PG, Hool LC. (2007) Transient exposure to hydrogen peroxide causes an increase in mitochondria-derived superoxide as a result of sustained alteration in L-type Ca2+ channel function in the absence of apoptosis in ventricular myocytes. Circ Res 100: 1036–1044.

34. Viola HM, Hool LC (2010) Qo site of mitochondrial complex III is the source of increased superoxide after transient exposure to hydrogen peroxide. J Mol Cell Cardiol 49: 875–885.

35. Korde AS, Yadav VR, Zheng YM, Wang YX (2011) Primary role of mitochondrial Rieske iron-sulfur protein in hypoxic ROS production in pulmonary artery myocytes. Free Radic Biol Med 50: 945–952.

36. Waypa GB, Marks JD, Guzy RD, Mungai PT, Schriewer JM, et al. (2013) Superoxide generated at mitochondrial complex III triggers acute responses to hypoxia in the pulmonary circulation. Am J Respir Crit Care Med 187: 424–432.

37. Diaz F, Enríquez JA, Moraes CT (2012) Cells lacking Rieske iron-sulfur protein have a reactive oxygen species-associated decrease in respiratory complexes I and IV. Mol Cell Biol 32: 415–429.

38. Nishihara M, Miura T, Miki T, Tanno M, Yano T, et al. (2007) Modulation of the mitochondrial permeability transition pore complex in GSK-3beta-mediated myocardial protection. J Mol Cell Cardiol 43: 564–570.

8

Sperm Flagellum Volume Determines Freezability in Red Deer Spermatozoa

José Luis Ros-Santaella[1,2]*, Álvaro Efrén Domínguez-Rebolledo[3], José Julián Garde[1]

1 SaBio, Instituto de Investigación en Recursos Cinegéticos (IREC), CSIC-UCLM-JCCM, Campus Universitario, Albacete, Spain, **2** Department of Animal Science and Food Processing, Faculty of Tropical AgriSciences, Czech University of Life Sciences, Prague, Czech Republic, **3** Instituto Nacional de Investigaciones Forestales, Agrícolas y Pecuarias (INIFAP), Mocochá, Yucatán, México

Abstract

The factors affecting the inter-individual differences in sperm freezability is a major line of research in spermatology. Poor sperm freezability is mainly characterised by a low sperm velocity, which in turn is associated with low fertility rates in most animal species. Studies concerning the implications of sperm morphometry on freezability are quite limited, and most of them are based on sperm head size regardless of the structural parts of the flagellum, which provides sperm motility. Here, for the first time, we determined the volumes of the flagellum structures in fresh epididymal red deer spermatozoa using a stereological method under phase contrast microscopy. Sperm samples from thirty-three stags were frozen and classified as good freezers (GF) or bad freezers (BF) at two hours post-thawing using three sperm kinetic parameters which are strongly correlated with fertility in this species. Fourteen stags were clearly identified as GF, whereas nineteen were BF. No significant difference in sperm head size between the two groups was found. On the contrary, the GF exhibited a lower principal piece volume than the BF (6.13 µm³ vs 6.61 µm³, respectively, p = 0.006). The volume of the flagellum structures showed a strong negative relationship with post-thawing sperm velocity. For instance, the volume of the sperm principal piece was negatively correlated with sperm velocity at two hours post-thawing (r = −0.60; p<0.001). Our results clearly show that a higher volume of the sperm principal piece results in poor freezability, and highlights the key role of flagellum size in sperm cryopreservation success.

Editor: Chris D. Wood, Universidad Nacional Autónoma de México, Mexico

Funding: JLRS was supported by I3P program (Spanish National Research Council, CSIC, Spain), post-doctoral fellowship no. 99830/1181/1822 (Czech University of Life Sciences, Prague, Czech Republic), and by the project CIGA 20145001 (Czech University of Life Sciences, Prague, Czech Republic). The funders had no role in study design, data collection and analysis, decision to publish, or preparation of the manuscript.

Competing Interests: The authors have declared that no competing interests exist.

* Email: joseluis.ros@irec.csic.es

Introduction

It is well known that the cryopreservation process negatively affects the viability and fertility of reproductive cells. There are several factors caused by cryopreservation protocols that alter sperm integrity, such as a change in the temperature of the diluents [1]; osmotic and toxic stresses induced by cryoprotectants [1,2]; formation/reshaping of intracellular ice during freezing and thawing [3]; and dissolution of ice in the extracellular environment [2]. All of these factors induce sperm volumetric changes, plasma membrane alterations, flagellum morphological defects, as well as decrease mitochondrial membrane potential, sperm motility, viability, and fertility [2,4–10]. As a generalisation, some 40–50% of the sperm population does not survive cryopreservation even with optimised protocols [2]. Thus, many of the frozen-thawed spermatozoa show a shorter life span and have difficulties in reaching the oocytes and penetrating their vestments after conventional artificial insemination [10]. Variations between individuals in sperm freezability have been reported in numerous animal species. Within this context, semen donors have routinely been categorized as "good freezers" (GF) or "bad freezers" (BF). When experiments involve comparing bulk samples before

freezing and after thawing, it is difficult to know which parts of the cryopreservation procedure may be causing problems and to detect differences between individual cells [11]. In boar sperm, it has been demonstrated that these consistent inter-individual variations in sperm freezability are genetically determined [12].

The understanding and prediction of the sperm functional response to cryopreservation is one of the major questions in sperm cryobiology [13]. For this reason, sperm morphometry classification has become an integral part of routine sperm analysis. In recent years, ASMA (Automated Sperm Morphometry Analysis) systems have been employed for determining morphometric parameters of the sperm head and midpiece to elucidate possible relations with sperm freezability [3,14]. These systems can only analyse the lengths and areas of the sperm head and midpiece, but they are unable, for example, to measure the sperm principal piece, which provides sperm motility. Thus, in order to study sperm midpiece morphometry with these systems, it is not easy to find a staining method which allows discerning this structure from the rest of the flagellum [15]. By contrast, with other software for analysis and image processing, it is possible to obtain the lengths of flagellum structures. Numerous studies argue the implications of sperm flagellum in several biological processes such as sperm velocity, male reproductive success, relationships with testicle size, and spermatogenesis [16–21]. On the other hand, studies concerning the role played by flagellum size on sperm freezability have not yet been reported, although there is evidence of the fragility of its internal and external structures when they are exposed to the cryopreservation process [5,7,22–24].

Curry et al. [25] reported that the sperm area and volume have great importance for determining optimal sperm cryopreservation protocols. For this purpose, stereology is able to accurately estimate surface areas and volumes for complex shapes from two-dimensional images [25]. Thereby, sperm volumes as well as the sperm head and flagellum structures can be examined. The sperm volumes of human and several mammalian species have been determined by stereological methods [19,25].

In the present study, we hypothesized that flagellum volume could predict freezability in red deer spermatozoa. For this reason, we focused on determining the volumes of the sperm flagellum, midpiece, and principal piece in fresh sperm using stereology under phase contrast microscopy. In this way, differences between males in sperm flagellum volume could be linked to sperm freezability, owing to the fragility of flagellum structures to the cryopreservation process [5,7,23], and the differences observed among red deer sperm to support cryopreservation [14]. In order to determine sperm freezability, sperm samples were classified as good or bad freezers using three kinetic parameters (VAP: average path velocity, VCL: curvilinear velocity, and VSL: straight linear velocity) assessed by a CASA (Computer Assisted Sperm Analysis) system at 2 hours post-thawing and with incubation at 37°C. Sperm viability, acrosomal status, and mitochondrial activity were also determined by flow cytometry.

Materials and Methods

Animals

The study was approved by the "Comité de Ética en Investigación de la Universidad de Castilla-La Mancha". All animal handling was done following the Spanish Animal Protection Regulation RD 53/2013, which conforms to the European Union Regulation 2003/65. Stags were legally culled and hunted in their natural habitat in accordance with the harvest plan of the game reserve. The harvest plans were made following the Spanish Harvest Regulation, Law 2/93 of Castilla-La Mancha, which conforms to European Union regulations. Thirty-three stags (age >4.5 years; body mass >130 kg) of red deer (*Cervus elaphus*) were used in this study. Landowners and managers of the red deer populations gave permission to the authors to use the samples.

Stags and testes collection

Testes were recovered from adult red deer culled during the 2008 hunting season in the south of Spain. Testes, together with the scrotum, were removed and transported at approximately 20–22°C to the laboratory. The time that elapsed between animal death and sperm recovery ranged from 3 to 6 hours, an adequate and reliable time interval for evaluating sperm parameters, because a decrease in the quality of sperm traits begins to take place 12 hours after the death of a male [26].

Chemicals and solutions

Unless otherwise stated, chemicals were obtained from Sigma-Aldrich (Madrid, Spain).

A Salamon's modified extender was prepared in two fractions, as previously described [27]. Fraction A contained: Tris (2.70%, w/v), fructose (1%, w/v), citric acid (1.4%, w/v), and clarified egg yolk (20%, v/v) (pH 6.8, osmolality 300 mOsm/Kg). Fraction B differed from the Fraction A in that water was replaced (12%, v/v) with the same volume of glycerol, with a final concentration = 6% (v/v). Glutaraldehyde solution was composed of 2% glutaraldehyde (v/v), and 0.165 mol/L cacodylate/HCl buffer (pH 7.3), as previously described [27]. Bovine gamete medium (BGM-3) was composed of 87 mmol/L NaCl, 3.1 mmol/L KCl, 2 mmol/L CaCl$_2$, 0.4 mmol/L MgCl$_2$, 0.3 mmol/L NaH$_2$PO$_4$, 40 mmol/L HEPES, 21.6 mmol/L sodium lactate, 1 mmol/L sodium pyruvate, 50 µg/mL kanamicine, 10 µg/mL phenol red, and 6 mg/mL BSA (Bovine Serum Albumine) (pH 7.5), as previously described [28].

Cryopreservation of epididymal spermatozoa

Post-mortem seminal recovery is the most practical option to obtain sperm samples from wild populations of red deer, with hunting representing a constant source from harvested animals [29]. Spermatozoa were collected from the cauda of the epididymis by repeated longitudinal and transverse cuts with a surgical scalpel and placed in 0.5 mL of fraction A. The contents from both epididymides of each individual were pooled for processing. Afterward, a routine sperm evaluation was made. Sperm concentration was determined using a Bürker counter chamber. The percentage of motile sperm and the quality of motility (QM) were subjectively evaluated, the latter using a scale from 0 (lowest: immobile) to 5 (highest: progressive and vigorous movement). Then, the sperm motility index (SMI) was calculated according to the formula [30]:

$$SMI = \frac{\% \, motile \, sperm + (quality \, of \, motility \times 20)}{2}$$

Only samples with at least 60% motile sperm were used for this study (Table S1). The sperm cryopreservation protocol was performed as previously described [27]. Briefly, the sperm mass was diluted at room temperature to 400×10^6 spermatozoa/mL with fraction A, and then to 200×10^6 spermatozoa/mL with fraction B. The diluted samples were refrigerated for approximately 10 min to reach 5°C and then equilibrated at the same temperature for 2 h. After equilibration, the suspended sperm was loaded into 0.25 mL plastic straws and frozen for 10 min in nitrogen vapours, 4 cm above the level of liquid nitrogen (−120°C). The straws were then immediately immersed into liquid nitrogen (−196°C) for storage.

Morphometry assessment of fresh sperm

Sperm samples were directly recovered from the cauda of both epididymides. Spermatozoa were fixed in glutaraldehyde solution. A sub-sample of 2 µl was used to prepare the smears. Semen smears were air-dried for one day, then immersed in the glutaraldehyde fixative solution for 5 min and immediately mounted, sealing the edges with dibutyl phthalate xylene (DPX). This method avoids floating cells on the slide (Figure 1A), which greatly helps sperm morphometry analysis. Sperm samples were photographed using a high-resolution camera DXM1200 (Nikon, Tokyo, Japan) under a phase-contrast microscopy using an Eclipse E600 microscope (Nikon, Tokyo, Japan), and a 40X objective (Nikon, Tokyo, Japan). The resolution of the pictures was 3840×3072 pixels (TIFF format). A scale of 10 µm (181 pixels) was used for the measurements. The pixel size was 0.055 µm in the horizontal and vertical axes. Sperm lengths were assessed using ImageJ software (National Institutes of Health, USA). The main structures of red deer spermatozoon are shown in Figure 1B. The following sperm morphometry parameters were determined: head width, head length, proximal midpiece width, distal midpiece width, midpiece length, flagellum length, and terminal piece length (Table S1). From these measurements, we calculated other morphometric parameters such as total sperm and principal piece lengths. The head area was calculated using the formula for the area of an ellipse [31,32]:

$$Area = \pi \times L \times W$$

Head perimeter was calculated using the Ramanujan's formula for calculating the perimeter of an ellipse [33]:

$$Perimeter \approx \pi \left[3 \times (L + W) - \sqrt{(3L + W) \times (L + 3W)} \right]$$

In both formulae, L and W are the semi-major and semi-minor axis of the sperm head, respectively. Twenty-five representative sperm were measured for each male as described by Malo et al. [16].

Stereology of the flagellum. Sperm flagellum volume and its structures were determined using a stereology method based on Anderson et al. [19] with some modifications. Anderson et al. [19] calculated the volumes of the sperm midpiece and flagellum using the formula for the volume of a cylinder. In this study, owing to the significant differences between the proximal and distal midpiece widths (0.94 μm and 0.74 μm, respectively; p<0.0001), we estimated sperm midpiece volume using the formula for the volume of a truncate cone:

$$Volume = \left(\frac{\pi \times L}{3} \right) \times \left(R^2 + r^2 + R \times r \right)$$

where L is the length of the midpiece, R is the half proximal

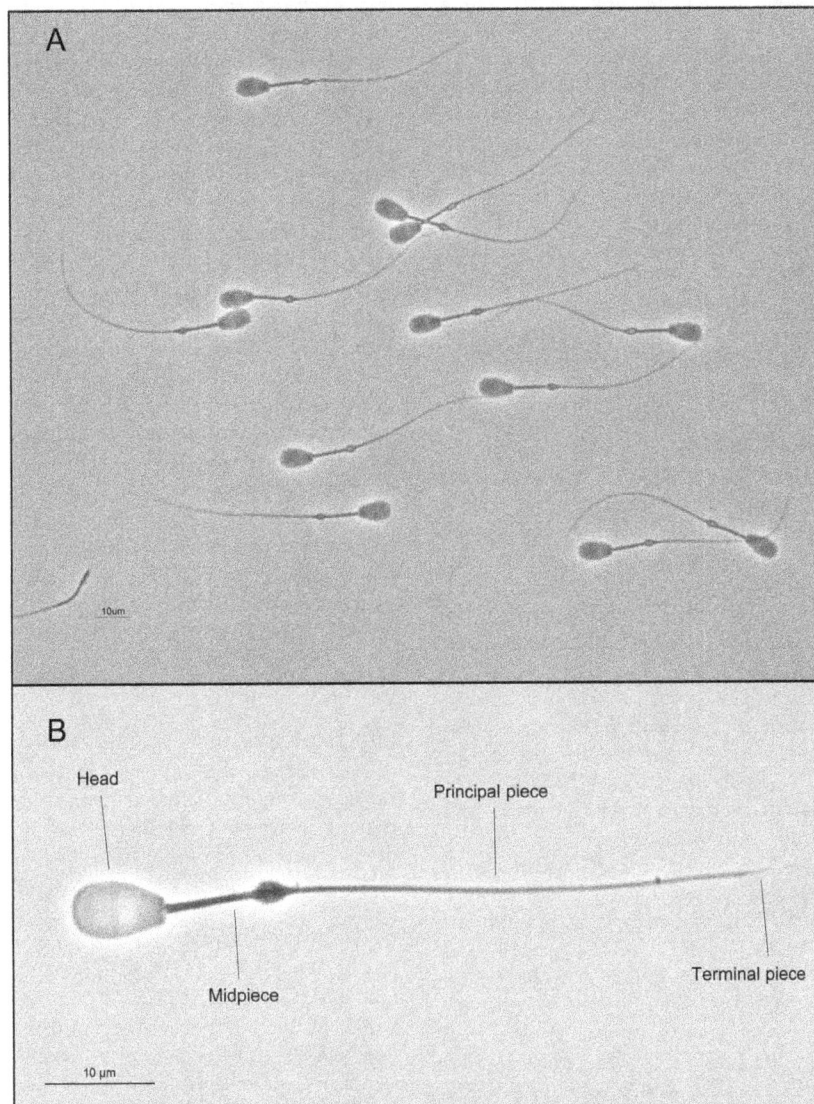

Figure 1. Red deer sperm. (A) It is remarkable how the method described in this study avoids floating cells. The picture was taken under phase contrast microscopy (40X objective). Scale bar, 10 μm. (B) Main structures of red deer spermatozoon. The picture was taken under phase contrast microscopy (100X objective) using the method described in the present study. Scale bar, 10 μm.

midpiece width, and r is the half distal midpiece width. On the other hand, the total flagellum and principal piece volumes were determined using the formula for the volume of a cone:

$$Volume = \frac{\pi \times R^2 \times L}{3}$$

where R is the half midpiece width (i.e., proximal or distal) and L is the length of the flagellum or the principal plus terminal piece.

Sperm thawing and sperm quality assessment

The sperm straws were thawed in a water bath with a saline solution at 37°C for 30 s, and each sample was poured in a tube. The samples were incubated at 37°C and analysed for motility (subjectively and with the CASA system), viability, acrosomal status, and mitochondrial status at 0 hours (after 10 min of incubation) and after 2 hours post-thawing (Table S2 and Table S3, respectively).

CASA analysis. Sperm were diluted down to 25–30×10^6 spermatozoa/mL with fraction A solution and loaded into a Makler counter chamber (Sefi-Medical instruments, Haifa, Israel) at 37°C. The CASA system consisted of a triocular optical phase contrast microscope Eclipse 80i (Nikon, Tokyo, Japan), equipped with a warming stage at 37°C and a Basler A302fs digital camera (Basler Vision Technologies, Ahrensburg, Germany). Images were captured and analysed using the Sperm Class Analyzer software (Microptic S.L., Barcelona, Spain). The analysis was carried out using a 10X negative phase-contrast objective (Nikon, Tokyo, Japan). A total of 4 descriptors of sperm motility were recorded analysing a minimum of 250 sperm per sample: average path velocity (VAP, μm/s), curvilinear velocity (VCL, μm/s), straight linear velocity (VSL, μm/s), and amplitude of lateral head displacement (ALH, μm). The standard parameters settings were as follows: 25 frames/s; 20 to 60 μm² for the head area.

Fluorescence probes for sperm viability, acrosomal status, and mitochondrial activity. Several physiological traits were assessed using fluorescent probes and flow cytometry, as previously described [28]. Briefly, the samples were diluted down to 10^6 spermatozoa/mL in BGM-3 solution and stained using four fluorophores. Sperm viability was assessed with 0.1 μmol/L YO-PRO-1 (Invitrogen, Barcelona, Spain) and 10 μmol/L PI (propidium iodide). Mitochondrial activity and acrosomal status were assessed with 0.1 μmol/L Mitotracker Deep Red (Invitrogen, Barcelona, Spain) and 4 μg/mL PNA-TRITC (peanut agglutinin), respectively. The spermatozoa stained in these two solutions were incubated 20 min in the dark before being run through a flow cytometer. The sperm populations shown in this work were: YO-PRO-1-/PI- (viable spermatozoa), MT+ (Mitotracker Deep Red, spermatozoa with active mitochondria), and PNA- (spermatozoa with intact acrosome).

Flow cytometry analysis. Samples were analysed as previously described [28]. Briefly, a Cytomics FC500 flow cytometer (Beckman Coulter, Brea, CA, USA) was utilized, with a 488-nm Ar-Ion laser (excitation for YOPRO-1, PNA-TRITC, and PI), and a 633-nm He-Ne laser (excitation for Mitotracker Deep Red). The FSC (forward-scattered light) and SSC (side-scattered light) signals were used to gate out debris (non-sperm events). Fluorescence from YO-PRO-1 was read using a 525/25BP filter, PNA-TRITC was read using a 575/20BP filter, PI was read using a 615DSP filter, and Mitotracker Deep Red was read using a 675/40BP filter. Fluorescence captures were controlled using the RXP software provided with the cytometer. All of the parameters were read using logarithmic amplification. For each sample, 5000 spermatozoa were recorded, saving the data in flow cytometry

standard (FCS) v. 2 files. The analysis of the flow cytometry data was carried out using WEASEL v. 2.6 (WEHI, Melbourne, Victoria, Australia).

Statistical analysis

All statistical analyses were performed using the SPSS 20.0 statistical software package (SPSS Inc, Chicago, IL, USA). Sperm samples from thirty-three (N = 33) red deer were used for statistical analysis. The Kolmogorov-Smirnov test was used to check the normal distribution of the data. A repeated measures one-way ANOVA test was used to compare subjective motility between fresh and thawed sperm (0 and 2 hours) using the Mauchly's sphericity test (Greenhouse-Geisser correction) to verify homogeneity of variance. The quality of motility (QM) was not normally distributed in some groups, therefore, non-parametric Friedman test (repeated measures) was used. On the other hand, to check differences in sperm parameters (kinetics-CASA and flow cytometry) between 0 and 2 hours post-thaw we used a paired-samples t-student test (repeated measures). In order to determine sperm freezability we performed a hierarchical clustering analysis using the Euclidean distance measure after determining automatically the number of conglomerates by cluster analysis in two phases. For this purpose, we used three sperm kinetic parameters (VAP, VCL, and VSL) because they are highly related with fertility in red deer [34] and are good indicators of post-thawing sperm quality [35]. Sperm freezability was determined at 2 hours post-thawing to test spermatozoa thermo-resistance. The independent-samples student-t test (Levene's test to verify homogeneity of variance) and the Mann-Whitney U test were used to check for differences in sperm parameters between the GF and the BF.

On the other hand, since the variables of sperm velocity (VAP, VCL, VSL, ALH, and SMI) were highly correlated among themselves, we reduced the number of predictor variables using principal component analysis (PCA) to obtain an overall sperm velocity at 0 and 2 hours post-thawing, respectively. Bartlett sphericity and Keiser-Meyer-Olkin tests were assessed as measures of sampling adequacy [36]. Pearson's correlation test and the linear regression model were used to assess the relationship between sperm velocity and morphometric parameters. We used the RMA software to reduced major axis regression [37]. The within male coefficient of variation (CV) in sperm morphometry was calculated to show the intra-male variability in sperm design.

Results

Effects of sperm cryopreservation

After the freezing-thawing process, a remarkable decrease in subjective motility parameters was observed between fresh and thawed sperm, but also between 0 and 2 hours post-thawing: motile sperm, $F(1.49, 47.55) = 94.27$, $p < 0.001$; QM, $\chi^2(2, N = 33) = 37.83$, $p < 0.001$; and SMI, $F(1.51, 48.22) = 116.02$, $p < 0.001$ (Table 1). In the same way, there were significant differences for all of the CASA kinetic parameters between 0 and 2 hours post-thawing: VAP, $t(32) = 9.77$, $p < 0.001$; VCL, $t(32) = 10.46$, $p < 0.001$; VSL, $t(32) = 10.29$, $p < 0.001$; and ALH, $t(32) = 10.18$, $p < 0.001$ (Table 2). Furthermore, sperm viability and organelle functionality showed a decrease after thawing, displaying highly significant differences between 0 and 2 hours post-thawing: YOPRO-1-/PI-, $t(32) = 11.47$, $p < 0.001$; MT+, $t(32) = 6.26$, $p < 0.001$; and PNA-, $t(32) = 11.18$, $p < 0.001$ (Table 2). For instance, the mean percentage of sperm with active mitochondria ranged from 51.13% to 28.63% during incubation (at 0 and 2 hours, respectively).

Table 1. Subjective motility of fresh and thawed red deer spermatozoa (N = 33).

	Sperm motility		
	Motile sperm (%)	QM (0–5)	SMI (%)
Fresh sperm	81.82±11.44[a]	2.24±0.48[a]	63.33±9.11[a]
0 hours post-thaw	55.91±16.32[b]	1.86±0.28[a]	46.51±9.96[b]
2 hours post-thaw	42.57±13.52[c]	1.43±0.21[b]	35.61±7.71[c]

Different superscript letters within the same column differ significantly (p<0.001). Data are shown as mean ± SD (standard deviation). QM, quality of motility; SMI, sperm motility index.

Deer sperm freezability (GF and BF)

The cluster dendrogram analysis of sperm freezability is shown in Figure 2. Fourteen stags were clearly identified as GF, whereas nineteen were BF. Not only did the GF and the BF show clear differences in the three parameters (VAP, VCL, and VSL) used for the cluster analysis, but they also showed differences in the other sperm functionality parameters. Indeed, at 0 hours post-thawing we found significant differences in the following sperm parameters: motile sperm, $t(31) = -2.23$, $p = 0.033$; VAP, $t(31) = -3.58$, $p = 0.001$; VCL, $t(31) = -4.18$, $p<0.001$; VSL, $t(31) = -2.66$, $p = 0.012$; ALH, $t(31) = -4.30$, $p<0.001$; YOPRO-1-/PI-, $t(27.97) = -2.82$, $p = 0.009$; and MT+, $t(26.75) = -2.30$, $p = 0.029$ (Table 3). The differences between the two groups of males in sperm quality were more evident across sperm incubation (Table 3). Thus, sperm kinetics, viability, and organelle functionality were significantly different between the GF and the BF at 2 hours post-thawing as follows: motile sperm, $t(31) = -3.87$, $p<0.001$; QM, $U = 66.00$, $p = 0.003$; SMI, $t(31) = -4.62$, $p<0.001$; VAP, $t(31) = -9.66$, $p<0.001$; VCL, $t(19.24) = -8.28$, $p<0.001$; VSL, $t(31) = -8.08$, $p<0.001$; ALH, $t(31) = -7.60$, $p<0.001$; YOPRO-1-/PI-, $t(28) = -2.67$, $p = 0.013$; MT+, $t(31) = -3.43$, $p = 0.002$; and PNA-, $t(31) = -2.69$, $p = 0.011$ (Table 3). On the other hand, the quality of motility, SMI, and acrosomal status were not significantly different between the GF and the BF at 0 hours post-thawing (Table 3).

Sperm quality before freezing

There were no significant differences between the GF and the BF before freezing in any of the subjective kinetics parameters

evaluated. Indeed, the GF and the BF showed similar subjective motility and, therefore, were expected to have the same freezability. Motility parameters in fresh spermatozoa were: motile sperm ($83.21±13.24$ vs $80.79±10.17$, $t(31) = -0.60$, $p = 0.556$), QM ($2.25±0.51$ vs $2.24±0.48$, $U = 127.50$, $p = 0.825$), and SMI ($64.11±10.63$ vs $62.76±8.07$, $t(31) = -0.41$, $p = 0.682$). Data are shown as the mean ± SD for the GF and the BF, respectively.

Sperm freezability and morphometry of fresh sperm

The descriptive statistics for the morphometric parameters are shown in Table 4. There were no differences between the GF and the BF in any of the sperm head measurements (Table 4). On the other hand, highly significant differences were observed between groups in regard to sperm flagellum morphometry (Table 4). Thus, the GF exhibited a lower mean principal piece volume than the BF ($6.13±0.42$ μm^3 vs. $6.61±0.49$ μm^3, $t(31) = 2.93$, $p = 0.006$). Also, the GF exhibited a smaller mean distal midpiece width than the BF ($0.73±0.02$ μm vs. $0.75±0.03$ μm, $t(31) = 2.37$, $p = 0.024$). The data are shown as the mean ± SD for the GF and the BF, respectively. Moreover, the flagellum and midpiece volumes together with some sperm lengths (e.g., total sperm and flagellum) showed lower values in the GF than the BF, although the differences were not statistically significant (Table 4). Sperm freezability according to the sperm principal piece volume is shown in Figure 3.

Table 2. Kinetics, viability, and organelle status of red deer spermatozoa (N = 33) at 0 and 2 hours post-thaw.

Assessed parameters	0 hours post-thaw		2 hours post-thaw	
	Mean ± SD	Range	Mean ± SD	Range
Sperm kinetics				
VAP (μm/s)	60.60±13.48[a]	26.08-87.21	40.09±10.41[b]	23.18–57.16
VCL (μm/s)	97.58±21.52[a]	55.49–144.36	66.60±14.14[b]	43.92–95.45
VSL (μm/s)	34.17±7.17[a]	15.56–49.03	22.17±5.00[b]	14.11–29.44
ALH (μm)	3.78±0.80[a]	2.49–5.47	2.69±0.43[b]	2.05–3.55
Sperm viability and organelle status				
Viability (YOPRO-1-/PI-, %)	35.81±8.91[a]	18.46–51.16	29.70±8.28[b]	13.20–41.32
Active mitochondria (MT+, %)	51.13±13.93[a]	26.42–78.51	28.63±19.00[b]	0.88–56.08
Intact acrosome (PNA-, %)	83.96±6.82[a]	67.40–93.96	69.67±11.16[b]	50.74–90.72

Different superscript letters within the same row differ significantly (p<0.001) between 0 and 2 hours post-thaw. SD, standard deviation; VAP, average path velocity; VCL, curvilinear velocity; VSL, straight linear velocity; ALH, amplitude of lateral head displacement.

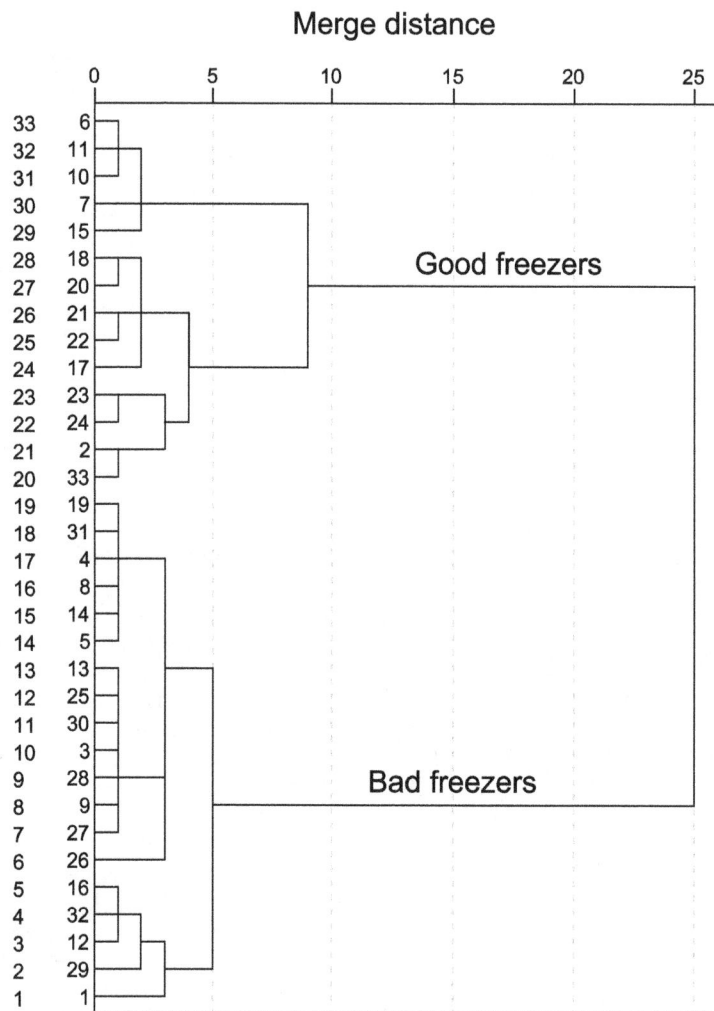

Figure 2. Cluster dendrogram analysis showing red deer sperm freezability. Fourteen males were identified as good freezers, whereas 19 as bad freezers.

Sperm velocity post-thaw and its relationship with morphometry of fresh sperm

Principal component analysis rendered only one component both at 0 and 2 hours post-thawing. The components explained 91.68% and 83.98% of the variance in sperm velocity at 0 and 2 hours post-thawing, respectively. The PCA are shown in Table 5.

None of the morphometric parameters of the sperm head were related with sperm velocity. In contrast, sperm velocity showed strong and negative relationships with sperm flagellum volumes, particularly with the sperm principal piece volume. Indeed, at 0 hours post-thawing, the principal piece volume was the only parameter showing a significant relationship with sperm velocity ($r = -0.36$; $p = 0.038$) (Figure 4A). In addition, such a relationship reached the highest values at 2 hours post-thawing ($r = -0.60$; $p < 0.001$) (Figure 4B). On the other hand, sperm velocity was also negatively correlated with distal midpiece width ($r = -0.55$; $p = 0.001$) (Figure 4C) and with midpiece and flagellum volumes ($r = -0.44$; $p = 0.011$ and $r = -0.36$; $p = 0.038$) at 2 hours post-thawing (Figures 4D–E, respectively).

Discussion

In this study, for the first time, the volumes of the flagellum structures in fresh sperm have been determined to predict sperm freezability in red deer. Our results clearly show that sperm with a higher principal piece volume freeze worse, that is, the GF have a lower principal piece volume than the BF. We also found that sperm velocity is strongly and negatively related with the volumes of the flagellum structures. Sperm velocity is crucial in the process of fertilization in a large number of taxa (including fish [38], birds [39], and mammals [34,40]). Moreover, VAP, VCL, and VSL have been proven to be good indicators of sperm freezability in red deer as previously described in canine sperm [35], and in turn, these parameters are closely related with fertility in red deer using thawed sperm [34].

The evaluation of sperm function throughout post-thawing and sperm incubation provides additional information about the quality of the spermatozoa [41,42] and is more closely related to sperm fertility than those sperm assessed immediately after thawing [43,44]. According to this assumption, we found more and stronger relationships between sperm velocity and the volumes of the flagellum structures at 2 hours than at 0 hours post-thawing. For example, we did not find any differences between the GF and

Table 3. Sperm kinetics, viability and organelle status in good freezers (GF, n = 14) and bad freezers (BF, n = 19) at 0 and 2 hours post-thaw.

Assessed parameters	0 hours post-thaw		2 hours post-thaw	
	GF	BF	GF	BF
Sperm kinetics				
VAP (µm/s)	68.96±12.40[a**]	54.44±10.83[b***]	50.43±5.38[a***]	32.47±5.21[b***]
VCL (µm/s)	112.40±20.13[a***]	86.65±15.30[b***]	80.37±9.70[a***]	56.45±5.56[b***]
VSL (µm/s)	37.71±7.02[a*]	31.56±6.24[b*]	26.90±2.27[a***]	18.70±3.25[b***]
ALH (µm)	4.34±0.78[a***]	3.36±0.53[b***]	3.08±0.32[a***]	2.40±0.19[b***]
Motile sperm (%)	62.86±15.53[a]	50.79±15.30[b*]	51.43±9.89[a***]	36.05±12.20[b***]
QM (0–5)	1.89±0.19[a]	1.83±0.33[a]	1.55±0.11[a**]	1.34±0.22[b**]
SMI (%)	50.36±9.14[a]	43.68±9.80[a]	41.25±5.07[a***]	31.45±6.63[b***]
Sperm viability and organelle status				
Viability (YOPRO-1-/PI-, %)	40.10±5.00[a**]	32.65±9.91[b**]	33.50±4.70[a]	26.91±9.29[b*]
Active mitochondria (MT+, %)	56.72±7.44[a*]	47.01±16.20[b*]	40.06±18.20[a**]	20.22±15.05[b***]
Intact acrosome (PNA-, %)	86.26±6.30[a]	82.26±6.84[a]	75.24±9.87[a*]	65.57±10.45[b*]

Different superscript letters within the same row differ significantly between GF and BF at 0 and 2 hours post-thaw, respectively (*p<0.05; **p<0.01; ***p<0.001). Data are shown as mean ± SD (standard deviation). GF, good freezers; BF, bad freezers; VAP, average path velocity; VCL, curvilinear velocity; VSL, straight linear velocity; ALH, amplitude of lateral head displacement; QM, quality of motility; SMI, sperm motility index.

the BF in acrosomal status at 0 hours post-thawing, but we did at 2 hours post-thawing. This is probably because if the sperm membrane or other structures (e.g. axoneme) were disrupted, this damage was not manifested immediately upon thawing, but occurred during post-thaw re-warming within specific temperatures [23]. Thus, the damage in this structure would be higher due to thermal stress during sperm incubation, inasmuch as sperm thawing is more deleterious than sperm freezing [23,45] and can result in more morphological damage [46].

In the present work, we did not find any significant differences between the GF and the BF in sperm head size, and also none of the morphometric parameters of this structure showed any relationship with sperm velocity. By contrast, Esteso et al. [14] found that sperm head size is related with sperm freezability in red deer (i.e., increased head size entails a poor sperm freezability). However, Esteso et al. [14] classified sperm donors as a GF or BF using the sperm motility index, instead of a CASA system, together with acrosomal status and membrane stability. Within this context, we found that the intact acrosome is negatively related with sperm head size at 0 hours post-thawing (Figure S1). On the other hand, the sperm principal piece volume and distal midpiece width showed significant differences between the GF and the BF, and

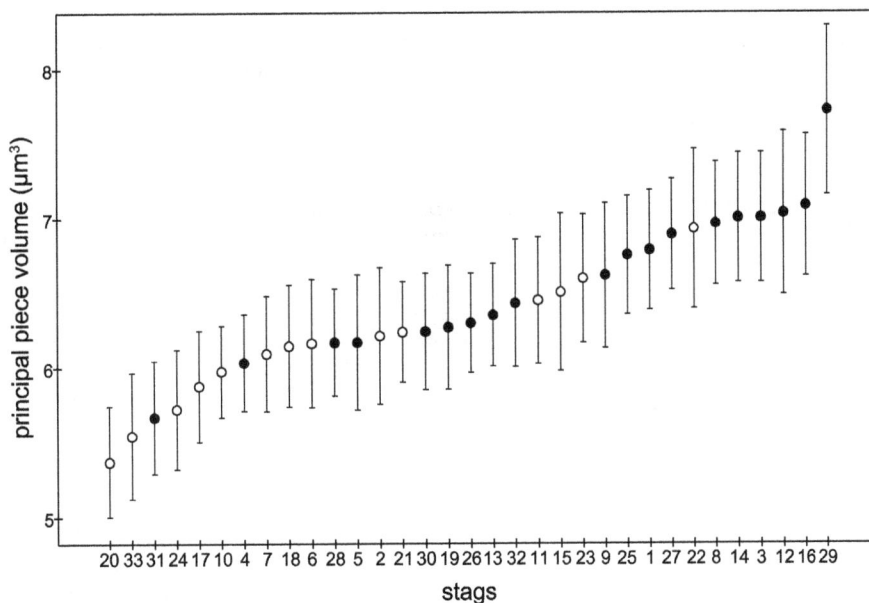

Figure 3. Classification of sperm principal piece volume according to red deer sperm freezability. Good freezers are shown in open circles and bad freezers in closed circles.

Table 4. Overall morphometry of fresh sperm (data derived from 825 spermatozoa from 33 red deer) and differences between good (GF) and bad freezers (BF) in sperm morphometry parameters.

Assessed parameters	Overall (N = 33)			GF (n = 14)	BF (n = 19)	P
	Mean ± SD	Range	CV	Mean ± SD	Mean ± SD	
Sperm head						
Width (μm)	5.17±0.12	4.82–5.35	2.35	5.17±0.11	5.16±0.13	0.766
Length (μm)	8.75±0.24	8.35–9.30	2.79	8.77±0.25	8.74±0.25	0.730
Area (μm²)	35.51±1.17	33.09–37.94	3.30	35.64±1.30	35.41±1.10	0.597
Perimeter (μm)	22.23±0.43	21.36–23.23	1.92	22.27±0.46	22.20±0.41	0.653
Flagellum and sperm length						
Midpiece width, proximal (μm)	0.94±0.03	0.85–1.03	3.56	0.93±0.05	0.94±0.02	0.382
Midpiece width, distal (μm)	0.74±0.03	0.68–0.80	3.63	0.73±0.02	0.75±0.03	**0.024**
Midpiece length (μm)	12.06±0.26	11.50–12.60	2.12	12.02±0.18	12.08±0.30	0.506
Principal piece length (μm)	41.47±1.26	38.92–44.39	3.04	41.07±1.42	41.76±1.07	0.122
Terminal piece length (μm)	2.68±0.26	2.05–3.21	9.89	2.61±0.28	2.72±0.25	0.248
Flagellum length (μm)	56.20±1.29	53.56–59.05	2.29	55.71±1.41	56.57±1.09	0.057
Midpiece volume (μm³)	6.73±0.48	5.68–7.70	7.09	6.57±0.56	6.85±0.38	0.097
Principal piece volume (μm³)	6.41±0.51	5.37–7.72	7.98	6.13±0.42	6.61±0.49	**0.006**
Flagellum volume (μm³)	12.95±1	10.36–15.22	7.70	12.67±1.24	13.16±0.73	0.161
Sperm length (μm)	64.96±1.29	62.35–67.52	1.99	64.48±1.34	65.31±1.17	0.068

Bold letters show significant differences between GF and BF. SD, standard deviation; CV, coefficient of variation.

they are also negatively correlated with sperm velocity. According to this fact, Peña et al. [3] found that midpiece width is a predictor of post-thawing boar sperm motility. In our study, the principal piece volume is mainly determined by the distal midpiece width (Figure S2), which can explain their similar relationships with sperm velocity, and also the differences found between the GF and the BF in these sperm measures. The damage caused by sperm freezing protocols to the sperm head and tail membranes may occur independently: an intact tail membrane does not necessarily indicate an intact sperm head membrane and vice versa [47]. For example, after the freezing-thawing process, the disruption of the sperm head membrane occurs more easily than the tail in human and ram sperm [23,47], whereas the flagellum membrane is more vulnerable to the cryopreservation process in equine sperm [7].

Sperm with an intact head membrane but a damaged flagellum are most likely immotile, explaining the low fertilization rates with frozen/thawed sperm [7,48]; therefore, these cells should be included in the dead rather than the alive category [48]. In our study, the sperm flagellum is likely more sensitive to the freezing/thawing process than the sperm head or, at least, a damaged flagellum negatively affects sperm kinetics more than a damaged sperm head.

The fragility of the sperm flagellum and its ability to withstand the freezing/thawing process has been reported in many studies. The addition of glycerol as a cryoprotectant agent alters sperm functionality, mainly during the post-thawing sperm incubation stage [49], showing an increase in the proportion of epididymal spermatozoa with axonemal vacuoles, damaged plasma mem-

Table 5. Results of principal component analysis (PCA) to determine overall sperm velocity at 0 and 2 hours post-thaw.

PCA	0 hours	2 hours
Variables	PC1	PC1
VAP	0.986	0.951
VCL	0.975	0.967
VSL	0.960	0.920
ALH	0.957	0.927
SMI	0.907	0.809
Variance explained (%)	91.676	83.978
Eigenvalue	4.584	4.199
Bartlett's test of sphericity	0.000	0.000
Kaiser-Meyer-Olkin test	0.795	0.744

VAP, average path velocity; VCL, curvilinear velocity; VSL, straight linear velocity; ALH, amplitude of lateral head displacement; SMI, sperm motility index.

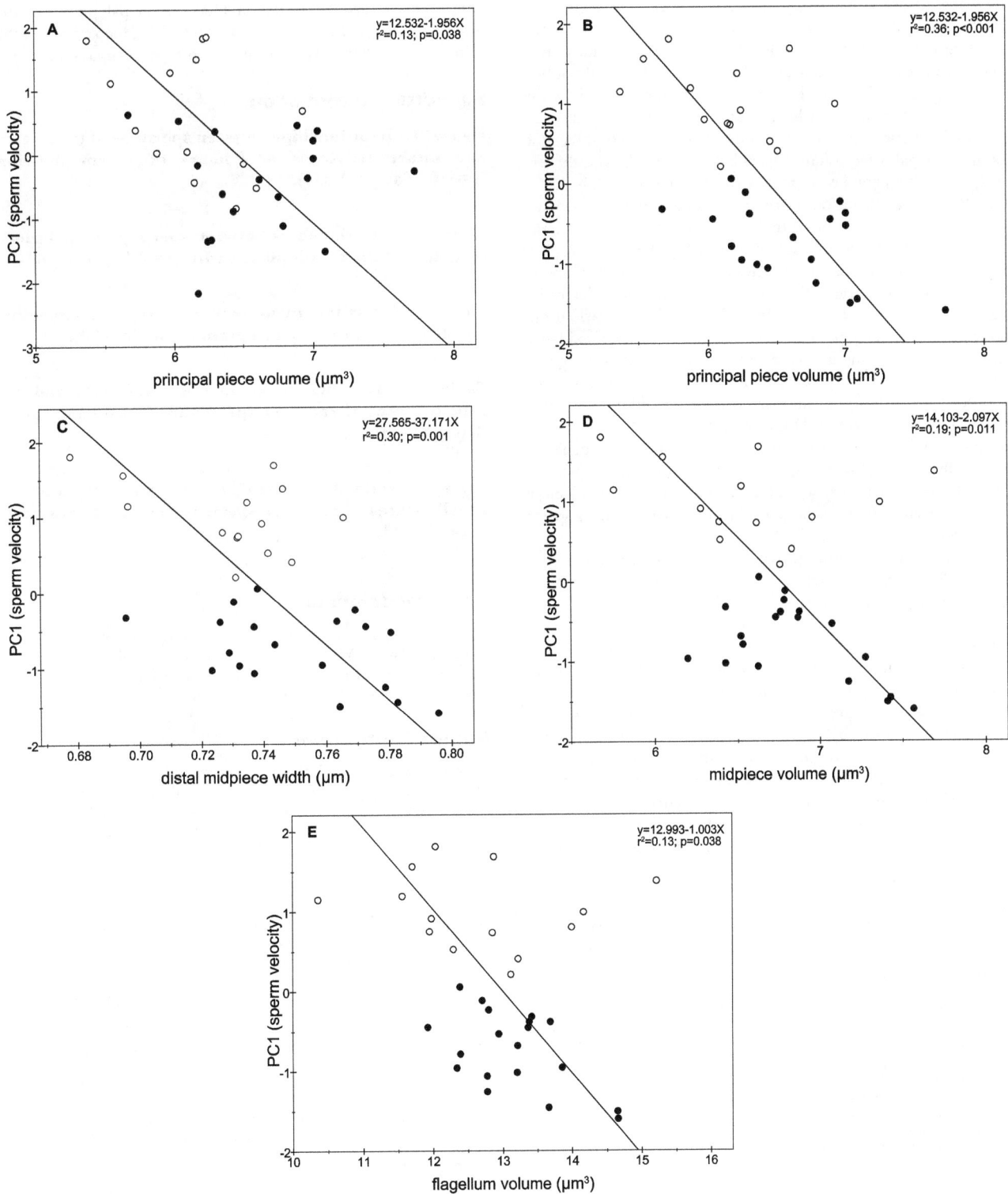

Figure 4. Relationships between sperm velocity and flagellum morphometry. Good freezers are shown in open circles and bad freezers in closed circles: A (0 hours post-thaw); B, C, D, and E (2 hours post-thaw).

branes, and abnormal mitochondria when compared to media without glycerol [50]. Furthermore, the cytoskeleton is responsible for the appropriate cell volume regulation and its stability is highly altered by the cryopreservation process, causing its proteins (e.g., F-actin) to become more fragile [13,24,51,52]. Thus, in red deer, a

higher volume of the sperm flagellum might result in an increased amount of ice crystals and increased formation of axonemal vacuoles during the freezing/thawing process, adversely affecting flagellum integrity and consequently, cell volume regulation and sperm velocity. Supporting this hypothesis, Correa et al. [53]

suggested a direct connection between cell volume regulation, flagellum morphology, motility, and the actin cytoskeleton in the sublethal damage that occurs during osmotic stress and, potentially, during cryopreservation. On the other hand, we did not find any significant differences between the GF and the BF in the midpiece or flagellum size which contain sperm mitochondria and the whole sperm axoneme, respectively. On the contrary, the sperm principal piece volume differs between the GF and the BF. Such a result might be related to the fibrous sheath, which is located along the principal piece and linked with sperm kinetics [54]. Thus, differences among males in the principal piece volume could differently affect their sperm freezability. Furthermore, it is thought that the fibrous sheath plays a mechanical role in sperm motility, providing a rigid support to the flagellum and determining its planar beat [55,56]. Additionally, glycolysis is carried out along the length of the principal piece and this, instead of oxidative phosphorylation in the midpiece, is the most important source of ATP for the tail [57]. At least one fibrous sheath protein may act to protect sperm from oxidative stress, which could interfere with sperm motility or cause DNA damage [58]. Therefore, the proteome and the general structure of the fibrous sheath could be potentially damaged by the sperm freezing and thawing process, particularly in those sperm with a higher sperm principal piece volume, causing a decrease in sperm function mainly during sperm incubation.

In conclusion, our results provide evidence that the volumes of the flagellum structures are a determinant to predict post-thaw sperm velocity in red deer, and the BF have a higher sperm principal piece volume than the GF. In contrast, sperm head size is not a good predictor of post-thaw sperm velocity in red deer spermatozoa. However, further studies, including additional analyses such as freeze-fracture electron microscopy observations of spermatozoa [50,59] and the evaluation of sperm tail membrane integrity by light microscopy [48] during sperm cooling and especially during sperm warming, are necessary in order to confirm our findings. On the other hand, our subsequent studies will be directed towards the use of electron microscopy, to assess sperm flagellum morphometry in more depth, and also to measure their internal structures (axoneme, fibrous sheath, etc.).

Our results clearly show that a higher principal piece volume results in poor sperm freezability, and highlights the key role of the volume of flagellum structures in sperm cryopreservation success.

Supporting Information

Figure S1   Relationships between sperm head perimeter and intact acrosome at 0 hours of sperm thawing ($r = -0.365$; $p = 0.037$).

Figure S2   Relationships between sperm principal piece volume and distal midpiece width ($r = 0.93$; $p < 0.0001$).

Table S1   Individual mean morphometry and subjective motility of fresh red deer spermatozoa ($N = 33$).

Table S2   Individual mean kinetics, viability, and organelle status of red deer spermatozoa at 0 hours post-thaw ($N = 33$).

Table S3   Individual mean kinetics, viability, and organelle status of red deer spermatozoa at 2 hours post-thaw ($N = 33$).

Acknowledgments

The assistance provided by Alfonso Bisbal, Enrique del Olmo, Mari Cruz Sotos, and Zandra Maulen is gratefully acknowledged. Eliana Pintus is acknowledged for her insightful comments. Landowners, managers, and rangers are acknowledged for facilitating access to samples.

Author Contributions

Conceived and designed the experiments: JLRS JJG. Performed the experiments: JLRS AEDR. Analyzed the data: JLRS. Contributed reagents/materials/analysis tools: JJG. Wrote the paper: JLRS. Sperm morphometry assessment: JLRS. Revised the manuscript: JLRS AEDR JJG.

References

1. Watson PF (1995) Recent developments and concepts in the cryopreservation of spermatozoa and the assessment of their post-thawing function. Reprod Fertil Dev 7: 871–891.
2. Watson PF (2000) The causes of reduced fertility with cryopreserved semen. Anim Reprod Sci 60–61: 481–492.
3. Peña FJ, Saravia F, Garcia-Herreros M, Nuñez-Martinez I, Tapia JA (2005) Identification of sperm morphometric subpopulations in two different portions of the boar ejaculate and its relation to postthaw quality. J Androl 26: 716–723.
4. Parks JE, Graham JK (1992) Effects of cryopreservation procedures on sperm membranes. Theriogenology 38: 209–222.
5. Billard R, Cosson J, Linhart O (2000) Changes in the flagellum morphology of intact and frozen/thawed Siberian sturgeon *Acipenser baerii* (Brandt) sperm during motility. Aquac Res 31: 283–287.
6. Holt WV (2000) Basic aspects of frozen storage of semen. Anim Reprod Sci 62: 3–22.
7. Domes U, Stolla R (2001) Staining of stallion spermatozoa with damaged tail membrane after freezing/thawing. Anim Reprod Sci 68: 329–330.
8. Devireddy RV, Swanlund DJ, Alghamdi AS, Duoos LA, Troedsson MH, et al. (2002) Measured effect of collection and cooling conditions on the motility and the water transport parameters at subzero temperatures of equine spermatozoa. Reproduction 124: 643–648.
9. Brouwers JF, Silva PF, Gadella BM (2005) New assays for detection and localization of endogenous lipid peroxidation products in living boar sperm after BTS dilution or after freeze-thawing. Theriogenology 63: 458–469.
10. Cremades T, Roca J, Rodriguez-Martinez H, Abaigar T, Vazquez JM, et al. (2005) Kinematic changes during the cryopreservation of boar spermatozoa. J Androl 26: 610–618.
11. Holt WV, Medrano A, Thurston LM, Watson PF (2005) The significance of cooling rates and animal variability for boar sperm cryopreservation: insights from the cryomicroscope. Theriogenology 63: 370–382.
12. Thurston LM, Siggins K, Mileham AJ, Watson PF, Holt WV (2002) Identification of amplified restriction fragment length polymorphism markers linked to genes controlling boar sperm viability following cryopreservation. Biol Reprod 66: 545–554.
13. Petrunkina AM, Gröpper B, Günzel-Apel AR, Töpfer-Petersen E (2004) Functional significance of the cell volume for detecting sperm membrane changes and predicting freezability in dog semen. Reproduction 128: 829–842.
14. Esteso MC, Soler AJ, Fernández-Santos MR, Quintero-Moreno AA, Garde JJ (2006) Functional significance of the sperm head morphometric size and shape for determining freezability in iberian red deer (*Cervus elaphus hispanicus*) epididymal sperm samples. J Androl 27: 662–670.
15. Tuset VM, Dietrich GJ, Wojtczak M, Słowińska M, de Monserrat J (2008) Comparison of three staining techniques for the morphometric study of rainbow trout (*Oncorhynchus mykiss*) spermatozoa. Theriogenology 69: 1033–1038.
16. Malo AF, Gomendio M, Garde J, Lang-Lenton B, Soler AJ (2006) Sperm design and sperm function. Biol Lett 2: 246–249.
17. Voordouw MJ, Koella JC, Hurd H (2008) Intra-specific variation of sperm length in the malaria vector *Anopheles gambiae*: males with shorter sperm have higher reproductive success. Malar J 7: 214.
18. Anderson MJ, Dixson AF (2002) Sperm competition: motility and the midpiece in primates. Nature 416: 496.
19. Anderson MJ, Nyholt J, Dixson AF (2005) Sperm competition and the evolution of sperm midpiece volume in mammals. J Zool 267: 135–142.
20. Parapanov RN, Nusslé S, Crausaz M, Senn A, Hausser J (2009) Testis size, sperm characteristics and testosterone concentrations in four species of shrews (Mammalia, Soricidae). Anim Reprod Sci 114: 269–278.

21. Ramm SA, Stockley P (2010) Sperm competition and sperm length influence the rate of mammalian spermatogenesis. Biol Lett 6: 219–221.
22. Keates RA (1980) Effects of glycerol on microtubule polymerization kinetics. Biochem Biophys Res Commun 97: 1163–1169.
23. Holt WV, North RD (1994) Effects of temperature and restoration of osmotic equilibrium during thawing on the induction of plasma membrane damage in cryopreserved ram spermatozoa. Biol Reprod 51: 414–424.
24. Felipe-Pérez YE, Valencia J, Juárez-Mosqueda Mde L, Pescador N, Roa-Espitia AL (2012) Cytoskeletal proteins F-actin and β-dystrobrevin are altered by the cryopreservation process in bull sperm. Cryobiology 64: 103–109.
25. Curry MR, Millar JD, Tamuli SM, Watson PF (1996) Surface area and volume measurements for ram and human spermatozoa. Biol Reprod 55: 1325–1332.
26. Garde JJ, Ortiz N, García AJ, Gallego L, Landete-Castillejos T (1998) Postmortem assessment of sperm characteristics of the red deer during the breeding season. Arch Androl 41: 195–202.
27. Fernández-Santos MR, Esteso MC, Montoro V, Soler AJ, Garde JJ (2006) Cryopreservation of Iberian red deer (Cervus elaphus hispanicus) epididymal spermatozoa: effects of egg yolk, glycerol and cooling rate. Theriogenology 66: 1931–1942.
28. Domínguez-Rebolledo AE, Fernández-Santos MR, Bisbal A, Ros-Santaella JL, Ramón M, et al. (2010) Improving the effect of incubation and oxidative stress on thawed spermatozoa from red deer by using different antioxidant treatments. Reprod Fertil Dev 22: 856–870.
29. Garde JJ, Martínez-Pastor F, Gomendio M, Malo AF, Soler AJ, et al. (2006) The application of reproductive technologies to natural populations of red deer. Reprod Domest Anim 41: 93–102.
30. Comizzoli P, Mauget R, Mermillod P (2001) Assessment of in vitro fertility of deer spermatozoa by heterologous IVF with zona-free bovine oocytes. Theriogenology 56: 261–274.
31. Hossain AM, Barik S, Kulkarni PM (2001) Lack of significant morphological differences between human X and Y spermatozoa and their precursor cells spermatids exposed to different prehybridization treatments. J Androl 22: 119–123.
32. Sanchez L, Petkov N, Alegre E (2005) Statistical approach to boar semen head classification based on intracellular intensity distribution. Computer Analysis of Images and Patterns, Proceedings 3691: 88–95.
33. Ramanujan S (1914) Modular equations and approximations to π. Quart J Pure Appl Math 45: 350–372.
34. Malo AF, Garde JJ, Soler AJ, Garcia AJ, Gomendio M (2005) Male fertility in natural populations of red deer is determined by sperm velocity and the proportion of normal spermatozoa. Biol Reprod 72: 822–829.
35. Nuñez-Martínez I, Morán JM, Peña FJ (2006) Two-step cluster procedure after principal component analysis identifies sperm subpopulations in canine ejaculates and its relation to cryoresistance. J Androl 27: 596–603.
36. Budaev SV (2010) Using principal components and factor analysis in animal behaviour research: caveats and guidelines. Ethology 116: 472–480.
37. Bohonak AJ, Linde K (2004) RMA: Software for reduced major axis regression, Java version. Website: http://www.kimvdlinde.com/professional/rma.html.
38. Gage MJ, Macfarlane CP, Yeates S, Ward RG, Searle JB (2004) Spermatozoal traits and sperm competition in Atlantic salmon: relative sperm velocity is the primary determinant of fertilization success. Curr Biol 14: 44–47.
39. Birkhead TR, Martinez JG, Burke T, Froman DP (1999) Sperm mobility determines the outcome of sperm competition in the domestic fowl. Proc R Soc B-Biol Sci 266: 1759–1764.

40. Holt WV, Shenfield F, Leonard T, Hartman TD, North RD (1989) The value of sperm swimming speed measurements in assessing the fertility of human frozen semen. Hum Reprod 4: 292–297.
41. Bollwein H, Fuchs I, Koess C (2008) Interrelationship between plasma membrane integrity mitochondrial membrane potential and DNA fragmentation in cryopreserved bovine spermatozoa. Reprod Domest Anim 43: 189–195.
42. Anel-López L, Alvarez-Rodríguez M, García-Álvarez O, Alvarez M, Maroto-Morales A, et al. (2012) Reduced glutathione and Trolox (vitamin E) as extender supplements in cryopreservation of red deer epididymal spermatozoa. Anim Reprod Sci 135: 37–46.
43. Saacke RG, White JM (1972) Semen quality tests and their relationship to fertility. Proceedings of the Fourth Technical Conference on Artificial Insemination and Reproduction National Association of Animal Breeders 2–7.
44. Del Olmo E, Bisbal A, Maroto-Morales A, García-Alvarez O, Ramon M, et al. (2013) Fertility of cryopreserved ovine semen is determined by sperm velocity. Anim Reprod Sci 138: 102–109.
45. Medrano A, Watson PF, Holt WV (2002) Importance of cooling rate and animal variability for boar sperm cryopreservation: insights from the cryomicroscope. Reproduction 123: 315–322.
46. Woolley DM, Richardson DW (1978) Ultrastructural injury to human spermatozoa after freezing and thawing. J Reprod Fertil 53: 389–394.
47. Zhu WJ, Liu XG (2000) Cryodamage to plasma membrane integrity in head and tail regions of human sperm. Asian J Androl 2: 135–138.
48. Nagy S, Házas G, Papp AB, Iváncsics J, Szász F (1999) Evaluation of sperm tail membrane integrity by light microscopy. Theriogenology 52: 1153–1159.
49. Macías García B, Ortega Ferrusola C, Aparicio IM, Miró-Morán A, Morillo Rodriguez A, et al. (2012) Toxicity of glycerol for the stallion spermatozoa: effects on membrane integrity and cytoskeleton, lipid peroxidation and mitochondrial membrane potential. Theriogenology 77: 1280–1289.
50. McClean RV, Holt WV, Johnston SD (2007) Ultrastructural observations of cryoinjury in kangaroo spermatozoa. Cryobiology 54: 271–280.
51. Petrunkina AM, Hebel M, Waberski D, Weitze KF, Töpfer-Petersen E (2004) Requirement for an intact cytoskeleton for volume regulation in boar spermatozoa. Reproduction 127: 105–115.
52. Petrunkina AM, Radcke S, Günzel-Apel AR, Harrison RAP, Töpfer-Petersen E (2004) Role of potassium channels, the sodium-potassium pump and the cytoskeleton in the control of dog sperm volume. Theriogenology 61: 35–54.
53. Correa LM, Thomas A, Meyers SA (2007) The macaque sperm actin cytoskeleton reorganizes in response to osmotic stress and contributes to morphological defects and decreased motility. Biol Reprod 77: 942–953.
54. Schlingmann K, Michaut MA, McElwee JL, Wolff CA, Travis AJ (2007) Calmodulin and CaMKII in the sperm principal piece: evidence for a motility-related calcium/calmodulin pathway. J Androl 28: 706–716.
55. Fawcett DW (1975) The mammalian spermatozoon. Dev Biol 44: 394–436.
56. Lindemann CB, Orlando A, Kanous KS (1992) The flagellar beat of rat sperm is organized by the interaction of two functionally distinct populations of dynein bridges with a stable central axonemal partition. J Cell Sci 102: 249–260.
57. Turner RM (2006) Moving to the beat: a review of mammalian sperm motility regulation. Reprod Fertil Dev 18: 25–38.
58. Fulcher KD, Welch JE, Klapper DG, O'Brien DA, Eddy EM (1995) Identification of a unique mu-class glutathione S-transferase in mouse spermatogenic cells. Mol Reprod Dev 42: 415–424.
59. Holt WV, North RD (1984) Partially irreversible cold-induced lipid phase transitions in mammalian sperm plasma membrane domains; freeze fracture study. J Exp Zool 230: 473–483.

9

Iron Overload and Apoptosis of HL-1 Cardiomyocytes: Effects of Calcium Channel Blockade

Mei-pian Chen[1], Z. Ioav Cabantchik[2], Shing Chan[1], Godfrey Chi-fung Chan[1]*, Yiu-fai Cheung[1]*

[1] Department of Pediatrics and Adolescent Medicine, The University of Hong Kong, Hong Kong, China, [2] Department of Biological Chemistry, Alexander Silberman Institute of Life Sciences, Hebrew University of Jerusalem, Safra Campus at Givat Ram, Jerusalem, Israel

Abstract

Background: Iron overload cardiomyopathy that prevails in some forms of hemosiderosis is caused by excessive deposition of iron into the heart tissue and ensuing damage caused by a raise in labile cell iron. The underlying mechanisms of iron uptake into cardiomyocytes in iron overload condition are still under investigation. Both L-type calcium channels (LTCC) and T-type calcium channels (TTCC) have been proposed to be the main portals of non-transferrinic iron into heart cells, but controversies remain. Here, we investigated the roles of LTCC and TTCC as mediators of cardiac iron overload and cellular damage by using specific Calcium channel blockers as potential suppressors of labile Fe(II) and Fe(III) ingress in cultured cardiomyocytes and ensuing apoptosis.

Methods: Fe(II) and Fe(III) uptake was assessed by exposing HL-1 cardiomyocytes to iron sources and quantitative real-time fluorescence imaging of cytosolic labile iron with the fluorescent iron sensor calcein while iron-induced apoptosis was quantitatively measured by flow cytometry analysis with Annexin V. The role of calcium channels as routes of iron uptake was assessed by cell pretreatment with specific blockers of LTCC and TTCC.

Results: Iron entered HL-1 cardiomyocytes in a time- and dose-dependent manner and induced cardiac apoptosis via mitochondria-mediated caspase-3 dependent pathways. Blockade of LTCC but not of TTCC demonstrably inhibited the uptake of ferric but not of ferrous iron. However, neither channel blocker conferred cardiomyocytes with protection from iron-induced apoptosis.

Conclusion: Our study implicates LTCC as major mediators of Fe(III) uptake into cardiomyocytes exposed to ferric salts but not necessarily as contributors to ensuing apoptosis. Thus, to the extent that apoptosis can be considered a biological indicator of damage, the etiopathology of cardiosiderotic damage that accompanies some forms of hemosiderosis would seem to be unrelated to LTCC or TTCC, but rather to other routes of iron ingress present in heart cells.

Editor: Alexander G. Obukhov, Indiana University School of Medicine, United States of America

Funding: This work was supported by Children's Thalassemia Foundation and Edward Sai Kim Hotung Pediatric Education and Research Fund (http://www.thalassaemia.org.hk). The funders had no role in study design, data collection and analysis, decision to publish, or preparation of the manuscript.

Competing Interests: The authors have declared that no competing interests exist.

* Email: gcfchan@hku.hk (GCFC); xfcheung@hku.hk (YFC)

Introduction

As an essential element for almost all living organisms, iron serves as a critical component in different metabolic processes including oxygen transport and storage, DNA, RNA and protein synthesis, and electron transport [1]. Tight regulation of iron concentrations is required for maintenance of cellular function, while excessive iron leads to generation of oxidative stress by increasing production of reactive oxygen species [2–4]. Of the different organs, the heart is particularly vulnerable to iron toxicity [5].

Iron overload cardiomyopathy (IOC) is well documented in patients with β-thalassemia major and is an important cause of morbidity and mortality [6–9]. Clinical manifestations include systolic and diastolic ventricular dysfunction, cardiac arrhythmias, and end-stage cardiomyopathy [5,8,10,11]. However, the mechanisms of iron-induced subclinical cardiac dysfunction and end-stage cardiomyopathy remain unclear. Progressive loss of cardio-myocytes, albeit at a low level, through apoptosis is believed to contribute to the remodeling process and ventricular dysfunction in heart failure [12–17]. There is, however, a paucity of data on the phenomenon of cardiomyocyte apoptosis and the pathway involved in the setting of iron overload.

Under physiologic condition, iron uptake into cardiomyocytes is mediated through transferrin-transferrin receptor-mediated endo-cytosis with negative feedback regulatory mechanisms [18]. However, under iron overloading conditions, transferrin becomes saturated and excess plasma iron will present as non-transferrin-bound iron (NTBI), which contributes to the intracellular labile iron pool and the generation of reactive oxygen species [9]. Reported mechanisms of NTBI entry into cardiomyocytes are nonetheless controversial [19]. While some studies have proposed L-type calcium channels (LTCC) to be a major pathway for NTBI entry [20–22], others suggest that T-type calcium channel (TTCC) may be the alternative portal of entry [23,24]. However, direct evidence for possible protective effects of calcium channel blockers against iron-induced cardiomyocyte apoptosis is lacking.

Using HL-1 cardiomyocytes, a spontaneously contracting cardiomyocyte cell line that expresses both LTCC and TTCC molecularly and functionally [25–27], together with the real-time technique tracing cellular iron uptake and flow cytometry, we explored (i) the phenomenon of and mechanisms involved in cardiomyocyte apoptosis induced by iron overload, (ii) the effects of LTCC and TTCC blockers on Fe(II) and Fe(III) entry into cardiomyocytes, and (iii) the potential protective effect on iron-induced cardiomyocyte apoptosis by calcium channel blockade.

Materials and Methods

Cell culture

HL-1 cardiomyocytes were kindly provided by Prof. W.C. Claycomb (Louisiana State University Health Science Center, New Orleans, LA, USA) who created the cell line [25]. HL-1 cells were established from the AT-1 mouse atrial cardiomyocyte tumor, and can be serially passaged while maintaining contractile phenotype. The cells were grown in culture vessels pre-coated with 0.02% gelatin (Difco, Fisher Scientific, Suwanee, GA, USA) - 5 µg/ml fibronectin (Sigma, St Louis, MO, USA) solution at 37°C in a humidified 5% CO_2 incubator, maintained in Claycomb Medium (SAFC Biosciences, Sigma) supplemented with 10% fetal bovine serum (Sigma), 0.1 mM norepinephrine (Sigma), 2 mM L-glutamine (Invitrogen, Life Technologies, Grand Island, NY, USA) and penicillin/streptomycin (100 U/ml:100 µg/ml) (Invitrogen). The medium was changed approximately 5 days per week.

Iron treatment and calcium channel blockade

For calcein green-acetomethoxy (CALG-AM) fluorescent assay, HL-1 cells were seeded at 6×10^4 cells/well in gelatin-fibronectin coated 96-well black CulturPlate (PerkinElmer, Waltham, Massachusetts, USA). Cells reached around 90% confluence after 24 hr culture. L-type calcium channel blockers including amlodipine (Cipla, India) and verapamil (Abbott, Ludwigshafen, Germany) and TTCC blocker, efonidipine (Sigma), were loaded at 0.1, 1, 10, 100 µM in assay buffer, which consisted of HEPES-buffered saline, pH 7.4 (HBS) supplemented with 0.5 mM probenecid (Sigma), 30 min before iron challenge, and the concentrations were maintained during the assay. $FeCl_3$ was loaded at 150, 300, 600 µM with and without 1 mM ascorbic acid in assay buffer, which has been indicated to represent Fe(II) and Fe(III) respectively [24,28,29]. Controls (with and without ascorbate) was defined as the conditions without calcium channel blockers and iron.

For flow cytometric assay, HL-1 cells were seeded at a density of 1.5×10^5 cells/ml in gelatin-fibronectin coated plates. After 24 hr incubation, culture medium was changed into norepinephrine-free medium containing 2% fetal bovine serum, 2 mM L-glutamine and penicillin/streptomycin (100 U/ml:100 µg/ml), and also 150, 300, 600 µM $FeCl_3$ with and without 1 mM ascorbic acid for test groups. Calcium channel blockers were pre-loaded at 1 µM 60 min before iron challenge without media change before treatment endpoint. For treatments with iron chelator deferiprone (Apotex, Toronto, Canada), 10 or 100 µM deferiprone was loaded 20 min after iron loading. Blank controls (with and without ascorbate) was defined as the conditions without calcium channel blockers, chelator and iron loading. After 72 hr of incubation, cells in the control group had confluency at around 90%, while cells in iron treatment groups had less. Cells were gently detached by 0.05% Trypsin-EDTA (Invitrogen) for flow cytometric assays.

CALG-AM fluorescent assay

To trace iron transport in live HL-1 cells, CALG-AM fluorescent assay was used [30]. Non-fluorescent CALG-AM is converted to green-fluorescent calcein once diffuses into live cells, going through acetoxymethyl ester hydrolysis by intracellular esterases. Cells were exposed to 0.25 µM CALG-AM (Molecular Probes, Life Technologies, Grand Island, NY, USA) at 37°C for 30 min in Claycomb Medium containing 10 mM Na-HEPES (Sigma). Cells were then rinsed with HBS, followed by the perfusion of assay buffer, HBS supplemented with 0.5 mM probenecid, which prevented leakage of anionic fluorescent probes from cells. Calcium channel blockers and ascorbic acid were added simultaneously under the conditions mentioned. Fluorescent intensity was measured using fluorescent plate reader Fusion (Packard, Perkin Elmer Life Sciences, Boston, MA, USA) at excitation/emission wavelength 485 nm/520 nm. Local average reading at 10 min after assay buffer loading was set as initial fluorescence level. $FeCl_3$ was loaded at 20 min after the first plate reading (Figure 1A). Calcein was quenched by intracellular labile iron, and hence, the fluorescence intensity was inversely proportional to the level of labile intracellular iron. Iron entry was terminated by adding 100 µM impermeant chelator diethylene-triamine-pentaacetic acid (DTPA) at 115 min after assay buffer loading. Identification of intracellular labile iron was verified by 100 µM permeant iron chelator deferasirox (Exjade, ICL670) at 136 min after assay buffer loading to reverse the calcein-Fe quenching. Control was defined as treatments without addition of calcium channel blockers, iron, DTPA and ICL670. Experiments were performed in triplicate. Each reading at any given time was normalized to the local initial fluorescence level.

Annexin V/PI assay

Fluorescein isothiocyanate (FITC) Annexin V Apoptosis Detection Kit (Becton Dickinson, Franklin Lakes, NJ, USA) was used according to manufacturer's instructions. Briefly, cells from cultures were collected and washed with cold PBS and then resuspended in annexin V binding buffer. After staining with annexin V-FITC and PI for 15 min at room temperature in the dark, cells suspended in annexin V binding buffer were tested by LSR II flow cytometer (Becton Dickinson). For each measurement, at least 10,000 cells were counted. Flow data were analyzed by FlowJo 8.8.4 (Tree Star). Only single cell events were gated out for analysis.

Activated caspase-3 assay

FITC Active Caspase-3 Apoptosis Kit (Becton Dickinson) was used according to manufacturer's instructions. Briefly, cells from culture were collected and washed with cold PBS, then fixed and permeabilized in BD Cytofix/Cytoperm solution for 20 min on ice. After washing with BD Perm/Wash buffer, cells were stained with FITC-conjugated anti- active caspase-3 antibody for 30 min at room temperature. With further wash with Perm/Wash buffer, cells suspended in Perm/Wash buffer were tested by LSR II flow cytometer. Flow cytometry was performed as aforementioned.

JC-1 assay

The mitochondrial membrane potential (Δψ) of HL-1 cardiomyocytes was evaluated by Flow Cytometry Mitochondrial Membrane Potential Detection Kit (Becton Dickinson). JC-1 (5,5',6,6'-tetrachloro-1,1',3,3'-tetraethylbenzimidazolcarbocyanine iodide) is a fluorochrome widely used to evaluate the status of Δψ. Mitochondria with normal Δψ increases JC-1 uptake, which leads to the formation of JC-1 aggregates that emit red

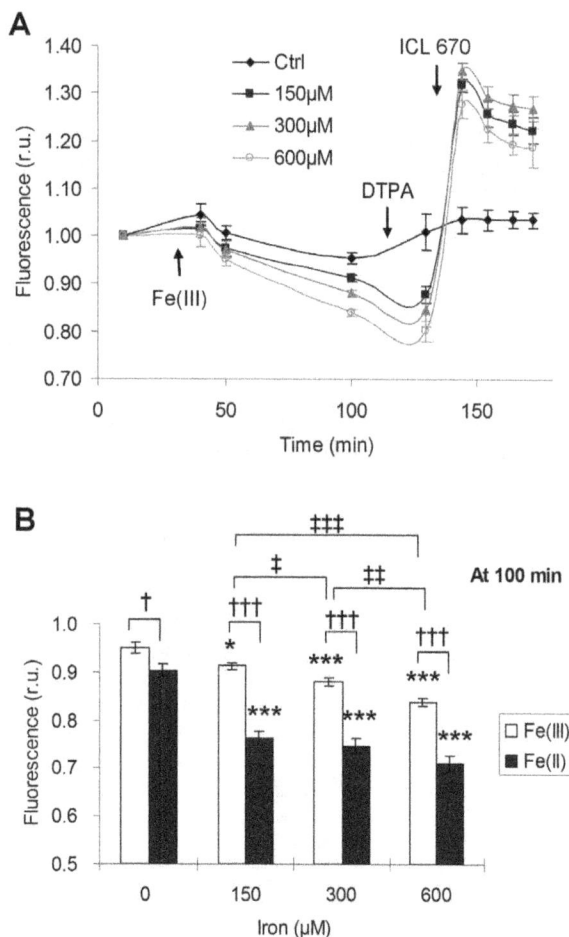

Figure 1. Exogenous iron entered cardiomyocytes in a time- and dose- dependent manner. (A) Fe(III) uptake by live HL-1 cells treated at 3 indicated doses, detected by CALG-AM fluorescent assay. Fluorescence intensity was carried out by fluorescent plate reader Fusion. Local average reading at 10 min was set as initial fluorescence level. Each reading at any given time was normalized to the local initial fluorescence level. FeCl$_3$ was load at 30 min. Impermeant chelator DTPA was loaded at 115 min; permeant iron chelator ICL 670 was loaded at 136 min. Control was defined as treatment without addition of iron, DTPA and ICL670. **(B)** Fe(III) and Fe(II) uptake at 100 min of the assessment time point indicated in (A), i.e. 70 min after iron loading. FeCl$_3$ loaded with ascorbate represented Fe(II) treatment. Both controls with and without ascorbate were shown. *, †, ‡, $p<0.05$; **, ††, ‡‡, $p<0.01$; ***, †††, ‡‡‡, $p<0.001$; * versus respective controls. The results represented as mean ± SEM of five independent triplicate experiments.

fluorescence at 590 nm. In depolarized mitochondria, low concentration of JC-1 inside would stay at monomer form, emitting green fluorescence maximally at 527 nm. The staining protocol followed manufacturer's instructions. Briefly, cells were collected and incubated in JC-1 solution for 15 min at 37°C in CO$_2$ incubator. After subsequent washes with Assay Buffer, cells were resuspended in Assay Buffer for flow cytometry by LSR II as aforementioned.

Statistical analysis

Data are presented as mean ± SEM. Statistical analysis was performed using one-way analysis of variance (ANOVA) with post test for multiple comparisons, and unpaired t test for comparisons of two groups by GraphPad Instat 3 (GraphPad Software, Inc.,

San Diego, CA, USA). A $p<0.05$ was regarded as statistically significant.

Results

Exogenous iron entered cardiomyocytes in a time- and dose- dependent manner

To detect intracellular labile iron, iron influx was visualized in real time by tracking the gradual decrease of fluorescence signals in the live HL-1 cardiomyocytes. Within the detection period from 10 to 70 min after iron loading (Figure 1A), we observed iron entering HL-1 cells in a time-dependent manner. With elimination of extracellular iron by addition of the impermeable chelator DTPA, the subsequent addition of permeable chelator ICL670 restored the calcein fluorescence quenched by labile iron significantly, confirming that CALG-AM assay could assess intracellular iron in HL-1 cardiomyocytes effectively.

Based on the difference of uptake rate at 70 min after iron challenge with or without ascorbate, Fe(II) was found to be significantly more permeable than Fe(III) ($p<0.001$) (Figure 1B). Fe(III) showed a dose-dependent acquisition at 150, 300, 600 µM loading. In contrast, Fe(II) achieved a near plateau loading at 150 µM (Figure 1B).

Iron loading induced cardiomyocyte apoptosis

Annexin V/Propidium Iodide (PI) flow cytometric assay was used to quantify the amount of apoptosis. Cells positive for annexin V but negative for PI represented those undergoing early apoptosis, while cells stained positive for both annexin V and PI represented the population undergoing late apoptosis or necrosis [31,32]. By quantifying the percentage of total annexin V positive cells (lower and upper right quadrant in the representative flow cytometry charts as shown in Figure 2A), we found a dose-dependent increase in apoptotic cell population when HL-1 cells were treated with FeCl$_3$ with or without ascorbic acid for 72 hr (Figure 2A) (pH of each condition changed within 7.4–7.8). Such increase in apoptosis was noted in cells treated with concentrations of FeCl$_3$ at ≥300 µM ($p<0.001$). At the concentration of 600 µM, Fe(II) induced significantly more apoptosis than Fe(III) ($p<0.01$).

To further define the underlying apoptotic mechanism of iron overload on HL-1 cardiomyocytes, caspase-3 activity and mitochondrial membrane potential change were also assessed. In line with the findings of annexin V/PI assay, iron overload induced a dose-dependent activation of caspase-3 (Figure 2B) and alteration of mitochondrial membrane potential (Figure 2C), which suggested an involvement of the intrinsic apoptotic pathway.

High-dose LTCC but not TTCC ameliorated Fe(III) entry under condition of iron load

The potential roles of LTCC and TTCC for iron entry into HL-1 cardiomyocytes were evaluated using CALG-AM fluorescent assay, with treatments with LTCC blockers, amlodipine and verapamil, and TTCC blocker, efonidipine, at 30 min prior to iron loading. The blockade effects for Fe(III) (Figure 3A) and Fe(II) (Figure 3B) treated at 150, 300, 600 µM were assessed at logarithmic increments of calcium channel blocker concentrations from 0.1 to 100 µM. The time point of assay was at 70 min after iron loading, which was approximately 100 min after administration of different calcium channel blockers. Fluorescent signal changes were normalized to respective negative controls of each treatment arm.

Compared with the increase in iron entry into cells under Fe(III) treatment alone with decreased fluorescent signals, pretreatment with 10 to 100 µM of amlodipine and verapamil significantly

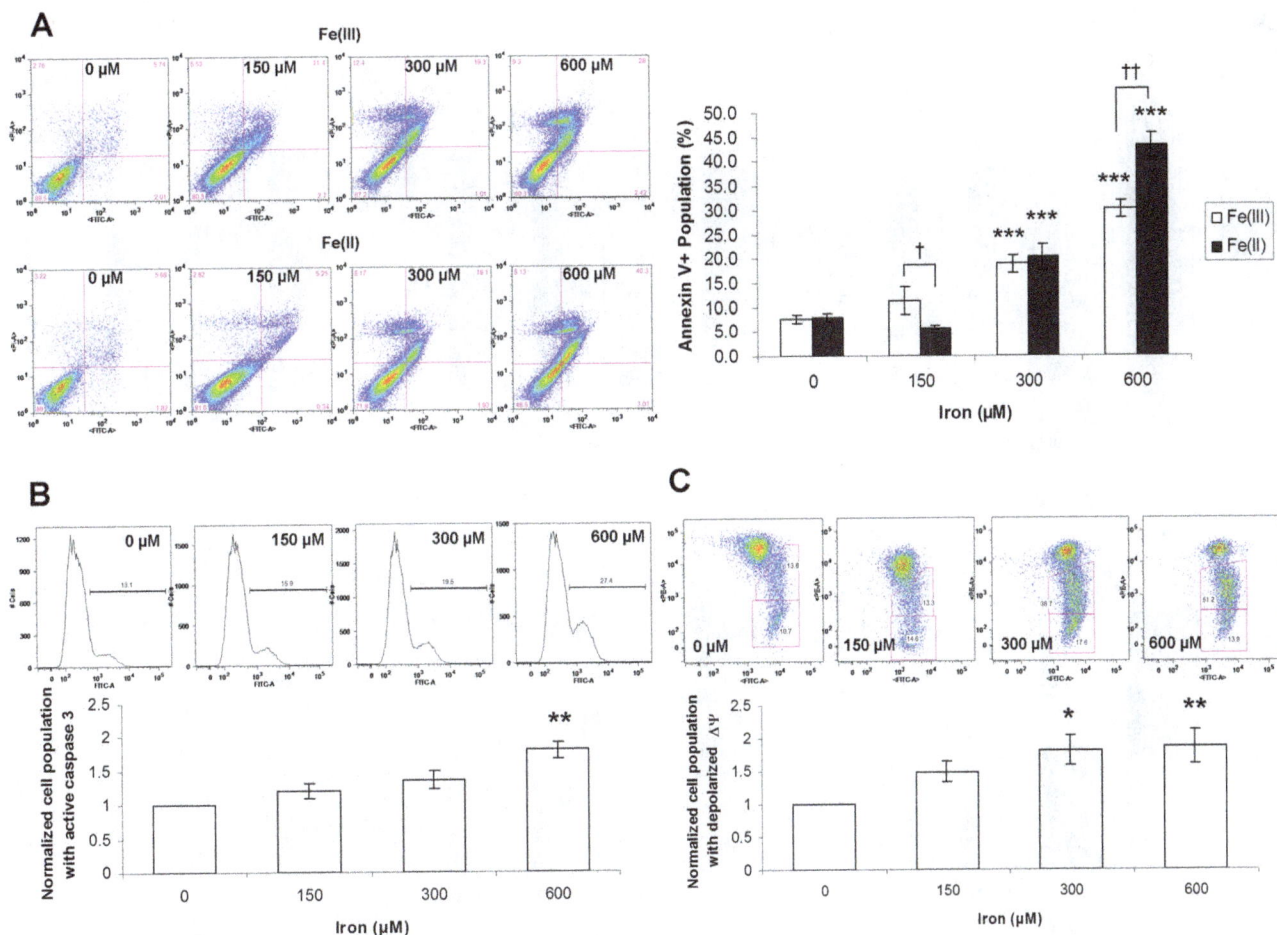

Figure 2. Iron overload induced cardiomyocyte apoptosis. HL-1 cells were treated with Fe(III) and Fe(II) for 72 hr, followed by (**A**) annexin V/PI flow cytometry assay, (**B**) active caspse-3 flow cytometry assay, and (**C**) JC-1 flow cytometry assay. *, †, $p<0.05$; **, ††, $p<0.01$; ***$p<0.001$; * versus respective controls. The results represented as mean $\pm$ SEM of five to six independent experiments.

increased normalized fluorescent signals (Figure 3A). The effect was more pronounced with 300 μM and 600 μM than 150 μM of Fe(III) load. These findings suggested blockade of Fe(III) entry by both LTCC blockers. However, efonidipine did not exert significant blocking effect on iron entry in Fe(III) overload.

Trend of LTCC and TTCC blockade of Fe(II) entry

With regard to Fe(II) loading condition, increased trends of fluorescent signals were observed with increased LTCC and TTCC blockade (Figure 3B). However, statistical significance was only found with pretreatment using 100 μM amlodipine.

Calcium channel blockers could not salvage cardiomyocytes from iron-induced apoptosis

To further explore whether calcium channel blockade could reduce HL-1 cardiomyocyte apoptosis induced by iron overload, annexin V/PI assay was performed on HL-1 cells loaded with Fe(III) and Fe(II) at different concentrations, with pretreatment of LTCC and TTCC blockers at a concentration of 1 μM. There was no significant decrease in apoptotic cell population, whether loaded with Fe(III) (Figure 4A) or Fe(II) (Figure 4B).

The findings suggested that calcium channel blockers at this concentration had no protective effects on HL-1 cells against iron-induced apoptosis. However, at the doses of 10 μM or 100 μM,

amlodipine or verapamil, which showed significant iron blockade effect on HL-1 cells (Figure 3), appeared to have high cellular toxicity (Figure 5A). Pretreatment of TTCC blockers in iron treated HL-1 cells led to similar or even worse effects.

By contrast, the commonly-used iron chelator deferiprone induced less toxic effect under non-iron overloaded condition (Figure 5A) and further showed protective effect on iron-induced apoptosis of cardiomyocytes (Figure 5B).

Discussion

The present study shows that i) iron induces apoptosis of HL-1 cardiomyocytes via the mitochondria-mediated caspase-3 dependent pathway, ii) blockade of LTCC but not TTCC prevented Fe(III) but not Fe(II) entry under iron overload condition and (iii) blockade of neither LTCC nor TTCC could salvage the cultured cardiomyocytes from iron overload induced apoptosis.

Iron-induced cardiomyocyte apoptosis

The levels of plasma NTBI in thalassemia patients under iron overload are variable, with an estimation suggested to be 0-25 μM [33]. For the proof of principle, comparable iron concentrations as previously reported were used in the current *in vitro* study [24,34]. The apoptotic effect of iron overload on HL-1 cells and its involvement of mitochondria-dependent pathway were suggested

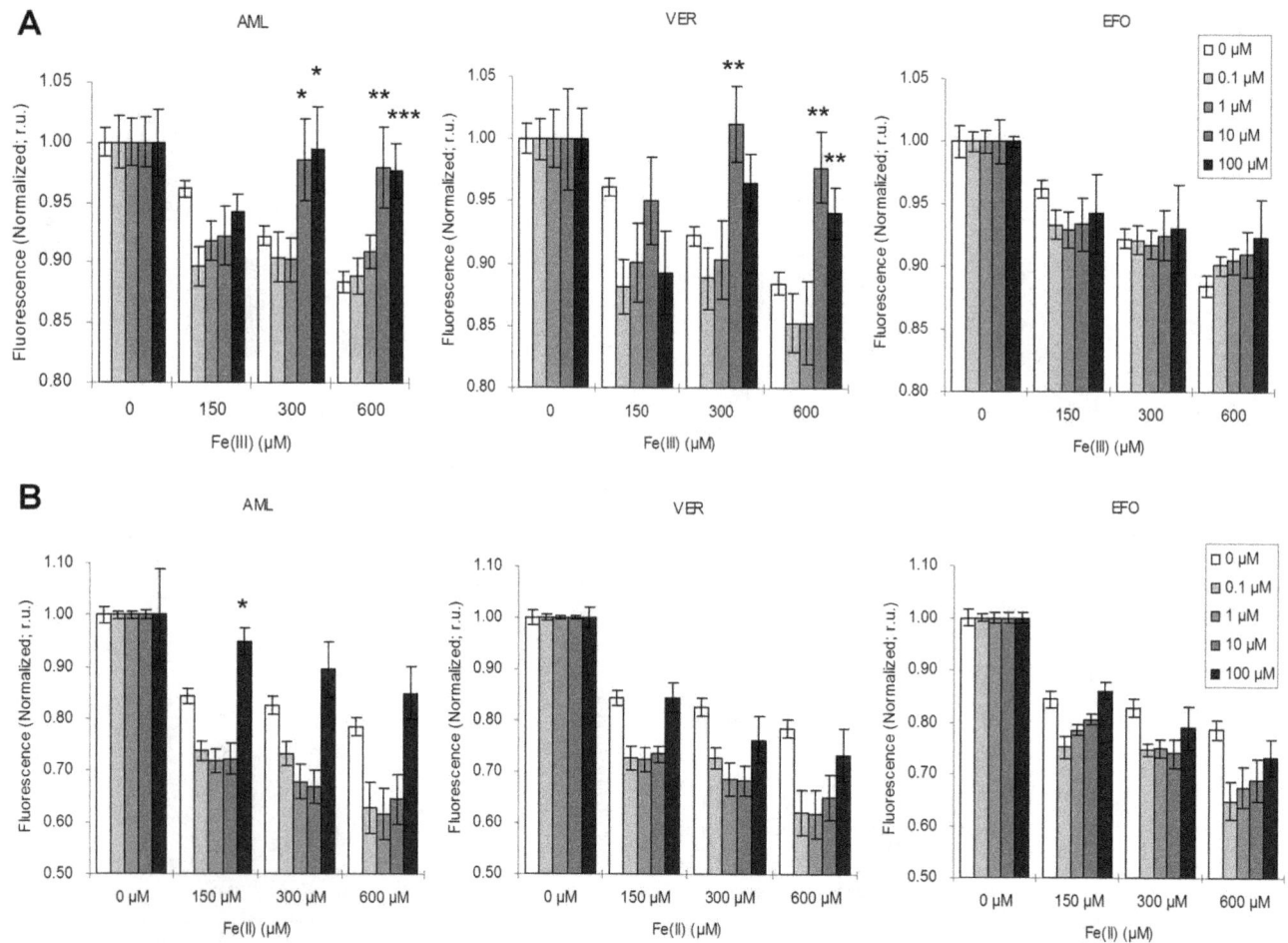

Figure 3. Iron blockade effects of LTCC and TTCC blockers on iron-overloaded cardiomyocytes. In this CALG-AM fluorescent assay, HL-1 cells were pretreated with LTCC blockers, amlodipine (AML) and verapamil (VER), and TTCC blocker, efonidipine (EFO), at logarithmic scale from 0.1 to 100 µM. 3 indicated doses of Fe(III) (**A**) and Fe(II) (**B**) were loaded 30 min after blocker treatment. Fluorescence readings were at 70 min after iron loading. Fluorescence signal changes were normalized to respective negative controls of each treatment arm. *p<0.05; **p<0.01; ***p<0.001. The results represented as mean ± SEM of four independent triplicate experiments.

by the findings of increase in phosphatidylserine exposure, increased caspase-3 activity, and a dose-dependent drop on mitochondrial membrane potential in iron-overloaded HL-1 cells. Our results are in agreement with the *in vivo* studies suggesting the cardiac apoptotic effect of iron overload on mice [20] and gerbils [35] as revealed by increased nucleic DNA fragmentation and caspase activity. Although other study suggesting the necrotic effect of iron overload on cardiomyocytes [36], more evidences will be of interest to the further mechanism behind, including the postulated cross link between apoptosis and necrosis in series or parallel [37], as well as the differences among experimental models.

Fe(II) and Fe(III) entry into cardiomyocytes

As both redox states of iron have been shown to form cardiac iron deposit [28], our study explored both ferric and ferrous irons. The results agree with those reported previously regarding the more permeative nature of ferrous iron, which is maintained with ascorbate as a reducing agent [24], as evaluated by kinetic parameters [28,38]. Previous studies have implicated either the LTCC or TTCC as the main candidate for NTBI entry into cardiomyocytes. The controversies have in part been related to different models and methods used.

The effect of LTCC blockade on iron entry

Calcium channels play an important role in myocardial contractility and remain open for long duration (>400 ms) in each contraction cycle [39]. Except for the primary transport of Ca^{2+}, LTCC also facilitate transport for many other divalent cations including Fe^{2+}, Co^{2+} and Zn^{2+} [22,40,41]. Previous studies suggest that LTCC is the major portal for iron uptake into cardiomyocytes in IOC [20–22]. For a further mechanism, we assessed the role of LTCC in iron-overloaded cardiomyocytes by the real-time approach.

Our results showed significant reduction of ferric iron ingress by both LTCC blockers at higher doses of iron treatment, 300 µM and 600 µM, but not at lower dose of iron at 150 µM. This phenomenon implicated the classic concept of iron delivery through transferrin at lower dose of iron treatment [9], while confirming the blockade effect from LTCC blockers toward excessive iron, Fe(III) from this result, uptake into cardiomyocytes [20–22]. It is worth noting, however, that LTCC blockers displayed their iron blockade effect only at concentrations of 10 and 100 µM, higher than the therapeutic serum levels of 0.1 to 1 µM [42,43]. Hence, the clinical translation of the use of LTCC blockers to prevent iron-induced cardiotoxicity remains uncertain.

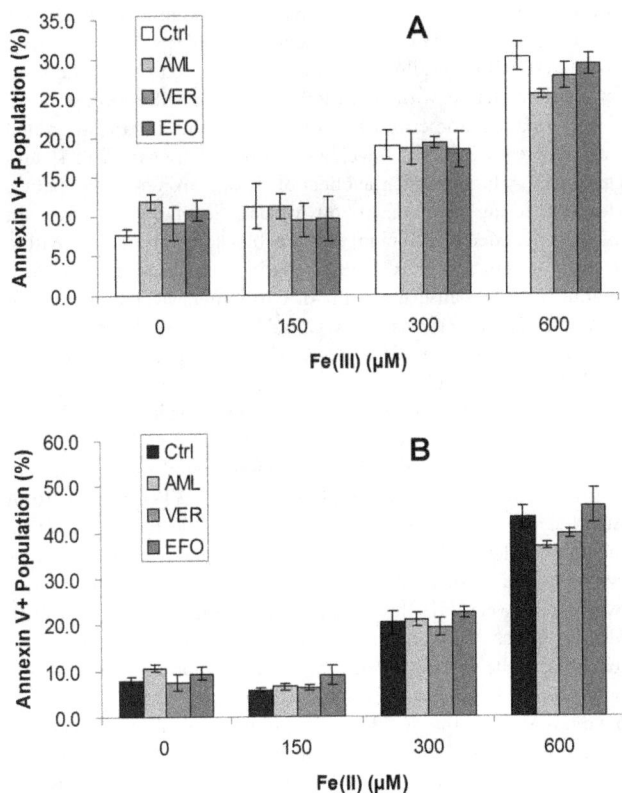

Figure 4. Calcium channel blockers could not salvage HL-1 cells from iron overload induced apoptosis. HL-1 cells were pretreated with LTCC blockers AML or VER, and TTCC blocker EFO for 1 hr, followed by Fe(III) (**A**) and Fe(II) (**B**) loading for 72 hr. Controls were defined as treatments without blockers. Apoptosis was determined by annexin V/PI flow cytometry assay. Total annexin V positive cell portion was counted. The results represented as mean ± SEM of three independent experiments.

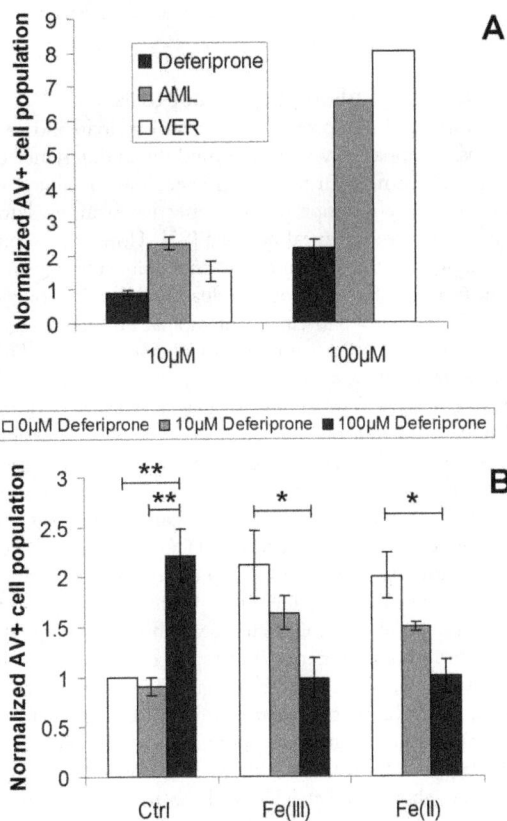

Figure 5. Cellular toxicity of LTCC blockers and the comparison with deferiprone. (A) Apoptotic effects of 10 or 100 μM AML, VER and deferiprone on HL-1 cells for 72 hr were assessed by annexin V/PI flow cytometry assay. (B) HL-1 cells were challenged with 300 μM Fe(III) or Fe(II), followed by treatments of 10 or 100 μM deferiprone 20 min after iron loading. Apoptosis was determined after 72 hr incubation by annexin V/PI assay. Data were shown as total annexin V positive cell portion with normalization to respective negative controls. * $p < 0.05$; ** $p < 0.01$. The results represented as mean ± SEM of three independent experiments, except 100 μM AML and VER (n = 1).

It is widely recognized that the promiscuous property of LTCC for the transport of other metals is limited to divalent, but not trivalent cations [22,40,41,44]. Interestingly, our data indicated a significant reduction of Fe(III) uptake, but only a trend to reduce Fe(II) uptake, at the presence of LTCC blockers. Together with evidence that a reduction of Fe(III) is required for NTBI uptake into cardiomyocytes [22,28], it raised the possibility that LTCC blockers achieve the effect on NTBI blockade not by stopping Fe(II) entry directly but through alternative mechanism. Recent studies provide an alternative explanation on the role of LTCC in NTBI entry. LTCC has been shown to contribute to the activation of endocytotic machinery in neuronal cells [45]. Interestingly, endocytosis has also been demonstrated to be a possible pathway for macromolecule-associated NTBI uptake into various cell types including cardiomyocytes [38,46]. As LTCC blockade interferes calcium-induced endocytosis, a subsequent interruption of Fe(III) uptake via such pathway can be a possible speculation.

The effect of TTCC blockade on iron entry

With abundant expression in embryonic cardiomyocytes, and subsequent suppression shortly after birth [47], TTCC has been shown to reappear in murine hearts with pathological abnormalities including hypertrophy [48], myocardial infarction [49] and also thalassemia [23,24]. Using efonidipine, the TTCC blocker, Kumfu et al. shows effective blockade of iron uptake both *in vitro*

and *in vivo* using the thalassemic mice model, together with the protection effects as assessed *in vivo*, while LTCC blockers appeared inferior [23,24]. However, in our present experimental model, with pretreatment of efonidipine, uptake of neither Fe(II) nor Fe(III) was significantly decreased in iron-overloaded HL-1 cardiomyocytes, implicating an insignificant role of TTCC in HL-1 cells for excessive iron uptake.

Differences in study models

The mechanisms and portal of iron entry into cardiomyocytes under iron overload condition have been controversial, in part being related to differences in experiment approaches, types of iron load models, and the nature of cardiomyocytes explored. In the present study, immortalized HL-1 atrial myocytes were used, which have the advantages of being the only cardiomyocyte cell line currently available that continuously divides and spontaneously contracts while retaining a differentiated adult cardiac phenotype [25,26]. Apart from the superior cardiac properties and cell purity compared with isolated primary cardiomyocytes, HL-1 cells express, from molecular and functional regards, both LTCC and TTCC *in vitro* [27]. In addition, atrial myocytes may provide a model for the study of cardiac iron toxicity, given that atrial dilation and dysfunction have been reported to be earlier markers than depressed ventricular function of cardiac iron toxicity in patients with thalassemia major [50].

LTCC blockade and cardiomyocyte apoptosis

For the therapy of IOC, protection of iron overload induced cardiac apoptosis is apparently crucial beyond the maintenance of regular iron metabolism. Such protection effect was presented in our *in vitro* study by deferiprone, the effective iron chelator commonly used in current clinical practice [51]. However, in our assessment, none of the calcium channel blockers showed significant protection effect on iron overload induced apoptosis, though LTCC blockers, in particular amlodipine, presented slight protection at 600 µM of ferric or ferrous iron challenge. This result is to a certain extent contrary to the previous finding that amlodipine and verapamil attenuate cardiac apoptosis in iron-overloaded mice evaluated by TUNEL assay [20]. One possible explanation is that NTBI initiates apoptosis of cardiomyocyte prior to its entry through cell membrane; and for the *in vivo* model, apart from the effect on NTBI blockade, it cannot rule out the possible contribution from the impacts of LTCC blockers on other physiological conditions which subsequently reduce such iron induced apoptosis. Furthermore, the different susceptibility to iron overload between atrial and ventricular cardiomyocytes should also be taken into consideration [52]. Despite the demonstrable ability of LTCC blockers to inhibit iron ingress into the cytosol of cardiomycytes, their apparent failure to protect them from apoptosis might be due to various properties associated with iron traffic within cells, particularly between cytosol and into mitochondria. As shown earlier [38,53,54], a major fraction of exogenously added iron can access mitochondria, by mechanism that seemingly by-pass the labile iron pool, which is sensed by the calcein probe, and it can even be refractory to some intracellular chelators [38]. While those features would imply that LTCC might provide a path for NTBI entry into cardiomyocytes, they also indicate that such paths might not be relevant for trafficking iron across cytosol to mitochondria, particularly in the pathophysiological context. Consequently, although the prevention of iron ingress into cardiomyocytes was observed in treatment with LTCC blockers at higher doses, due to their toxicity, at least shown *in vitro*, further studies would be of importance for their protective roles in iron-overloaded cardiomyocytes, and also for a better understanding of the etiology of IOC.

Clinical implications

Apoptosis is rare in normal human heart. In all reported cases, including those in failing hearts, apoptosis levels are substantially lower than 1% as revealed by TUNEL assay [55]. Due to the poor regenerative capacity of cardiomyocytes, a constant, albeit low, level of apoptosis can have serious consequence. Apart from limited studies showing the potential anti-apoptotic effect of deferasirox [35] and taurine [4] in myocardium of iron-overloaded murine model, little is known about the anti-apoptotic approach for iron overload. Further studies on the mechanism of iron induced apoptosis would provide novel targets for advanced therapy against IOC.

Limitations

Several limitations to this study warrant discussion. Firstly, the findings of the present *in vitro* study may reflect perhaps a relatively acute effect of iron load on cardiomyocytes. Ideally, the experimental protocols should be extended to longer duration with

lower iron levels. However, given the technical constraints including the confounding influence of cell proliferation with prolonged culture on fluorescent assay of iron entry and the need for medium change with alteration in iron concentrations, we have elected to adopt the current methodology. With regard to animal studies, previous works have been done on mouse [20,24] and gerbil [35], which mimic the effect of chronic iron overload better, although results remained controversial. Secondly, we have not assessed the effects of calcium channel blockade on cellular beating in the present study. Calcium channel blockade may reduce beating rate or cause cessation of cardiomyocyte contraction *in vitro* [56,57]. Nonetheless, LTCC and TTCC have been shown to remain functional in HL-1 cells without apparent contraction [27,58]. The effect of cardiomyocytes beating rate on iron uptake, however, requires further studies for its clarification. Thirdly, although HL-1 cells are the only cardiomyocyte cell line that retains contractile phenotype with differentiated cardiac characteristics [25,26], they are established from AT-1 mouse atrial cardiomyocyte tumor lineage. The different electrical properties, including calcium kinetics, between atrial and ventricular myocytes [52] may potentially lead to differences in response to iron overload between HL-1 cells and ventricular cardiomyocytes merit further studies. With advances in the induced pluripotent stem cell technology, the use of human ventricular cardiomyocytes may be a better model to study the effects of iron cardiotoxicity. Finally, we have not assessed the detailed pro-apoptotic signaling pathways in the present study. In mesenchymal stem cells [59,60], hepatocytes [61], neuroblastoma cells [62] and gerbil [63], p38 and JNK are activated under iron overload conditions. This would undoubtedly be important when designing future studies.

Conclusions

In summary, our current study illustrated the patterns of iron entry in HL-1 atrial myocytes under ferric or ferrous iron overload condition. The blockade of LTCC but not TTCC was identified to prevent labile ferric iron entry. The uptake of ferrous iron probably involves other mechanism. As expected, iron overload was shown to induce cardiac apoptosis via mitochondria-mediated caspase-3 dependent pathways. However, LTCC blockers have very limited protective effect toward iron induced apoptosis. Our study provided a better understanding to the role of LTCC and TTCC on NTBI uptake into cardiomyocytes, contributing to the conceptual framework in the development of advanced therapeutic strategy for IOC in combination with the current chelation therapy.

Acknowledgments

We thank Prof. William C. Claycomb (Louisiana State University Health Science Center, LA, USA) for HL-1 cardiomyocytes, and Dr. Wing Keung Chan (St. Jude Children's Research Hospital, TN, USA) for technical advice on experiments and data analysis. We also thank Prof. George J. Kontoghiorghes (Postgraduate Research Institute, Limassol,Cyprus) for comment on iron chelation experiment.

Author Contributions

Conceived and designed the experiments: MPC GCFC YFC. Performed the experiments: MPC. Analyzed the data: MPC GCFC YFC. Contributed reagents/materials/analysis tools: MPC ZIC SC. Wrote the paper: MPC ZIC GCFC YFC.

References

1. Lieu PT, Heiskala M, Peterson PA, Yang Y (2001) The roles of iron in health and disease. Mol Aspects Med 22: 1–87.
2. Esposito BP, Breuer W, Sirankapracha P, Pootrakul P, Hershko C, et al. (2003) Labile plasma iron in iron overload: redox activity and susceptibility to chelation. Blood 102: 2670–2677.
3. Hershko CM, Link GM, Konijn AM, Cabantchik ZI (2005) Iron chelation therapy. Curr Hematol Rep 4: 110–116.
4. Oudit GY, Trivieri MG, Khaper N, Husain T, Wilson GJ, et al. (2004) Taurine supplementation reduces oxidative stress and improves cardiovascular function in an iron-overload murine model. Circulation 109: 1877–1885.
5. Gujja P, Rosing DR, Tripodi DJ, Shizukuda Y (2010) Iron overload cardiomyopathy: better understanding of an increasing disorder. J Am Coll Cardiol 56: 1001–1012.
6. Kremastinos DT, Tiniakos G, Theodorakis GN, Katritsis DG, Toutouzas PK (1995) Myocarditis in beta-thalassemia major. A cause of heart failure. Circulation 91: 66–71.
7. Kremastinos DT, Flevari P, Spyropoulou M, Vrettou H, Tsiapras D, et al. (1999) Association of heart failure in homozygous beta-thalassemia with the major histocompatibility complex. Circulation 100: 2074–2078.
8. Muhlestein JB (2000) Cardiac abnormalities in hemochromatosis. In: Barton JC, Edwards CQ, editors. Hemochromatosis: genetics, pathophysiology, diagnosis, and treatment Cambridge University Press. pp. 297–310.
9. Murphy CJ, Oudit GY (2010) Iron-overload cardiomyopathy: pathophysiology, diagnosis, and treatment. J Card Fail 16: 888–900.
10. Olivieri NF, Nathan DG, MacMillan JH, Wayne AS, Liu PP, et al. (1994) Survival in medically treated patients with homozygous beta-thalassemia. N Engl J Med 331: 574–578.
11. Horwitz LD, Rosenthal EA (1999) Iron-mediated cardiovascular injury. Vasc Med 4: 93–99.
12. Narula J, Haider N, Virmani R, DiSalvo TG, Kolodgie FD, et al. (1996) Apoptosis in myocytes in end-stage heart failure. N Engl J Med 335: 1182–1189.
13. Olivetti G, Abbi R, Quaini F, Kajstura J, Cheng W, et al. (1997) Apoptosis in the failing human heart. N Engl J Med 336: 1131–1141.
14. Kang PM, Izumo S (2000) Apoptosis and heart failure: A critical review of the literature. Circ Res 86: 1107–1113.
15. Wencker D, Chandra M, Nguyen K, Miao W, Garantziotis S, et al. (2003) A mechanistic role for cardiac myocyte apoptosis in heart failure. J Clin Invest 111: 1497–1504.
16. Foo RS, Mani K, Kitsis RN (2005) Death begets failure in the heart. J Clin Invest 115: 565–571.
17. Lee Y, Gustafsson AB (2009) Role of apoptosis in cardiovascular disease. Apoptosis 14: 536–548.
18. Hentze MW, Muckenthaler MU, Andrews NC (2004) Balancing acts: molecular control of mammalian iron metabolism. Cell 117: 285–297.
19. Chattipakorn N, Kumfu S, Fucharoen S, Chattipakorn S (2011) Calcium channels and iron uptake into the heart. World J Cardiol 3: 215–218.
20. Oudit GY, Sun H, Trivieri MG, Koch SE, Dawood F, et al. (2003) L-type Ca2+ channels provide a major pathway for iron entry into cardiomyocytes in iron-overload cardiomyopathy. Nat Med 9: 1187–1194.
21. Oudit GY, Trivieri MG, Khaper N, Liu PP, Backx PH (2006) Role of L-type Ca2+ channels in iron transport and iron-overload cardiomyopathy. J Mol Med (Berl) 84: 349–364.
22. Tsushima RG, Wickenden AD, Bouchard RA, Oudit GY, Liu PP, et al. (1999) Modulation of iron uptake in heart by L-type Ca2+ channel modifiers: possible implications in iron overload. Circ Res 84: 1302–1309.
23. Kumfu S, Chattipakorn S, Chinda K, Fucharoen S, Chattipakorn N (2012) T-type calcium channel blockade improves survival and cardiovascular function in thalassemic mice. Eur J Haematol 88: 535–548.
24. Kumfu S, Chattipakorn S, Srichairatanakool S, Settakorn J, Fucharoen S, et al. (2011) T-type calcium channel as a portal of iron uptake into cardiomyocytes of beta-thalassemic mice. Eur J Haematol 86: 156–166.
25. Claycomb WC, Lanson NA Jr, Stallworth BS, Egeland DB, Delcarpio JB, et al. (1998) HL-1 cells: a cardiac muscle cell line that contracts and retains phenotypic characteristics of the adult cardiomyocyte. Proc Natl Acad Sci U S A 95: 2979–2984.
26. White SM, Constantin PE, Claycomb WC (2004) Cardiac physiology at the cellular level: use of cultured HL-1 cardiomyocytes for studies of cardiac muscle cell structure and function. Am J Physiol Heart Circ Physiol 286: H823–829.
27. Xia S, Salata JJ, Figueroa DJ, Lawlor AM, Liang HA, et al. (2004) Functional expression of L- and T-type Ca2+ channels in murine HL-1 cells. J Mol Cell Cardiol 36: 111–119.
28. Parkes JG, Olivieri NF, Templeton DM (1997) Characterization of Fe2+ and Fe3+ transport by iron-loaded cardiac myocytes. Toxicology 117: 141–151.
29. Randell EW, Parkes JG, Olivieri NF, Templeton DM (1994) Uptake of non-transferrin-bound iron by both reductive and nonreductive processes is modulated by intracellular iron. J Biol Chem 269: 16046–16053.
30. Glickstein H, El RB, Shvartsman M, Cabantchik ZI (2005) Intracellular labile iron pools as direct targets of iron chelators: a fluorescence study of chelator action in living cells. Blood 106: 3242–3250.
31. Lecoeur H, Melki MT, Saidi H, Gougeon ML (2008) Analysis of apoptotic pathways by multiparametric flow cytometry: application to HIV infection. Methods Enzymol 442: 51–82.
32. Oancea M, Mazumder S, Crosby ME, Almasan A (2006) Apoptosis assays. Methods Mol Med 129: 279–290.
33. Kontoghiorghes GJ (2006) Iron mobilization from transferrin and non-transferrin-bound-iron by deferiprone. Implications in the treatment of thalassemia, anemia of chronic disease, cancer and other conditions. Hemoglobin 30: 183–200.
34. Nday CM, Malollari G, Petanidis S, Salifoglou A (2012) In vitro neurotoxic Fe(III) and Fe(III)-chelator activities in rat hippocampal cultures. From neurotoxicity to neuroprotection prospects. J Inorg Biochem 117: 342–350.
35. Wang Y, Wu M, Al-Rousan R, Liu H, Fannin J, et al. (2011) Iron-induced cardiac damage: role of apoptosis and deferasirox intervention. J Pharmacol Exp Ther 336: 56–63.
36. Munoz JP, Chiong M, Garcia L, Troncoso R, Toro B, et al. (2010) Iron induces protection and necrosis in cultured cardiomyocytes: Role of reactive oxygen species and nitric oxide. Free Radic Biol Med 48: 526–534.
37. Whelan RS, Kaplinskiy V, Kitsis RN (2010) Cell death in the pathogenesis of heart disease: mechanisms and significance. Annu Rev Physiol 72: 19–44.
38. Shvartsman M, Kikkeri R, Shanzer A, Cabantchik ZI (2007) Non-transferrin-bound iron reaches mitochondria by a chelator-inaccessible mechanism: biological and clinical implications. Am J Physiol Cell Physiol 293: C1383–1394.
39. Catterall WA, Striessnig J (1992) Receptor sites for Ca2+ channel antagonists. Trends Pharmacol Sci 13: 256–262.
40. Winegar BD, Kelly R, Lansman JB (1991) Block of current through single calcium channels by Fe, Co, and Ni. Location of the transition metal binding site in the pore. J Gen Physiol 97: 351–367.
41. Atar D, Backx PH, Appel MM, Gao WD, Marban E (1995) Excitation-transcription coupling mediated by zinc influx through voltage-dependent calcium channels. J Biol Chem 270: 2473–2477.
42. Hamann SR, Blouin RA, McAllister RG Jr (1984) Clinical pharmacokinetics of verapamil. Clin Pharmacokinet 9: 26–41.
43. Mak IT, Weglicki WB (1990) Comparative antioxidant activities of propranolol, nifedipine, verapamil, and diltiazem against sarcolemmal membrane lipid peroxidation. Circ Res 66: 1449–1452.
44. Lansman JB, Hess P, Tsien RW (1986) Blockade of current through single calcium channels by Cd2+, Mg2+, and Ca2+. Voltage and concentration dependence of calcium entry into the pore. J Gen Physiol 88: 321–347.
45. Rosa JM, Nanclares C, Orozco A, Colmena I, de Pascual R, et al. (2012) Regulation by L-Type Calcium Channels of Endocytosis: An Overview. J Mol Neurosci.
46. Sohn YS, Ghoti H, Breuer W, Rachmilewitz E, Attar S, et al. (2012) The role of endocytic pathways in cellular uptake of plasma non-transferrin iron. Haematologica 97: 670–678.
47. Yasui K, Niwa N, Takemura H, Opthof T, Muto T, et al. (2005) Pathophysiological significance of T-type Ca2+ channels: expression of T-type Ca2+ channels in fetal and diseased heart. J Pharmacol Sci 99: 205–210.
48. Martinez ML, Heredia MP, Delgado C (1999) Expression of T-type Ca(2+) channels in ventricular cells from hypertrophied rat hearts. J Mol Cell Cardiol 31: 1617–1625.
49. Huang B, Qin D, Deng L, Boutjdir M, Nabil ES (2000) Reexpression of T-type Ca2+ channel gene and current in post-infarction remodeled rat left ventricle. Cardiovasc Res 46: 442–449.
50. Li W, Coates T, Wood JC (2008) Atrial dysfunction as a marker of iron cardiotoxicity in thalassemia major. Haematologica 93: 311–312.
51. Kolnagou A, Kleanthous M, Kontoghiorghes GJ (2011) Efficacy, compliance and toxicity factors are affecting the rate of normalization of body iron stores in thalassemia patients using the deferiprone and deferoxamine combination therapy. Hemoglobin 35: 186–198.
52. Grandi E, Pandit SV, Voigt N, Workman AJ, Dobrev D, et al. (2011) Human atrial action potential and Ca2+ model: sinus rhythm and chronic atrial fibrillation. Circ Res 109: 1055–1066.
53. Shvartsman M, Fibach E, Cabantchik ZI (2010) Transferrin-iron routing to the cytosol and mitochondria as studied by live and real-time fluorescence. Biochem J 429: 185–193.
54. Shvartsman M, Ioav Cabantchik Z (2012) Intracellular iron trafficking: role of cytosolic ligands. Biometals 25: 711–723.
55. Chiong M, Wang ZV, Pedrozo Z, Cao DJ, Troncoso R, et al. (2011) Cardiomyocyte death: mechanisms and translational implications. Cell Death Dis 2: e244.
56. Wang T, Hu N, Cao J, Wu J, Su K, et al. (2013) A cardiomyocyte-based biosensor for antiarrhythmic drug evaluation by simultaneously monitoring cell growth and beating. Biosens Bioelectron 49: 9–13.
57. Jonsson MK, Wang QD, Becker B (2011) Impedance-based detection of beating rhythm and proarrhythmic effects of compounds on stem cell-derived cardiomyocytes. Assay Drug Dev Technol 9: 589–599.
58. Rao F, Deng CY, Wu SL, Xiao DZ, Huang W, et al. (2013) Mechanism of macrophage migration inhibitory factor-induced decrease of T-type Ca(2+) channel current in atrium-derived cells. Exp Physiol 98: 172–182.

59. Lu WY, Zhao MF, Chai X, Meng JX, Zhao N, et al. (2013) [Reactive oxygen species mediate the injury and deficient hematopoietic supportive capacity of umbilical cord derived mesenchymal stem cells induced by iron overload]. Zhonghua Yi Xue Za Zhi 93: 930–934.

60. Lu WY, Zhao MF, Sajin R, Zhao N, Xie F, et al. (2013) [Effect and mechanism of iron-catalyzed oxidative stress on mesenchymal stem cells]. Zhongguo Yi Xue Ke Xue Yuan Xue Bao 35: 6–12.

61. Dai J, Huang C, Wu J, Yang C, Frenkel K, et al. (2004) Iron-induced interleukin-6 gene expression: possible mediation through the extracellular signal-regulated kinase and p38 mitogen-activated protein kinase pathways. Toxicology 203: 199–209.

62. Salvador GA, Oteiza PI (2011) Iron overload triggers redox-sensitive signals in human IMR-32 neuroblastoma cells. Neurotoxicology 32: 75–82.

63. Al-Rousan RM, Paturi S, Laurino JP, Kakarla SK, Gutta AK, et al. (2009) Deferasirox removes cardiac iron and attenuates oxidative stress in the iron-overloaded gerbil. Am J Hematol 84: 565–570.

10

Metabolic Profiling and Flux Analysis of MEL-2 Human Embryonic Stem Cells during Exponential Growth at Physiological and Atmospheric Oxygen Concentrations

Jennifer Turner[1,2,9], Lake-Ee Quek[3,9], Drew Titmarsh[1,2], Jens O. Krömer[3,4], Li-Pin Kao[2], Lars Nielsen[3,4], Ernst Wolvetang[2]*, Justin Cooper-White[1,5]*

1 Tissue Engineering and Microfluidics Laboratory, The Australian Institute for Bioengineering and Nanotechnology, The University of Queensland, St Lucia, Queensland, Australia, 2 Stem Cell Engineering Group, The Australian Institute for Bioengineering and Nanotechnology, The University of Queensland, St Lucia, Queensland, Australia, 3 Centre for Systems and Synthetic Biotechnology, The Australian Institute for Bioengineering and Nanotechnology, The University of Queensland, St Lucia, Queensland, Australia, 4 Metabolomics Australia, The Australian Institute for Bioengineering and Nanotechnology, The University of Queensland, St Lucia, Queensland, Australia, 5 School of Chemical Engineering, The University of Queensland, St Lucia, Queensland, Australia

Abstract

As human embryonic stem cells (hESCs) steadily progress towards regenerative medicine applications there is an increasing emphasis on the development of bioreactor platforms that enable expansion of these cells to clinically relevant numbers. Surprisingly little is known about the metabolic requirements of hESCs, precluding the rational design and optimisation of such platforms. In this study, we undertook an in-depth characterisation of MEL-2 hESC metabolic behaviour during the exponential growth phase, combining metabolic profiling and flux analysis tools at physiological (hypoxic) and atmospheric (normoxic) oxygen concentrations. To overcome variability in growth profiles and the problem of closing mass balances in a complex environment, we developed protocols to accurately measure uptake and production rates of metabolites, cell density, growth rate and biomass composition, and designed a metabolic flux analysis model for estimating internal rates. hESCs are commonly considered to be highly glycolytic with inactive or immature mitochondria, however, whilst the results of this study confirmed that glycolysis is indeed highly active, we show that at least in MEL-2 hESC, it is supported by the use of oxidative phosphorylation within the mitochondria utilising carbon sources, such as glutamine to maximise ATP production. Under both conditions, glycolysis was disconnected from the mitochondria with all of the glucose being converted to lactate. No difference in the growth rates of cells cultured under physiological or atmospheric oxygen concentrations was observed nor did this cause differences in fluxes through the majority of the internal metabolic pathways associated with biogenesis. These results suggest that hESCs display the conventional Warburg effect, with high aerobic activity despite high lactate production, challenging the idea of an anaerobic metabolism with low mitochondrial activity. The results of this study provide new insight that can be used in rational bioreactor design and in the development of novel culture media for hESC maintenance and expansion.

Editor: Martin Pera, University of Melbourne, Australia

Funding: The authors would like to acknowledge the financial support of the Australian Research Council Special Research Initiative Stem Cells Australia (SR110001002), and the support of the Australian Stem Cell Centre's Core hESC Laboratories (Stem Core) for providing cell culture and support services. The authors would also like to acknowledge the Queensland Brain Institute Flow Cytometry Facility for assistance sorting cells. The funders had no role in study design, data collection and analysis, decision to publish, or preparation of the manuscript.

Competing Interests: The authors have declared that no competing interests exist.

* Email: e.wolvetang@uq.edu.au (EW); j.cooperwhite@uq.edu.au (JCW)

9 JT and LQ are co-first authors on this work.

Introduction

The pluripotent nature of human embryonic stem cells (hESCs) along with their capacity for unlimited self-renewal makes them ideal candidates for use in regenerative medicine. However, before this potential can truly be realised expansion of hESCs to clinically relevant numbers must be based on a more detailed understanding of their metabolic and growth characteristics compared with workhorse lines such as CHO cells [1]. This has led to research and exploration into hESC expansion [2,3], bioreactor platforms [4–7] and the development of maintenance media [8–13] and has led to an explosion in biological research to understand the molecular mechanisms governing hESC behaviour. There has been very little exploration however into the fundamental metabolic requirements necessary to support cell expansion in a pluripotent state. Such data would enable the rational design of hESC expansion systems. This work, which describes an in-depth study of hESC metabolism during the exponential growth phase, addresses this deficit.

Human ESC cultures are generally considered to be highly metabolically active, with energy substrates such as glucose being rapidly consumed and waste products such as lactate and ammonia rapidly produced [14]. This provides an explanation for why daily medium changes are necessary in routine hESC culture, a constraint highlighted in a recent study that demonstrated that lactate levels of 25 mM or above inhibit proliferation of hESCs [14] and effects the pluripotent state as indicated by a reduction in Tra-1-60 expression [14].

In addition to being highly metabolically active, previous work has suggested that ESCs are also highly glycolytic [14–16], with lactate production to glucose consumption ratios reported to be between 1.8 and 2.8 [14,15] of a theoretical maximum of 2. This indicates that the majority of the pyruvate generated from glucose metabolism is converted to lactate, rather than entering the mitochondria and the citric acid (TCA) cycle. In contrast, adult mammalian cells typically exhibit lactate to glucose ratios of 1.5 to 1.7 [17]. hESCs are also known to have fewer mitochondria than terminally differentiated cells [18], and possess mitochondria that appear immature and lack normal cristae [16,19]. The highly glycolytic nature of hESCs combined with the immature structure of their mitochondria has led to the proposition that the mitochondria in hESCs are less active than in differentiated cells, and that energy generation by oxidative phosphorylation in the mitochondria plays no or only a minor role in hESCs [16]. It is interesting to note however, that a recent study found that expression of TCA cycle genes is higher in hESCs than in differentiated cells [16].

Since hESCs are isolated from the inner cell mass of a day 5 blastocyst prior to implantation [20] and *in vivo* reside in a hypoxic environment [21], the effect of oxygen tension on hESC behaviour has been an area of intense research. Physiological oxygen concentrations of 1–5% have been shown to improve human and mouse embryonic stem cell ESC survival [22,23] and enhance the maintenance of pluripotency [24–26], when compared with atmospheric oxygen concentrations of 20% typically used in routine hESC maintenance. While there is data available in the literature on amino acid [14], glucose [14,16], lactate [14,16] and ammonium [14] uptake and production rates for hESCs cultured at atmospheric oxygen concentrations, to date little has been done to determine how the metabolic requirements differ when cultured at physiologically-relevant oxygen concentrations (e.g. 1–5%). In addition, current studies limit themselves to uptake and production rates of only a few key metabolites and have not investigated the activity of internal metabolic pathways that are known to be highly interconnected.

In order to gain greater understanding of hESC metabolism, we have employed metabolic profiling and flux analysis techniques to investigate the metabolic requirements and to predict the activity of the internal metabolic pathways at both physiological (2%) and atmospheric (20%) oxygen concentrations. The metabolic flux analysis (MFA) model developed herein consists of over 2000 reactions, which should realistically approximate hESC metabolism. Quantitative analysis shows that the model accounts for all major carbon sources, amino acids, metabolic by-products, total DNA content and total protein content measured during the exponential growth phase of hESC culture.

To our knowledge, even though it is currently only available for one hESC line, this is the first detailed metabolic flux analysis of hESCs under varying oxygen conditions, and the resultant data the most in depth for hESC metabolism to date. The results of this study will provide a useful resource for researchers interested in probing and understanding hESC metabolism, improving rational bioreactor design and the development of novel culture media for hESC maintenance and expansion.

Methods

Cell culture

The hESCs used within this study were of the MEL-2 hESC line, previously characterised by the International Stem Cell Initiative [27]. All cells used within this study were obtained from the Australian Stem Cell Centre, Queensland Node Core hESC Laboratory (StemCore). In order to accurately enumerate cell number for the metabolic flux analysis, it was necessary to be able to passage hESCs as a single cell suspension. Adapted hESCs for passaging as a single-cell suspension (using TrypLE express (Invitrogen)) were provided by StemCore, following protocols adapted from Costa et al 2008 [28]. Briefly, cells were moved from passaging as pieces (W), to bulk enzymatic passaging (Y), to passaging as a single cell suspension on a mouse embryonic fibroblast (MEF) layer (Y), and finally passaging as a single cell suspension in a feeder-free system (Z). Passage numbers were designated as pW+X+Y+Z. Cells received from StemCore were p18+2+9−13. They were then passaged as a single cell suspension for a further two passages on BD Matrigel hESC-qualified matrix (BD Biosciences), used at 1/8 of the recommended concentration, in mTeSR-1 (StemCell Technologies) prior to use in all experiments.

Cell sorting, culture initiation and sampling

Cell stocks were harvested as a single cell suspension and sorted into well plates for metabolism experiments. Prior to cell detachment the culture medium was supplemented with 10 μM of small molecule Y-27632 dihydrochloride monohydrate (ROCK inhibitor) (Sigma) and incubated for 2 hours. To detach cells in a single cell suspension for sorting, cells were rinsed twice with warm PBS and treated with TrypLE Express (Invitrogen) for 5 minutes at 37°C. Cells were spun down and resuspended in DMEM F12 (Invitrogen) +1% (v/v) Penicillin/Streptomycin (Invitrogen) + 10 μM Y-27632 + propidium iodide (PI).

Cells were sorted with either a Cytopeia Influx or a BD FACS Aria Instrument at the Queensland Brain Institute Flow Cytometry Facility, University of Queensland. Cells were sorted into 6-well plates that had been pre-coated with BD Matrigel hESC-qualified matrix (BD Biosciences), at 1/8 of the recommended concentration at 4°C overnight, and filled with 2 mL mTeSR-1 (StemCell Technologies) supplemented with 1% (v/v) Penicillin/Streptomycin (Invitrogen) and 10 μM Y-27632 (Sigma). 120,000 cells were sorted into each well, resulting in an even distribution of cells. Only viable single cells were sorted into wells utilising doublet exclusion (using forward- and side-scatter height and width parameters) and exclusion of cells staining positive for PI. Once sorted the cells were placed in the appropriate incubator, either a Binder CB160 variable oxygen incubator at 2% O_2 and 5% CO_2 or a standard Sanyo CO_2 incubator set at 20% O_2 and 5% CO_2. The cells were left overnight to attach. The cell culture medium was then exchanged with 2 mL fresh mTeSR-1 supplemented with 1% (v/v) Penicillin/Streptomycin (Invitrogen) per well to remove Y-27632 and any unattached cells from the system. Note that media exchange was performed using media preconditioned to match oxygen concentration, that is, either preconditioned hypoxic media or normoxic media. This media change marked the commencement of the experiment and was considered time = 0 hours.

Samples were then collected approximately every eight hours for 4–5 days. For each biological replicate at each timepoint, the following procedure was conducted: 1) Phase-contrast images were taken using an Olympus IX81 inverted microscope; 2) Cell culture supernatant was collected, filtered with a 0.22 μm syringe filter and frozen at $-20°C$ until analysis; and 3) Cells were harvested using TrypLE express (Invitrogen) and fixed for 10 minutes in filtered ice-cold 70% ethanol and stored in PBS +2% foetal bovine serum (FBS) (Invitrogen) at 4°C until cell number enumeration analysis.

Cell imaging

Phase contrast images were taken of each biological replicate at time = 0 hours and each subsequent timepoint. Images were taken using an Olympus IX81 inverted microscope with a 4x objective lens. Pre-programmed stage positions, using Cell-R software, were utilised to ensure that images were taken at the same position within the culture well at each timepoint.

Cell enumeration by flow cytometry

Flow cytometry was used to quantify cell numbers throughout the study. 35 μL of a known concentration of Flow-Count Fluorospheres (Beckman Coulter) was added to a fixed cell sample and vortexed to mix. Approximately 30 minutes prior to flow cytometry analysis, 10 ug/mL Hoechst 33342 (Invitrogen, Molecular Probes) was added to each sample to obtain a cell cycle profile to allow cells to be readily discriminated from debris. Samples were analysed on a BD LSR II flow cytometer at the Queensland Brain Institute Flow Cytometry Facility. Counting beads were identified with a FITC (530/30) detector, and Hoechst positive cells were identified with the (450/50) detector, see Figure S1 for more detail. Multiplets could be discerned from the Hoechst histogram as cells brighter than the G2/M peak. Events were recorded until 1000 beads had been counted. Cell number could then be determined by Equation 1: Total cells = ((volume of beads added x bead concentration)/(#beads counted)) x (# cells counted + # doublets).

Biochemical analysis

Frozen cell culture supernatant samples were thawed, vortex-mixed and subsequently analysed with a Bioprofile FLEX Chemistry Analyser (Nova Biomedical, Waltham). This analysis provided data for ammonia concentration, pH and osmolality.

High-Performance Liquid Chromatography

Extracellular metabolite concentrations were analysed by high performance liquid chromatography (HPLC). All samples were deproteinated via ultrafiltration (<3 kDa) prior to analysis. Amino acid analysis was performed as described previously [29], except that cysteine was not quantified. Organic acids and glucose were quantified with UV and RI detection, respectively. Separation of compounds was achieved on a Rezex RHM-monosaccharide column (300×7.8 mm, 8 μm, Phenomenex) at 70°C and 0.6 mL min^{-1} of 4 mM H_2SO_4 in water.

Determining total cellular protein content

Cells were sorted into 6-well plates as described above. Cells were cultured at 20% oxygen, as it was assumed that oxygen concentration does not affect total protein. Cells were harvested at timepoints approximately 40 hours, 70 hours and 100 hours after the initial media exchange. At each timepoint cells were harvested with TrypLE express (Invitrogen) and then resuspended in 1 mL of DMEM-F12 (Invitrogen). A known number of viable single cells, either 1×10^5 or 2×10^5 cells, were then sorted into a tube using a Cytopeia Influx cell sorter as described above. Once sorted, the cells were spun down and a known volume of liquid supernatant removed. This was then replaced with 150 μL of RIPA buffer (100 mM EDTA, 50 mM Tris-HCL, pH 8, 150 mM

NaCl, 1% Triton, 0.5% sodium deoxycholate, 0.1% SDS) + 1 x complete EDTA-free protease inhibitor cocktail (ROCHE Applied Science). The tube weight was measured before and after sorting, this combined with controlling the volume of liquid in the tube allowed the cell number per mL to be calculated with a high degree of accuracy. This step was critical to ensuring that our cell numbers were consistent throughout all experiments, allowing for closure of the flux analysis. The protein concentration was determined using a BCA protein assay kit (Pierce Thermo Scientific) according to the manufacturer's instructions. All samples and standards were diluted with a mixture of PBS and RIPA buffer to a comparable concentration. The plate was read on a Spectromax plate reader. Finally, the protein concentration per cell was determined by dividing the protein concentration by the total number of cells in the sample. Three biological replicates were included per experiment and the experiment repeated twice.

Determining total DNA content

Cellular DNA content was ascertained by adding 10 ug/mL Hoechst 33342 (Invitrogen, Molecular Probes) to samples taken at various timepoints for cell enumeration. Staining was detected using a BD LSRII flow cytometer equipped with a UV laser and 450/50 nm detector. Flow cytometric data were analysed in WEASEL v2.7.4 software (Walter and Eliza Hall Institute for Medical Research, Melbourne, Australia). Multiplets were excluded from analysis by gating on forward- and side-scatter area, height, and width parameters. Cells in different stages of the cell cycle (G0/G1, S, G2/M) were estimated by using the Curve Fit - Cell Cycle function in Weasel. This function fits Gaussian distributions to the G0/G1 and G2/M peaks identified in the DNA histogram, and calculates percentages of cells in each compartment. Samples to which curves could not be fitted were discarded. Cell cycle profiles were determined for multiple (n = 4–6) biological replicates at each timepoint.

Metabolic Flux Analysis

The intracellular fluxes of hESC cultures were determined by linear programming using the measured extracellular metabolite concentrations and cell numbers as constraints [30]. The mouse genome scale model was used because it adequately represents core mammalian metabolism and can be directly applied to cell culture flux experiments [31]. Briefly, intracellular fluxes (v) can be calculated using the metabolite balancing constraints $S \cdot v = 0$, whereby S is the stoichiometric matrix derived from the metabolic model, and that cell metabolism is assumed to be at a pseudo-steady state. The constraint $v \geq 0$ is imposed on all irreversible reactions, while the lower and upper boundary values of measured fluxes were specified using the measured cell-specific consumption or production rates and the estimated standard error ($v_{measured} \pm SE_{measured}$). The maximum ATP yield objective function is used in order to generate a version of flux distributions that is energetically most efficient. Flux calculations were done in MATLAB (The Mathworks) using a third-party LP solver (Gurobi Optimizer and Gurobi Mex).

The biomass composition of hESC was approximated using literature values of hybridoma cell lines. It was assumed that RNA content is 3 times of DNA, and that lipid and carbohydrate contents are 1/7 and 1/10 of protein, respectively [32]. The relative fractions of amino acids, nucleotides and lipids within each subgroup were also used (Sheikh 2005). The biomass composition was further refined using measured cellular protein and DNA content of hESC. This was accomplished by adjusting the absolute amount (mmol per cell) of each biomass component such that the weights of the total protein, DNA, RNA, lipid and carbohydrate are met, while maintaining the same relative fractions of amino acids, nucleotides and lipids within each subgroup.

Blocking mitochondrial complex I and II

MEL2 hESCs were seeded at 200,000 cells per well in a 6-well plate coated with BD Matrigel hESC qualified matrix (BD Biosciences). Cells were cultured for 60 hours to ensure that they were in the exponential phase of growth. Cell culture medium was exchanged with either fresh 2 mL of mTeSR-1 (StemCell Technologies) or mTeSR-1 supplemented with 50 µM α-tocopherol succinate (α-TOS) (Sigma) to block mitochondrial respiratory complex II [33], or 0.5 µg/mL rotenone (Sigma) to block mitochondrial complex I [34], or 0.01% ethanol as a carrier control for α-TOS, or 0.01% DMSO as a carrier control for rotenone. The cells were incubated for a further 24 hours prior to harvesting with TrypLE express (Invitrogen) and cell enumeration. All experiments were performed in triplicate.

Western blot analysis

For HIF-1α protein quantification cells were lysed at 80 hours during the exponential growth phase in RIPA buffer (150 mM sodium chloride, 1% Triton X−100, 1% sodium deoxycholate, 0.1% SDS, 50 mM Tris-HCl, pH 7.5, and 2 mM EDTA) supplemented with 1x EDTA-free protease inhibitor cocktail (Roche Applied Science), as recommended by the manufacturer. Cell lysates were then loaded to 6% polyacrylamide gel and electrotransferred onto nitrocellulose membranes. Membranes were blocked with 5% skim milk in TBST (10 mM Tris-HCl buffer, pH 7.6, 150 mM NaCl, and 0.1% Tween-20) for 1 h. Membranes were then incubated with a primary antibody against HIF1α (610958, BD Transduction Laboratories) diluted 1:1000 in TBST and β-tubulin antibody (088K4795, Sigma) diluted 1:10000 in TBST, respectively at 4°C overnight. Primary antibody was detected with a horseradish peroxidase-conjugated anti-mouse secondary antibody (12–348 and 12–349, Millipore) following a 1 hour incubation at RT using the ECL Western blotting detection system (GE Healthcare). Moreover, we detected two bands, the expected ~120 kDa band for HIF1α, and a smaller size band of protein of ~80 kDa molecular weight, corresponding to a lower molecular weight splice variant of HIF1α that has previously been reported in human cells. For the purposes of analysis and comparison, the relative HIF1-α protein concentrations were determined from the sum of both bands using Image J following protocols described in the literature [35] and normalised to β-tubulin protein expression determined using the same protocol. The apparent molecular weights of the proteins detected were 120, 80 and 55 kDa, for HIF1α, HIF1α splice variant (sv) and β-tubulin, respectively. Three biological replicates were measured at both 2% and 20% oxygen.

Quantitative real-time PCR (q-PCR)

Total RNA was extracted from cells at 80 hours during the exponential growth phase at 2% and 20% oxygen with a Qiagen RNeasy Mini RNA extraction kit (Qiagen) and quantified with a NanoDrop 1000 spectrophotometer. cDNA was synthesized, using 1.5 µg of total RNA and 1 µl d(T)$_{20}$ (500 µg/ml) (Geneworks) mixed with RNase-free water to a final volume of 12 µl. The reaction was incubated at 65°C for 5 min, placed on ice, and then 4 µl 5× first-strand buffer and 2 µl 0.1 M DTT (Invitrogen) were added and incubated at 37°C for 2 min. Finally, 1 µl (200 U) MMLV reverse transcriptase (Invitrogen) was added to the reaction and incubated at 37°C for 50 min, followed by incubation at 70°C for 15 min. The resulting cDNA was used for quantitative real-time PCR (qPCR) analysis.

Expression of the genes, HIF1-α, HIF2-α, and β-actin were quantified by qPCR using an ABI 7500 detection system (Applied Biosystems), with fluorescein as an internal passive reference dye for normalization of well-to-well optical variation. PCR amplification was performed in a total volume of 10 µl containing 5 µl 2x SYBR Green supermix (Applied Biosystems), 0.2 µl primers (10 µM each), 0.2 µl cDNA and DNase-free water (Invitrogen, Gibco). The reaction conditions were as follows: 95°C for 1 min, followed by 40 cycles of 95°C for 30 sec, and 52°C for 30 sec, with a final dissociation step to generate a melting curve for verification of amplification product specificity. Real-time qPCR was monitored and analysed with ABI 7500 fast optical system software. The primers used are as follows; HIF-1α, F: 5′-GTAGTTGTG-GAAGTTTATGCTAATATTGTGT-3′, R: 5′-CTTGTTTA-CAGTCTGCTCA-AAATATCTT-3′; β-actin F: 5′-GCTGTG-CTACGTCGCCCTG-3′, R 5′- GGAGGAGCTGGAAGCAG-CC-3′. Each reaction was performed in triplicate, and amplification in the presence of a single primer was performed as a negative control. Relative mRNA levels were calculated using the comparative CT method according to the Applied Biosystems manual and normalized to β-actin mRNA. The fold change in expression of each target mRNA relative to β-actin was calculated by the 2-Δ(ΔCT) method [36,37]. To calculate PCR efficiency, standard curves were generated with serial dilutions of cDNA from experiments performed in triplicate, enabling the determination of CT values and PCR efficiencies for individual assays and variations between individual assays. The PCR efficiency (E) was calculated using the equation E = (10 [1/−slope]−1)×100). Thus, E is between 110% and 90% when the slope falls between −3.1 and −3.6. The slope was calculated by plotting the fold-dilution of cDNA versus the CT value [38].

Flow cytometry analysis of pluripotency marker expression

Cell samples at time = 0 and the endpoint of an experiment were analysed by flow cytometry for expression of pluripotent markers. Samples were blocked with 3% bovine serum albumin (BSA) (Sigma-Aldrich). Samples were incubated for 30 mins at RT with 100 µL of primary antibody against TG30 (1:400) (MAB4427, Millipore), TRA-1-60 (1:400) (MAB4360, Millipore) or Oct-4 (1:400) (MAB4401, Millipore). Samples were washed twice with 500 µL PBS, then incubated for 30 mins at RT with 100 µL of AlexaFluor 488-conjugated secondary antibodies against mouse IgG (H+L) or mouse IgM (µ chain) (1:500) (Invitrogen). 30 minutes prior to analysis 10 ug/mL Hoechst 33342 (Invitrogen, Molecular Probes) was added to each sample to aid in gating the cell population. Multiplets and debris were excluded from the marker expression analysis, as described above.

Karyotype analysis

Regular karyotype analysis of hESC lines is performed by StemCore as part of their routine quality procedures. The MEL-2 hESCs used within this study were karyotyped at p18+2+6 and p18+2+15.

Results

The characterisation of the metabolism of MEL-2 human embryonic stem cells involved three stages. First, the cell growth profile, cell DNA and protein content, as well as the uptake and production rates of key metabolites were measured by experimentation. The second stage utilised the experimental data to develop a fluxomic model of hESC metabolism capable of resolving fluxes through internal metabolic pathways. The third stage involved a

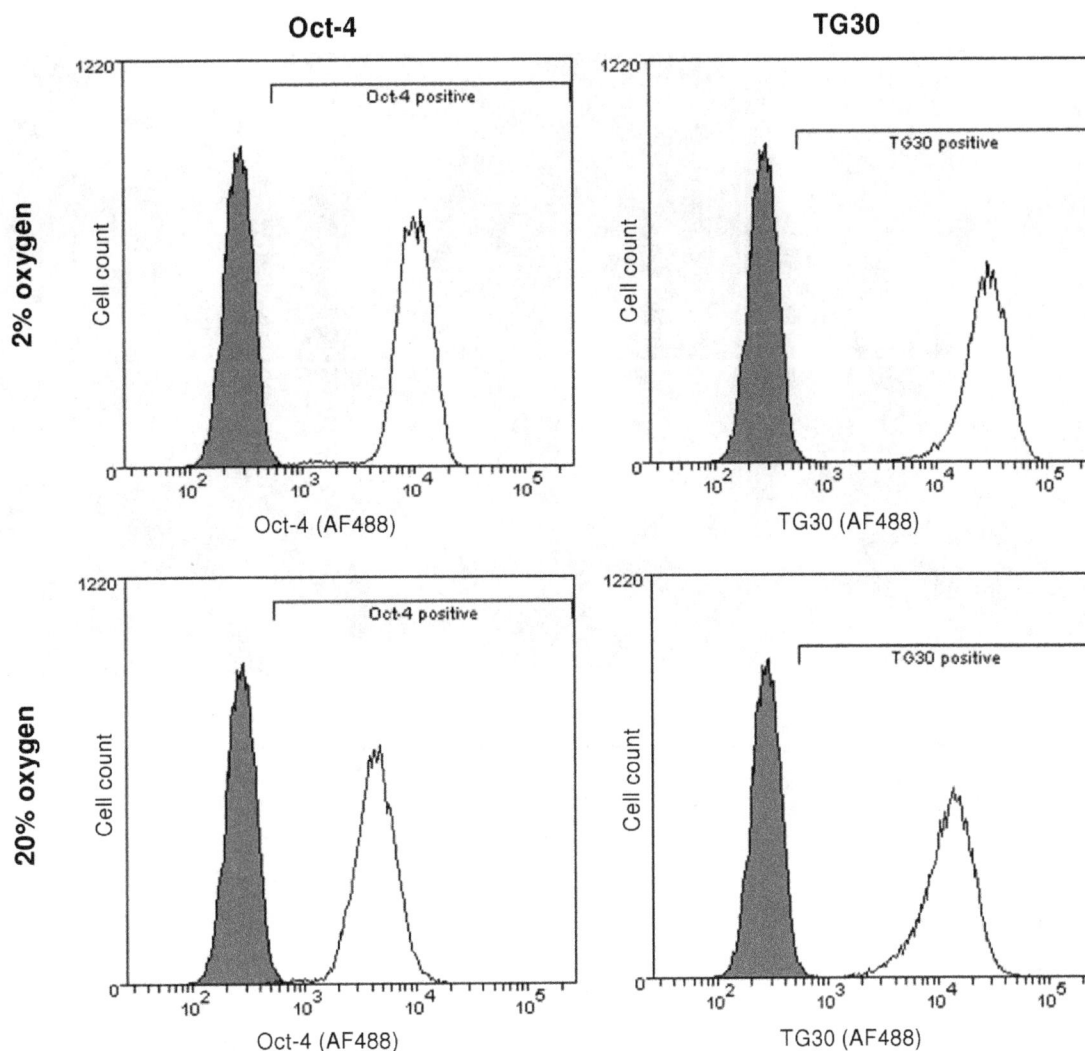

Figure 1. Pluripotent marker expression of human embryonic stem cells. Pluripotent marker expression of hESCs cultured under physiological, 2%, and atmospheric, 20%, oxygen concentrations were tested by FACS analysis after the completion of each experiment to ensure that the cells remained pluripotent throughout the 100 hour experiment. Representative FACS plots showing the shift in marker expression from the negative controls at the endpoint of an experiment are shown. All cell samples were found to be 92–99% positive for both TG30 and Oct-4 after each experiment at both 2% and 20% oxygen.

preliminary analysis of the results generated by the model to highlight key features of hESC metabolism. All stages of the study were conducted at physiological (2%) and atmospheric (20%) oxygen concentrations for comparison.

Pluripotency, control of cell culture conditions and cell DNA and protein measurements

The experiments were performed with MEL-2 hESCs of passage 2 in a single cell adapted state. Karyotype analysis of the cell stocks was performed at passage 15 in a single cell adapted state. No karyotype abnormalities were detected (Figure S2). The cells were stained for pluripotency markers, Oct-4 and TG30, and analysed by flow cytometry before and after each experiment (Figure 1). Cells were found to be 92–99% positive for both Oct-4 and TG30 after each experiment at both 2% and 20% oxygen.

The pH and osmolality of the cell culture medium supernatant was measured at each time point throughout the experiment (Figure S3). The pH and osmolality were within limits to promote

normal cell growth, as defined in Ludwig et al. (2006) [8], throughout the exponential growth phase.

The total protein content of hESCs was measured to be 156 ± 31.9 pg/cell. The total DNA content was estimated from cell cycle data to be 9.39 ± 0.78 pg/cell. These parameters were used to estimate other cell parameters, such as RNA content, using rules of thumb outlined by Bonarius et al. 1996 [32]. See materials and methods for a full list of assumptions.

hESC exponential growth profiles are similar at physiological and atmospheric O_2 concentrations

Throughout this study MEL-2 hESCs were passaged as a single cell suspension using TrypLE, instead of the more traditional enzymatic or manual culture techniques which typically transfer hESCs in clumps of cells to a new culture vessel. Our seeding protocol resulted in a greater control over cell number and allowed for a consistently uniform initial seeding density, which in turn ensured uniform progression through the phases of cell growth within a specific culture condition (Figure 2). The initial lag phase

A

B

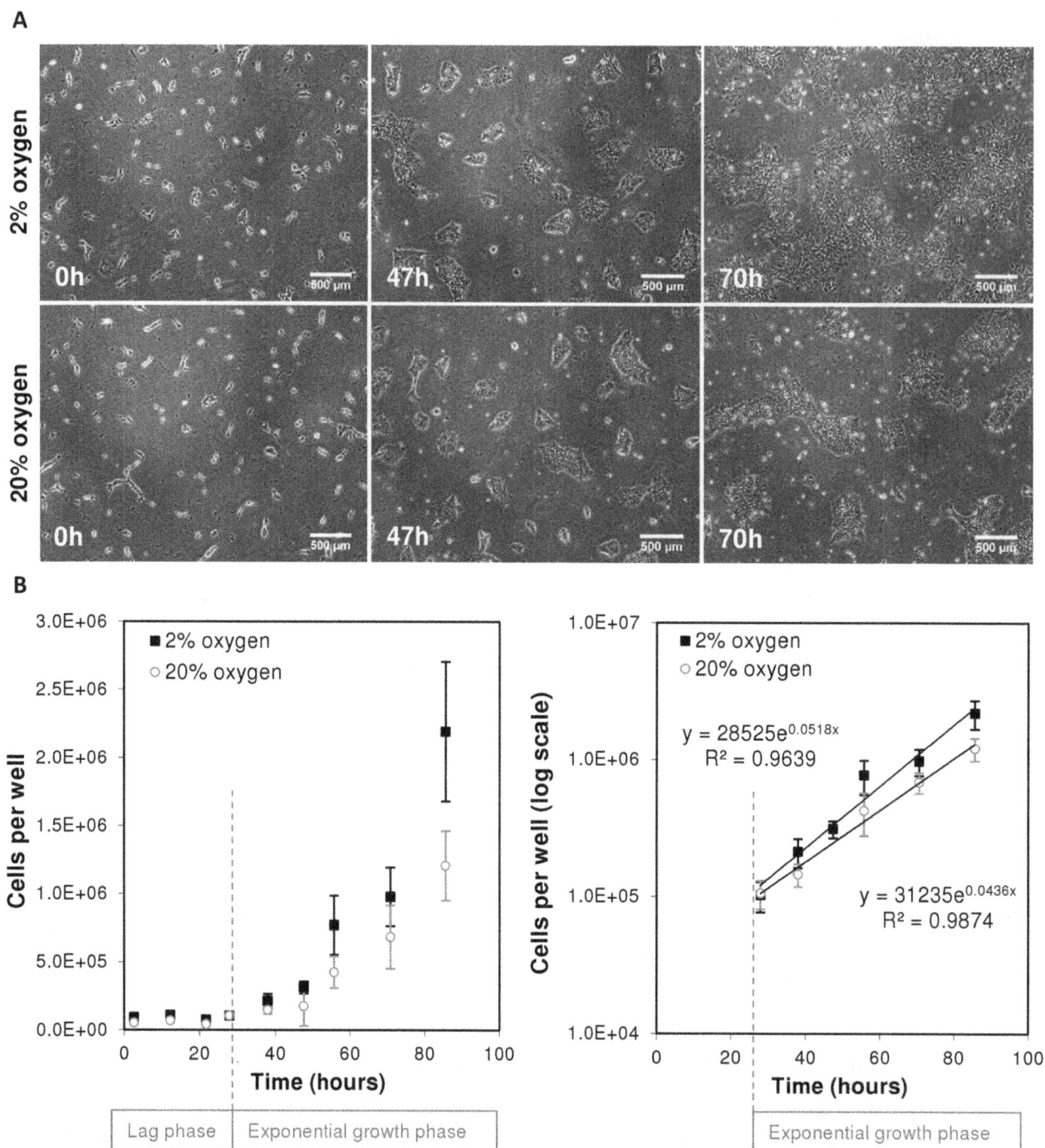

Figure 2. Growth profile of human embryonic stem cells cultured at physiological and atmospheric oxygen concentrations. hESCs were harvested as a single cell suspension and then FACS sorted into 6-well plates to ensure a uniform seeding distribution across all wells. The cells were then cultured at physiological, 2%, and atmospheric, 20%, oxygen concentrations. At each time point cells were harvested and counted using flow cytometry techniques. A) The phase contrast images represent the spatial distribution of cells in the well and colony morphology at time equals 0, 47 and 70 hours after the experiment initialisation. The cell seeding methodologies employed allowed for consistent uniform seeding which translated to consistent initial cell numbers. Scale bar represents 500 μm. B) Growth profile of hESCs showing the lag and exponential growth phase on the semi-log plot to determine the specific growth rates. Values are means $\pm$ standard deviation, n = 6.

lasted approximately 25 hours after the commencement of the experiment, and was observed to be independent of oxygen concentration (Figure 2). The following exponential growth phase was used to calculate the growth rates. The growth rate was found to be 0.051 ± 0.0025 h^{-1} and 0.043 ± 0.0024 h^{-1} (values $\pm$ SE) at

2% and 20% oxygen, respectively (Figure 2). The growth rates were not significantly different, p-value 0.05. Only data from the exponential growth phase was used in constructing the metabolism profile and flux analysis.

Table 1. Flux of major metabolites consumed and produced by human embryonic stem cells cultured at physiological and atmospheric oxygen concentrations.

Consumed metabolites	Specific Rates (mmol h^{-1} 112 cells^{-1})		Produced metabolites	Specific Rates (mmol h^{-1} 112 cells^{-1})	
	2% Oxygen	20% Oxygen		2% Oxygen	20% Oxygen
Glucose**	569.91±56.85	307.45±23.03	Lactate**	1316.94±121	712.94±56.62
Aspartate*	2.54±0.56	0.57±0.43	Ammonia**	20.51±1.56	35.32±3.84
Asparagine*	3.27±0.39	1.69±0.40	Glutamate	5.84±0.71	4.63±0.59
Serine	17.92±1.51	12.89±1.18	Alanine**	25.20±2.52	8.20±0.81
Glutamine**	15.62±6.02	68.12±12.68			
Histidine	2.00±0.32	1.79±0.35			
Glycine*	3.82±0.74	0.73±0.76			
Threonine	5.08±0.92	4.18±1.02			
Arginine	26.84±2.33	19.38±2.09			
Tyrosine	2.79±0.48	2.54±0.52			
Valine	7.27±1.03	6.10±1.09			
Methionine	2.83±0.30	2.33±0.30			
Tryptophan	0.91±0.14	1.00±0.14			
Phenylalanine	3.50±0.50	3.06±0.54			
Isoleucine	9.02±1.03	6.77±1.03			
Leucine	10.59±1.18	8.08±1.15			
Lysine	7.55±1.23	4.92±1.18			
Proline	−3.31±1.14	−0.23±0.71			

Note: All data measured by HPLC analysis, except for ammonia, which was measured using the Bioflex analyser. Values are mean ± standard error, n = 6. Significance between the values measured at 20% and 2% oxygen were determined using a student's t-test where ** indicates a p-value<0.05, deemed statistically significant and *indicates a p-value<0.08, deemed statistically significant.

Differences in metabolite uptake and production rates seen with differing O$_2$ concentrations

Metabolite concentrations in the cell culture supernatant were measured at each experimental time point to generate concentration profiles (See Figure S4). The concentration profiles were then used in conjunction with cell numbers to determine the specific uptake and production rates (Table 1). Glucose was consumed and lactate produced throughout the culture at 2% and 20% oxygen. The glucose and lactate consumption and production rates were calculated and found to be statistically different at 2% and 20% oxygen, with rates 1.8 times higher at 2% oxygen (Table 1). Despite the difference in rates, the lactate production to glucose consumption ratios ($Y_{Lac/Glc}$) were calculated to be 2.3, at both oxygen concentrations during the exponential phase of growth (Table 2). In contrast to glucose and lactate, ammonia rates were found to be 1.6 times higher at 20% oxygen than at 2% oxygen (Table 1).

The amino acid (AA) profile, in terms of the AA's produced and consumed, was the same regardless of oxygen concentration. There were however some statistically significant differences in the rates. Alanine was produced, and aspartate, asparagine and glycine were consumed at statistically higher rates at 2% oxygen than at 20% oxygen (Table 1). Glutamine, however, was consumed at a statistically higher rate at 20% oxygen compared with 2% oxygen (Table 1). While $Y_{Lac/Glc}$ was the same at both oxygen concentrations, $Y_{Amm/Gln}$ (ammonia production to glutamine consumption ratio) was determined to be 1.31 and 0.52 at 2% and 20% oxygen respectively (Table 2). No AA's were found to be limiting during the culture.

Table 2. Biological ratios of human embryonic stem cells cultured at physiological and atmospheric oxygen concentrations.

Biological ratios *	2% Oxygen	20% Oxygen
Lactate production to glucose consumption ratio ($Y_{Lac/Glc}$)	2.31±0.31	2.32±0.25
Ammonia production to glutamine consumption ratio ($Y_{NH3/Gln}$)	1.31±0.52	0.52±0.11
Glucose to glutamine consumption ratio ($Y_{Glc/Gln}$)	36.49±14.53	4.51±0.91
Respiratory quotient (RQ = $Y_{CO2/O2}$)	1.04±0.01	1.09±0.01
Ratio cytoplasmic ATP production to ATP production in the mitochondria	0.83±0.09	0.42±0.06

* Note: The specific consumption and production rates used to calculate the biological ratios for lactate, glucose, ammonia and glutamine were measured while the specific rates for carbon dioxide, oxygen and ATP were calculated from the metabolic flux analysis results.

Table 3. ATP production.

	Total flux ATP produced	Total flux ATP produced by Oxidative phosphorylation	Total flux ATP produced by Glycolysis
2% oxygen	219–251 mmol/10^6 cells	116–141 mmol/10^6 cells * (53–56% total ATP production)	103–110 mmol/10^6 cells * (47–44% total ATP production)
20% oxygen	207–252 mmol/10^6 cells	144–178 mmol/10^6 cells * (69–71% total ATP production)	64–74 mmol/10^6 cells * (31–29% total ATP production)

*Note: The range of values reported for ATP flux were calculated from the upper and lower bounds of the resulting fluxes from the MFA as reported in Table S1.

Metabolic Flux Analysis shows a difference in energy pathways at physiological and atmospheric O$_2$ concentrations

The metabolic flux analysis (MFA) model used within this study consisted of over 2000 internal metabolic reactions. The measured data detailed above was used as inputs to the model to provide constraints on the theoretical solution space, allowing for the mass balance to be closed and a solution to be generated. The output of this model shows that of the 2000 reactions, 288 are considered to be essential for hESC metabolism (Table S1). Of the reactions considered essential, 61 reactions show a significant difference in the flux between hESCs cultured at physiological, 2%, and atmospheric, 20%, oxygen concentrations, with the flux being greater at physiological oxygen in 70% of these reactions (Table S1). At first glance, this suggests that hESCs are more metabol-ically active at physiological oxygen concentrations compared with atmospheric oxygen concentrations.

Interestingly, overall the fluxes through pathways involved in biogenesis are not statistically different between the two oxygen concentrations (Table S1). This is in keeping with the observation that the growth rates of MEL-2 hESCs at these two oxygen concentrations were not statistically different.

Another similarity is that the results of the model show no statistically significant difference in the total amount of ATP produced by MEL-2 hESCs at physiological and atmospheric oxygen (see Table 3). There is however a significant difference in the flux through the reactions used to generate ATP (see Figure 3).

At physiological oxygen concentrations, the flux through the glycolysis pathway is much greater than that at atmospheric oxygen concentrations (Figures 3 and 4). It is interesting to note however, that unlike most terminally differentiated cells, in MEL-2 hESCs, glycolysis is completely disconnected from the citric acid

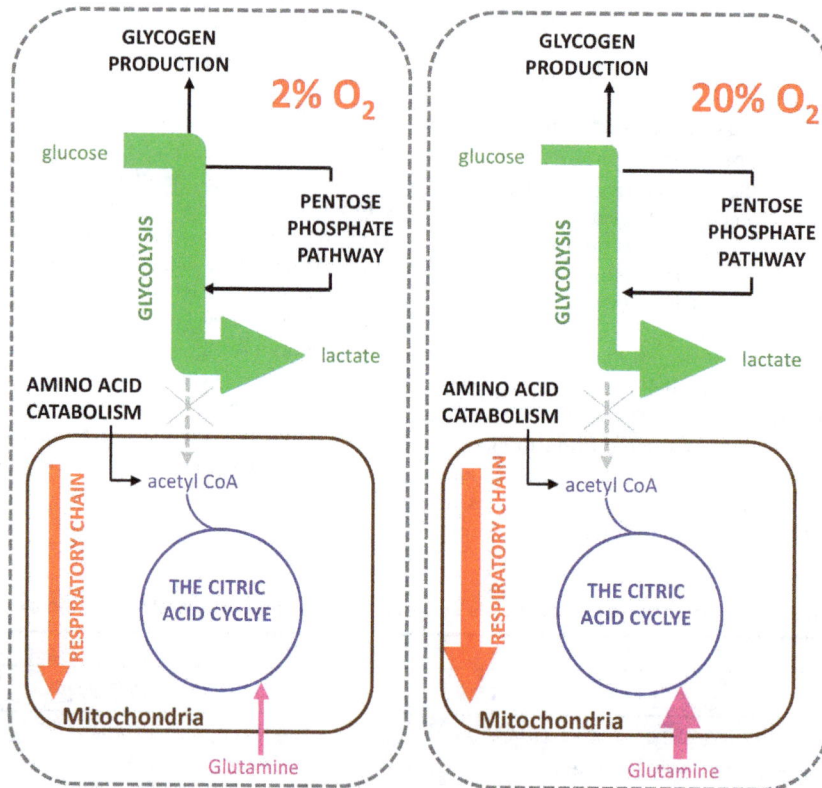

Figure 3. Activity of human embryonic stem cells metabolic pathways cultured at physiological and atmospheric oxygen concentrations. A visual representation of the flux through key metabolic pathways of hESCs cultured at both physiological and atmospheric oxygen concentrations. At both oxygen concentrations glycolysis is disconnected from the TCA cycle. The flux through the glycolysis pathways is greater at physiological oxygen concentrations while the uptake of glutamine and the flux through the respiratory chain is greater at atmospheric oxygen concentrations.

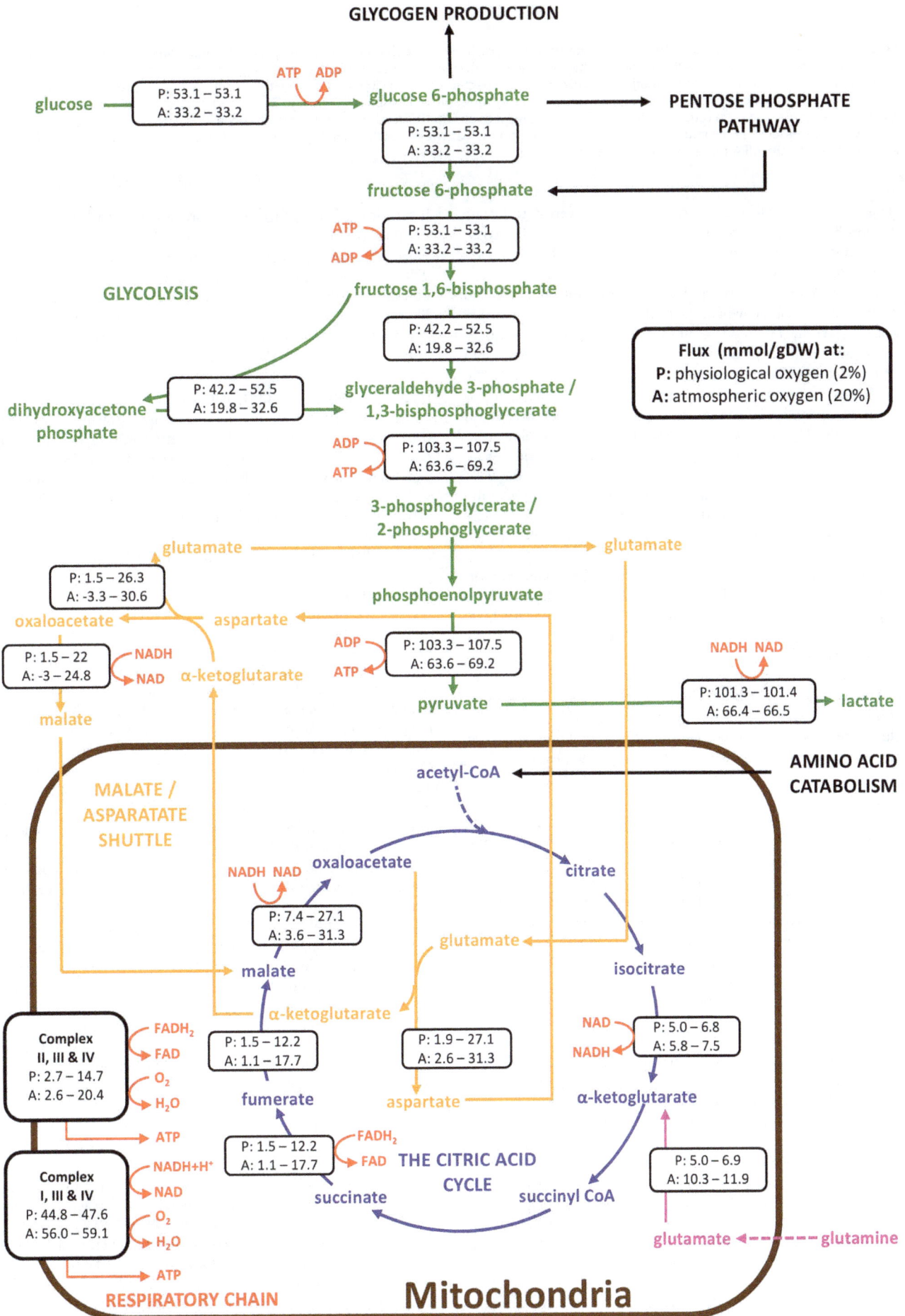

GLYCOGEN PRODUCTION

glucose — P: 53.1 – 53.1 / A: 33.2 – 33.2 — ATP ADP — glucose 6-phosphate → **PENTOSE PHOSPHATE PATHWAY**

P: 53.1 – 53.1 / A: 33.2 – 33.2

fructose 6-phosphate

ATP ADP — P: 53.1 – 53.1 / A: 33.2 – 33.2

GLYCOLYSIS

fructose 1,6-bisphosphate

P: 42.2 – 52.5 / A: 19.8 – 32.6

Flux (mmol/gDW) at:
P: physiological oxygen (2%)
A: atmospheric oxygen (20%)

dihydroxyacetone phosphate — P: 42.2 – 52.5 / A: 19.8 – 32.6 — glyceraldehyde 3-phosphate / 1,3-bisphosphoglycerate

ADP ATP — P: 103.3 – 107.5 / A: 63.6 – 69.2

3-phosphoglycerate / 2-phosphoglycerate

glutamate — glutamate

P: 1.5 – 26.3 / A: -3.3 – 30.6

oxaloacetate ← aspartate ← phosphoenolpyruvate

P: 1.5 – 22 / A: -3 – 24.8 — NADH NAD — α-ketoglutarate — ADP ATP — P: 103.3 – 107.5 / A: 63.6 – 69.2 — NADH NAD — P: 101.3 – 101.4 / A: 66.4 – 66.5 → lactate

malate — pyruvate

MALATE / ASPARATATE SHUTTLE

acetyl-CoA ← **AMINO ACID CATABOLISM**

NADH NAD — oxaloacetate — citrate

P: 7.4 – 27.1 / A: 3.6 – 31.3

glutamate

malate — isocitrate

α-ketoglutarate — P: 1.5 – 12.2 / A: 1.1 – 17.7 — glutamate — P: 1.9 – 27.1 / A: 2.6 – 31.3 — NAD NADH — P: 5.0 – 6.8 / A: 5.8 – 7.5

Complex II, III & IV
P: 2.7 – 14.7
A: 2.6 – 20.4
FADH₂ → FAD / O₂ → H₂O / → ATP

fumerate — aspartate — α-ketoglutarate

P: 1.5 – 12.2 / A: 1.1 – 17.7 — FADH₂ FAD — **THE CITRIC ACID CYCLE**

P: 5.0 – 6.9 / A: 10.3 – 11.9

Complex I, III & IV
P: 44.8 – 47.6
A: 56.0 – 59.1
NADH+H⁺ → NAD / O₂ → H₂O / → ATP

succinate — succinyl CoA

glutamate ◄---- glutamine

RESPIRATORY CHAIN

Mitochondria

Figure 4. Metabolic pathway fluxes of human embryonic stem cells cultured at physiological and atmospheric oxygen concentrations. Overview of the fluxes through key metabolic pathways of hESCs cultured at physiological, 2%, and atmospheric, 20%, oxygen concentrations. At both oxygen concentrations glycolysis is disconnected from the citric acid cycle with majority of the glucose being converted to lactate, with some being diverted to glycogen production and the pentose phosphate pathway. The mitochondria is metabolically active with glutamine serving as the main carbon source for ATP aerobic respiration. The TCA cycle within the mitochondria is driven by the catabolism of amino acids. Values represent the minimum and maximum flux through the pathway at both physiological (P) and atmospheric (A) oxygen concentrations as determined by the MFA model developed within this study.

(TCA) cycle, with all of the pyruvate being converted to lactate (Figures 3 and 4). This phenomenon is observed at both oxygen concentrations, and further, is supported by the high $Y_{Lac/Glc}$ measured (see Table 2).

This MFA model however shows that the mitochondria and TCA cycle are still active and facilitates the catabolism of amino acids. In fact the TCA cycle is thus mostly driven by catabolism of amino acids, and there is no net oxidation of glucose carbon into CO_2 via the TCA cycle. In addition, glutamine is fed into the TCA cycle, following transamination to α-ketoglutarate, acting as a carbon source for ATP production by oxidative phosphorylation (OXPHOS) (Figure 4). Consumption of glutamine dominated in the 20% cells, while consumption of arginine, glutamine and serine had comparable contributions in the 2% cells. In the 20% cells, the catabolism of glutamine accounted for up to 79% of the oxygen consumed (Calculated from data given in Table S1). For the 2% cells, glutamine catabolism required at most 28% of the total oxygen consumed. For the same amount of ATP produced, 20% cells needs 3 times more O_2 (Calculated from data given in Table S1), i.e., TCA cycle activity is 3 times greater. Without oxygen, 2% cells can make 56 mmol/gDW/hr ATP, while 20% cells can make 17 mmol/gDW/hr ATP (Calculated from data given in Table S1).

The ratio of cytoplasmic to mitochondrial ATP production of hESCs is twice as high at physiological oxygen compared to atmospheric oxygen (see Table 2). This is driven by a glucose to glutamine consumption ratio, $Y_{Glc/Gln}$, that is eight times higher at physiological oxygen compared with atmospheric (see Table 2).

Oxygen uptake

The oxygen uptake rate, OUR, of MEL-2 hESCs was calculated from the results of the MFA to be 369–375 mmol h^{-1} 10^{12} cells^{-1} when cultured at 2% oxygen and 397–403 mmol h^{-1} 10^{12} cells^{-1} when cultured at 20% oxygen. The oxygen uptake rates (OUR) were used to determine the maximum number of cells that could be supported in this experimental set-up before oxygen became limiting using Equation 2 (Equation 2: Total cells = $((C-C_0) \times D_{OW} \times SA)/(OUR \times h))$ and setting C, the oxygen concentration at the cell layer, to 0. The diffusivity of oxygen in cell culture medium was assumed to be the same as the diffusivity of oxygen in water, D_{OW} for the purposes of this calculation. D_{OW} at 37 degrees Celsius is 3.55×10^{-5} cm^{2}s^{-1} [39]. The surface area, SA, of a well of a Nunc 6-well plate is 9.6 cm^2 and the height of liquid was 0.21 cm. The concentration of oxygen at the surface, C_0, is equal to the solubility multiplied by the partial pressure. The solubility of oxygen in water at 37 degrees Celsius is 0.00021 mmol/cm^3 [40].

The maximum number of cells that can be supported before oxygen becomes limiting is 6.6×10^4–6.71×10^4 and 61.4×10^4–62.4×10^4 cells at 2% and 20% oxygen, respectively. This means that oxygen becomes limiting at the start and during the exponential growth phase at 2% oxygen and 20% oxygen, respectively, in this experimental set-up.

Inhibition of mitochondrial complexes I and II

Mitochondrial complex I and II were blocked with rotenone to block mitochondrial complex I and α-tocopherol succinate (α-TOS) to block mitochondrial respiratory complex II to get an indication of the importance of the respiratory chain in hESCs. Blocking the complexes caused an arrest in cell growth (Figure 5). From this result, it can be inferred that the respiratory chain is essential to MEL-2 hESC expansion.

Hypoxia inducible factor, HIF-1α

HIF-1α expression during the exponential growth phase (at the protein level) was measured by western blot and quantified by densitometry analysis. As expected HIF-1α was found to be expressed at a statistically higher level at 2% oxygen compared with 20% oxygen (Figure 6). No statistical difference was found in HIF-1α expression at the RNA level under the same conditions, in keeping with the notion that the expression of this protein is controlled by protein stability and not mRNA expression.

Discussion

Human embryonic stem cells are considered to be highly metabolically active with a highly glycolytic nature [14–16]. They are also known to have fewer mitochondria than terminally differentiated cells [18] and possess mitochondria that appear immature and lack normal cristae [16,19]. Together this has led to the preposition that the mitochondria are more inactive in hESCs than in differentiated cells and that energy generation by OXPHOS in the mitochondria is also reduced in hESCs [16]. The metabolic flux analysis conducted in this study reaffirms the highly glycolytic nature of hESCs but also shows an active citric acid cycle within the mitochondria and opportunistic use of energy substrates to maximise ATP production.

At first glance MEL-2 hESCs appear to be more metabolically active at physiological oxygen concentrations with 70% of fluxes showing significant differences between 2% and 20% oxygen being greater at 2% oxygen. In particular, large differences were noted for those fluxes associated with glycolysis, with glucose uptake being 1.8 times greater at physiological oxygen. Glycolysis is completely disconnected from the TCA cycle at both oxygen concentrations, as indicated by lactate to glucose ratios of 2.3, of a theoretical maximum of 2. The results of the MFA further support this by indicating that all of the pyruvate is converted to lactate, rather than entering the mitochondria. This phenomenon is often seen in highly proliferative cells, such as cancer cells [41–43] and during embryogenesis [41].

Hypotheses put forward in the literature to explain the disconnection of glycolysis from the TCA cycle in highly proliferative cells range from the phenomena being a more rapid way to produce ATP despite being less efficient [19] to it being a protection mechanism from the production of reactive oxygen species (ROS) during ATP production by OXPHOS. What is often not investigated within these studies, however, is the activity of the mitochondria and OXPHOS. Our metabolic flux analysis uniquely allowed us to reveal that mitochondrial oxidative phosphorylation in fact significantly contributes to cellular ATP

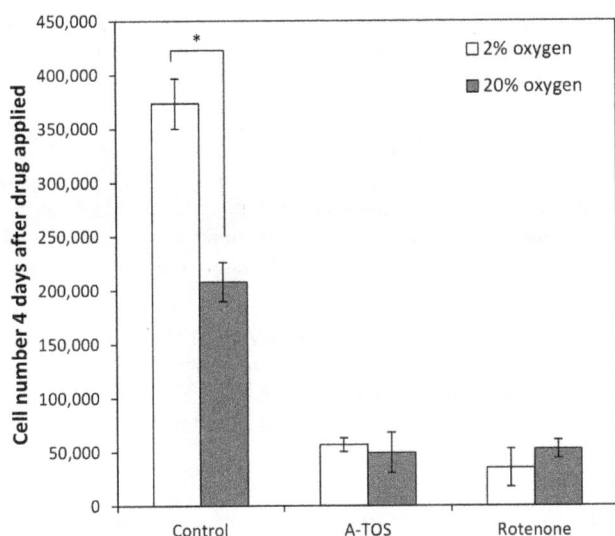

Figure 5. Effect of blocking complex I and II on human embryonic stem cell growth. Effect of blocking complex I and complex II on hESC growth at 20% and 2% oxygen. Plot shows the cell number per well 4 days after the addition of rotenone or D-α-tocopherol succinate (A-TOS) or to the cell culture medium block complex I and II respectively. Control has no drug added. Values are averages ± standard deviation, n = 3. * indicates a p-value <0.5, deemed statistically significant.

oxidation is increased by the mitochondrial uncoupling protein UCP2 [44,45]. UCP2 has been shown to be repressed during hESC differentiation [46] lending further support to the idea that UCP expression may play a central role in shaping the metabolic make-up of hESC and establishing this aerobic glycolysis state.

The results of the MFA showed no statistical difference in the total ATP produced by MEL-2 hESCs but rather a difference in the flux through the internal pathways contributing to the total ATP production at 2% and 20% oxygen. The results show a greater flux through glycolysis pathways at physiological oxygen and a greater flux through OXPHOS at atmospheric oxygen concentrations. This is supported by the greater uptake of glucose at 2% oxygen and a greater uptake of glutamine at 20% oxygen. The question then is what is the molecular mechanism governing the use of different internal metabolic pathways? While answering this question will require additional studies, it is possible to suggest some potential hypotheses based on the results of this study.

In general, metabolic homeostasis is controlled by the supply of metabolites or cellular demand. All of the metabolites measured were found to be in excess throughout the culture, including glucose and glutamine and so the uptake of either energy substrate is not limited by the supply to the cell. Oxygen however was found to be limiting in cultures under both physiological and atmospheric oxygen concentrations, but becomes limiting earlier at physiological oxygen. It is then reasonable to hypothesise that oxygen supply controls the flux through OXPHOS, and that to compensate and meet the total cellular energy demand, hESCs may then increase the flux through glycolysis pathways. A requirement of an increase in flux through glycolysis would be an increase in glucose uptake. Glucose is transported into the cell by facilitated transport via the GLUT family of proteins. One of the key regulators of GLUT expression, and in fact of many proteins required for glycolysis, is HIF-1α [47]. Indeed, greater levels of HIF1-α protein were found in hESCs cultured under physiological oxygen compared with atmospheric oxygen. HIF-1α

production by utilising glutamine as the main carbon source. In confirmation of this result, when complex I and II, necessary for OXPHOS, were blocked there was a reduction in cell proliferation, indicating that OXPHOS is contributing to the energy demand required for the rapid proliferation of MEL-2 hESCs observed in this study. Interestingly in other cell types glutamine

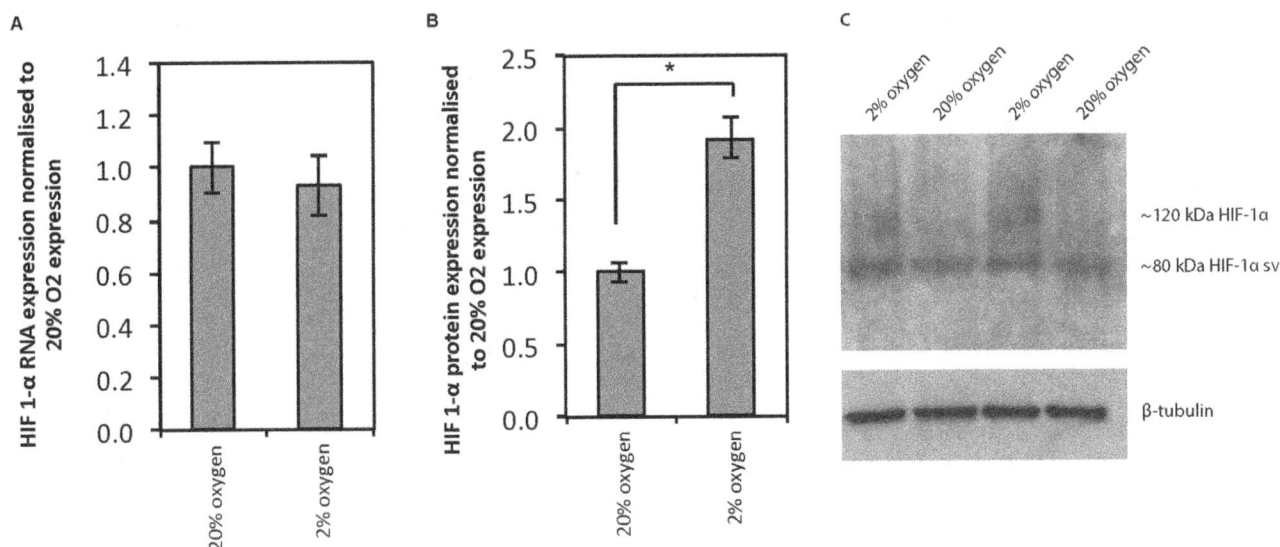

Figure 6. HIF-1α expression of human embryonic stem cells cultured at physiological and atmospheric oxygen concentrations. HIF-1α expression during the exponential phase of growth for hESCs cultured at 20% and 2% oxygen. A) HIF-1α RNA expression normalised to RNA expression at 20% oxygen. Values are averages ± standard deviation, n = 3. No significant difference was found between HIF-1α expression at the RNA level at 20% oxygen compared with 2% oxygen. B) Total HIF-1α protein expression normalised to protein expression at 20% oxygen. Protein expression measured by densitometry analysis of a western blot. Values are averages ± standard deviation, n = 2. * indicates a p-value <0.5, deemed statistically significant. C) Western blot showing protein probed for HIF-1α (both 120 kDa form and the 80 kDa splice variant (sv)) and β-tubulin as a loading control.

is in turn regulated by oxygen concentration within the cell and is up regulated under hypoxic conditions.

While major differences were found in the flux through energy pathways, no statistical difference was found between the growth rates of hESCs at physiological and atmospheric oxygen concentrations. Previously this point has been contested in the literature, with some groups reporting no dependence of growth rates on oxygen concentration [48] while others report that oxygen does effect cell growth rate [24]. It should be noted that in the past, however, growth rates have typically been measured indirectly by measuring expansion in colony or EB diameter. While this methodology allows for the maintenance of pluripotent cells, it does not allow for precise control of initial seeding densities. The protocols developed within this paper build on the published single cell passaging techniques and couple them with flow cytometry techniques for cell counting to allow for precise cell enumeration and accurate measurements of cellular growth rates. The similarity in growth rates observed within this study were further supported by the results of the MFA, which indicated no significant differences in the internal fluxes for reactions associated with biosynthesis at physiological and atmospheric oxygen concentrations.

In summary, the results presented in this paper, even though at present based only on a single well known hESC line, MLE-2, provide the most detailed metabolic profile of hESCs to date, providing an invaluable resource for understanding hESC metabolism. The results indicate that hESCs alter the flux through energy pathways, including OXPHOS, to maximise ATP production depending on the culture conditions. To do this hESCs utilise not only glycolysis but also OXPHOS in the mitochondria utilising glutamine as a carbon source. The results also showed no difference in the growth rates of cells cultured under physiological or atmospheric oxygen concentrations. The results may be used in the development of novel culture mediums for hESC maintenance and expansion, as well as to improve rational bioreactor design to ensure that metabolites are delivered and waste products are removed in an effective fashion. In addition, the growth profiles presented may be used to predict the fold expansion in cultures under the conditions described within this paper.

Supporting Information

Figure S1 Cell enumeration by flow cytometry strategy.
Cell enumeration by flow cytometry strategy. A known volume of Flow count flurospheres of a known concentration was added to a known volume of cell suspension. The sample was analysed on a BD LSR II flow cytometer. Analysis was performed using the following regions. First, the bead population was selected on the FITC 530/30 channel, shown in red. Next, the cell population was selected based on Hoechst staining on the 450/50 channel, shown in green. Finally, the multiplets were selected on a histogram of 450/50 gated to select the cell population as the population after the G2/M peak, this region is shown in blue on the dot plots.

Figure S2 Karyotype analysis of human embryonic stem cell line. Karyotype analysis was conducted on the

MEL-2 hESC cell stocks at p18+2+15 after experiments were conducted. Female karyotype with no abnormalities was detected for 25 cells tested.

Figure S3 Osmolality and pH of the cell culture supernatant. The pH (A) and the osmolality (B) of the cell culture medium were measured at each time point throughout the experiment and found to be within the limits to promote normal hESC growth. Values are averages ± standard deviation, n = 6.

Figure S4 Metabolite concentration profiles. Concentration of metabolites in the cell culture media at time points throughout the experiment at physiological, 2%, and atmospheric, 20% oxygen concentrations. Values are means ± standard deviation, n = 6.

Table S1 Flux through reactions considered essential for hESC metabolism. Results of the metabolic flux analysis - The MFA model consisted of over 2000 reactions. Of these reactions 288 were considered essential for hESC metabolism. The metabolic reactions modelled within the MFA can be broken into six categories:

- Biosynthesis – reactions directly involved with synthesising biomass;"
- Amino acid catabolism;"
- Central pathway – metabolism pathways present in all three domains of life;"
- Energy – reactions involved with the production of ATP;"
- Transport – transport of metabolites within the cell, eg from cytoplasm to the mitochondria; and"
- Exchange – transport of metabolites into and out of the cell."

The flux through these reactions in hESCs cultured at both physiological and atmospheric oxygen concentrations are given in Table S1. Note: Reaction ID or component followed by '_mt' indicates that the reaction takes place or the reactant is located within the mitochondria.

Acknowledgments

The authors would like to acknowledge the Australian Stem Cell Centre's Core hESC Laboratories (Stem Core) for providing cell culture and support services, as well as the Australian Stem Cell Centre for financial support. The authors would also like to acknowledge the Queensland Brain Institute Flow Cytometry Facility for assistance sorting cells.

Author Contributions

Conceived and designed the experiments: JT LQ DT JK LN EW JCW. Performed the experiments: JT LQ DT JK L-PK. Analyzed the data: JT LQ DT JK L-PK LN EW JCW. Contributed reagents/materials/analysis tools: LN EW JCW. Wrote the paper: JT LQ DT JK LN EW JCW.

References

1. Oh SKW, Choo ABH (2006) Human embryonic stem cells: Technological challenges towards therapy. Clinical and Experimental Pharmacology and Physiology 33: 489–495.
2. Hernandez D, Ruban L, Mason C (2011) Feeder-Free Culture of Human Embryonic Stem Cells for Scalable Expansion in a Reproducible Manner. Stem Cells and Development 20: 1089–1098.
3. Zweigerdt R, Olmer R, Singh H, Haverich A, Martin U (2011) Scalable expansion of human pluripotent stem cells in suspension culture. Nature Protocols 6: 689–700.
4. Fong WJ, Tan HL, Choo A, Oh SKW (2005) Perfusion cultures of human embryonic stem cells. Bioprocess and Biosystems Engineering 27: 381–387.

5. Oh SKW, Chen AK, Mok Y, Chen XL, Lim UM, et al. (2009) Long-term microcarrier suspension cultures of human embryonic stem cells. Stem Cell Research 2: 219–230.

6. Serra M, Brito C, Sousa MFQ, Jensen J, Tostoes R, et al. (2010) Improving expansion of pluripotent human embryonic stem cells in perfused bioreactors through oxygen control. Journal of Biotechnology 148: 208–215.

7. Krawetz R, Taiani JT, Liu SY, Meng GL, Li XY, et al. (2010) Large-Scale Expansion of Pluripotent Human Embryonic Stem Cells in Stirred-Suspension Bioreactors. Tissue Engineering Part C-Methods 16: 573–582.

8. Ludwig TE, Levenstein ME, Jones JM, Berggren WT, Mitchen ER, et al. (2006) Derivation of human embryonic stem cells in defined conditions. Nature Biotechnology 24: 185–187.

9. Wang L, Schuiz TC, Sherrer ES, Dauphin DS, Shin S, et al. (2007) Self-renewal of human embryonic stem cells requires insuhn-like growth factor-1 receptor and ERBB2 receptor signaling. Blood 110: 4111–4119.

10. Lu J, Hou RH, Booth CJ, Yang SH, Snyder M (2006) Defined culture conditions of human embryonic stem cells. Proceedings of the National Academy of Sciences of the United States of America 103: 5688–5693.

11. Li Y, Powell S, Brunette E, Lebkowski J, Mandalam R (2005) Expansion of human embryonic stem cells in defined serum-free medium devoid of animal-derived products. Biotechnology and Bioengineering 91: 688–698.

12. Akopian V, Andrews PW, Beil S, Benvenisty N, Brehm J, et al. (2010) Comparison of defined culture systems for feeder cell free propagation of human embryonic stem cells. In Vitro Cellular & Developmental Biology-Animal 46: 247–258.

13. Furue MK, Na J, Jackson JP, Okamoto T, Jones M, et al. (2008) Heparin promotes the growth of human embryonic stem cells in a defined serum-free medium. Proceedings of the National Academy of Sciences of the United States of America 105: 13409–13414.

14. Chen X, Chen A, Woo TL, Choo ABH, Reuveny S, et al. (2010) Investigations into the metabolism of two-dimensional colony and suspended microcarrier cultures of human embryonic stem cells in serum-free medium. Stem Cells and Development 19.

15. Fernandes TG, Fernandes-Platzgummer AM, da Silva CL, Diogo MM, Cabral JMS (2010) Kinetic and metabolic analysis of mouse embryonic stem cell expansion under serum-free conditions. Biotechnology Letters 32: 171–179.

16. Varum S, Rodrigues AS, Moura MB, Momclovic O, Easley CA, et al. (2011) Energy metabolism in human pluripotent stem cells and their differentiated counterparts. Plos One 6: e20914–e20914.

17. Zeng AP, Hu WS, Deckwer WD (1998) Variation of stoichiometric ratios and their correlation for monitoring and control of animal cell cultures. Biotechnology Progress 14: 434–441.

18. Facucho-Oliveira JM, St John JC (2009) The Relationship Between Pluripotency and Mitochondrial DNA Proliferation During Early Embryo Development and Embryonic Stem Cell Differentiation. Stem Cell Reviews and Reports 5: 140–158.

19. Jezek P, Plecita-Hlavata L, Smolkova K, Rossignol R (2010) Distinctions and similarities of cell bioenergetics and the role of mitochondria in hypoxia, cancer, and embryonic development. International Journal of Biochemistry & Cell Biology 42: 604–622.

20. Thomson JA, Itskovitz-Eldor J, Shapiro SS, Waknitz MA, Swiergiel JJ, et al. (1998) Embryonic stem cell lines derived from human blastocysts. Science 282: 1145–1147.

21. Fischer B, Bavister BD (1993) Oxygen-tension in the oviduct and uterus of rhesus-monkeys, hamsters and rabbits. Journal of Reproducition and Fertility 99: 673–679.

22. Forsyth NR, Musio A, Vezzoni P, Simpson A, Noble BS, et al. (2006) Physiologic oxygen enhances human embryonic stem cell clonal recovery and reduces chromosomal abnormalities. Cloning and Stem Cells 8: 16–23.

23. Wang F, Thirumangalathu S, Loeken MR (2006) Establishment of new mouse embryonic stem cell lines is improved by physiological glucose and oxygen. Cloning and Stem Cells 8: 108–116.

24. Ezashi T, Das P, Roberts RM (2005) Low O2 tensions and the prevention of differentiation of hES cells. Proceedings of the National Academy of Sciences of the United States of America 102: 4783–4788.

25. Prasad SM, Czepiel M, Cetinkaya C, Smigielska K, Weli SC, et al. (2009) Continuous hypoxic culturing maintains activation of Notch and allows long-term propagation of human embryonic stem cells without spontaneous differentiation. Cell Proliferation 42: 63–74.

26. Forristal CE, Wright KL, Hanley NA, Oreffo ROC, Houghton FD (2010) Hypoxia inducible factors regulate pluripotency and proliferation in human embryonic stem cells cultured at reduced oxygen tensions. Reproduction 139: 85–97.

27. Adewumi O, Aflatoonian B, Ahrlund-Richter L, Amit M, Andrews PW, et al. (2007) Characterization of human embryonic stem cell lines by the International Stem Cell Initiative. Nature Biotechnology 25: 803–816.

28. Costa M, Sourris K, Hatzistavrou T, Elefanty AG, Stanley EG (2008) Expansion of human embryonic stem cells in vitro. Current Protocols in Stem Cell Biology 1: 1C.1.1–1C.1.7.

29. Krömer JO, Fritz M, Heinzle E, Wittmann C (2005) In vivo quantification of intracellular amino acids and intermediates of the methionine pathway in Corynebacterium glutamicum. Analytical Biochemistry 340: 171–173.

30. Quek L, Dietmair S, Krömer J, Nielsen L (2010) Metabolic flux analysis in mammalian cell culture. Metabolic Engineering Journal 12: 161–171.

31. Quek L, Nielsen L (2008) On the reconstruction of the Mus musculus genome-scale metabolic network model. Genome Information 21: 89–100.

32. Bonarius HPJ, Hatzimanikatis V, Meesters KPH, deGooijer CD, Schmid G, et al. (1996) Metabolic flux analysis of hybridoma cells in different culture media using mass balances. Biotechnology and Bioengineering 50: 299–318.

33. Dong LF, Low P, Dyason JC, Wang XF, Prochazka L, et al. (2008) alpha-tocopheryl succinate induces apoptosis by targeting ubiquinone-binding sites in mitochondrial respiratory complex II. Oncogene 27: 4324–4335.

34. Li B, Chauvin C, De Paulis D, De Oliveira F, Gharib A, et al. (2012) Inhibition of complex I regulates the mitochondrial permeability transition through a phosphate-sensitive inhibitory site masked by cyclophilin D. Biochimica Et Biophysica Acta-Bioenergetics 1817: 1628–1634.

35. Miller L (2010) Analyzing gels and western blots with ImageJ

36. Schmittgen TD, Livak KJ (2008) Analyzing real-time PCR data by the comparative C(T) method. Nat Protoc 3: 1101–1108.

37. Livak KJ, Schmittgen TD (2001) Analysis of relative gene expression data using real-time quantitative PCR and the 2(-Delta Delta C(T)) Method. Methods 25: 402–408.

38. Bustin SA, Benes V, Garson JA, Hellemans J, Huggett J, et al. (2009) The MIQE guidelines: minimum information for publication of quantitative real-time PCR experiments. Clin Chem 55: 611–622.

39. (1999) Perry's chemical engineering handbook; Perry RH, Green DW, editors. New York: McGraw Hill.

40. (2003) OxyMicro User Manual - Fibre-optic oxygen measurement systems with microsensors. In: Instruments WP, editor.

41. Fritz V, Fajas L (2010) Metabolism and proliferation share common regulatory pathways in cancer cells. Oncogene 29: 4369–4377.

42. Heiden MGV, Locasale JW, Swanson KD, Sharfi H, Heffron GJ, et al. (2010) Evidence for an Alternative Glycolytic Pathway in Rapidly Proliferating Cells. Science 329: 1492–1499.

43. Argiles JM, Lopezsoriano FJ (1990) Why do cancer-cells have such a high glycolytic rate? Medical Hypotheses 32: 151–155.

44. Pecqueur C, Bui T, Gelly C, Hauchard J, Barbot C, et al. (2008) Uncoupling protein-2 controls proliferation by promoting fatty acid oxidation and limiting glycolysis-derived pyruvate utilization. FASEB Journal 22: 9–18.

45. Pecqueur C, Alves-Guerra C, Ricquier D, Bouillaud F (2009) UCP2, a metabolic sensor coupling glucose oxidation to mitochondrial metabolism? IUBMB Life 61: 762–767.

46. Zhang J, Khvorostov I, Hong JS, Oktay Y, Vergnes L, et al. (2011) UCP2 regulates energy metabolism and differentiation potential of human pluripotent stem cells. EMBO Journal 30: 4860–4873.

47. Airley RE, Mobasheri A (2007) Hypoxic regulation of glucose transport, anaerobic metabolism and angiogenesis in cancer: Novel pathways and targets for anticancer therapeutics. Chemotherapy 53: 233–256.

48. Kurosawa H, Kimura M, Noda T, Amano Y (2006) Effect of oxygen on In Vitro differentiation of mouse embryonic stem cells. Journal of Bioscience and Engineering 101: 26–30.

11

Identification of a Common Non-Apoptotic Cell Death Mechanism in Hereditary Retinal Degeneration

Blanca Arango-Gonzalez[1]*[9], **Dragana Trifunović**[1][9], **Ayse Sahaboglu**[1], **Katharina Kranz**[2], **Stylianos Michalakis**[3], **Pietro Farinelli**[1,4], **Susanne Koch**[3], **Fred Koch**[3], **Sandra Cottet**[5], **Ulrike Janssen-Bienhold**[2], **Karin Dedek**[2], **Martin Biel**[3], **Eberhart Zrenner**[1,6], **Thomas Euler**[1,6], **Per Ekström**[4], **Marius Ueffing**[1], **François Paquet-Durand**[1]*

1 Institute for Ophthalmic Research, University of Tuebingen, Tuebingen, Germany, 2 Department of Neurobiology, University of Oldenburg, Oldenburg, Germany, 3 Center for Integrated Protein Science Munich and Department of Pharmacy - Center for Drug Research, Ludwig-Maximilians-University Munich, Munich, Germany, 4 Division of Ophthalmology, Department of Clinical Sciences, University of Lund, Lund, Sweden, 5 Institute for Research in Ophthalmology, Sion, Switzerland, 6 Centre for Integrative Neuroscience, University of Tuebingen, Tuebingen, Germany

Abstract

Cell death in neurodegenerative diseases is often thought to be governed by apoptosis; however, an increasing body of evidence suggests the involvement of alternative cell death mechanisms in neuronal degeneration. We studied retinal neurodegeneration using 10 different animal models, covering all major groups of hereditary human blindness (*rd1, rd2, rd10, Cngb1* KO, *Rho* KO, S334ter, P23H, *Cnga3* KO, *cpfl1, Rpe65* KO), by investigating metabolic processes relevant for different forms of cell death. We show that apoptosis plays only a minor role in the inherited forms of retinal neurodegeneration studied, where instead, a non-apoptotic degenerative mechanism common to all mutants is of major importance. Hallmark features of this pathway are activation of histone deacetylase, poly-ADP-ribose-polymerase, and calpain, as well as accumulation of cyclic guanosine monophosphate and poly-ADP-ribose. Our work thus demonstrates the prevalence of alternative cell death mechanisms in inherited retinal degeneration and provides a rational basis for the design of mutation-independent treatments.

Editor: Olaf Strauß, Eye Hospital, Charité, Germany

Funding: This work was supported by the Kerstan Foundation, Deutsche Forschungsgemeinschaft [DFG PA1751/1-1, 4-1], Alcon Research Institute, European Commission [DRUGSFORD: HEALTH-F2-2012-304963; PANOPTES: NMP4-SL-2010-246180], German Ministry of Education and Research [BMBF HOPE2 - FKZ 01GM1108A], Centre for Integrative Neuroscience [CIN pool project 2009-20], Kronprinsessan Margaretas Arbetsnämnd (KMA), Stiftelsen Olle Engkvist Byggmästare, The Swedish Research Council 2009-3855, and Stiftelsen för Synskadade i f.d. Malmöhus län. The funders had no role in study design, data collection and analysis, decision to publish, or preparation of the manuscript.

Competing Interests: The authors declare that they have no conflict of interest.

* Email: francois.paquet-durand@klinikum.uni-tuebingen.de (FDP); blanca.arango-gonzalez@klinikum.uni-tuebingen.de (BAG)

9 These authors are joint first authors on this work.

Introduction

Apoptosis is a programmed cell death mechanism that is often invoked for neurodegenerative diseases. The classical apoptotic pathway starts with a BAX dependent permeabilisation of mitochondrial membranes, cytochrome c leakage to the cytoplasm and subsequent activation of initiator and executioner caspases [1]. Inherited neurodegenerative diseases of the retina are also generally thought to be governed by apoptotic cell death [2,3], which has given rise to numerous attempts to use anti-apoptotic strategies for therapy development [4–6]. Unfortunately, these approaches were generally unsuccessful and efficient neuroprotective therapies for hereditary retinal degenerations (RD) such as retinitis pigmentosa (RP), Leber's congenital amaurosis (LCA), or Stargardt's disease are still missing. Recent findings suggest alternative, non-apoptotic cell death mechanisms for photoreceptor degeneration [7,8]. Hence, we decided to systematically re-

evaluate the situation in the retina using a variety of markers for both classical apoptosis and non-apoptotic cell death.

The retina harbours two general types of photoreceptors, rods, responsible for vision under dim-light conditions (*i.e.* at night), and cones, responsible for vision during bright daylight. In addition, the retina hosts a variety of different 2nd and 3rd order neurons, responsible for relaying photoreceptor output to the brain. For studies into hereditary degenerative mechanisms in the retina a large number of human homologous animal models are available [9], faithfully reproducing the photoreceptor degeneration phenotype. Two major categories of mutations and diseases can be distinguished: primary *rod* photoreceptor degeneration, which usually entails secondary cone death and complete blindness, and is characteristic of human diseases such as RP, LCA, or Usher syndrome. Primary *cone* photoreceptor degeneration, which leaves rods mostly unaffected but nevertheless causes a severe loss of visual acuity and daylight vision and typifies human diseases, such as cone-dystrophy, Stargardt's disease or age-related macular degeneration [10,11].

In the present study, we asked the question whether there was a common mechanism governing photoreceptor cell death independent of the initial causative genetic defect, since this could open up for broadly applicable therapies. To address the heterogeneity of hereditary photoreceptor degeneration, we employed ten different animal models RD (Figure 1), eight models for primary rod degeneration, as seen in autosomal dominant RP (P23H and S334ter transgenic rats) and autosomal recessive RP (rd1, rd2, rd10, Cngb1 KO, Rho KO mice), as well as in LCA (Rpe65 KO mice). In addition, we also included two animal models for primary cone death (cpfl1, Cnga3 KO mice).

Surprisingly, our single cell resolution analysis of metabolic changes at the peak of cell death suggested that hereditary photoreceptor death was predominantly non-apoptotic, with only a marginal role, if any, for apoptosis. Instead, our study delineated a non-apoptotic cell death pathway and highlighted the general importance of this pathway for photoreceptor neurodegeneration. This finding has major ramifications for future therapy developments.

Materials and Methods

Animals

Animals were housed under standard white cyclic lighting, had free access to food and water, and were used irrespective of gender. Ten different mouse lines (C3H or C57Bl6 background) either wild-type or carrying naturally occurring mutations or engineered genetic deletions were used together with three different rat lines (CD background) expressing different rhodopsin transgenes (see Table 1). Day of birth was considered as postnatal day (P) 0. All procedures were approved by the respective local ethics and animal protection authorities and performed in compliance with the ARVO statement for the use of animals in

Ophthalmic and Visual Research. Specifically, procedures performed in Tübingen (concerning C3H wt, C57Bl6 wt, rd1, rd2, rd10, cpfl1, CD wt, S334ter, and P23H animals) were reviewed and approved by the Tuebingen University "Einrichtung für Tierschutz, Tierärztlichen Dienst und Labortierkunde". Procedures performed in Munich (on Cngb1 KO and Cnga3 KO animals) were reviewed and approved by the "Regierung von Oberbayern". Procedures performed in Oldenburg (on Rho KO animals) were reviewed and approved by the Oldenburg University animal welfare committee. Procedures performed in Sion (on Rpe65 KO animals) were reviewed and approved by the Veterinary Service of the State of Valais (Switzerland). Procedures performed in Lund (rd1, rd2 animals) adhered to permit # M220/09 issued by the local animal ethics committee. All efforts were made to minimize the number of animals used and their suffering.

Histology, immunohistochemistry, and immunofluorescence

Animals were sacrificed in the morning (10–11 am), their eyes enucleated and fixed in 4% paraformaldehyde (PFA) in 0.1 M phosphate buffer (pH 7.4) for 45 min at 4°C. PFA fixation was followed by cryoprotection in graded sucrose solutions (10, 20, 30%). Unfixed eyecups were directly embedded in cryomatrix (Tissue-Tek, Leica, Bensheim, Germany). Sagittal 12 µm sections were obtained and stored at −20°C.

Sections were incubated overnight at 4°C with primary antibodies (Table 2). Immunostaining was performed employing the avidin-biotin-peroxidase technique (Vectastain ABC system, Vector laboratories, Burlingame, CA). Immunofluorescence was performed using Alexa Fluor 488-conjugated secondary antibodies (Molecular Probes, Inc. Eugene, USA). Negative controls were carried out by omitting the primary antibody. Sections were mounted with Vectashield (Vectorlabs, Burlingame, CA, USA) for imaging.

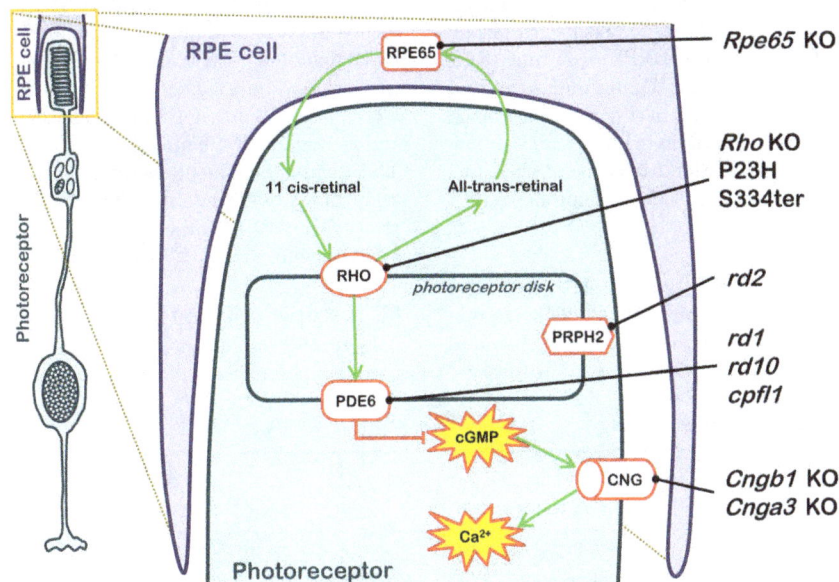

Figure 1. RD animal models used and their genetic defects. The cartoon illustrates the anatomical localization and metabolic consequences of the causative genetic mutations in the ten different RD animal models used in this study. RD causing mutations in these animal models interfere with the various stages of the phototransduction cascade, from the 11-cis-retinal recycling enzyme RPE65 (Rpe65 KO), via the light-sensitive Rhodopsin (Rho KO, P23H, S334ter), cGMP-hydrolyzing phosphodiesterase-6 (PDE6; rd1, rd10, cpfl1), the structural protein Peripherin (Prph2; rd2), to the cyclic-nucleotide-gated (CNG; Cngb1 KO, Cnga3 KO) channel that allows for Ca^{2+}-influx.

Table 1. List of animals used, genes affected, and original references (where applicable).

Background/Line	Mutant	Gene	Reference
C3H	wild-type	-	[51]
C3H	rd1	Pde6b	[52]
C3H	rd2	Prph2	[53]
C57Bl/6J	wild-type	-	-
C57Bl/6	Rho KO	Rho	[54]
C57Bl/6N	Cngb1 KO	Cngb1	[55]
C57Bl/6N	Cnga3 KO	Cnga3	[56]
C57Bl/6J	cpfl1	Pde6c	[57]
C57Bl/6	Rpe65 KO	Rpe65	[49]
C57Bl/6J	rd10	Pde6b	[58]
Crl: CD(SD)	wild-type	-	-
Crl: CD(SD)	P23H tg	Rho	[59]
Crl: CD(SD)	S334ter tg	Rho	[59]

Italic fonts indicate mutant name or affected gene.

TUNEL Assay

Terminal deoxynucleotidyl transferase dUTP nick end labelling (TUNEL) assay was performed using an *in situ* cell death detection kit (Fluorescein or TMR; Roche Diagnostics GmbH, Mannheim, Germany). For controls terminal deoxynucleotidyl transferase enzyme was either omitted from the labelling solution (negative control), or sections were pre-treated for 30 min with DNAse I (Roche, 3 U/ml) in 50 mM Tris-HCl, pH 7.5, 1 mg/ml BSA to induce DNA strand breaks (positive control). While negative control gave no staining, positive control stained all nuclei in all layers of the retina [12].

Calpain *in situ* activity assay

Calpain activity was investigated with an enzymatic *in situ* assay [13]. Briefly, unfixed cryosections were incubated for 15 min in calpain reaction buffer (CRB; 25 mM HEPES, 65 mM KCl, 2 mM MgCl^2, 1,5 mM CaCl^2, 2 mM DTT) and then incubated at 35°C for 1 h in the dark in CRB with 2 mM fluorescent calpain substrate 7-amino-4-chloromethylcoumarin, t-BOC-Leucyl-L-methionine amide (CMAC, t-BOC-Leu-Met; Molecular Probes, Inc. Eugene, USA). Fluorescence was uncaged by calpain-dependent cleavage of t-Boc-Leu-Met-CMAC.

Poly-ADP-ribose polymerase (PARP) *in situ* activity assay

Unfixed cryosections were incubated in an avidin/biotin blocking kit (Vector Laboratories, Burlingame, USA), followed by incubation at 37°C for 2 h in PARP reaction mixture containing 10 mM MgCl$_2$, 1 mM DTT, 5 μM biotinylated NAD$^+$ (Trevigen, Gaithersburg, USA) in 100 mM Tris buffer with 0.2% Triton X-100 (pH 8.0). Biotin incorporation was detected by avidin - Alexa Fluor 488 conjugate (1:800, 1 h at room temperature). For controls biotinylated NAD+ was omitted from the reaction mixture [14].

HDAC *in situ* activity assay

HDAC activity assays were performed on retinal cryosections obtained from 4% PFA fixed eyes. Retinal sections were exposed to 200 μM Fluor de Lys-SIRT2 deacetylase substrate (Biomol, Hamburg, Germany) and 500 μM NAD+ (Biomol) in assay buffer (50 mM Tris/HCl, pH 8.0; 137 mM NaCl; 2.7 mM KCl; 1 mM MgCl2) and incubated for 2 h at 37°C. The tissue sections were then washed three times for 5 min in PBS and subsequently fixed in Methanol at −20°C, for 20 min. After refixation, the sections were washed once again for 5 min in PBS, then incubated in 1x Developer II (Biomol) in assay buffer and immediately coversliped and viewed under the microscope. The inclusion of either 100 μM TSA (Sigma, Steinheim, Germany) or 2 mM NAM (Sigma) in the assay allows to distinguish between HDAC activities coming from class I, II or IV (inhibited by TSA) or from class III (sirtuin-type HDACs, inhibited by NAM) [15].

Microscopy, cell counting, and statistical analysis

Light and fluorescence microscopy were usually performed at room temperature on an Axio Imager Z.1 ApoTome Microscope,

Table 2. List of antibodies used in this study.

Antigen	Source/Cat. Number	Dilution IF/IHC	Reference
BAX (clone 6A7)	Sigma/B8429	1:20	[60]
Cleaved Caspase-3 (Asp175) (clone 5A1E)	Cell Signalling/9664	1:300	[61]
Cleaved Caspase-9 (Asp353) (rabbit, polyclonal)	Abcam/ab52298	1:100	[16]
Cytochrome C (clone 7H8.2C12)	Abcam/mab13575	1:2000	[62]
cGMP (sheep, polyclonal)	Prof. Harry Steinbusch, Maastricht University, The Netherlands	1:500	[63]
PAR (clone 10H)	Enzo/ALX-804-220	1:200	[12]

equipped with a Zeiss Axiocam MRm digital camera. Images were captured using Zeiss Axiovision 4.7 software; representative pictures were taken from central areas of the retina using a 20x/ 0,8 Zeiss Plan-APOCHROMAT objective. Adobe Photoshop CS3 (Adobe Systems Incorporated, San Jose, CA) was used for primary image processing.

For quantifications, pictures were captured on three entire sagittal sections for at least three different animals for each genotype and age using Mosaic mode of Axiovision 4.7 at 20x magnification. The average area occupied by a photoreceptor cell (*i.e.* cell size) for each genotype and age was determined by counting DAPI-stained nuclei in 9 different areas (50×50 μm) of the retina. The total number of photoreceptor cells was estimated by dividing the outer nuclear layer (ONL) area by this average cell size. The number of positively labelled cells in the ONL was counted manually. To be able to compare the various markers in the different genotypes, we considered cells as positively labelled only if they showed a strong staining of either the photoreceptor nuclei or perinuclear areas. Since some markers actually stained predominantly the photoreceptor inner and/or outer segments (*i.e.* BAX, cGMP in *Cngb1* KO retina) these may thus in the present study have been systematically underestimated. Values obtained are given as fraction of total cell number in ONL (*i.e.* as percentage) and expressed as mean ± standard error of the mean (SEM). For statistical comparisons the unpaired Student t-test as implemented in Prism 5 for Windows (GraphPad Software, La Jolla, CA) was employed.

Results

In RD models the peak of cell death varied depending on severity of genetic insult

To study the cell death mechanisms governing RD, we first performed a detailed analysis of the temporal progression of the degeneration for each of the 10 animal models used (Figure 1). We used the TUNEL assay to label dying cells at different postnatal ages and quantified the percentages of TUNEL-positive cells in the outer nuclear layer (ONL), *i.e.* the photoreceptor layer (Figure 2).

In all RD models, once the degeneration sets in, the TUNEL assay detected a moderate to strong elevation of dying cells when compared to the respective wild-type, depending on degeneration speed and whether rods or cones were affected. In each RD animal model the peak of cell death was identified (Figure 2) and all following experiments were performed at this time-point to increase the chances of detecting characteristic cell death processes. From previous experiments [12,15–17], we know that the peak of TUNEL also corresponds to a strong activation of critical cell death processes; both for apoptotic and non-apoptotic cell death (*cf.* Figure S1). For the different animal models these time-points were: *rd1* = Postnatal day 13 (P13), *rd10* = P18, *rd2* = P18, *Cngb1* KO = P24, *Rho* KO = P42, *Rpe65* KO = P16, *cpfl1* = P24, *Cnga3* KO = P35, S334ter = P12, P23H = P15 (Data for *rd1*, *cpfl1*, S334ter, and P23H adapted from [16–18], respectively).

Since photoreceptor cell death is often seen as an apoptotic process [2,3], we initially focused our analysis on detecting characteristic markers for apoptosis, and then extended our investigation to also include metabolic processes involved in alternative cell death mechanisms. To assess the extent to which apoptotic or non-apoptotic cell death mechanisms were active in the different animal models, we compared the number of cells displaying a specific metabolic activity with the number of TUNEL-positive cells in both mutant and wild-type retina (Table S1 and S2).

Apoptosis was restricted to degenerating S334ter retina

We looked for increased expression, localization, or activation of Bcl-2–associated X protein (BAX), cytochrome c, cleaved, activated caspase-9 and -3 (Figure 3, quantification in Table S1 and S2). Increases in these apoptotic markers were found only in the S334ter model when compared to the corresponding wild-type.

Classical apoptosis starts with an activation of BAX [1]. Although early studies have already ruled out an involvement of BAX in RD [19], a recent study reported on the apparent activation of BAX in *rd1*, P23H, and Rho KO mice [20]. Nevertheless, in our hands a significant BAX activation (using the same antibody as in [20], Table 2) was observed only in S334ter retina. Here, prominent BAX staining was observed near mitochondria, in particular in individual photoreceptor inner segments, synaptic terminals, and occasionally around nuclei (Figure 3, Figure S2). This staining pattern in S334ter ONL is consistent with the reported role of BAX in the formation of the mitochondrial permeability transition pore [1].

Consequently, cytochrome c release from mitochondria was observed as an increased staining of individual photoreceptor cells in the S334ter ONL (Figure 3). A relative increase of cytochrome c leakage was found in *cpfl1* retina, however, this was not statistically significant (Table S1). Increased caspase-9 staining was present in S334ter retina only, with a peri-nuclear staining predominantly in the lower part of the ONL. A very similar staining pattern was found using an antibody specific for activated, cleaved caspase-3, again exclusively in S334ter retinal sections. These data are in line with previous studies [16,21].

Thus, whereas large numbers of TUNEL-positive cells were detected in all analysed RD models, clear evidence for apoptosis was only detected in S334ter rats. This suggested the execution of alternative, non-apoptotic cell death mechanisms.

Non-apoptotic cell death in photoreceptor degeneration

We have previously shown that rod photoreceptor degeneration in *rd1* mice is characterized by accumulation of cyclic guanosine monophosphate (cGMP), increased activities of histone deacetylases (HDAC), poly-ADP-ribose polymerases (PARP), and calpains [13,15,22].

cGMP accumulation in phosphodiesterase-6 mutants (*rd1*, *rd10*, *cpfl1*) is a direct consequence of the lack of phosphodiesterase activity that normally hydrolyses cGMP. Surprisingly, significant cGMP accumulation was observed also in all other analysed mouse and rat models (Figure 4, Table S1) except for *Rpe65* KO retina, where the initial causative defect does not reside in photoreceptors themselves but in retinal pigment epithelial cells. However, the patterns of cGMP accumulation varied between different RD models (Figure 4). In the case of *rd1*, *rd10*, *rd2*, *Cnga3* and *cpfl1* cGMP was visible in cell bodies as well as in photoreceptor inner/outer segments, whereas in *Cngb1* KO retina the signal was more prominent in inner/outer segments. For methodological reasons, we only quantified cGMP positive cell bodies. As a consequence, most likely the true number of photoreceptors showing elevated cGMP levels is higher in *Cngb1* KO retina than assessed here. P23H and S334ter rat retinas were characterized by diffuse cGMP accumulation in the ONL, contrary to *Rho* KO mice in which only very few nuclei were cGMP-positive.

The HDAC assay revealed significantly increased activity in all the analysed mutants when compared to corresponding wild-type (Figure 4). The number of nuclei stained with the HDAC assay varied between different mutants (Table S1 and S2) with more

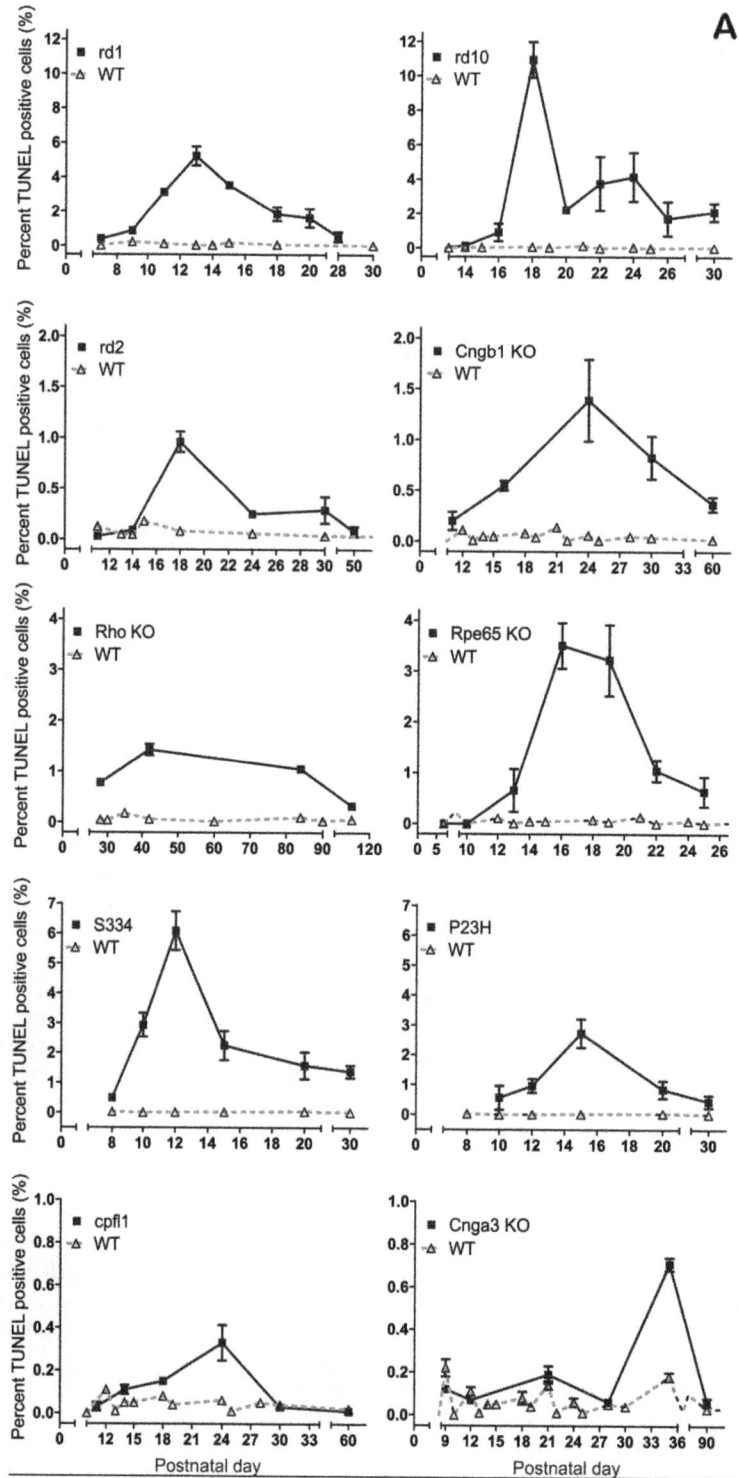

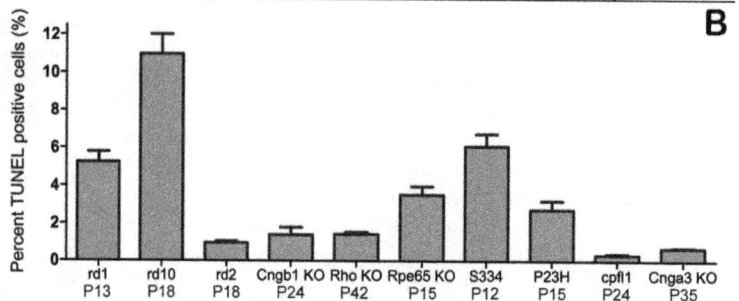

Figure 2. Progression of cell death in inherited RD models. Depending on the causative genetic insult, the temporal development of retinal degeneration is highly variable in the different animal models. The quantification of dying, TUNEL-positive photoreceptor cells in the outer nuclear layer (ONL) allowed determination of the evolution and the peak of photoreceptor death for each of these animal models (**A**). The peak was taken as reference point for the ensuing analysis of cell death mechanisms. The bar graph (**B**) shows a comparison of maximum peak heights for all ten RD models studied. Note the different scales in line graphs. Values are mean ± SEM from at least three different animals. See also Table S1 and S2.

cells showing HDAC activity in the case of *rd1*, *rd10*, and S334ter and less positive cells in the case of *Cngb1* KO and *Rpe65* KO.

To determine if poly-ADP-ribosylation, as an additional epigenetic process, was involved in photoreceptor degeneration, we looked for increased PARP *in situ* activity as well as for accumulation of poly-ADP-ribosylated proteins (PAR), *i.e.* the products of PARP activity. Nuclear staining of both PARP activity and PAR followed the patterns observed for HDAC activity. Mutants characterized by a high number of TUNEL-positive cells (*rd1*, *rd10*, S334ter and *Rpe65* KO) also displayed comparatively higher numbers of both PARP and PAR stained cells compared to models with low degeneration rates. The *in situ* staining for calpain activity was also significantly increased in all analysed RD models (Figure 4, quantification in Table S1).

To compare the different cell death processes, we related the numbers of positive cells detected by each individual assay to the numbers of TUNEL positive cells. To match the various RD models and their very different degeneration kinetics with each other, all values were expressed as logarithm to base 10. Since the TUNEL values were defined as 100%, its logarithm was 2. This comparative analysis highlighted the fact that non-apoptotic processes were clearly dominant for photoreceptor degeneration in all RD models (Figure 5). This was also true for the S334ter model which, interestingly, showed the additional involvement of apoptotic cell death. We also analysed the relative contribution of apoptotic and non-apoptotic processes to developmental cell death in wild-type retina (P13-P42). Here, the relative contributions of

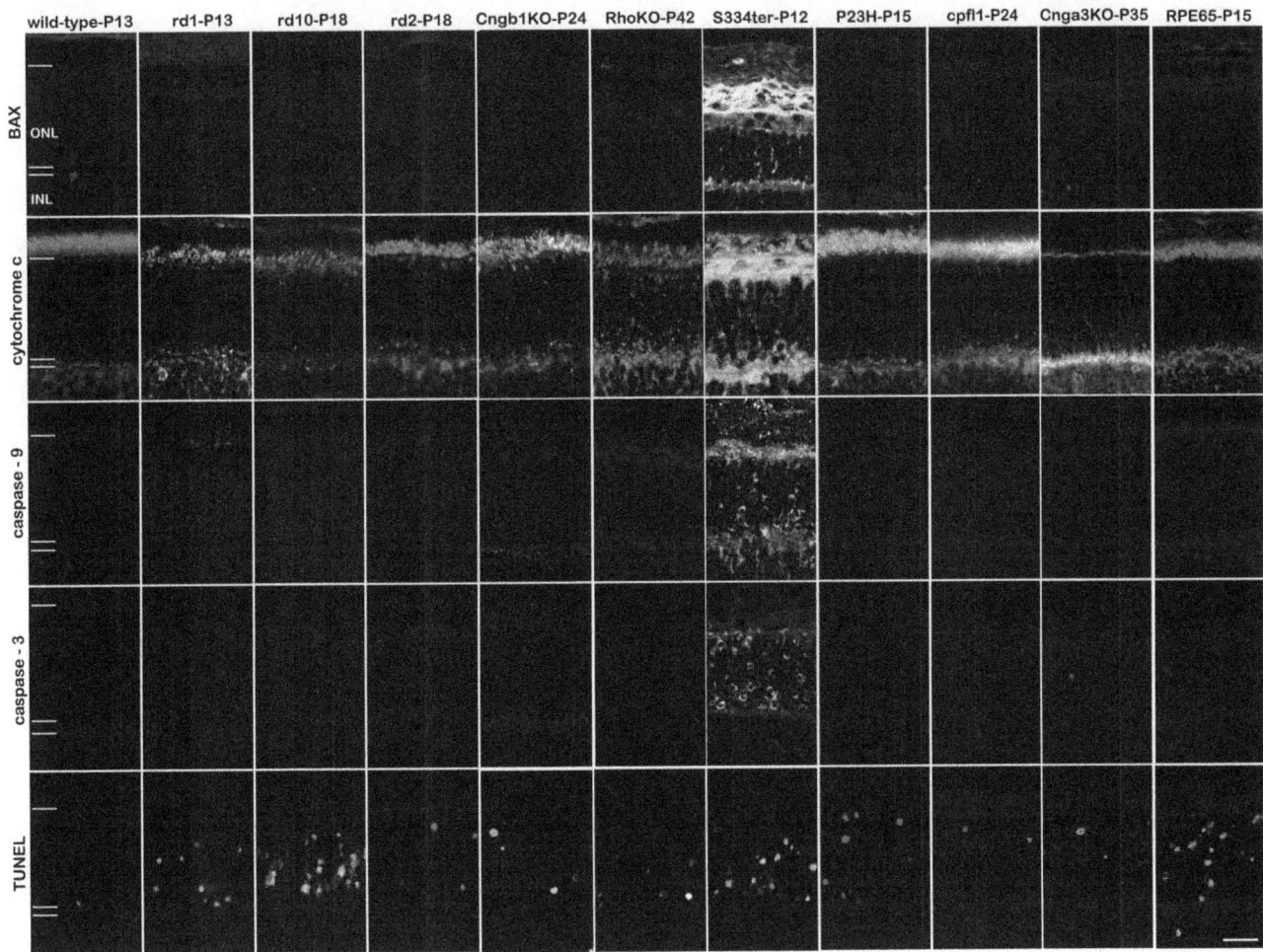

Figure 3. Apoptosis in the retina is restricted to the S334ter rat model. The analysis of BAX expression, mitochondrial cytochrome c release, activation of caspase-9 and -3 shows essentially no positive detection in 9 out of 10 animal models for hereditary retinal degeneration. The notable exception was the S334ter transgenic rat which harbours a mutation in the rhodopsin gene leading to a truncated protein and in which many photoreceptors were positive for apoptosis. In all other animal models, while there were cells displaying clear evidence for apoptosis, their numbers were within the wild-type levels, indicating that this was related to physiological, developmental cell death, which is characteristic for the postnatal rodent retina. Importantly, the numbers of apoptotic cells did not match the numbers of mutation-induced dying cells as evidenced by the TUNEL assay. Scale bar 20 µm.

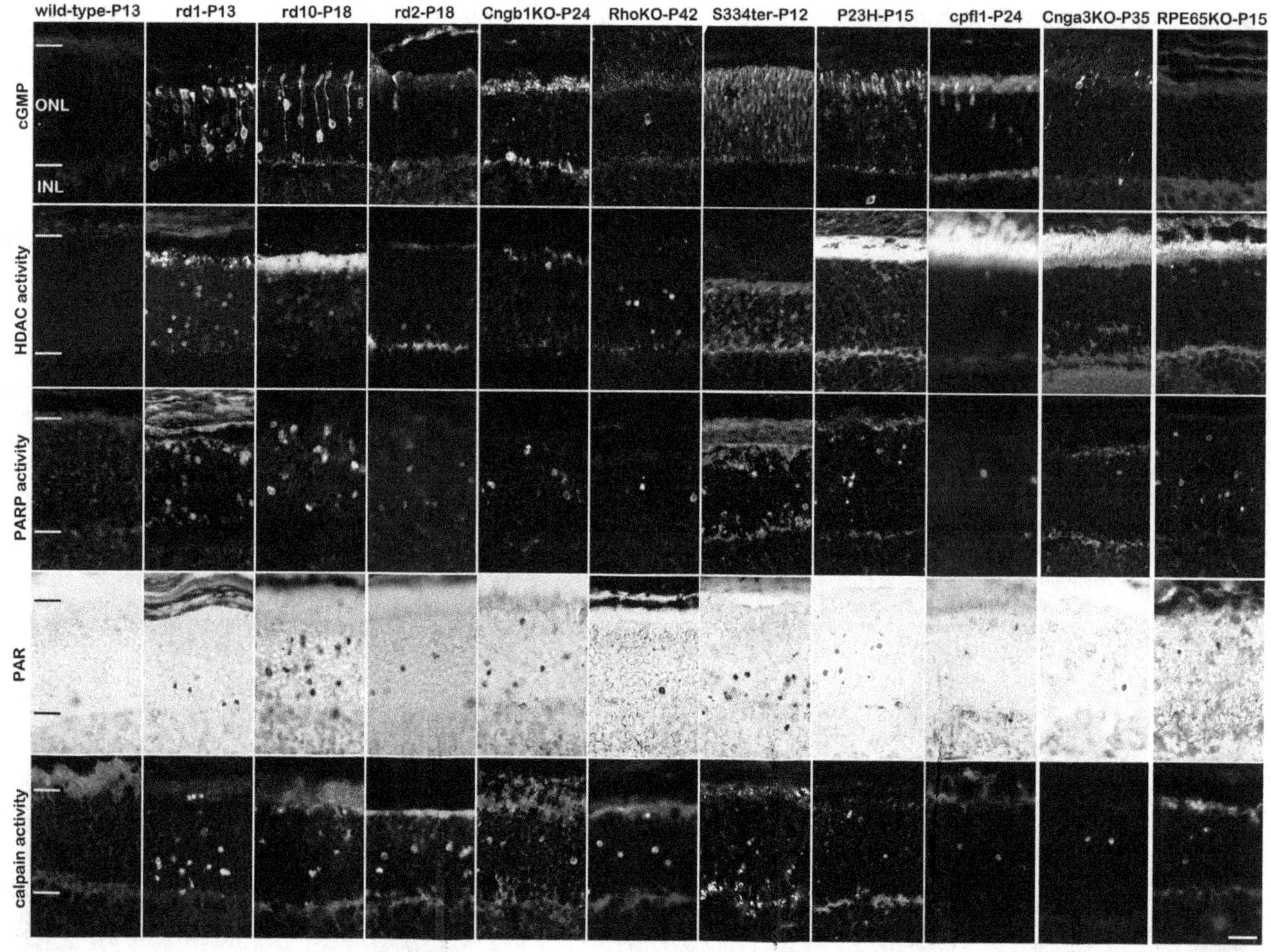

Figure 4. Cell death in hereditary retinal degeneration is predominantly non-apoptotic. In 10 out of 10 animal models for hereditary retinal degeneration, large numbers of photoreceptors display cGMP accumulation, HDAC and PARP activity, PAR accumulation, and calpain activity, respectively. Intriguingly, these non-apoptotic markers are prominent even in the S334ter retina, concomitant with this also showing signs of apoptosis. This suggests that in S334ter retina two different cell death mechanisms may run in parallel while in all other studied RD models the mutation-induced cell death followed a non-apoptotic mechanism. Scale bar 20 μm.

apoptotic and non-apoptotic cell death mechanisms appeared to be equally important (Figure S3).

Discussion

Our study provides a detailed and comprehensive overview of the temporal progression and the kinetics of cell death in ten different, commonly used RD animal models. These RD models harbour genetic defects mostly affecting the phototransduction cascade but include also such which are disturbing the visual cycle (*Rpe65* KO) and the structural integrity of the outer segment (*rd2*). As a result, the comparative analysis of characteristic cell death processes for the first time highlights the over-riding importance of a common, alternative mechanism for photoreceptor degeneration. Contrary to previous studies on retinal degeneration mechanisms [23,24] our study focused on the elevated activity and presence of key enzymes and/or metabolites, respectively, and thus may be seen as a first attempt to assess the so called reactome or metabolome (www.reactome.org; [25]) of photoreceptor degeneration at the level of the individual dying cell.

To put our report in a perspective, many studies on cell death in the retina and other parts of the central nervous system have

previously resorted to tissue based methods (*e.g.* micro-array, western blot) [23,24]. Such methods are particularly useful in conditions where there is a homogenous cell population and a highly synchronized onset of cell death and are thus ideal, for instance, for cell culture. However, in a complex neuronal tissue such as the retina, with >50 different neuronal cell types among which only one – the rod photoreceptor – undergoes non-synchronized primary degeneration, with cell death of individual photoreceptors spread out over a time of weeks to many years, tissue based analysis runs the risk of suffering from very low detection rates and overall poor signal-to-noise ratio. For our analysis, we thus focussed on methods that afforded cellular resolution to be able to unequivocally attribute cell death related processes to primary photoreceptor death and to distinguish these processes from secondary or tertiary events.

Apoptosis during retinal degeneration

Previous studies on cell death in hereditary retinal degeneration have often suggested apoptosis as the main degenerative mechanism [2,3,26]. These earlier studies, however, based their conclusion on analysis methods now known not to discriminate between apoptosis and other forms of cell death. For instance, the

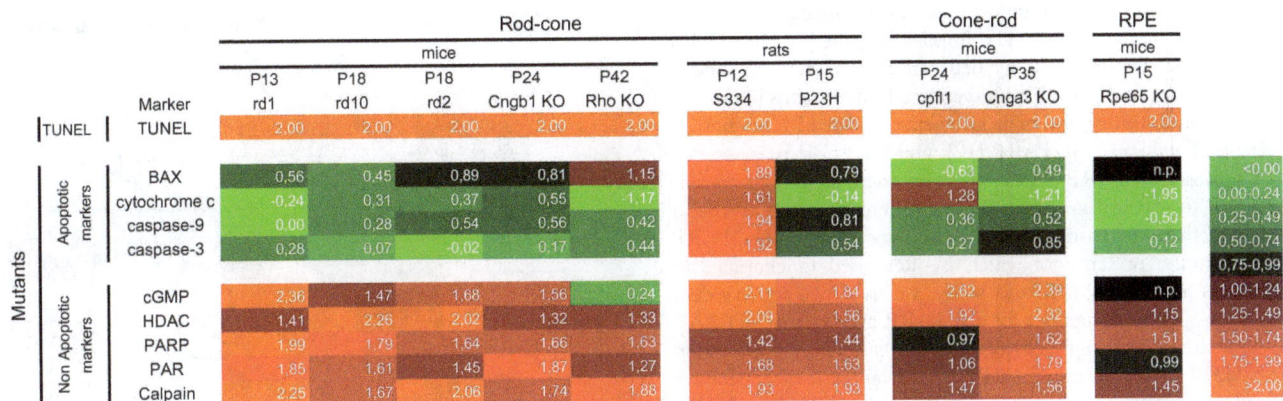

	Marker	Rod-cone							Cone-rod		RPE
		mice					rats		mice		mice
		P13 rd1	P18 rd10	P18 rd2	P24 Cngb1 KO	P42 Rho KO	P12 S334	P15 P23H	P24 cpfl1	P35 Cnga3 KO	P15 Rpe65 KO
TUNEL	TUNEL	2,00	2,00	2,00	2,00	2,00	2,00	2,00	2,00	2,00	2,00
Apoptotic markers	BAX	0,56	0,45	0,89	0,81	1,15	1,89	0,79	-0,63	0,49	n.p.
	cytochrome c	-0,24	0,31	0,37	0,55	-1,17	1,61	-0,14	1,28	-1,21	-1,95
	caspase-9	0,00	0,28	0,54	0,56	0,42	1,94	0,81	0,36	0,52	-0,50
	caspase-3	0,28	0,07	-0,02	0,17	0,44	1,92	0,54	0,27	0,85	0,12
Non Apoptotic markers	cGMP	2,36	1,47	1,68	1,56	0,24	2,11	1,84	2,62	2,39	n.p.
	HDAC	1,41	2,26	2,02	1,32	1,33	2,09	1,56	1,92	2,32	1,15
	PARP	1,99	1,79	1,64	1,66	1,63	1,42	1,44	0,97	1,62	1,51
	PAR	1,85	1,61	1,45	1,87	1,27	1,68	1,63	1,06	1,79	0,99
	Calpain	2,25	1,67	2,06	1,74	1,88	1,93	1,93	1,47	1,56	1,45

Legend: <0,00 | 0,00-0,24 | 0,25-0,49 | 0,50-0,74 | 0,75-0,99 | 1,00-1,24 | 1,25-1,49 | 1,50-1,74 | 1,75-1,99 | >2,00

Figure 5. Heat map representing metabolic activities in different RD models. The RD models were grouped according to the peak of degeneration, the cell type affected by the mutation (rod, cone, RPE), and species (mouse, rat). The number of TUNEL-positive cells in each model was normalized to 100, expressed as logarithm, and compared with the number of positively labelled cells for each marker. The heat map clearly illustrates the prevalence of non-apoptotic vs. apoptotic cell death in 9 out of 10 RD models. The S334ter rhodopsin mutant was unique, showing concurrent activation of both cell death pathways. n.p.: null positive. See also Table S1 and S2.

TUNEL assay, originally thought to be a marker for apoptosis [27], generally labels all kinds of dying cells, including necrotic cells [28].

Apoptosis may be defined as an active process resulting in orderly self-disintegration of a cell. Hallmark features of apoptosis include an up-regulation of pro-apoptotic genes and proteins, such as the transcription factor c-fos and in particular Bcl-2 family proteins such as BAX, which participate in forming the mitochondrial permeability transition pore (MPTP), allowing mitochondrial proteins including cytochrome c to enter the cytoplasm. Cytoplasmic cytochrome c aggregates with apoptotic protease-activating factor (APAF) and caspase-9 to form a multimeric protein complex termed the apoptosome [1]. This complex then cleaves and activates down-stream executioner caspases such as caspase-3.

Classical apoptosis occurs during retinal development until about 3–4 weeks post-natal [29]. Indeed, developmental apoptosis temporally coincides, at least partially, with mutation-induced cell death [7]. This introduces a confounding factor which may explain some of the contradictory reports in the literature. Our study demonstrates that wild-type photoreceptors are capable of executing apoptosis at least until P42; by contrast, however, we see that mutant photoreceptors normally take a non-apoptotic route as a means for orderly self-destruction.

Importantly, therapeutic strategies based on the inhibition of the apoptotic cascade have had little success or produced conflicting findings. For instance, neither the pharmacological inhibition of the caspase cascade [5], nor the genetic manipulation of Bcl-2 and Bcl-XL [30], c-fos [31], or caspase-3 [6] promoted long-term photoreceptor survival. On the other hand BAX KO may delay rod but not cone death in the *Rpe65* KO animals [4].

Recently, increased BAX activation was suggested to be connected to retinal degeneration in *rd1*, *Rho* KO, and P23H mice [20]. At present it is not clear whether these findings relate in part to developmental cell death (see above) or would have been interpreted differently if the study [20] had also included observations of a model with a much stronger BAX response, such as the S334ter rat investigated by us. At any rate, our results do not show any evidence for major BAX activation in degenerating retina, with the notable exception of S334ter photoreceptors. This model thus constitutes a "positive control"

for BAX and further apoptotic markers, lending additional credit to our findings in all other mutants.

Alternative cell death mechanisms

In recent years a growing body of evidence has suggested the activity of alternative cell death mechanisms in RD [8,32–34]. The analysis of such mechanisms faces the major obstacle of identifying alternative and causative metabolic processes. In a number of previous studies, we showed activation of the cGMP targets protein kinase G (PKG) and cyclic nucleotide-gated (CNG) channel [22,35] in degenerating *rd1* photoreceptors. Excessive cGMP signalling was associated with a strong increase in enzymatic activities of calpain-type proteases [13], PARP [12], and HDAC [15], which we found to be causally involved in photoreceptor cell death. Calpain activation, which was also seen by others in different RD models [20], is a well-established phenomenon in necrosis and alternative cell death mechanisms [21,36]. While HDAC and PARP enzymes are ubiquitously expressed and involved in epigenetic gene regulation and DNA repair [37], respectively, their excessive activation has repeatedly been connected to alternative mechanisms of neuronal cell death [38–40].

We found that all these processes were also involved in RD caused by the different mutations, in various genes and in both mouse and rat. Importantly, the cellular resolution afforded by the used assays allowed clear distinction between cells dying an apoptotic death and cells dying through an alternative pathway. In this alternative pathway the activities of calpain and PARP activity co-localize to a large extent with the TUNEL assay [12,17], while cGMP detection and HDAC activity do not [15,41]. This could suggest that the latter two relate to early metabolic processes in the execution of cell death.

Together with other earlier data [8,16,42,43] our present findings prompt us to propose a potential pathway for cGMP-induced cell death: Elevated levels of cGMP activate CNG channels and/or PKG to cause excessive Ca^{2+}-influx and protein phosphorylation, respectively. As a possible consequence of the latter, PKG dependent phosphorylation could trigger HDAC activation [44], down-stream of which PARP can be activated [15]. Ca^{2+}-influx might on the other hand, and in parallel, cause calpain activation [13,35]. Both routes (Figure 6) act in unison to drive a photoreceptor cell to its demise, but, surprisingly, this alternative form of cGMP-induced cell death appears to be 4–6 times slower than apoptosis [41]. Importantly, the presence of this pathway and the connections between the different metabolic processes were confirmed by interventional experiments in the *rd1* mouse demonstrating the neuroprotective effects of inhibition of PKG [22], calpain [13], PARP [12], and HDAC [15].

The observed PARP activity deserves some additional considerations: In classical apoptosis the PARP enzyme is cleaved and inactivated by caspases, resulting in a specific 85 kDa PARP fragment, the presence of which is often used to characterize apoptosis as such [45]. In our study, we used two independent methods – immunostaining for the PARP activity product PAR and direct *in situ* PARP activity detection based on incorporation of NAD$^+$ – to demonstrate PARP over-activation. Hence, what we found in mutant photoreceptors is the exact opposite of what would happen in apoptosis, which thus provides further evidence for a non-apoptotic photoreceptor cell death, an alternative cell death mechanism that could share some features with PARthanatos [40].

The fact that photoreceptors use a non-apoptotic mechanism when in principle they are capable of executing apoptosis raises the question as to what the physiological and even evolutionary advantage of this mechanism may be. Apoptosis is a process that requires energy in the form of ATP [1]. The insult caused by a genetic mutation may exhaust such energy resources to the point that apoptosis can no longer be executed. Necrosis on the other hand would result in inflammation and could cause additional extensive tissue damage. Hence, it may make sense for a cell to execute the slow, alternative and probably ATP-independent pathway laid out here to limit the damage to the surrounding neuronal tissue.

Perspectives for mutation-independent RD treatment

An important consequence of the high genetic heterogeneity of retinal degenerations is that for any pathogenic mutation there may be only a very low number of patients [10,11]. This calls for the development of mutation-independent treatments that could address larger groups of RD patients. The finding that the same non-apoptotic mechanism was the prevalent mode of cell death in 9/10 RD models strongly increases the chances to find neuroprotective treatments that are independent of the initial causal mutation. In the context of rare retinal diseases, such treatments appropriate for a large number of patients may dramatically improve the perspectives for both a successful clinical translation and the commercial viability of corresponding drugs.

We found that the alternative cell death mechanism described above was active in all investigated animal models. Of particular importance for this mechanism may be the observed accumulation of cGMP in mutant photoreceptors. While this was already known for retina suffering from mutations in *Pde6b* and *Pde6c* (i.e. rd1, cpfl1; [18,46]), *Prph2* (i.e.rd2 [22]), *Cngb1* and *Cnga3* [35,42], our work also showed cGMP accumulation in retina suffering for three different types of rhodopsin mutations (*Rho* KO, S334ter, P23H). A potential explanation for this remarkable phenomenon in rhodopsin mutants could be either the longer life-times of activated rhodopsin resulting in a stimulation of cGMP synthesis and an increase in net cGMP [47] or a failure to activate downstream PDE6 in cases where rhodopsin is absent (i.e. in *Rho* KO).

While these findings highlight cGMP-signalling for the development of novel neuroprotective treatments, there is one exception: in *Rpe65* KO retina, we did not find elevations of cGMP. Indeed, here, unliganded opsin was proposed to cause a constitutive activation of phototransduction and hence low cGMP-

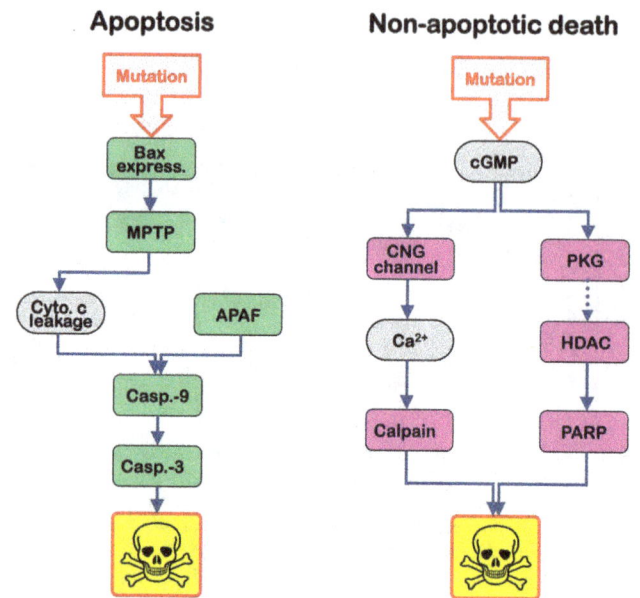

Figure 6. Two routes to cell death. Classical apoptosis, such as it occurs in S334ter transgenic photoreceptors, involves a mutation-induced up-regulation and translocation of BAX protein to form the mitochondrial permeability transition pore (MPTP). This leads to leakage of cytochrome c from the mitochondria to the cytoplasm, where it combines with apoptotic protease activating factor (APAF) and caspase-9 to form the apoptosome, which in turn activates down-stream executioner caspases, including caspase-3. In 9/10 RD animal models investigated here, photoreceptor death followed a different route: mutation-induced up-regulation of cGMP on the one hand causes activation of the CNG channel, leading to Ca2+ influx and calpain activation. On the other hand cGMP-dependent activation of protein kinase G (PKG) is associated with histone deacetylase (HDAC) and poly-ADP-ribose-polymerase (PARP) activation. Importantly, this alternative, non-apoptotic cell death mechanism offers a number of novel targets for neuroprotection of photoreceptors.

levels [48]. On the other hand, since all further down-stream processes appear to be the same in all mutants investigated, a disruption of the visual cycle by *Rpe65* KO [49] might cause minor elevations of cGMP – perhaps below the detection levels of our immunohistological methods – and still trigger cell death.

Mutations in the same gene may potentially trigger distinct degenerative processes [16]. Our study more extensively shows how intragenic variability of RD mutations may initiate different cell death mechanisms: The recessive *rd1* and *rd10* mutations in the *Pde6b* gene result in activation of the same non-apoptotic pathways. This is also true for the recessive *Rho* KO and the dominant P23H mutation, but not for the dominant S334ter mutation. While all three mutations reside in the rhodopsin gene, the concurrent activation of apoptotic and non-apoptotic cell death observed in the S334ter situation suggests that human patients with similar mutations may need combination therapy targeting both degenerative pathways simultaneously. Likewise, since we found that photoreceptors (wild-type) are in principle able to execute apoptosis, we cannot exclude the possibility that under circumstances in which non-apoptotic cell death is blocked, the cell may switch to apoptosis. This possibility needs further investigation and might also require the development of combination therapies.

Another question, that will be important to address in the future, relates to the fact that all mutant photoreceptors carry a genetic defect that will eventually destroy them. Yet, the timepoint at which a mutant photoreceptor dies appears to be entirely random, and, in the human situation, the time from the first to the last photoreceptors' death may cover many decades [10]. The exact reasons for this phenomenon are unknown but could be explained by stochastic effects similar to what is seen in the decay of radioactive elements [50]. This opens the possibility that even a minor shift in the dynamics of these stochastic processes – such as interference with processes like those studied here – could improve photoreceptor survival dramatically.

In conclusion, this work demonstrates the existence of a common, non-apoptotic cell death mechanism for hereditary photoreceptor degeneration. The tentative cell death pathway laid out here (Figure 6) provides a number of novel targets for neuroprotective treatment approaches [12,13,15,16,22] and, importantly, a unifying principle for RD caused by a variety of different mutations in different genes. As such, this common cell death pathway may be of major importance for future RD therapy developments and possibly for also other neurodegenerative diseases.

Supporting Information

Figure S1 Correlation of selected cell death markers to loss of photoreceptors, related to Figure 1. Percentage of labelled ONLcells (left y-axis) and number of surviving photoreceptor rows (right y-axis) for (A) *rd1* mice, (B) P23H, and (C) S334ter transgenic rats. In all three models, calpain activation peaked together with the TUNEL assay, and correlated with the strongest loss in the number of photoreceptor rows. The grey area indicates the loss of photoreceptors. Throughout the retinal degeneration, activation of caspase-3 was ***absent*** in *rd1* and P23H retina, but ***present*** in S334ter retina. Values are mean from at least three different animals.

Figure S2 Expression of activated BAX in wild-type, rd1 and S334ter retina. In wild-type mouse retina at P11 (left panel), a mouse monoclonal antibody directed against activated BAX (clone 6A7) detected positive cells only rarely, but then in all layers of the retina. The white arrowhead indicates a cell positive for activated BAX in the ganglion cell layer (GCL). In *rd1* mouse retina at P11 – the onset of RD in this model – activated BAX is detected only very rarely, with BAX detection levels very similar to age-matched wild-type (middle panel; cf. Table S2). In contrast to this, in the outer nuclear layer (ONL) of P12 S334ter rat retina, the BAX antibody immunodecorates mitochondria, in particular in individual photoreceptor inner segments, synaptic terminals, and perinuclear areas (right panel). This mitochondria specific staining pattern in S334ter retina is consistent with the reported role of BAX in the formation of the mitochondrial permeability transition pore and the initiation of apoptosis. Images are representative for immunostainings obtained from at least three different animals for each genotype. Note that use of secondary anti-mouse antibodies

led to an unspecific IGG decoration in inner retinal blood vessels in mouse tissues (see asterisks in wild-type, *rd1*). INL = inner nuclear layer.

Figure S3 Cell death markers in wild-type mouse retina. Well-type retina occasionally showed cells positive for both apoptotic and non-apoptotic cell death markers (A). As the number of positive cells is rather small, please note that the pictures shown are selected not as the representative but somewhat an exaggeration of the real number of dying cells. Heat map representing metabolic activities in corresponding wild-types (B), similarly as in Figure 5 for RD mutants, shows that cell death during wild-type retina development displayed activation of both apoptotic and non-apoptotic pathways. Scale bar 20 μm. n.p.: null positive. See also Table S2.

Table S1 Quantification of cell death processes in 10 different RD animals related to Figures 1 and 4. Numbers given represent mean values for the percentages of positive cells for each marker, followed by standard error of the mean (SEM), and p-values for comparisons with corresponding, age-matched WT. Green label indicates statistically significant p-values (p<0.05); red label indicates non-significance. Significant differences between RD mutants and WT were found almost only for non-apoptotic processes, with the notable exception of the S334ter mutant where also apoptotic processes were significantly activated. Note that in contrast to Fig. 4, here, values were not normalized to the numbers of TUNEL positive, dying cells.

Table S2 Quantification of labelled photoreceptors in different RD models related to Figures 1 and 4. For each genotype, at the respective peak of degeneration, the percentage of cells positively labelled for the various cell death processes is given as mean value, followed by SEM, and number (n) of different specimens analysed. To assess the relative importance of these processes for retinal degeneration the percentage of TUNEL positive cells is also given.

Acknowledgments

We thank M.M. LaVail for transgenic rats; T. Schubert, J. Kaur, and W. Haq for helpful comments and discussions; S. Bernhard-Kurz, K. Masarini, N. Rieger, M. Dierstein, B. Sandström, and H. Abdshill for skilful technical assistance.

Author Contributions

Conceived and designed the experiments: BAG FPD. Performed the experiments: BAG DT AS KK PF FPD. Analyzed the data: BAG DT AS KK SC PF KD EZ TE MU FPD. Contributed reagents/materials/analysis tools: BAG SM SK FK SC UJB KD MB TE PE FPD. Wrote the paper: BAG DT FPD.

References

1. Orrenius S, Zhivotovsky B, Nicotera P (2003) Regulation of cell death: the calcium-apoptosis link. Nat Rev Mol Cell Biol 4: 552–565.
2. Chang GQ, Hao Y, Wong F (1993) Apoptosis: final common pathway of photoreceptor death in rd, rds, and rhodopsin mutant mice. Neuron 11: 595–605.
3. Marigo V (2007) Programmed cell death in retinal degeneration: targeting apoptosis in photoreceptors as potential therapy for retinal degeneration. Cell Cycle 6: 652–655.
4. Hamann S, Schorderet DF, Cottet S (2009) Bax-induced apoptosis in Leber's congenital amaurosis: a dual role in rod and cone degeneration. PLoS One 4: e6616. doi: 10.1371/journal.pone.0006616

5. Yoshizawa K, Kiuchi K, Nambu H, Yang J, Senzaki H, et al. (2002) Caspase-3 inhibitor transiently delays inherited retinal degeneration in C3H mice carrying the rd gene. Graefes Arch Clin Exp Ophthalmol 240: 214–219.
6. Zeiss CJ, Neal J, Johnson EA (2004) Caspase-3 in postnatal retinal development and degeneration. Invest Ophthalmol Vis Sci 45: 964–970.
7. Sancho-Pelluz J, Arango-Gonzalez B, Kustermann S, Romero FJ, van Veen T., et al. (2008) Photoreceptor cell death mechanisms in inherited retinal degeneration. Mol Neurobiol 38: 253–269.
8. Trifunovic D, Sahaboglu A, Kaur J, Mencl S, Zrenner E, et al. (2012) Neuroprotective Strategies for the Treatment of Inherited Photoreceptor Degeneration. Curr Mol Med 12(5): 598–612.

9. Dalke C, Graw J (2005) Mouse mutants as models for congenital retinal disorders. Exp Eye Res 81: 503–512.

10. Hamel C (2006) Retinitis pigmentosa. Orphanet J Rare Dis 1: 40.

11. Hamel CP (2007) Cone rod dystrophies. Orphanet J Rare Dis 2: 7.

12. Paquet-Durand F, Silva J, Talukdar T, Johnson LE, Azadi S, et al. (2007) Excessive activation of poly(ADP-ribose) polymerase contributes to inherited photoreceptor degeneration in the retinal degeneration 1 mouse. J Neurosci 27: 10311–10319.

13. Paquet-Durand F, Sanges D, McCall J, Silva J, van Veen T, et al. (2010) Photoreceptor rescue and toxicity induced by different calpain inhibitors. J Neurochem 115: 930–940.

14. Sahaboglu A, Bolz S, Lowenheim H, Paquet-Durand F (2014) Expression of Poly(ADP-Ribose) Glycohydrolase in Wild-Type and PARG-110 Knock-Out Retina. Adv Exp Med Biol 801: 463–469.

15. Sancho-Pelluz J, Alavi M, Sahaboglu A, Kustermann S, Farinelli P, et al. (2010) Excessive HDAC activation is critical for neurodegeneration in the rd1 mouse. Cell Death & Disease 1: 1–9.

16. Kaur J, Mencl S, Sahaboglu A, Farinelli P, van Veen T, et al. (2011) Calpain and PARP Activation during Photoreceptor Cell Death in P23H and S334ter Rhodopsin Mutant Rats. PLoS ONE 6: e22181. doi:10.1371/journal.pone.0022181

17. Paquet-Durand F, Azadi S, Hauck SM, Ueffing M, van Veen T, et al. (2006) Calpain is activated in degenerating photoreceptors in the rd1 mouse. J Neurochem 96: 802–814.

18. Trifunovic D, Dengler K, Michalakis S, Zrenner E, Wissinger B, et al. (2010) cGMP-dependent cone photoreceptor degeneration in the cpfl1 mouse retina. J Comp Neurol 518: 3604–3617.

19. Mosinger OJ, Deckwerth TL, Knudson CM, Korsmeyer SJ (1998) Suppression of developmental retinal cell death but not photoreceptor degeneration in Bax-deficient mice. Invest Ophthalmol Vis Sci 39: 1713–1720.

20. Comitato A, Sanges D, Rossi A, Humphries MM, Marigo V (2014) Activation of bax in three models of retinitis pigmentosa. Invest Ophthalmol Vis Sci 55(6):3555–62.

21. Shinde VM, Sizova OS, Lin JH, LaVail MM, Gorbatyuk MS (2012) ER stress in retinal degeneration in S334ter Rho rats. PLoS One 7: e33266. doi:10.1371/journal.pone.0033266

22. Paquet-Durand F, Hauck SM, van Veen T, Ueffing M, Ekstrom P (2009) PKG activity causes photoreceptor cell death in two retinitis pigmentosa models. J Neurochem 108: 796–810.

23. Azadi S, Paquet-Durand F, Medstrand P, van Veen T, Ekstrom PA (2006) Up-regulation and increased phosphorylation of protein kinase C (PKC) delta, mu and theta in the degenerating rd1 mouse retina. Mol Cell Neurosci 31: 759–773.

24. Hauck SM, Ekstrom PA, Ahuja-Jensen S, Suppmann S, Paquet-Durand F, et al. (2006) Differential modification of phosducin protein in degenerating rd1 retina is associated with constitutively active Ca2+/calmodulin kinase II in rod outer segments. Mol Cell Proteomics 5: 324–336.

25. Rubakhin SS, Lanni EJ, Sweedler JV (2013) Progress toward single cell metabolomics. Curr Opin Biotechnol 24: 95–104.

26. Portera-Cailliau C, Sung CH, Nathans J, Adler R (1994) Apoptotic photoreceptor cell death in mouse models of retinitis pigmentosa. Proc Natl Acad Sci U S A 91: 974–978.

27. Gavrieli Y, Sherman Y, Ben-Sasson SA (1992) Identification of programmed cell death in situ via specific labeling of nuclear DNA fragmentation. J Cell Biol 119: 493–501.

28. Grasl-Kraupp B, Ruttkay-Nedecky B, Koudelka H, Bukowska K, Bursch W, et al. (1995) In situ detection of fragmented DNA (TUNEL assay) fails to discriminate among apoptosis, necrosis, and autolytic cell death: a cautionary note. Hepatology 21: 1465–1468.

29. Young RW (1984) Cell death during differentiation of the retina in the mouse. J Comp Neurol 229: 362–373.

30. Joseph RM, Li T (1996) Overexpression of Bcl-2 or Bcl-XL transgenes and photoreceptor degeneration. Invest Ophthalmol Vis Sci 37: 2434–2446.

31. Hafezi F, Abegg M, Grimm C, Wenzel A, Munz K, et al. (1998) Retinal degeneration in the rd mouse in the absence of c-fos. Invest Ophthalmol Vis Sci 39: 2239–2244.

32. Lohr HR, Kuntchithapautham K, Sharma AK, Rohrer B (2006) Multiple, parallel cellular suicide mechanisms participate in photoreceptor cell death. Exp Eye Res 83: 380–389.

33. Punzo C, Kornacker K, Cepko CL (2009) Stimulation of the insulin/mTOR pathway delays cone death in a mouse model of retinitis pigmentosa. Nat Neurosci 12: 44–52.

34. Doonan F, Donovan M, Cotter TG (2003) Caspase-independent photoreceptor apoptosis in mouse models of retinal degeneration. J Neurosci 23: 5723–5731.

35. Paquet-Durand F, Beck S, Michalakis S, Goldmann T, Huber G, et al. (2011) A key role for cyclic nucleotide gated (CNG) channels in cGMP-related retinitis pigmentosa. Hum Mol Genet 20: 941–947.

36. McCall K (2010) Genetic control of necrosis - another type of programmed cell death. Curr Opin Cell Biol 22: 882–888.

37. Bai P, Canto C (2012) The role of PARP-1 and PARP-2 enzymes in metabolic regulation and disease. Cell Metab 16: 290–295.

38. Bardai FH, Price V, Zaayman M, Wang L, D'Mello SR (2012) Histone deacetylase-1 (HDAC1) is a molecular switch between neuronal survival and death. J Biol Chem 287: 35444–35453.

39. Jaumann M, Dettling J, Gubelt M, Zimmermann U, Gerling A, et al. (2012) cGMP-Prkg1 signaling and Pde5 inhibition shelter cochlear hair cells and hearing function. Nat Med 18: 252–259.

40. Wang Y, Kim NS, Haince JF, Kang HC, David KK, et al. (2011) Poly(ADP-ribose) (PAR) binding to apoptosis-inducing factor is critical for PAR polymerase-1-dependent cell death (parthanatos). Sci Signal 4(167): ra20.

41. Sahaboglu A, Paquet-Durand O, Dietter J, Dengler K, Bernhard-Kurz S, et al. (2013) Retinitis Pigmentosa: Rapid neurodegeneration is governed by slow cell death mechanisms. Cell Death & Disease 4:e488.

42. Michalakis S, Muhlfriedel R, Tanimoto N, Krishnamoorthy V, Koch S, et al. (2010) Restoration of Cone Vision in the CNGA3(-/-) Mouse Model of Congenital Complete Lack of Cone Photoreceptor Function. Mol Ther. 18(12):2057–63.

43. Sahaboglu A, Tanimoto N, Kaur J, Sancho-Pelluz J, Huber G, et al. (2010) PARP1 gene knock-out increases resistance to retinal degeneration without affecting retinal function. PLoS ONE 5: e15495.

44. Hao N, Xu N, Box AC, Schaefer L, Kannan K, et al. (2011) Nuclear cGMP-dependent kinase regulates gene expression via activity-dependent recruitment of a conserved histone deacetylase complex. PLoS Genet 7: e1002065. doi:10.1371/journal.pgen.1002065.

45. Lazebnik YA, Kaufmann SH, Desnoyers S, Poirier GG, Earnshaw WC (1994) Cleavage of poly(ADP-ribose) polymerase by a proteinase with properties like ICE. Nature 371: 346–347.

46. Farber DB, Lolley RN (1974) Cyclic guanosine monophosphate: elevation in degenerating photoreceptor cells of the C3H mouse retina. Science 186: 449–451.

47. Gross OP, Pugh EN Jr, Burns ME (2012) Calcium feedback to cGMP synthesis strongly attenuates single-photon responses driven by long rhodopsin lifetimes. Neuron 76: 370–382.

48. Woodruff ML, Wang Z, Chung HY, Redmond TM, Fain GL, et al. (2003) Spontaneous activity of opsin apoprotein is a cause of Leber congenital amaurosis. Nat Genet 35: 158–164.

49. Redmond TM, Yu S, Lee E, Bok D, Hamasaki D, et al. (1998) Rpe65 is necessary for production of 11-cis-vitamin A in the retinal visual cycle. Nat Genet 20: 344–351.

50. Clarke G, Collins RA, Leavitt BR, Andrews DF, Hayden MR, et al. (2000) A one-hit model of cell death in inherited neuronal degenerations. Nature 406: 195–199.

51. Sanyal S, Bal AK (1973) Comparative light and electron microscopic study of retinal histogenesis in normal and rd mutant mice. Z Anat Entwicklungsgesch 142: 219–238.

52. Keeler CE (1924) The Inheritance of a Retinal Abnormality in White Mice. Proc Natl Acad Sci U S A 10: 329–333.

53. Sanyal S, De RA, Hawkins RK (1980) Development and degeneration of retina in rds mutant mice: light microscopy. J Comp Neurol 194: 193–207.

54. Humphries MM, Rancourt D, Farrar GJ, Kenna P, Hazel M, et al. (1997) Retinopathy induced in mice by targeted disruption of the rhodopsin gene. Nat Genet 15: 216–219.

55. Huttl S, Michalakis S, Seeliger M, Luo DG, Acar N, et al. (2005) Impaired channel targeting and retinal degeneration in mice lacking the cyclic nucleotide-gated channel subunit CNGB1. J Neurosci 25: 130–138.

56. Biel M, Seeliger M, Pfeifer A, Kohler K, Gerstner A, et al. (1999) Selective loss of cone function in mice lacking the cyclic nucleotide-gated channel CNG3. Proc Natl Acad Sci U S A 96: 7553–7557.

57. Chang B, Hawes NL, Hurd RE, Davisson MT, Nusinowitz S, et al. (2001) A new mouse model of cone photoreceptor function loss (cpfl1). Invest Ophthalmol Vis Sci 42(4): S527.

58. Chang B, Hawes NL, Hurd RE, Davisson MT, Nusinowitz S, Heckenlively JR (2002) Retinal degeneration mutants in the mouse. Vision Research 42: 517–525.

59. Steinberg RH, Flannery JG, Naash M, Oh P, Matthes MT, et al. (1996) Transgenic rat models of inherited retinal degeneration caused by mutant opsin genes. Invest Ophthalmol Vis Sci 37: S698.

60. Hsu YT, Youle RJ (1997) Nonionic detergents induce dimerization among members of the Bcl-2 family. J Biol Chem 272: 13829–13834.

61. Liu C, Li Y, Peng M, Laties AM, Wen R (1999) Activation of caspase-3 in the retina of transgenic rats with the rhodopsin mutation s334ter during photoreceptor degeneration. J Neurosci 19: 4778–4785.

62. Tiwari S, Hudson S, Gattone VH, Miller C, Chernoff EA, et al. (2013) Meckelin 3 is necessary for photoreceptor outer segment development in rat Meckel syndrome. PLoS One 8: e59306. doi:10.1371/journal.pone.0059306.

63. De Vente J, Steinbusch HW, Schipper J (1987) A new approach to immunocytochemistry of 3′,5′-cyclic guanosine monophosphate: preparation, specificity, and initial application of a new antiserum against formaldehyde-fixed 3′,5′-cyclic guanosine monophosphate. Neuroscience 22: 361–373.

12

Enterovirus 71 Induces Mitochondrial Reactive Oxygen Species Generation That is Required for Efficient Replication

Mei-Ling Cheng[1,2,3], Shiue-Fen Weng[4], Chih-Hao Kuo[4], Hung-Yao Ho[2,4,5]*

1 Department of Biomedical Sciences, College of Medicine, Chang Gung University, Tao-Yuan, Taiwan, 2 Healthy Aging Research Center, Chang Gung University, Tao-Yuan, Taiwan, 3 Metabolomics Core Laboratory, Chang Gung University, Tao-Yuan, Taiwan, 4 Department of Medical Biotechnology and Laboratory Science, College of Medicine, Chang Gung University, Tao-Yuan, Taiwan, 5 Office of Research and Development, Chang Gung Memorial Hospital, Tao-Yuan, Taiwan

Abstract

Redox homeostasis is an important host factor determining the outcome of infectious disease. Enterovirus 71 (EV71) infection has become an important endemic disease in Southeast Asia and China. We have previously shown that oxidative stress promotes viral replication, and progeny virus induces oxidative stress in host cells. The detailed mechanism for reactive oxygen species (ROS) generation in infected cells remains elusive. In the current study, we demonstrate that mitochondria were a major ROS source in EV71-infected cells. Mitochondria in productively infected cells underwent morphologic changes and exhibited functional anomalies, such as a decrease in mitochondrial electrochemical potential $\Delta\Psi_m$ and an increase in oligomycin-insensitive oxygen consumption. Respiratory control ratio of mitochondria from infected cells was significantly lower than that of normal cells. The total adenine nucleotide pool and ATP content of EV71-infected cells significantly diminished. However, there appeared to be a compensatory increase in mitochondrial mass. Treatment with mito-TEMPO reduced eIF2α phosphorylation and viral replication, suggesting that mitochondrial ROS act to promote viral replication. It is plausible that EV71 infection induces mitochondrial ROS generation, which is essential to viral replication, at the sacrifice of efficient energy production, and that infected cells up-regulate biogenesis of mitochondria to compensate for their functional defect.

Editor: Hsin-Chih Lai, Chang-Gung University, Taiwan

Funding: This project is supported by grants from Chang Gung University (CMRPD190443, CMRPD1A0562, CMRPD1C0751, CMRPD1A0563, CMRPD1A0521, CMRPD1A0522, CMRPD391683, CMRPD1C0441, and CMRPD1C0761), National Science Council of Taiwan (NSC99-2320-B-182-021-MY3 and NSC101-2320-B-182-024-MY3), and the Ministry of Education of Taiwan (EMRPD1C0271 and EMRPD1D0241). The funders had no role in study design, data collection and analysis, decision to publish, or preparation of the manuscript.

Competing Interests: The authors have declared that no competing interests exist.

* Email: hoh01@mail.cgu.edu.tw

Introduction

Enterovirus 71 (EV71), a member of the family *Picornaviridae*, is a non-enveloped RNA virus [1]. Since its identification in 1969 [2], episodes of EV71 outbreak occurred periodically throughout the world [3–6]. Clinical manifestation of EV71 infection includes febrile illness, acute respiratory disease, hand-foot-and-mouth disease, herpangia, myocarditis, aseptic meningitis, acute flaccid paralysis, brainstem and/or cerebellar encephalitis, Guillain-Barre syndrome, or combinations of these clinical features [7,8]. Though hand-foot-and-mouth disease is a benign disease, the neurologic and systemic complications are more severe [9]. Children aged below 5 years are susceptible to development of permanent neurologic sequelae, or even succumb to such complications [10]. To date, the largest epidemic of EV71 occurred in China in 2008 and 2009. Nearly 490000 cases, of which 126 deaths occurred, were reported; over a million cases of hand-foot-and-mouth disease were reported in 2009 [4,5]. EV71 infection recurs every 2 or 3 years in Asia-Pacific region.

It is evident that redox environment is a factor affecting host-microbe interactions. The susceptibility of host cells to HIV, coxsackievirus and influenza virus is affected by redox microenvironment [11–15]. We have recently found that increased oxidative stress in host cells enhances EV71 infection [16]. It is intriguing that viral infection can itself induce oxidative stress in host cells. Increased superoxide production has been observed in pneumonia caused by influenza A virus [17]. Herpes simplex virus (HSV) infection of microglia cells elicits oxidative stress, which probably causes the neurotoxicity [18]. Rhinovirus and respiratory syncytial virus induce ROS production in fibroblasts and epithelial cells [19,20]. We have recently found that EV71 infection can lead to increased reactive oxygen species (ROS) generation [16]. However, the mechanism for ROS production in EV71-infected cells remains elusive.

Mitochondria are important organelle for cellular energy metabolism, and have been recently implicated in responses of host cell to viral infection. They lie at the crossroad of metabolism, redox homeostasis, apoptosis, and innate immune signaling, and probably form a multi-functional signaling platform termed mitoxosome [21]. Mitochondria-associated adaptors such as MAVS are involved in modulation of innate immunity [22]. Additionally, intermembrane space proteins, such as cytochrome c and Smac/DIABLO, and outer mitochondrial membrane proteins, such as Bcl-2, are involved in initiation and regulation of apoptosis [23,24]. Viral infection is associated with changes in mitochondrial functions. Hepatitis C virus (HCV)-infected cells have impaired oxidative phosphorylation and increased ROS generation [25,26]. Expression of HCV protein represses mitochondrial membrane potential and mitochondrial coupling efficiency, and leads to increased ROS production [27]. Likewise, expression of human immunodeficiency virus (HIV) Tat protein, hepatitis B virus X protein, severe acute respiratory syndrome (SARS) coronavirus non-structural protein 10 results in mitochondrial depolarization [28–30]. These proteins are known to induce oxidative stress [30–32]. It is currently unknown whether EV71-induced oxidative stress is associated with mitochondrial dysfunction.

In the present study, we investigated the mechanistic aspect of ROS generation induced by EV71 infection. We examined the changes in mitochondria in EV71-infected cells. Our findings show that EV71 infection induces mitochondrial dysfunction and anomalous changes in their morphology and subcellular distribution. These mitochondria represent the sites of ROS generation. Such changes are associated with altered cellular energy metabolism. Mitochondrial ROS are essential to viral replication process. The increase in mitochondrial mass in infected cells may compensate for energy deficit caused by abnormal mitochondria.

Materials and Methods

Cell culture and virological techniques

SF268 glioblastoma cells (National Cancer Institute Center for Cancer Research ID: 59) were cultured in Dulbecco's modified Eagle's medium (DMEM) supplemented with 10% fetal calf serum (FCS), 2 mM glutamine, 2 mM non-essential amino acids, 100 U/ml penicillin, 0.1 mg/ml streptomycin, and 0.25 μg/ml amphotericin at 37°C in a humidified atmosphere of 5% CO_2 [33,34]. EV71 (BrCr strain; ATCC VR784) was propagated in Vero cells as described previously [16]. Vero cells (ATCC CCL-81) were cultured in modified Eagle's medium (MEM) supplemented with 10% FCS, 100 U/ml penicillin, 0.1 mg/ml streptomycin and 0.25 μg/ml amphotericin in an atmosphere of 5% CO_2 at 37°C. Quantitative PCR analysis of the copy number of EV71 genomic DNA was performed as described previously [16].

Detection of ROS and $\Delta\Psi_m$ by confocal microscopy

Visualization of ROS generation was performed by confocal microscopy as described previously [35]. The fluorescence of dichlorofluorescein (DCF) is derived from oxidation of its fluorogenic precursor [35]. In brief, cells were grown in glass-bottomed culture dish, and infected with virus at m.o.i. of 1.25. The infected cells were stained with 100 nM MitoTracker Red and 5 μM H_2DCFDA for 20 min at 37°C. and counterstained with with 5 μg/ml of Hoechst 33342. The specimens were examined with Zeiss LSM 510 Meta system (Carl Zeiss MicroImaging GmbH, Heidelberg, Germany). Confocal fluorescence images of labeled cells were acquired using Plan-Apochromat 100×1.40 NA oil immersion objective. To scan for DCF signal, we used the 488 nm excitation line of argon laser, beam splitter (HFT 405/488/561/633/KP720), and an emission window set at 505–550 nm. To scan for MitoTracker Red signal, we used the 561 nm excitation line of DPSS laser and an emission window set at 575–615 nm. For scanning of Hoechst dye, we used 405 nm excitation line of diode laser and an emission an emission window set at 420–480 nm. To scan for Hoechst dye signal, we

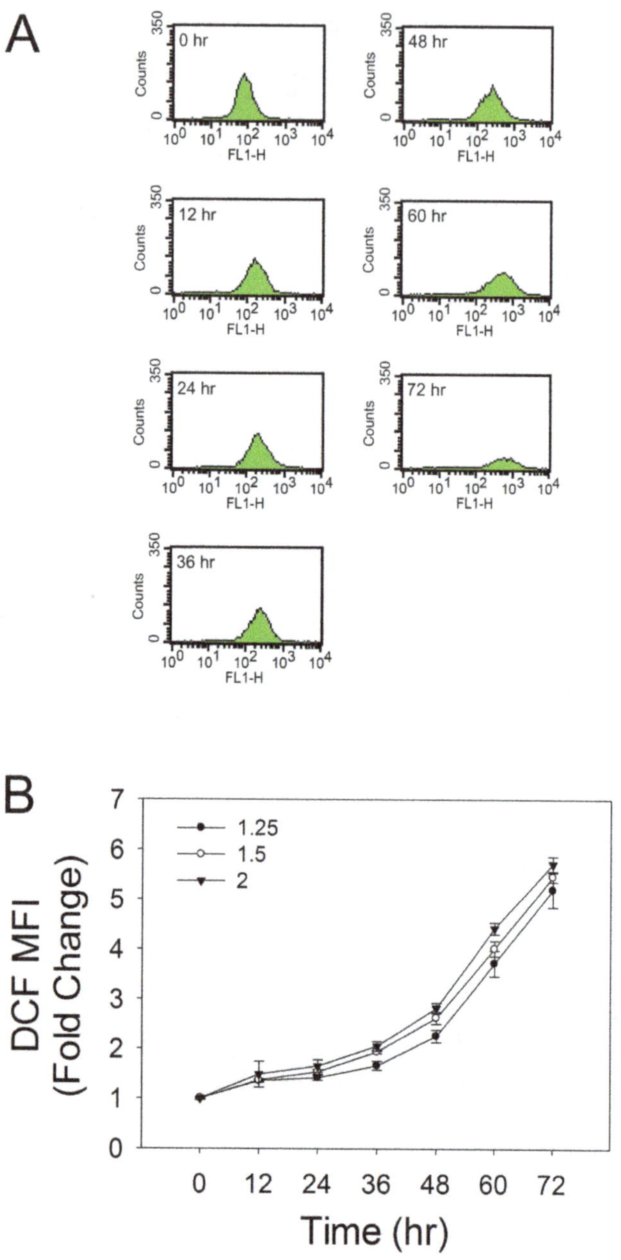

Figure 1. EV71 infection induces ROS in neural cells in a time-dependent manner. SF268 cells were infected with EV71 at m.o.i. of 1.25, 1.5 and 2 for 0, 12, 24, 36, 48, 60 and 72 hr, and were subject to H_2DCFDA staining and flow cytometric analysis. (A) Representative histograms of cell counts (*counts*) vs. DCF fluorescence (*FL1-H*) for cells infected at an m.o.i. of 1.25 at indicated times are shown. (B) The mean fluorescence intensity (MFI) of DCF of infected cells is expressed as fold change relative to that of uninfected cells. The results are presented as mean±SD of three separate experiments.

used the 405 nm excitation line of diode laser and an emission an emission window set at 420–480 nm. We analyzed all images using Zeiss Zen software package (Carl Zeiss MicroImaging GmbH, Heidelberg, Germany).

Confocal microscopic analysis of $\Delta\Psi_m$ was performed as described previously [36]. The mitochondrial membrane potential was determined using the cationic, lipophilic dye JC-1 (5,5',6,6'-tetrachloro-1,1',3,3'-tetraethylbenz- imidazolocarbocyanine io-

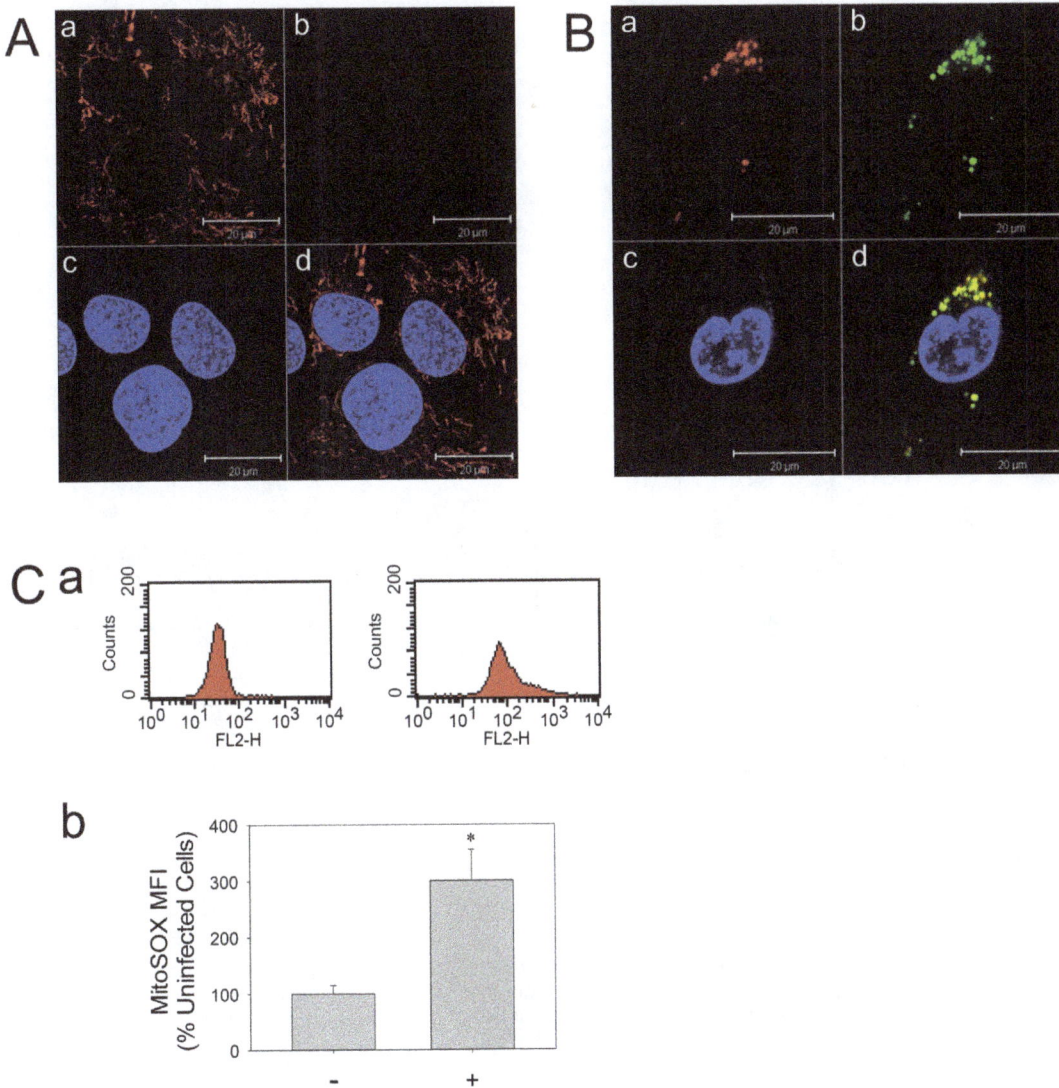

Figure 2. EV71 infection causes mitochondrial ROS production. (A) SF268 cells were mock- (A) or infected with (B) EV71 at an m.o.i. of 1.25 for 54 hr, and were subject to H₂DCFDA and Mitotracker Red staining and confocal microscopic examination. The MitoSOX-stained mitochondria (a), DCF-stained ROS generation sites (b), and Hoechst 33342-stained nuclei (c) are shown. The corresponding images are overlaid (d). The photographs shown here are representative of three experiments. Scale bar, 20 µm. (C) SF268 cells were mock- or infected with EV71 at an m.o.i. of 1.25 for 48 hr, and were subject to MitoSOX Red staining and flow cytometric analysis. (a) Representative histograms of cell counts (*counts*) vs. MitoSOX fluorescence (*FL2-H*) for un- (*left panel*) and infected cells (*right panel*) are shown. (b) The mean fluorescence intensity (MFI) of MitoSOX of mock- (−) and infected (+) cells is expressed as the percentage of that of uninfected cells. The results are presented as mean±SD, n = 3. *p<0.05 vs. uninfected cells.

dide) (Invitrogen, CA, USA). JC-1 monomers emit at 527 nm, and J-aggregates emit at 590 nm. Infected cells were loaded with 0.5 µM of JC-1 in culture medium at 37°C for 60 min, and counterstained with 5 µg/ml of Hoechst 33342 prior to examination with Leica TCS SP2 MP system (Leica Microsystems, Mannheim, Germany). Confocal fluorescence images of JC-1 labeled cells were obtained using HCX PL-APO CS 100×1.40 NA oil immersion objective. To scan for signal of JC-1 monomeric form, we used the 488 nm excitation line of argon laser, triple-dichroic beam splitter (TD 488/543/633) and an emission window set at 500–540 nm. To scan for signal of JC-1 aggregated form, we used the 543 nm excitation line of He-Ne laser, triple-dichroic beam splitter (TD 488/543/633) and an emission window set at 580–620 nm. To scan for Hoechst dye signal, we employed a fiber coupling system equipped with a

Ti:Sapphire laser (model Millenia/Tsunami; Spectra-Physics; Mountain View, CA, USA), and selected the wavelength of 800 nm for illumination. Image analysis was performed with Leica LCS software packages.

Cytometric analysis of ROS, mitochondrial mass, and $\Delta\Psi_m$

ROS formation was analyzed quantitatively by cytometric analysis. In brief, infected cells were loaded with 2 µM MitoSOX red or 5 µM H₂DCFDA for 20 min at 37°C, washed twice with PBS, and trypsinized for flow cytometric analysis as previously described [37]. The mean fluorescence intensity (MFI) of the fluorescence of oxidized MitoSOX red or MFI of DCF fluorescence was quantified using CellQuest Pro software (Becton Dickinson, CA, USA). Likewise, for determination of mitochon-

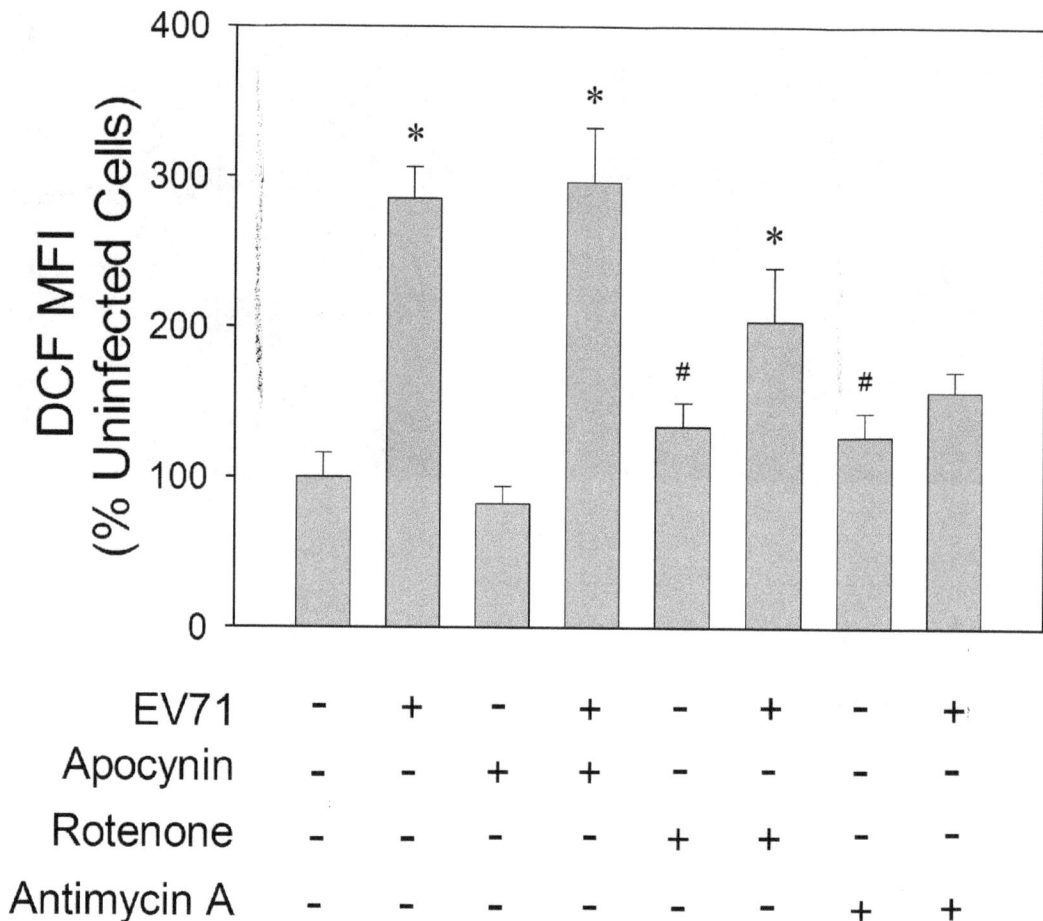

Figure 3. Mitochondrial electron transport-dependent ROS generation. SF268 cells were mock- or infected with EV71 at an m.o.i. of 1.25, and treated without or with apocyanin, rotenone or antimycin A. Forty-eight hours later, cells were stained with H_2DCFDA and analyzed by flow cytometry. The MFI of DCF of infected cells is expressed as the percentage of that of uninfected cells. The results are presented as mean±SD, n = 3. *p<0.05 vs. untreated, uninfected cells; #p<0.05, drug-treated vs. untreated cells.

drial mass, cells were stained with 100 nM Mitotracker Red 580FM (Invitrogen, CA, USA), and analyzed according to manufacturer's instruction. MFI was quantified using CellQuest Pro software. For analysis of $\Delta\Psi_m$, infected cells were loaded with 0.5 µM JC-1 as described above. The cell monolayer was washed three times with PBS, and trypsinized for cytometric analysis. Cells were analyzed for JC-1 monomer fluorescence (FL1 channel) and J-aggregate fluorescence (FL2 channel) with CELLQuest software. The ratio of the mean fluorescence intensity of FL2 channel to that of FL1 channel was calculated.

Measurement of oxygen consumption

SF268 cells were infected with EV71 at m.o.i. of 1.25 for indicated times, washed twice with PBS, and trypsinized. The cells were pelleted, and then resuspended in respiration buffer (20 mM NaKPO4 (pH 7.2), 65 mM KCl, 125 mM sucrose, 2 mM $MgCl_2$). The cell suspension was transferred to the respiratory chamber of Mitocell equipped with Clark-type electrode, which was connected to Strathkelvin 928 6-Channel Oxygen System (Strathkelvin Instruments, Glasgow, UK). Oxygen consumption rate was monitored. In some experiments, oligomycin was added to cell suspension at 2 µg/ml. Data are expressed as µg O_2 per 5×10^5 cells per min. For data normalization with respect to mitochondrial mass, the oxygen consumption rate was normalized

to mitochondrial mass determined by Mitotracker Red 580FM Staining as described above. Data are expressed as µg O_2 per mitochondrial mass unit (MMU) per min. A similar approach for determination of oxygen consumption rate per unit of mitochondrial mass was previously described [38].

Isolation of mitochondria and measurement of respiratory control ratio (RCR)

Cells were infected with EV71 at m.o.i. of 1.25, washed twice with PBS, and scrapped in ice-cold SHE buffer (0.25 M sucrose, 1 mM EGTA, 3 mM HEPES). The cells were homogenized. The homogenate was centrifuged at 1000×g at 4°C for 10 min, and subsequently centrifuged at 10000×g at 4°C for 10 min. Finally, the mitochondrial pellet was suspended in a suitable volume of SHE buffer.

For measurement of oxygen consumption of mitochondrial preparation, 0.5 mg of mitochondria was added to 300 µl of assay buffer (20 mM K_2HPO_4/KH_2PO_4 (pH 7.2), 125 mM sucrose, 65 mM KCl, 2 mM $MgCl_2$), and transferred to the Mitocell chamber equipped with Clark-type electrode. Malate and glutamate were added to chamber at the final concentration of 5 mM. ADP was added at a concentration of 0.5 mM to stimulate oxygen consumption. The respiratory control ratio (RCR) is defined as rate of ADP-stimulated state 3 oxygen consumption

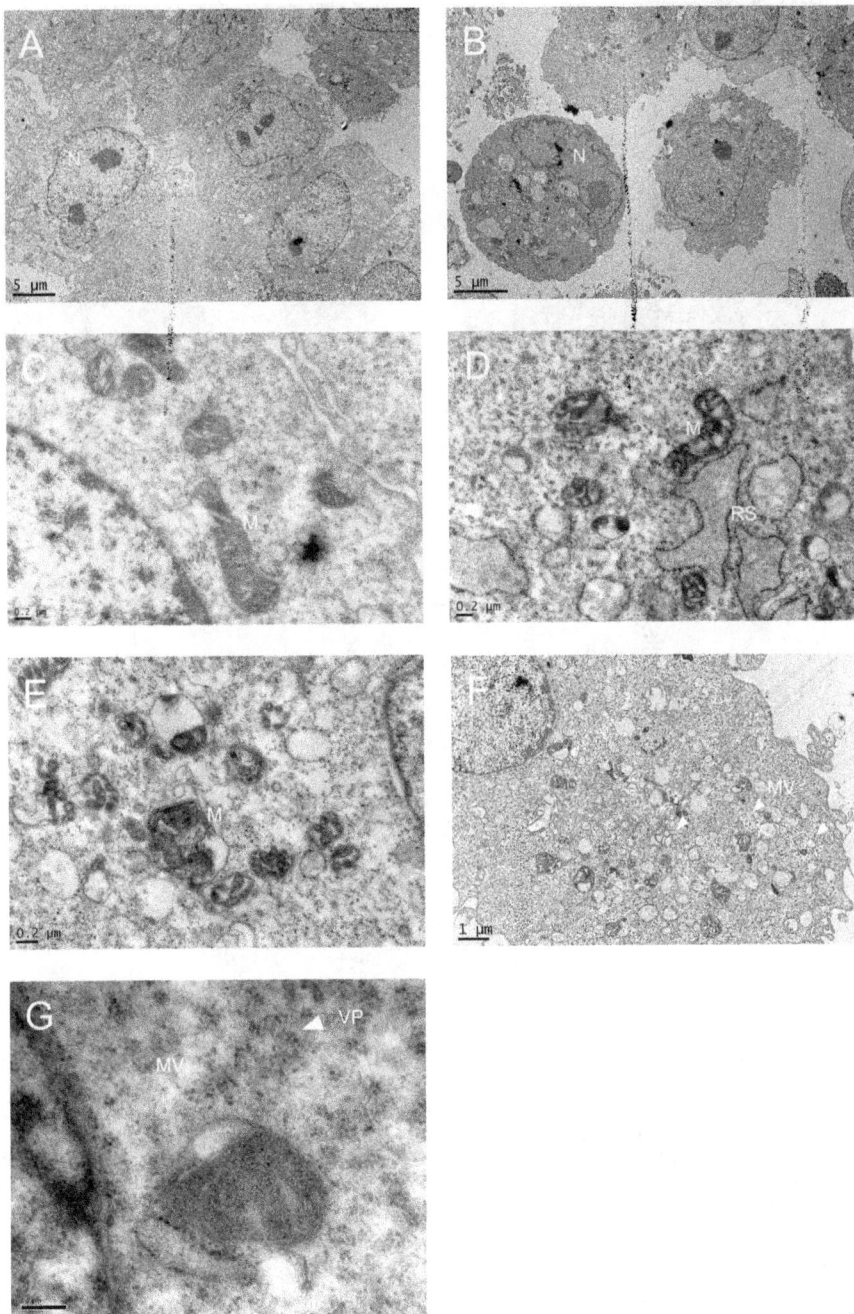

Figure 4. EV71 infection results in morphological changes in mitochondria. SF268 cells were mock- (A & C) or infected with (B, D–G) EV71 at an m.o.i. of 1.25 for 48 hr, and processed for electron microscopic examination. The mock-infected cells had typical nucleus (N) and mitochondria (M). In EV71-infected cells, a number of mitochondria underwent changes in morphology, characterized by deranged cristae (D & E). The developing viral replication site (RS) was lined with ribosomes and was in proximity to mitochondria (D). Numerous single or double membrane-bound vesicles (MV) developed in EV71-infected cells, and some contained virus particles (VP). For A & B, bar represents 5 µm; for F, bar represents 1 µm; for C, D, E & G, bar represents 0.2 µm.

divided by the rate of oxygen consumption (State 4_o) determined in the presence of 2 µg/ml oligomycin [39]. The state 4o is a specific indicator of mitochondrial proton movement being limited by leakiness of inner mitochondrial membrane to proton. RCR was calculated as state 3 oxygen consumption rate divided by that of state 4_o.

Transmission electron microscopy

The transmission electron microscopy was performed as previously described with slight modifications [40]. Briefly stated, the infected cells were washed three times with PBS; trypsinized; placed in a BEEM capsule; and centrifuged to form a pellet with a size of 0.5–1 mm^3. The pellet was fixed with fresh 2.5% glutaraldehyde/PBS at 4°C for 2 hr. After fixation, it was rinsed with PBS thrice for 10 min each. One ml of 1% OsO_4 was added

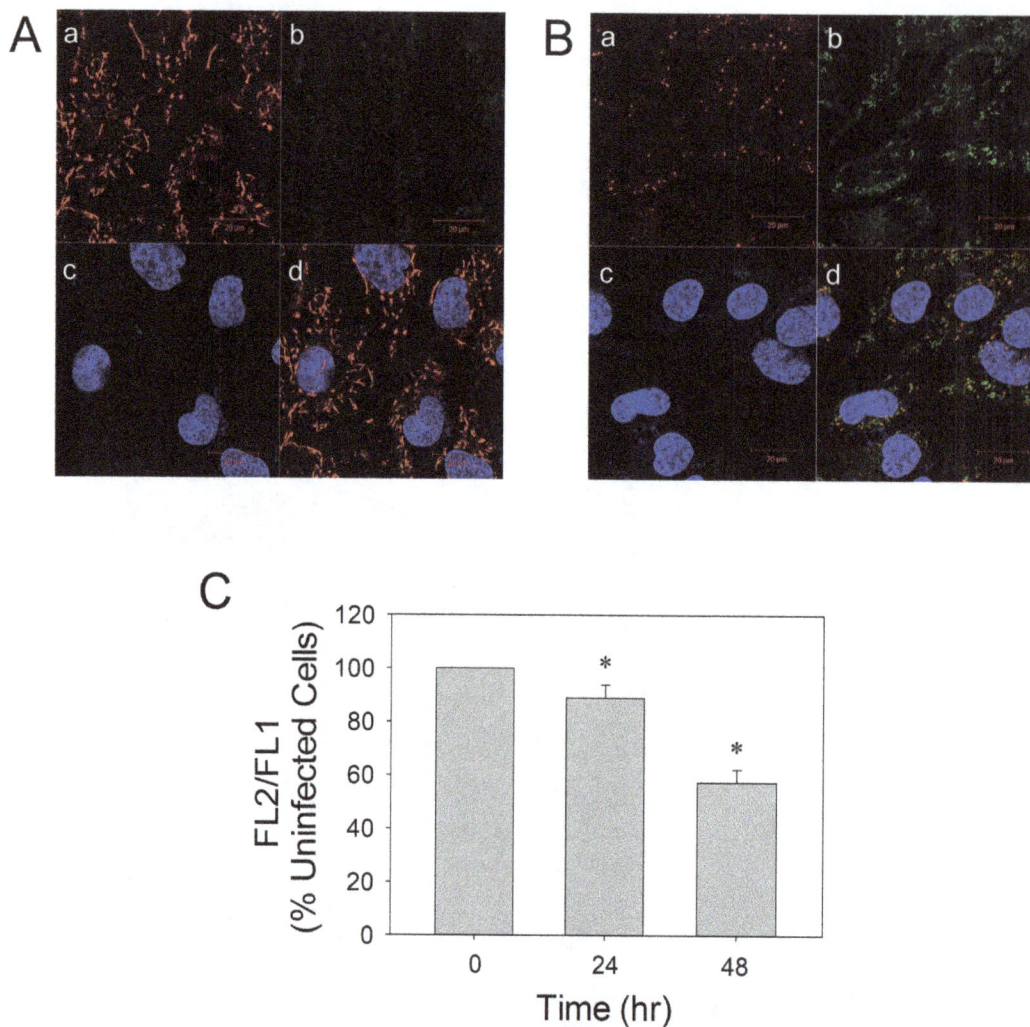

Figure 5. EV71 infection causes decline in $\Delta\Psi_m$. SF268 cells were mock- (A) or infected with (B) EV71 at an m.o.i. of 1.25 for 48 hr. Cell were stained with JC-1 and Hoechst 33342 dyes, and examined by confocal microscopy. Intracellular distribution of JC-1 J-aggregate (a) and monomer (b) is indicative of $\Delta\Psi_m$ in cells. Nuclei of these cells are shown (c). The corresponding images are overlaid (d). The photographs shown here are representative of three experiments. Scale bar, 20 μm. (C) SF268 cells were mock- or infected with EV71 at an m.o.i. of 1.25 for indicated times, and were subject to JC-1 staining and flow cytometric analysis. The ratio of MRI of FL2 channel to that of FL1 channel (FL2/FL1) was calculated, and is expressed relative to that of uninfected cells. Results are mean ± SD, n = 3. *p<0.05 vs. uninfected cells.

to the cell pellet for 1 hr. The pellet was washed with PBS thrice for 10 min each, and then with 50% ethanol thrice for 10 min each. The specimen was then stained *en bloc* with 0.5% uranyl acetate/50% ethanol at 4°C overnight. It was dehydrated in ethanol with graded concentrations (70%, 80%, 90%, 95%, 100%, 100%, 100%, 100%) for 15 min each. The specimen was successively infiltrated with ethanol/Spurr's resin (50%:50%), ethanol/Spurr's resin (25%:75%), and pure Spurr's resin overnight for each treatment. It was then heated at 70°C for 72 hr. Thin sections with thickness of 70 nm were obtained with the use of ultramicrotome, and picked on TEM grids. Sections were stained with 2% uranyl acetate for 30 min, followed by 6 min staining with 0.4% lead citrate. The sections were rinsed with water, blotted dry, and examined on a JEOL JEM-1230 electron microscope.

Determination of adenine nucleotides

Levels of ATP, ADP and AMP were determined using UPLC under conditions as previously described for measurement of nicotinamide nucleotides [41].

Western blotting and silver staining

SDS-PAGE and western blotting were performed as previously described [42]. The chemiluminescent signal was detected with X-ray film (Fujifilm, Japan). The silver staining was performed as previously described [43].

Results

EV71 induces mitochondrial ROS generation

Infection of neural SF268 cells with EV71 induces oxidative stress in time-dependent manner. The levels of ROS, as indicated by DCF fluorescence, in cells infected at multiplicity of infection (m.o.i.) of 2 were 3- and 6-fold that of uninfected control at 48 and 72 hr respectively (Figure 1). As mitochondria are major sites of

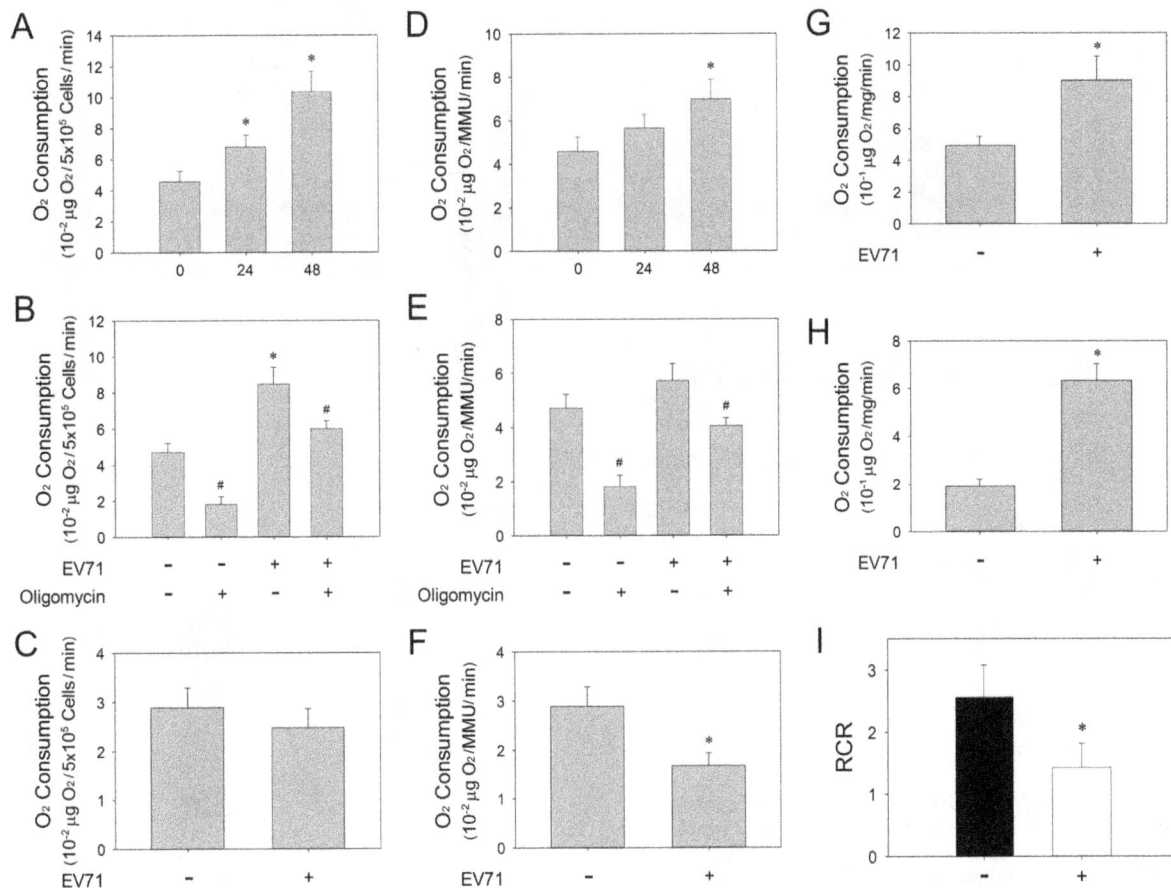

Figure 6. EV71 infection-induced oxygen consumption is associated with a reduction in respiratory efficiency. (A) SF268 cells were mock- or infected with EV71 at an m.o.i. of 1.25 for indicated times. Oxygen concentration was assayed with Clark oxygen electrode, and oxygen consumption rate (10^{-2} μg $O_2/5 \times 10^5$ cells/min) was calculated accordingly. Results are mean $\pm$ SD, n=6. *p<0.05 vs. uninfected cells. (B) SF268 cells were mock- or infected with EV71 at an m.o.i. of 1.25, and were treated without or with oligomycin. Oxygen concentration was assayed with Clark oxygen electrode, and oxygen consumption rate (10^{-2} μg $O_2/5 \times 10^5$ cells/min) was calculated. Results are mean $\pm$ SD, n=6. *p<0.05 vs. uninfected cells; #p<0.05, oligomycin-treated vs. untreated cells. (C) Oligomycin-sensitive oxygen consumption rate was calculated as the difference in the absence and presence of oligomycin. (D & E) The oxygen consumption was measured as described in (A) and (B). Data were normalized to the relative mitochondrial mass unit (MMU) of control and infected cells. Results are mean $\pm$ SD, n=6. *p<0.05 vs. uninfected cells. (F) Oligomycin-sensitive oxygen consumption rate was calculated as described in (C), and normalized to the relative mitochondrial mass unit (MMU) of control and infected cells. (G–I) SF268 cells were mock- (−) or infected (+) with EV71 at an m.o.i. of 1.25 for 48 hr, and mitochondria were isolated and assayed for oxygen consumption rates (10^{-1} μg O_2/mg mitochondrial protein/min) during state 3 (G) and 4_o (H) respiration as described in *Materials and Methods*. RCRs were calculated accordingly, and are shown (I). Results are mean $\pm$ SD, n=6. *p<0.05 vs. uninfected cells.

ROS generation, we test if EV71 induces ROS generation in mitochondria. The EV71 infected-SF268 cells were stained at 54 hr post-infection (p.i.) with Mitotracker Red and H_2DCFDA, and examined under confocal microscope. As expected, the DCF fluorescence increased in EV71-infected cell but not in the uninfected cells (Figures 2A & B). More important, the sites of ROS production coincided with mitochondria in EV71-infected cells (Figure 2B). It was consistent with staining of the infected cells with MitoSOX Red, a mitochondrion-specific probe for ROS (Figure 2C). The infected cells showed 3-fold increase in mitochondrial ROS generation. At infection time greater than 60 hr, the infected cells exhibited a great loss of viability. To study the host cell responses to viral infection, we focused on analyses of cells at early or mid-phase of infection in subsequent studies.

The mitochondrial production of ROS is validated by pharmacological approach. Treatment of EV71-infected cells with complex I inhibitor rotenone and complex III inhibitor antimycin A significantly inhibited ROS generation (Figure 3). Unlike respiratory complex inhibitors, treatment of cells with apocyanin,

a NADPH oxidase (NOX) inhibitor, did not affect EV71-induced ROS generation. These findings suggest that NOX is not involved in ROS generation in infected SF268 cells, and that ROS is generated at a site of mitochondrial electron transport chain distal to complex III.

ROS production is associated with mitochondrial morphological anomalies

Increased ROS production is suggestive of mitochondrial anomalies. To test such possibility, we examined the subcellular structure of infected SF268 cells using electron microscope. Forty-eight hour after infection, numerous SF268 cells were rounded. A number of membrane bound vesicle appeared in cytoplasm (Figures 4B & F). In some cells, replication sites of EV71 developed from rough endoplasmic reticulum, and were characterized by the presence of ribosomes at its periphery and accumulation of mitochondria in its proximity (Figure 4D). Smooth membrane bound vesicles were also observed. The replication sites and the virus-induced membrane vesicles were

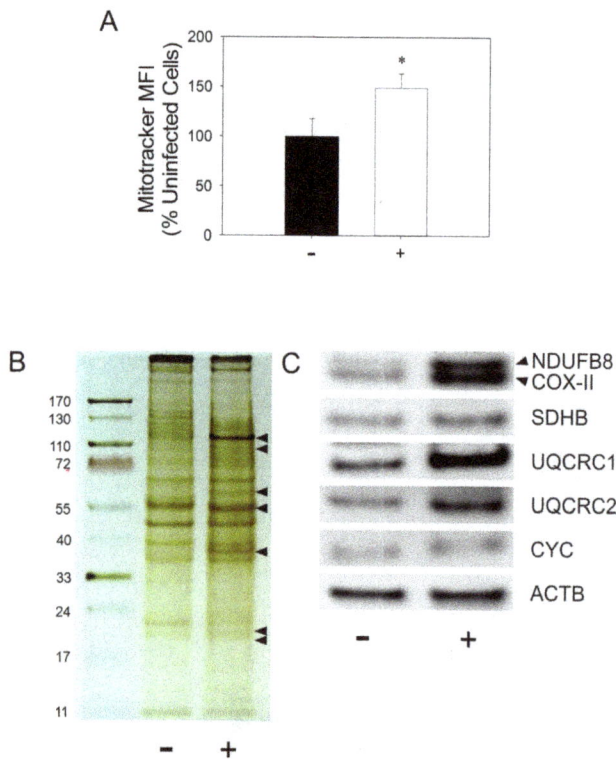

Figure 7. Mitochondrial mass increases and expression of mitochondrial proteins changes in response to EV71 infection. (A) SF268 cells were mock- (−) or infected (+) with EV71 at an m.o.i. of 1.25 for 48 hr, and were subject to Mitotracker dye staining and flow cytometric analysis as described in *Materials and Methods*. The MFI of the stained cells is expressed relative to that of control cells. Results are mean ± SD, n = 3. *p<0.05 vs. uninfected cells. (B & C) Cells were un- or infected under the similar condition, and mitochondria were isolated for SDS-PAGE electrophoresis and silver staining (B). In the silver-stained gel, the leftmost lane corresponds to protein markers with respective molecular weights indicated alongside the bands. (C) Cells similarly infected were harvested for western blotting with indicated antibodies. A representative experiment out of three is shown here.

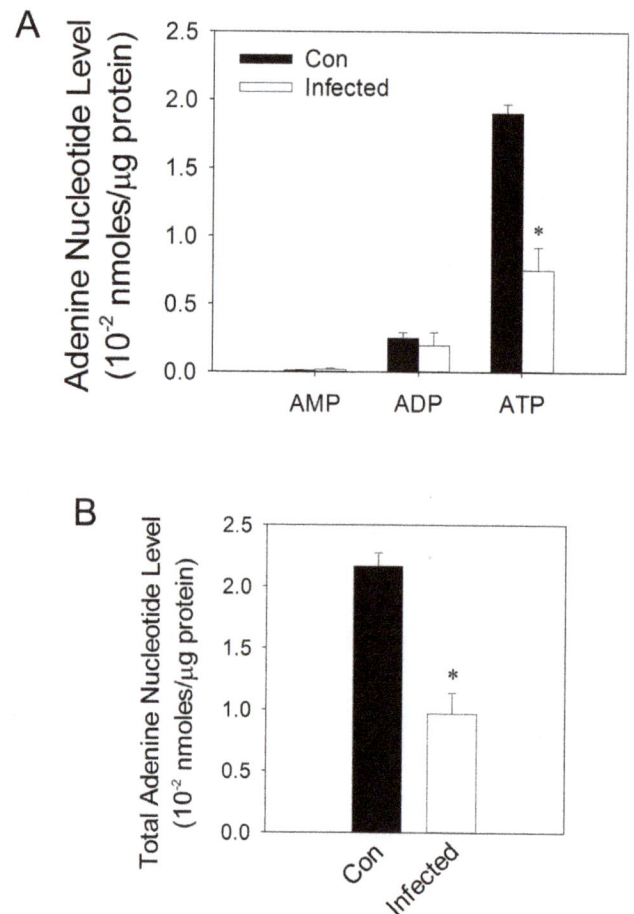

Figure 8. Levels of ATP, ADP and AMP in EV71-infected cells. SF268 cells were mock- (*Con*) or infected (*Infected*) with EV71 at an m.o.i. of 1.25 for 48 hr, and were harvested for UPLC-based analyses of ATP, ADP and AMP. These adenine nucleotides are normalized to cellular protein content. Levels of ATP, ADP and AMP (A) and total adenine nucleotides (B) are shown. Results are mean ± SD, n = 3. *p<0.05 vs. uninfected cells.

absent from the uninfected cells. A significant percentage of atypical mitochondria were observed. They were more electron-dense in infected cells as compared with control cells (Figure 4C & D). The crista arrangement in mitochondria of the infected cells became largely distorted (Figure 4D). The cristae were swollen and appeared as "holes" in electron micrograph. In some mitochondria, the crista structure was lost. The outer membrane was no longer closely apposed to the inner membrane, and the intermembrane space was enlarged (Figure 4E). Inner membrane was less conspicuous and even discontinuous in some mitochondria. These findings suggest that EV71 induces morphological anomalies in mitochondria.

EV71 causes decline in mitochondrial electrochemical potential $\Delta\Psi_m$

It is plausible that EV71 induces mitochondrial dysfunction in EV71-infected cells. A functional parameter of mitochondria is electrochemical potential $\Delta\Psi_m$. We studied $\Delta\Psi_m$ in infected cells using JC-1 staining. Mitochondria are heterogeneous in morphology, ranging from round, ovoid structure to long, interconnected structure. In uninfected cells, most mitochondria appeared red and represent energized mitochondria (Figure 5A). A large percentage

of the energized mitochondria were elongated tubular or oblong structure. They formed a network distributed more evenly throughout cytoplasm. Upon infection, there were changes in mitochondria. Most mitochondria showed green fluorescence, and were depolarized or de-energized (Figure 5B). They appeared to cluster in the perinuclear region (Figures 2B & 5B). Some of mitochondria were swollen, and a majority of the remainder took a short ovoid form. Additionally, we applied a flow cytometric analysis to quantify $\Delta\Psi_m$ in infected cells. Infection of cells with EV71 causes about 10 and 40% reduction in $\Delta\Psi_m$ at 24 and 48 hr p.i., respectively. These findings suggest that EV71 infection induces significant drop in electrochemical potential.

EV71-induced ROS generation and reduction in $\Delta\Psi_m$ are associated with defective electron transport

It is likely that increase in ROS generation and reduction in $\Delta\Psi_m$ are causally related to anomalous electron transport. Amperometric method was applied to measure the oxygen consumption rate, which reflects the function of electron transport chain. As shown in Figure 6A, the oxygen consumption rate of SF268 cells increased after EV71 infection. The rate (normalized to cell number) was elevated about 2.3-fold at 48 hr p.i. However, the oligomycin-insensitive oxygen consumption (i.e. oxygen

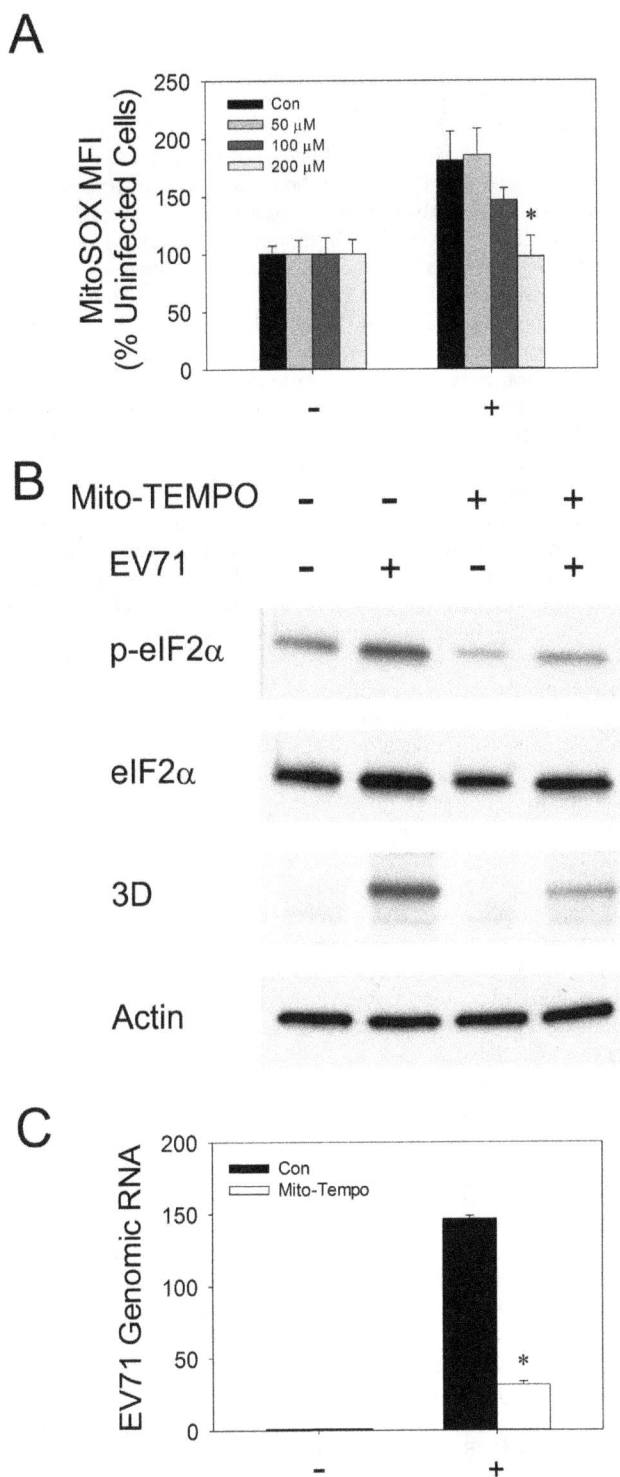

Figure 9. Mitochondrial ROS are essential to EV71 replication. (A) SF268 cells were mock- (−) or infected (+) with EV71 at an m.o.i. of 1.25, and treated without (Con) or with indicated concentrations of Mito-TEMPO. Forty-eight hours later, cells were subject to MitoSOX Red staining and flow cytometric analysis. The mean fluorescence intensity (MFI) of MitoSOX of mock- and infected cells is expressed as the percentage of that of uninfected cells. The results are presented as mean ± SD, n = 3. *p<0.05 vs. infected Con group. (B) SF268 cells were mock- or infected with EV71 at an m.o.i. of 1.25, and treated without or with 200 μM of Mito-TEMPO. Forty-eight hours later, cells were harvested for western blotting with antibodies to phosphorylated eIF2α and total eIF2α, viral protein 3D, and actin. A representative

experiment out of three is shown here. (C) SF268 cells were mock- (−) or infected (+) with EV71 at an m.o.i. of 1.25, and treated without (Con) or with 200 μM of Mito-TEMPO. Forty-eight hours later, cells were analyzed for levels of EV71 genomic RNA. The results are presented as means ± SD n = 3. *p<0.05 vs. infected Con group.

consumption in the presence of oligomycin), which represents that unrelated to ATP synthesis, accounted for a large percentage of oxygen consumption in infected cells. Oligomycin-insensitive oxygen consumption rate was about 38% of total oxygen consumption in control cells, while it increased about 3.3-fold and was up to 70% of total oxygen consumption in infected cells (Figure 6B). Oligomycin-sensitive oxygen consumption rates (i.e. difference in rates of oxygen consumption measured in the absence and presence of oligomycin, which represents oxygen used for oxidative phosphorylation) of control and infected cells were not significantly different (Figure 6C). A different picture emerged when the oxygen consumption rates were normalized to the relative mitochondrial mass of control and infected cells (Figure 6D–F). The oligomycin-sensitive oxygen consumption rate of infected cells was 42% lower than that of control cells (Figure 6F). Furthermore, we measured state 3 and 4_o respiration of isolated mitochondria from control and EV71-infected cells (Figure 6G–I), and calculated the respective respiratory control ratio (RCR), which is considered an index of electron transport coupling. As shown in Figure 6I, it was 44% lower for mitochondria from infected cells than for those from control cells.

EV71 infection causes an increase in mitochondrial mass and alteration in protein expression

There appears to be a discrepancy between oxygen consumption rates normalized to cell number and mitochondrial mass. It is possible that EV71-infected cells have a compensatory increase in mitochondrial mass. To test such possibility, we stained control and infected cells with mitochondrion-specific dye, and analyzed them cytometrically for quantification of mitochondrial mass. As shown in Figure 7A, mitochondrial mass of EV71-infected cells at 48 hr p.i. was about 50% higher than that of control. It was accompanied by an increased expression of mitochondrial proteins. Figure 7B shows the silver-stained SDS-PAGE of mitochondrial preparations from control and infected cells. Intensities of a number of bands increased in infected cells, which is probably indicative of enhanced expression of mitochondrial proteins. Additionally, some proteins were differentially expressed in control and infected cells. Furthermore, we determined the expression of specific mitochondrial proteins using immunoblotting. Expression of such mitochondrial proteins as NADH dehydrogenase (ubiquinone) 1 β subcomplex 8 (NDUFB8), ubiquinol-cytochrome C reductase core protein I (UQCRC1), ubiquinol-cytochrome C reductase core protein II (UQCRC2), and cytochrome c oxidase II (COX-II) was elevated. Other proteins, for example succinate dehydrogenase complex subunit B iron sulfur protein (SDHB) and cytochrome C (CYC), remained largely unchanged in their expression. These findings suggest that EV71 infection induces mitochondrial proliferation and changes in expression of mitochondrial proteins.

Mitochondrial dysfunction is associated with anomalous metabolism of adenine nucleotides

It is possible that mitochondrial dysfunction affects ATP supply. We measured the levels of ATP, ADP and AMP in control and infected cells. As shown in Figure 8, intracellular level of ATP was reduced by about 60% in infected cells, while those of ADP and

AMP remained largely unchanged. Total adenine nucleotides decreased in infected cells. The energy charge ($\frac{[ATP]+\frac{1}{2}[ADP]}{[ATP]+[ADP]+[AMP]}$) of control and infected cells were 0.94 and 0.87, respectively. These findings suggest that EV71 infection results in lower energy status of infected cells.

Mitochondrial ROS generation is essential to EV71 replication

We have previously shown that ROS are conducive to viral replication. It is possible that mitochondrion-generated ROS act as signal to initiate molecular events associated with EV71 replication. We treated control and uninfected cells with mitochondrion-specific antioxidant mito-TEMPO, and studied its consequences on EV71 infection. To determine the optimal mito-TEMPO concentration for experiments, we treated control and EV71-infected cells with increasing concentrations of mito-TEMPO, and studied its effect on ROS generation. Mito-TEMPO reduced EV71-induced mitochondrial ROS generation in a dose-dependent manner (Figure 9A). At 200 μM, mito-TEMPO completely suppressed mitochondrial ROS production in infected cells. This concentration was used in subsequent experiments. It is known that phosphorylation of eIF2α signifies the initiation of viral protein synthesis during enteroviral infection [44,45]. We treated control and EV71-infected cells with 200 μM mito-TEMPO, and examined the effect of mito-TEMPO on eIF2α phosphorylation. As shown in Figure 9B, the level of eIF2α phosphorylation increased significantly in infected cells. Treatment with mito-TEMPO reduced both basal and EV71-induced eIF2α phosphorylation. It was associated with significant reduction in expression of viral 3D protein and viral replication (Figures 9B & C). At 48 hr p.i., the copy number of viral genome in cells treated with mito-TEMPO was about 80% lower than that of untreated cells (Figure 9C). These findings suggest that mitochondrial ROS are essential to EV71 replication.

Discussion

In the present study, we demonstrate that EV71 induces mitochondrial dysfunction and ROS generation, which acts to promote replication. There appears to be increases in mitochondrial mass, which probably represents a mechanism to compensate for mitochondrial defects.

EV71 induces generation of ROS, which in turn promotes viral replication [16]. The co-localization of fluorescence signals of Mitotracker Red and DCF as well as the increase in fluorescence of MitoSox Red in infected cells suggest that ROS generated in EV71-infected cells are of mitochondrial origin. Previous study has shown that EV71 activates Rac1-dependent NADPH oxidase [46]. The findings that treatment of infected cells with rotenone and antimycin A but not with apocyanin inhibited ROS generation are suggestive of ROS generation through inadvertent reaction between electron passing down electron transport chain and oxygen. Moreover, ROS appear to be generated at a point of electron transport chain downstream of complex III. For instance, cytochrome C can transfer electron directly to p66^Shc, a redox enzyme, to generate ROS [47]. The heme center of cytochrome C may also act to enhance ROS production [48].

The mechanism for EV71-induced ROS production is currently enigmatic. A number of viral proteins interact with mitochondria or their proteins, and causes mitochondrial dysfunction and ROS generation. For example, hepatitis B virus X protein (HBx) interacts with mitochondrial heat shock protein 60 and 70, and VDAC3 [29,49], and promotes oxidative stress [50]. HCV core protein binds to mitochondria and increases oxidative stress [51–53]. Influenza A virus PB1-F2 protein targets to mitochondria and induces their anomalies [54,55]. Our preliminary results have recently shown that EV71 proteins associate with mitochondria and elicit oxidative stress.

The functional significance of ROS in EV71 remains unclear. It is undisputable that ROS generation has functional consequence on viral infection. Rabies virus can induce oxidative stress in dorsal root ganglion neurons, which diminishes axonal growth [56]. Oxidative stress also promotes S-glutathionylation of cellular proteins, which regulates activities of various cellular proteins [57]. It has been shown that de-glutathionylation of interferon regulatory factor 3 is required for its efficient interaction with CBP and transactivation of interferon genes [58]. Treatment with mito-TEMPO that reduces mitochondrial ROS generation inhibits eIF2α phosphorylation, a hallmark event in EV71 infection. It is probable that ROS play a regulatory role in EV71 infection.

Mitochondria play a prominent role in energy metabolism. Electron transport through respiratory complexes and the resulting proton motive force are critical to oxidative phosphorylation. Virus-induced mitochondrial dysfunction is not specific for EV71 and can be observed for other viruses. In most cases, viral infection induces mitochondrial dysfunction, and reduction in proton motive force and cellular ATP content. Hepatitis C virus (HCV)-infected cells show mitochondrial dysfunction and decrease in $\Delta\Psi_m$ [59]. Infection of mouse neural cells with Sindbis virus causes mitochondrial dysfunction [60]. The lowered mitochondrial respiratory efficiency of EV71-infected cells is consistent with change in cristae. It has been recently shown that cristae shape affects the interaction between respiratory complexes and the respiratory efficiency [61]. Decrease in RCR is accompanied by decrease in $\Delta\Psi_m$. A drop in $\Delta\Psi_m$ causes a deficit in oxidative phosphorylation, which is indicated by decreases in energy charge and ATP/ADP ratio. Moreover, it has been recently shown that dissipation of proton motive force leads to defective MAVS signaling and hence crippled antiviral defense [62,63].

The subcellular distribution of mitochondria in EV71-infected cells changes after infection. After infection, the mitochondria are converted from the interconnecting tubular form to short ovoid form, and are clustered near nuclei. Perinuclear clustering of mitochondria has also been observed for HBV and HCV [64,65]. It is reasoned that re-distribution of mitochondria in infected cells serves the purpose either to meet the energy requirement of viral infection process or to cordon off mitochondria to prevent pro-apoptotic mediators.

Energy is needed for synthesis of viral protein and nucleic acid. Changes in metabolism in host cells have been implicated in viral pathogenesis [66]. The depletion of ATP during EV71 infection is attributed to its direct incorporation into newly synthesized viral genome, or to its involvement in energy-requiring processes. Decrease in RCR has much impact on cellular metabolism. To ensure the continual ATP supply to support viral replication process, the infected cells may put a strain on "shop-worn" but still working mitochondria for energy production, and may activate glycolytic pathway to meet such need. Our preliminary results have recently shown that glycolytic pathway may be enhanced in EV71-infected cells. The decrease in respiratory efficiency of mitochondria, together with the low ATP-generating efficiency of glycolysis, results in reduction in energy charge. Moreover, mitochondria are involved in biosynthesis of such biomolecules as lipids. Picornaviral replication machinery is assembled on membrane, which is probably derived from endoplasmic reticulum [67]. Such process depends on significant alteration of membrane lipids. Alteration of mitochondria during infection may shift the lipid metabolic reactions, such as phosphatidylserine decarboxylation, cytochrome P450-dependent steroid metabolism or acetyl-CoA/citrate metabolism, to fulfill the metabolic need associated with viral replication. It has been recently shown that poliovirus induces long chain acyl-CoA synthetase 3 activity [68], the mitochondrial outer membrane protein that regulates fatty acid import for formation of replication complex. It is envisaged that impaired mitochondria are functionally altered to sustain viral replication.

As mentioned in preceding paragraphs, functional alterations of mitochondria may play vital roles during EV71 infection. Mitochondria undergo adaptive changes to compensate for their reduced respiratory efficiency and diminished functions. A non-exclusive view is that different pools of mitochondria exist [69,70]. Some mitochondria may experience different extent of dysfunctional change. The damaged but still-functional mitochondria may undergo biogenesis and make up for the deficit in mitochondrial functions. The increases in mitochondrial mass may represent such adaptive response. It is not unprecedented that virus induces mitochondrial biogenesis. For instance, human cytomegalovirus infection causes biogenesis of mitochondria [71]. Accumulation of mitochondria in the vicinity of viral assembly factories in African swine fever virus-infected cells is accompanied by *de novo* synthesis of mitochondria mass [72]. Increased mitochondrial mass in EV71-infected cells is accompanied by differential increases in levels of mitochondrial proteins. Mitochondrial proteins, such as COX-II, are specifically elevated. Increase in level of COX-II is consistent with increase in its transcription [73]. Interestingly, levels of mitochondrial proteins are differentially changed during infection. It is speculative whether such differential changes in protein expression contribute to mitochondrial respiratory functions. It has been recently suggested that the respiratory chain complexes are not arranged simply in a linear order of enzymatic complexes. These complexes display different activities toward one another [74]. Subunits of respiratory complexes differ in their correlation between protein expression and enzymatic activity [75]. It is exemplified by the 8 kDa subunit of complex I, for which one-thirds of protein level correlates with 70% of complex I activity. It follows that differential increases in mitochondrial proteins can contribute to maintenance of respiratory functions.

Previous studies have shown that EV71 infection elicits apoptosis in SF268 cells [76]. Conspicuous apoptosis is detected at 72 hr post-infection. However, apoptosis is atypical in the sense that Bid is not cleaved in EV71-infected SF268 cells. tBid, the cleavage product of Bid, is involved in permeabilization of mitochondrial outer membrane [77]. It is currently unknown whether mitochondrial membrane permeabilization is involved in EV71-induced death of SF268 cells. It is likely that mitochondria in these cells differ in some ways from those of infected non-neural cells, which exhibit typical apoptosis. Moreover, our studies focus on the early or mid-phase of infection. It is conceivable that mitochondria can undergo biogenesis in SF268 cells at early or mid-phase of infection, whilst they experience destructive changes during the late phase of infection.

Vesicular structures in picornavirus-infected cells represent the site of replication. These membranous structures are thought to be derived from endoplasmic reticulum (ER). Synthesis of picornaviral protein can activate ER stress response. It has been recently shown that EV71 induces ER stress response and causes eIF2α phosphorylation through activation of PKR and PERK [44]. Phosphorylation of eIF2α suppresses cap-dependent translation, causing a shift in translation to that of internal ribosome-entry site (IRES)-containing cellular mRNA and enteroviral RNA [78]. Interestingly, ROS are implicated in activation of PKR and PERK [79–83], raising the possibility that mitochondrial ROS may play signaling role in the process.

Antioxidants may be used as therapeutic intervention of enteroviral infection. We have previously shown that *N*-acetylcysteine and epigallocatechin gallate can suppress EV71 replication [16,35]. The antiviral activities of epigallocatechin gallate and related compounds correlate with their antioxidative activities. The fact that mito-TEMPO has antiviral effect pinpoints mitochondrial ROS as target for therapeutic intervention.

Taken together, EV71 causes alterations in mitochondria to induce ROS generation. The mitochondrial ROS are essential to viral replication. The "dysfunction" of mitochondria, such as decline in respiratory functions, may represent virus-induced functional alterations, which are optimized for viral replication. Mitochondria may undergo adaptive or compensatory increase in mitochondrial mass to serve such purpose.

Author Contributions

Conceived and designed the experiments: HYH MLC. Performed the experiments: HYH MLC SFW CHK. Analyzed the data: HYH MLC. Contributed reagents/materials/analysis tools: HYH MLC. Contributed to the writing of the manuscript: HYH MLC.

References

1. Racaniello VR (2001) Picornaviridae: the viruses and their replication. In: Fields BN, Knipe DM, Howley PM, Griffin DE, editors. Fields' Virology. Philadelphia: Lippincott Williams & Wilkins. 685–722.
2. Schmidt NJ, Lennette EH, Ho HH (1974) An apparently new enterovirus isolated from patients with disease of the central nervous system. J Infect Dis 129: 304–309.
3. Ho M, Chen ER, Hsu KH, Twu SJ, Chen KT, et al. (1999) An epidemic of enterovirus 71 infection in Taiwan. Taiwan Enterovirus Epidemic Working Group. N Engl J Med 341: 929–935.
4. Yang F, Ren L, Xiong Z, Li J, Xiao Y, et al. (2009) Enterovirus 71 outbreak in the People's Republic of China in 2008. J Clin Microbiol 47: 2351–2352.
5. Yang F, Zhang T, Hu Y, Wang X, Du J, et al. (2011) Survey of enterovirus infections from hand, foot and mouth disease outbreak in China, 2009. Virol J 8: 508.
6. McMinn PC (2002) An overview of the evolution of enterovirus 71 and its clinical and public health significance. FEMS Microbiol Rev 26: 91–107.
7. Ishimaru Y, Nakano S, Yamaoka K, Takami S (1980) Outbreaks of hand, foot, and mouth disease by enterovirus 71. High incidence of complication disorders of central nervous system. Arch Dis Child 55: 583–588.
8. Lin TY, Chang LY, Hsia SH, Huang YC, Chiu CH, et al. (2002) The 1998 enterovirus 71 outbreak in Taiwan: pathogenesis and management. Clin Infect Dis 34 Suppl 2: S52–57.
9. Ooi MH, Wong SC, Lewthwaite P, Cardosa MJ, Solomon T (2010) Clinical features, diagnosis, and management of enterovirus 71. Lancet Neurol 9: 1097–1105.
10. Huang CC, Liu CC, Chang YC, Chen CY, Wang ST, et al. (1999) Neurologic complications in children with enterovirus 71 infection. N Engl J Med 341: 936–942.
11. Beck MA, Handy J, Levander OA (2000) The role of oxidative stress in viral infections. Ann N Y Acad Sci 917: 906–912.
12. Beck MA, Levander OA, Handy J (2003) Selenium deficiency and viral infection. J Nutr 133: 1463S–1467S.
13. Cai J, Chen Y, Seth S, Furukawa S, Compans RW, et al. (2003) Inhibition of influenza infection by glutathione. Free Radic Biol Med 34: 928–936.
14. Beck MA, Shi Q, Morris VC, Levander OA (2005) Benign coxsackievirus damages heart muscle in iron-loaded vitamin E-deficient mice. Free Radic Biol Med 38: 112–116.
15. Aquaro S, Muscoli C, Ranazzi A, Pollicita M, Granato T, et al. (2007) The contribution of peroxynitrite generation in HIV replication in human primary macrophages. Retrovirology 4: 76.
16. Ho HY, Cheng ML, Weng SF, Chang L, Yeh TT, et al. (2008) Glucose-6-phosphate dehydrogenase deficiency enhances enterovirus 71 infection. J Gen Virol 89: 2080–2089.

17. Oda T, Akaike T, Hamamoto T, Suzuki F, Hirano T, et al. (1989) Oxygen radicals in influenza-induced pathogenesis and treatment with pyran polymer-conjugated SOD. Science 244: 974–976.

18. Schachtele SJ, Hu S, Little MR, Lokensgard JR (2010) Herpes simplex virus induces neural oxidative damage via microglial cell Toll-like receptor-2. J Neuroinflammation 7: 35.

19. Kaul P, Biagioli MC, Singh I, Turner RB (2000) Rhinovirus-induced oxidative stress and interleukin-8 elaboration involves p47-phox but is independent of attachment to intercellular adhesion molecule-1 and viral replication. J Infect Dis 181: 1885–1890.

20. Mochizuki H, Todokoro M, Arakawa H (2009) RS virus-induced inflammation and the intracellular glutathione redox state in cultured human airway epithelial cells. Inflammation 32: 252–264.

21. Tal MC, Iwasaki A (2011) Mitoxosome: a mitochondrial platform for cross-talk between cellular stress and antiviral signaling. Immunol Rev 243: 215–234.

22. Ohta A, Nishiyama Y (2011) Mitochondria and viruses. Mitochondrion 11: 1–12.

23. Kuwana T, Newmeyer DD (2003) Bcl-2-family proteins and the role of mitochondria in apoptosis. Curr Opin Cell Biol 15: 691–699.

24. Brunelle JK, Letai A (2009) Control of mitochondrial apoptosis by the Bcl-2 family. J Cell Sci 122: 437–441.

25. Piccoli C, Quarato G, Ripoli M, D'Aprile A, Scrima R, et al. (2009) HCV infection induces mitochondrial bioenergetic unbalance: causes and effects. Biochim Biophys Acta 1787: 539–546.

26. Wang T, Weinman SA (2006) Causes and consequences of mitochondrial reactive oxygen species generation in hepatitis C. J Gastroenterol Hepatol 21 Suppl 3: S34–37.

27. Piccoli C, Scrima R, Quarato G, D'Aprile A, Ripoli M, et al. (2007) Hepatitis C virus protein expression causes calcium-mediated mitochondrial bioenergetic dysfunction and nitro-oxidative stress. Hepatology 46: 58–65.

28. Lecoeur H, Borgne-Sanchez A, Chaloin O, El-Khoury R, Brabant M, et al. (2012) HIV-1 Tat protein directly induces mitochondrial membrane permeabilization and inactivates cytochrome c oxidase. Cell Death Dis 3: e282.

29. Rahmani Z, Huh KW, Lasher R, Siddiqui A (2000) Hepatitis B virus X protein colocalizes to mitochondria with a human voltage-dependent anion channel, HVDAC3, and alters its transmembrane potential. J Virol 74: 2840–2846.

30. Li Q, Wang L, Dong C, Che Y, Jiang L, et al. (2005) The interaction of the SARS coronavirus non-structural protein 10 with the cellular oxido-reductase system causes an extensive cytopathic effect. J Clin Virol 34: 133–139.

31. Lachgar A, Sojic N, Arbault S, Bruce D, Sarasin A, et al. (1999) Amplification of the inflammatory cellular redox state by human immunodeficiency virus type 1-immunosuppressive tat and gp160 proteins. J Virol 73: 1447–1452.

32. Waris G, Huh KW, Siddiqui A (2001) Mitochondrially associated hepatitis B virus X protein constitutively activates transcription factors STAT-3 and NF-kappa B via oxidative stress. Mol Cell Biol 21: 7721–7730.

33. Rutka JT, Giblin JR, Dougherty DY, Liu HC, McCulloch JR, et al. (1987) Establishment and characterization of five cell lines derived from human malignant gliomas. Acta Neuropathol 75: 92–103.

34. Kornblith PL, Szypko PE (1978) Variations in response of human brain tumors to BCNU in vitro. J Neurosurg 48: 580–586.

35. Ho HY, Cheng ML, Weng SF, Leu YL, Chiu DT (2009) Antiviral effect of epigallocatechin gallate on enterovirus 71. J Agric Food Chem 57: 6140–6147.

36. Ho HY, Cheng ML, Chiu HY, Weng SF, Chiu DT (2008) Dehydroepiandrosterone induces growth arrest of hepatoma cells via alteration of mitochondrial gene expression and function. Int J Oncol 33: 969–977.

37. Cheng ML, Shiao MS, Chiu DT, Weng SF, Tang HY, et al. (2011) Biochemical disorders associated with antiproliferative effect of dehydroepiandrosterone in hepatoma cells as revealed by LC-based metabolomics. Biochem Pharmacol 82: 1549–1561.

38. Moran M, Rivera H, Sanchez-Arago M, Blazquez A, Merinero B, et al. (2010) Mitochondrial bioenergetics and dynamics interplay in complex I-deficient fibroblasts. Biochim Biophys Acta 1802: 443–453.

39. Kristian T, Hopkins IB, McKenna MC, Fiskum G (2006) Isolation of mitochondria with high respiratory control from primary cultures of neurons and astrocytes using nitrogen cavitation. J Neurosci Methods 152: 136–143.

40. Schrand AM, Schlager JJ, Dai L, Hussain SM (2010) Preparation of cells for assessing ultrastructural localization of nanoparticles with transmission electron microscopy. Nat Protoc 5: 744–757.

41. Cheng ML, Ho HY, Lin HY, Lai YC, Chiu DT (2013) Effective NET formation in neutrophils from individuals with G6PD Taiwan-Hakka is associated with enhanced NADP(+) biosynthesis. Free Radic Res 47: 699–709.

42. Lin CJ, Ho HY, Cheng ML, You TH, Yu JS, et al. (2010) Impaired dephosphorylation renders G6PD-knockdown HepG2 cells more susceptible to H(2)O(2)-induced apoptosis. Free Radic Biol Med 49: 361–373.

43. Shevchenko A, Wilm M, Vorm O, Mann M (1996) Mass spectrometric sequencing of proteins silver-stained polyacrylamide gels. Anal Chem 68: 850–858.

44. Jheng JR, Lau KS, Tang WF, Wu MS, Horng JT (2010) Endoplasmic reticulum stress is induced and modulated by enterovirus 71. Cell Microbiol 12: 796–813.

45. Hanson PJ, Zhang HM, Hemid MG, Ye X, Qiu Y, et al. (2013) Viral Replication Strategies: Manipulation of ER Stress Response Pathways and Promotion of IRES-Dependent Translation. In: Rosas-Acosta G, editor. Viral Replication. Croatia: Intech. 103–126.

46. Tung WH, Hsieh HL, Lee IT, Yang CM (2011) Enterovirus 71 induces integrin beta1/EGFR-Rac1-dependent oxidative stress in SK-N-SH cells: role of HO-1/CO in viral replication. J Cell Physiol 226: 3316–3329.

47. Giorgio M, Migliaccio E, Orsini F, Paolucci D, Moroni M, et al. (2005) Electron transfer between cytochrome c and p66Shc generates reactive oxygen species that trigger mitochondrial apoptosis. Cell 122: 221–233.

48. Akopova OV, Kolchinskaya LI, Nosar VI, Bouryi VA, Mankovska IN, et al. (2012) Cytochrome C as an amplifier of ROS release in mitochondria. Fiziol Zh 58: 3–12.

49. Zhang SM, Sun DC, Lou S, Bo XC, Lu Z, et al. (2005) HBx protein of hepatitis B virus (HBV) can form complex with mitochondrial HSP60 and HSP70. Arch Virol 150: 1579–1590.

50. Anand SK, Tikoo SK (2013) Viruses as modulators of mitochondrial functions. Adv Virol 2013: 738794.

51. Koike K (2007) Hepatitis C virus contributes to hepatocarcinogenesis by modulating metabolic and intracellular signaling pathways. J Gastroenterol Hepatol 22 Suppl 1: S108–111.

52. Paracha UZ, Fatima K, Alqahtani M, Chaudhary A, Abuzenadah A, et al. (2013) Oxidative stress and hepatitis C virus. Virol J 10: 251.

53. Schwer B, Ren S, Pietschmann T, Kartenbeck J, Kaehlcke K, et al. (2004) Targeting of hepatitis C virus core protein to mitochondria through a novel C-terminal localization motif. J Virol 78: 7958–7968.

54. Gibbs JS, Malide D, Hornung F, Bennink JR, Yewdell JW (2003) The influenza A virus PB1-F2 protein targets the inner mitochondrial membrane via a predicted basic amphipathic helix that disrupts mitochondrial function. J Virol 77: 7214–7224.

55. Yamada H, Chounan R, Higashi Y, Kurihara N, Kido H (2004) Mitochondrial targeting sequence of the influenza A virus PB1-F2 protein and its function in mitochondria. FEBS Lett 578: 331–336.

56. Jackson AC, Kammouni W, Fernyhough P (2011) Role of oxidative stress in rabies virus infection. Adv Virus Res 79: 127–138.

57. Grek CL, Zhang J, Manevich Y, Townsend DM, Tew KD (2013) Causes and consequences of cysteine S-glutathionylation. J Biol Chem 288: 26497–26504.

58. Prinarakis E, Chantzoura E, Thanos D, Spyrou G (2008) S-glutathionylation of IRF3 regulates IRF3-CBP interaction and activation of the IFN beta pathway. EMBO J 27: 865–875.

59. Quarato G, Scrima R, Agriesti F, Moradpour D, Capitanio N, et al. (2013) Targeting mitochondria in the infection strategy of the hepatitis C virus. Int J Biochem Cell Biol 45: 156–166.

60. Silva da Costa L, Pereira da Silva AP, Da Poian AT, El-Bacha T (2012) Mitochondrial bioenergetic alterations in mouse neuroblastoma cells infected with Sindbis virus: implications to viral replication and neuronal death. PLoS One 7: e33871.

61. Cogliati S, Frezza C, Soriano ME, Varanita T, Quintana-Cabrera R, et al. (2013) Mitochondrial cristae shape determines respiratory chain supercomplexes assembly and respiratory efficiency. Cell 155: 160–171.

62. Koshiba T, Yasukawa K, Yanagi Y, Kawabata S (2011) Mitochondrial membrane potential is required for MAVS-mediated antiviral signaling. Sci Signal 4: ra7.

63. Sasaki O, Yoshizumi T, Kuboyama M, Ishihara T, Suzuki E, et al. (2013) A structural perspective of the MAVS-regulatory mechanism on the mitochondrial outer membrane using bioluminescence resonance energy transfer. Biochim Biophys Acta 1833: 1017–1027.

64. Kim S, Kim HY, Lee S, Kim SW, Sohn S, et al. (2007) Hepatitis B virus x protein induces perinuclear mitochondrial clustering in microtubule- and Dynein-dependent manners. J Virol 81: 1714–1726.

65. Nomura-Takigawa Y, Nagano-Fujii M, Deng L, Kitazawa S, Ishido S, et al. (2006) Non-structural protein 4A of Hepatitis C virus accumulates on mitochondria and renders the cells prone to undergoing mitochondria-mediated apoptosis. J Gen Virol 87: 1935–1945.

66. El-Bacha T, Menezes MM, Azevedo e Silva MC, Sola-Penna M, Da Poian AT (2004) Mayaro virus infection alters glucose metabolism in cultured cells through activation of the enzyme 6-phosphofructo 1-kinase. Mol Cell Biochem 266: 191–198.

67. Miller S, Krijnse-Locker J (2008) Modification of intracellular membrane structures for virus replication. Nat Rev Microbiol 6: 363–374.

68. Nchoutmboube JA, Viktorova EG, Scott AJ, Ford LA, Pei Z, et al. (2013) Increased long chain acyl-Coa synthetase activity and fatty acid import is linked to membrane synthesis for development of picornavirus replication organelles. PLoS Pathog 9: e1003401.

69. Parekh AB (2003) Mitochondrial regulation of intracellular Ca2+ signaling: more than just simple Ca2+ buffers. News Physiol Sci 18: 252–256.

70. Hollander JM, Thapa D, Shepherd DL (2014) Physiological and Structural Differences in Spatially-Distinct Subpopulations of Cardiac Mitochondria: Influence of Pathologies. Am J Physiol Heart Circ Physiol.

71. Kaarbo M, Ager-Wick E, Osenbroch PO, Kilander A, Skinnes R, et al. (2011) Human cytomegalovirus infection increases mitochondrial biogenesis. Mitochondrion 11: 935–945.

72. Rojo G, Chamorro M, Salas ML, Vinuela E, Cuezva JM, et al. (1998) Migration of mitochondria to viral assembly sites in African swine fever virus-infected cells. J Virol 72: 7583–7588.

73. Shih SR, Stollar V, Lin JY, Chang SC, Chen GW, et al. (2004) Identification of genes involved in the host response to enterovirus 71 infection. J Neurovirol 10: 293–304.

74. Claus C, Schonefeld K, Hubner D, Chey S, Reibetanz U, et al. (2013) Activity increase in respiratory chain complexes by rubella virus with marginal induction of oxidative stress. J Virol 87: 8481–8492.

75. Rossignol R, Faustin B, Rocher C, Malgat M, Mazat JP, et al. (2003) Mitochondrial threshold effects. Biochem J 370: 751–762.

76. Chang SC, Lin JY, Lo LY, Li ML, Shih SR (2004) Diverse apoptotic pathways in enterovirus 71-infected cells. J Neurovirol 10: 338–349.

77. Korytowski W, Basova LV, Pilat A, Kernstock RM, Girotti AW (2011) Permeabilization of the mitochondrial outer membrane by Bax/truncated Bid (tBid) proteins as sensitized by cardiolipin hydroperoxide translocation: mechanistic implications for the intrinsic pathway of oxidative apoptosis. J Biol Chem 286: 26334–26343.

78. Harding HP, Calfon M, Urano F, Novoa I, Ron D (2002) Transcriptional and translational control in the Mammalian unfolded protein response. Annu Rev Cell Dev Biol 18: 575–599.

79. Liu ZW, Zhu HT, Chen KL, Dong X, Wei J, et al. (2013) Protein kinase RNA-like endoplasmic reticulum kinase (PERK) signaling pathway plays a major role in reactive oxygen species (ROS)-mediated endoplasmic reticulum stress-induced apoptosis in diabetic cardiomyopathy. Cardiovasc Diabetol 12: 158.

80. Verfaillie T, Rubio N, Garg AD, Bultynck G, Rizzuto R, et al. (2012) PERK is required at the ER-mitochondrial contact sites to convey apoptosis after ROS-based ER stress. Cell Death Differ 19: 1880–1891.

81. Joshi M, Kulkarni A, Pal JK (2013) Small molecule modulators of eukaryotic initiation factor 2alpha kinases, the key regulators of protein synthesis. Biochimie 95: 1980–1990.

82. Donnelly N, Gorman AM, Gupta S, Samali A (2013) The eIF2alpha kinases: their structures and functions. Cell Mol Life Sci 70: 3493–3511.

83. Ito T, Yang M, May WS (1999) RAX, a cellular activator for double-stranded RNA-dependent protein kinase during stress signaling. J Biol Chem 274: 15427–15432.

13

Heart Mitochondrial Proteome Study Elucidates Changes in Cardiac Energy Metabolism and Antioxidant PRDX3 in Human Dilated Cardiomyopathy

Esther Roselló-Lletí[1], Estefanía Tarazón[1], María G. Barderas[2], Ana Ortega[1], Manuel Otero[3], Maria Micaela Molina-Navarro[1], Francisca Lago[3], Jose Ramón González-Juanatey[3], Antonio Salvador[4], Manuel Portolés[5], Miguel Rivera[1]*

1 Cardiocirculatory Unit, Health Research Institute Hospital La Fe, Valencia, Spain, 2 Department of Vascular Physiopathology, Hospital Nacional de Parapléjicos, SESCAM, Toledo, Spain, 3 Cellular and Molecular Cardiology Research Unit, Department of Cardiology and Institute of Biomedical Research, University Clinical Hospital, Santiago de Compostela, Spain, 4 Cardiology Service, Hospital La Fe, Valencia, Spain, 5 Cell Biology and Pathology Unit, Health Research Institute Hospital La Fe, Valencia, Spain

Abstract

Background: Dilated cardiomyopathy (DCM) is a public health problem with no available curative treatment, and mitochondrial dysfunction plays a critical role in its development. The present study is the first to analyze the mitochondrial proteome in cardiac tissue of patients with DCM to identify potential molecular targets for its therapeutic intervention.

Methods and Results: 16 left ventricular (LV) samples obtained from explanted human hearts with DCM (n = 8) and control donors (n = 8) were extracted to perform a proteomic approach to investigate the variations in mitochondrial protein expression. The proteome of the samples was analyzed by quantitative differential electrophoresis and Mass Spectrometry. These changes were validated by classical techniques and by novel and precise selected reaction monitoring analysis and RNA sequencing approach increasing the total heart samples up to 25. We found significant alterations in energy metabolism, especially in molecules involved in substrate utilization (ODPA, ETFD, DLDH), energy production (ATPA), other metabolic pathways (AL4A1) and protein synthesis (EFTU), obtaining considerable and specific relationships between the alterations detected in these processes. Importantly, we observed that the antioxidant PRDX3 overexpression is associated with impaired ventricular function. PRDX3 is significantly related to LV end systolic and diastolic diameter (r = 0.73, *p* value< 0.01; r = 0.71, *p* value<0.01), fractional shortening, and ejection fraction (r = −0.61, *p* value<0.05; and r = −0.62, *p* value< 0.05, respectively).

Conclusion: This work could be a pivotal study to gain more knowledge on the cellular mechanisms related to the pathophysiology of this disease and may lead to the development of etiology-specific heart failure therapies. We suggest new molecular targets for therapeutic interventions, something that up to now has been lacking.

Editor: Vincenzo Lionetti, Scuola Superiore Sant'Anna, Italy

Funding: This work was supported by grants from the NIH "Fondo de Investigaciones Sanitarias del Instituto Carlos III," (RD12/0042/0003; FIS Project PI10/00275; PI13/00100) and Health Research Institute Hospital La Fe (Project 07/2013). The funders had no role in study design, data collection and analysis, decision to publish, or preparation of the manuscript.

Competing Interests: The authors have declared that no competing interests exist.

* Email: miguelrivera492@gmail.com

Introduction

Heart failure (HF), a major and growing public health problem, is a current worldwide pandemic with an unacceptable high level of morbidity and mortality in industrialized countries and with no curative treatment currently available. Dilated cardiomyopathy (DCM), one of the most frequent causes of HF, is a severe pathology of unknown etiology characterized by impaired systolic function with increased ventricular mass, volume, and wall thickness [1,2]. The mechanisms underlying the development of this cardiomyopathy are multiple, complex, and not well understood.

Mitochondria are the major energy production sites within cells [3]. Cardiac energy deficits have been reported in the failing heart, with convincing evidence of the important effect of mitochondrial dysfunction in the development and progression of HF in human and animal models resulting from its central role in energy production, metabolism, calcium homeostasis, oxidative stress, and cell death [4–8]. Some studies identify mitochondria as both the target and origin of major pathogenic pathways that cause myocardial dysfunction [9]. Nevertheless, the mitochondria-specific role and the proteins contributing to HF are unclear. In earlier studies, this organelle has been studied using experimental models and classic biochemical methods [10–12]. These studies usually focused on only one particular protein rather than the whole cardiac mitochondrial proteome, despite the fact that methods designed to enrich and purify the mitochondria represent one of the most long-standing examples of proteome subfractionation [13–15]. Thus, characterization of the mitochondrial proteome could provide new insight into cardiac dysfunction and suggest new molecular targets for the therapeutic intervention of DCM. However, the mitochondrial proteome has not been analyzed in pathological human hearts.

Here, we isolate mitochondria from left ventricular (LV) samples of explanted human hearts with DCM and use a proteomic approach to investigate the variations in mitochondrial protein expression. Our results identify the overexpression of several proteins involved mainly in energy metabolism but also in stress response and protein synthesis in dilated human hearts. We focus on seven representative mitochondrial proteins with different expressions in control (CNT) and diseased hearts validated by different classical techniques as well as novel and precise selected reaction monitoring (SRM) analysis and RNA sequencing (RNAseq) approach. We find that some proteins involved in the different components of cardiac energy metabolism and protein biosynthesis could have an important role in this cardiomyopathy. LV dysfunction is directly related with the antioxidant PRDX3 expression in DCM.

Materials and Methods

Ethics statement

The project was approved by the Ethics Committee of Hospital La Fe, Valencia, and all participants gave their written, informed consent. The study was conducted in accordance with the guidelines of the Declaration of Helsinki [16].

Tissue sources

The experiments were performed using LV samples from explanted human hearts from Caucasian patients with DCM undergoing cardiac transplantation. Clinical history, hemodynamic study, electrocardiography, and Doppler echocardiography data were available from all of these patients. Non-ischemic DCM was diagnosed when patients had LV systolic dysfunction (ejection fraction, <40%) with a dilated non-hypertrophic LV (LV diastolic diameter, >55 mm) on echocardiography. Moreover, none of the patients had existing primary valvular disease or a familial history of DCM. All of the patients were functionally classified according to the New York Heart Association (NYHA) criteria and were receiving medical treatment following the guidelines of the European Society of Cardiology [17].

Non-diseased donor hearts were used as CNT samples. The hearts were initially considered for transplantation but were subsequently deemed unsuitable either because of blood type or size incompatibility. The cause of death was cerebrovascular or motor vehicle accident. All donors had normal LV function and no history of myocardial disease or active infection at the time of the transplantation.

Transmural samples were taken from the region around the apex of the left ventricle and stored at 4°C for a maximum of 6 h from the time of coronary circulation loss. The samples were stored at −80°C until the mitochondrial isolation was performed. Of the 25 heart samples, 16 were used in the proteomic analysis (DCM, n = 8; CNT, n = 8). The 25 heart samples were used in the validation to improve the numerical base with a higher number of patients (DCM, n = 17; CNT, n = 8).

Mitochondrial isolation and proteomic analysis

Mitochondrial isolation was performed using standard homogenization, protease digestion, and differential centrifugation methods as previously described by Imahashi et al. [15]. The isolation was made from 50 mg of left ventricular tissue, obtaining a final protein concentration of 10 μg/μl.

2D–DIGE

Protein samples were precipitated using a 2-D Clean-Up Kit (GE Healthcare) as per the manufacturer's protocol. Samples were labeled using CyDye DIGE Fluor Minimal Dyes (GE Healthcare) according to the manufacturer's recommendations. The samples were resuspended in 20 μL of labeling buffer containing 7 M urea, 2 M thiourea, 4% CHAPS, and 30 mM Tris. The pH was then checked in every sample to ensure an optimal labeling reaction (pH 8.0–9.0). Aliquots (50 μg protein in 10 μL) were separated into individual tubes and a pooled internal standard was generated by mixing of equal amounts of all samples included in the experiment. A total of 400 pmol of the appropriate CyDye was added to each sample according to the experimental protocol. This experimental design included 2 experimental groups and 8 samples per group, and thus 16 samples were labeled with either Cy3 or Cy5. The labeling reactions were developed for 30 min on ice in the dark and then quenched by incubation with 1 μL of 10 mM lysine for 10 min in the dark. The labeled samples were combined according to the experimental design and loaded onto the same immobilized pH-gradient (IPG) strip. We randomized the CyDye assignments and sample combinations in each gel, which also included one aliquot of the pooled internal standard labeled with Cy2.

Two-dimensional electrophoresis

The combined samples were diluted with a rehydration solution containing 8 M urea, 4% CHAPS, 20 mM DTT, and 1% IPG buffer (pH 3–11 NL; GE Healthcare) to a final volume of 450 μL. The IPG strips (24 cm; pH 3–11 NL) were allowed to rehydrate overnight in a re-swelling tray (GE Healthcare). After the strips were passively rehydrated, they were placed in an Ettan IPGphor Manifold ceramic tray (GE Healthcare) and isoelectric focusing (IEF) was performed in an Ettan IPGphor 3 unit (GE Healthcare) at 20°C according to the following program: 1-h gradient to 1000 V, 1.5-h gradient to 1500 V, 2-h gradient to 3500 V, 4-h constant at 3500 V, 2-h gradient to 6000 V, and, finally, constant at 6000 V to constitute a total of 60 kV/h. After IEF, the strips were equilibrated for 20 min in 1.5 M Tris (pH 8.8) buffer containing 6 M urea, 30% glycerol, 2% sodium dodecyl sulfate (SDS), and bromophenol blue to which 1% DTT had been added, followed by equilibration for 20 min more in the same buffer to which 2.5% iodoacetamide had been added. SDS–polyacrylamide gel electrophoresis (SDS-PAGE) was performed overnight as described by Laemmli [18] by using the Ettan DALTsix Electrophoresis System (GE Healthcare) at 1 W/gel.

Image acquisition and analysis

After SDS-PAGE, the gels were scanned with a Typhoon 9400 fluorescence gel scanner (GE Healthcare, Piscataway, NJ) using appropriate individual excitation and emission wavelengths, filters and photomultiplier (PTM) values that are sensitive for each of the Cy3, Cy5 and Cy2 dyes (PTM values: 480 nm, 490 nm, 500 nm, respectively). Relative protein quantification was performed on AS and healthy valves with DeCyder software v6.5 (GE Healthcare) and the multivariate statistical module EDA (Extended data analysis). The Differential in-gel analysis (DIA) module co-detected the 3 images of a gel (the internal standard and the two samples), measured the spot abundance in each image, and expressed these values as Cy3/Cy2 and Cy5/Cy2 ratios. These DIA datasets were then analysed using the Biological Variation Analysis module (BVA), which enabled the spot maps to be matched and the Cy3/Cy2 and Cy5/Cy2 ratios to be compared. Only protein spots with >1.5-fold differences in abundance were considered for the analysis. A statistical analysis was then carried out to determine the changes in protein species, with P-values below 0.05 accepted as significant when the Students t-test was applied. The gels were then re-stained with a silver staining kit (GE-Healthcare).

In-gel protein digestion

Protein spots were excised manually and then automatically digested using the Ettan Digester (GE Healthcare). We used the digestion protocol previously described by Schevchenko et al. [19] with minor variations: the gel plugs were reduced using 10 mM DTT (Sigma-Aldrich; St. Louis, MO, USA) in 50 mM ammonium bicarbonate (99% purity; Scharlau) and alkylated using 55 mM iodoacetamide (Sigma-Aldrich) in 50 mM ammonium bicarbonate. The gel pieces were then rinsed with 50 mM ammonium bicarbonate in 50% methanol (gradient, high-performance liquid chromatography [HPLC] grade; Scharlau) and acetonitrile (gradient, HPLC grade; Scharlau) and dried in a Speedvac. Modified porcine trypsin (sequencing grade; Promega, Madison, WI, USA) at a final concentration of 20 ng/μL in 20 mM ammonium bicarbonate was added to the dry gel pieces, and the digestion was allowed to proceed at 37°C overnight. Finally, 60% aqueous acetonitrile and 0.5% trifluoroacetic acid (99.5% purity; Sigma-Aldrich) were added to extract peptides.

MALDI-MS(/MS) and database searching

A total of 0.5 μL of each digestion solution was deposited using the thin-layer method onto a 384 Opti-TOF 123×81 mm MALDI plate (Applied Biosystems) and allowed to dry at RT. The same volume of matrix (3 mg/mL α-cyano-4-hydroxycinnamic acid (Sigma-Aldrich) in 60% acetonitrile/0.5% trifluoroacetic acid) was applied on every sample in the MALDI plate. MALDI-MS(/MS) data were obtained in an automated analysis loop by using a 4800 Plus MALDI TOF/TOF Analyzer (Applied Biosystems). Spectra were acquired in the reflector positive-ion mode by using an Nd:YAG 355-nm wavelength laser at 200 Hz laser frequency, and 1,000–2,000 individual spectra were averaged. The spectra were acquired uniformly by using fixed laser intensity. For the MS/MS 1-kV analysis mode, precursors were accelerated to 8 kV in Source 1 and selected using a relative resolution of 200 (FWHM) and metastable suppression. Fragment ions generated by collision with air in a CID chamber were further accelerated by 15 kV in Source 2. The mass data were automatically analyzed using the 4000 Series Explorer Software, Version 3.5.3 (Applied Biosystems). The MALDI-TOF mass spectra were internally calibrated using 2 trypsin autolysis ions with m/z = 842.510 and 2211.105. In the case of MALDI-MS/MS, calibrations were performed using the fragment-ion spectra obtained for Glub-fibrinopeptide (4700 Cal Mix; Applied Biosystems). MALDI-MS and MS/MS data were combined in the GPS Explorer Software Version 3.6 to search a non-redundant protein database (Swiss-Prot 2012_08) by using the Mascot software, version 2.2 (Matrix Science) [20], featuring 50 ppm precursor tolerance, 0.6-Da MS/MS fragment tolerance, carbamidomethyl cysteine as the fixed modification, and oxidized methionine as the variable modification, and allowing for 1 missed cleavage. The MALDI-MS (/MS) spectra and database search results were manually inspected in detail using the aforementioned software. In the case of the combined MS and MS/MS data,

identifications were accepted when the confidence interval (CI%) calculated using the GPS software was ≥95%. Because Protein Scores and Ion Scores obtained from distinct searches cannot be directly compared, the GPS software calculates the CI% to combine the results of MS and MS/MS database searches. This coefficient value refers to a <5% probability of the observed match being a random event. In the case of the PMF spectra, identifications were accepted when the CI% was ≥99%.

Gel electrophoresis and western blot analysis

Protein samples for the detection of pyruvate dehydrogenase E1 component subunit α, somatic form (ODPA), electron transfer flavoprotein-ubiquinone oxidoreductase (ETFD), dihydrolipoyl dehydrogenase (DLDH), delta-1-pyrroline-5-carboxylate dehydrogenase (AL4A1), ATP synthase subunit α (ATPA), elongation factor Tu (EFTU) and thioredoxin-dependent peroxide reductase (PRDX3) were separated using Bis-Tris electrophoresis on 4–12% polyacrylamide gels under reducing conditions. After the electrophoresis, the proteins were transferred from the gel to a polyvinylidene difluoride membrane using an iBlot Dry Blotting System (Invitrogen Ltd., UK) for western blot analyses. The primary detection antibodies used were anti-pyruvate dehydrogenase E1-alpha subunit mouse monoclonal antibody (1:400), anti-EFTDH mouse monoclonal antibody (1:400), anti-lipoamide dehydrogenase rabbit monoclonal antibody (1:1000), anti-ATP5A mouse monoclonal antibody (1:300), anti-TUFM rabbit polyclonal antibody (1:2000), and anti-peroxiredoxin 3 mouse monoclonal antibody (1:1000) (all obtained from Abcam, Cambridge, UK); and anti-ALDH4A1 mouse monoclonal antibody (1:500) from Sigma-Aldrich. Anti-COX IV rabbit polyclonal antibody (1:200) (Thermo Scientific, Rockford, IL, USA) was used as a loading control.

The bands were visualized using an acid phosphatase-conjugated secondary antibody and nitro blue tetrazolium/5-bromo-4-chloro-3-indolyl phosphate (NBT/BCIP, Sigma-Aldrich) substrate system. Finally, the bands were digitalized using an image analyzer (DNR Bio-Imagining Systems, Israel) and quantified using the GelQuant Pro (v12.2) program.

Fluorescence microscopy

Human myocardial LV samples were fixed in 4% formalin, embedded in paraffin, cut into 5-μm sections, and mounted on superfrost glass slides. Sections were kept at 60°C overnight, deparaffinized with xylol followed by washing in 100%, 96%, 80%, and 70% ethanol. The samples were then blocked with phosphate buffered saline (PBS) containing 1% bovine serum albumin (BSA) for 15 min at RT. After blocking, the sections were incubated for 120 min at RT with the primary antibodies (described in the Western Blot Analysis section above) in the same buffer solution, and then with Alexa-conjugated secondary antibody (Invitrogen, USA) for 60 min at RT [21]. Finally, the sections were rinsed in PBS, mounted in Vectashield-conjugated 4′,6-diamidino-2-phenylindole (DAPI) for identifying the nucleous (Vector Laboratories, Burlingame, CA, USA), and examined under an Olympus BX50 fluorescence microscope (Tokyo, Japan). The images were processed by ImageJ (v. 1.4.3.67) software.

Immunocytochemistry and electron microscopy

Myocardial samples (size 1 mm^3) from the LV were fixed in a solution of 1.5% glutaraldehyde and 1% formaldehyde in 0.05 M cacodylate buffer (pH 7.4) for 1 h at 4°C. The samples were then post-fixed in 1% OsO$_4$ for 1 h at 4°C, dehydrated in ethanol, and embedded in Epon 812. Ultra-thin sections measuring 80 nm were obtained and mounted on nickel grids and counter-stained with 2% uranyl acetate for 20 min and 2.7% lead citrate for 3 min [22,23].

For immunogold labeling, ultra-thin sections were floated for 30 min on 0.1% BSA-Tris buffer (20 mM Tris-HCl, 0.9% NaCl [pH 7.4] containing 0.1% BSA, type V) and 2 h in a moist chamber at RT on sodium metaperiodate [24]. After being rinsed with bi-distilled water, the sections were incubated for 5 min with 3% H_2O_2. The grids were rinsed again with bi-distilled water and incubated separately in a moist chamber overnight at RT with the primary antibodies (described in the Western Blot Analysis section above) in the 0.1% BSA-Tris buffer. After being rinsed with 0.1% BSA-Tris buffer, the sections were incubated in a moist chamber for 1 h at 37°C with 0.1% BSA-Tris buffer (containing 0.05% Tween-20) and a goat anti-rabbit IgG-gold antibody (10 nm, 1:10 dilution; Sigma) for DLDH and EFTU and a goat anti-mouse IgG-gold antibody (5 nm, 1:10 dilution; Sigma) for ODPA, EFTD, AL4A1, ATPA, and PRDX3.

After rinses with 0.1% BSA-Tris buffer and bi-distilled water, the sections were air dried and counterstained first with uranyl acetate for 30 min and then with lead citrate for 5 s. Finally, the grids were air dried completely. For the electron microscopy observation, a Philips CM-100 was used, with magnifications ranging X4500–15000.

Selected reaction monitoring (SRM)

Protein samples were reduced by incubating them with 100 mM DTT (Sigma Aldrich) in 50 mM ammonium bicarbonate (99% purity; Scharlau) for 30 min at 37°C. After reduction, alkylation with 55 mM iodoacetamide (Sigma Aldrich) in 50 mM ammonium bicarbonate was conducted for 20 min at RT. Next, we added 50 mM ammonium bicarbonate, 15% acetonitrile (LCMS grade, Scharlau) and, finally, sequencing-grade modified porcine trypsin (Promega) at a final ratio of 1 µg trypsin: 50 µg protein. After digestion at 37°C overnight, 2% formic acid (99.5% purity; Sigma Aldrich) was added and samples were cleaned using Pep-Clean spin columns (Pierce) according to the manufacturer's instructions. Tryptic digests were dried in a Speedvac and resuspended in 2% acetonitrile/2% formic acid prior to MS analysis.

The LC-MS/MS system consisted of a TEMPO nano LC system (Applied Biosystems) combined with a nano LC Autosampler and coupled to a modified triple quadrupole (Applied Biosystems 4000 QTRAP LC/MS/MS System). Three replicate injections (4 µL containing 8 µg of protein) were made for each sample (except 2 samples with only 1 injection per sample) by using mobile phase A (2% ACN/98% water, 0.1% FA) at a flow rate of 10 µL/min for 5 min. Peptides were loaded onto a µ-Precolumn Cartridge (Acclaim Pep Map 100 C18; 5 µm, 100Å; 300 µm i.d. ×5 mm, LC Packings) to preconcentrate and desalt samples. Reversed-phase LC was performed on a C18 column (Onyx Monolithic C18; 150×0.1 mm i.d., Phenomenex) in a gradient of phase A and phase B (98% ACN/2% water, 0.1% FA). Peptides were eluted at a flow rate of 900 nL/min by following these steps: 2–15% B for 2 min, 15–30% B for 18 min, 30–50% B for 5 min, 50–90% B for 2 min, and, finally, 90% B for 3 min. The column was then regenerated with 2% B for another 15 min. Both TEMPO nano LC and 4000 QTRAP system were controlled using the Analyst Software, v.1.4.5. Theoretical SRM transitions were designed using MRMpilot software v1.1 (ABSciex), with the following settings: Enzyme = trypsin, missed cleavages = 0; modifications in peptide ≤3; charge states = +1 from 300 to 600 Da, +2 from 500 to 2000 Da, +3 from 900 to 3000 Da, +4 from 1600 to 4000 Da, +5 from 2400 to 10,000 Da; studied modification = -none; fixed modifications = carboxyamidomethylation; variable

modifications = none; min. number of amino acids ≥5; max. number of amino acids ≤30; ignore multiple modification sites; 3 transitions per peptide (Table_S2 in File S1). A pool containing a mixture of all the samples was digested as described previously and analyzed in the 4000QTrap using a MIDAS acquisition method that included the theoretical transitions. Transitions were selected when the three co-eluting peaks (corresponding to the three transitions of the same peptide) had a signal-to-noise ratio over 5 and the MS/MS data matched the theoretical spectrum for that peptide.

The mass spectrometer was set to operate in the positive-ion mode with an ion-spray voltage of 2800 V and a nanoflow interface heater temperature of 150°C. Source gas 1 and curtain gas were set to 20 and 20 psi, respectively, and nitrogen was applied as both curtain and collision gases. Collision energy was optimized to obtain maximal transmission efficiency and sensitivity for each SRM transition. A total of 42 MRM transitions (3 per peptide) were monitored during the analysis of each sample and were acquired at unit resolution in both Q1 and Q3, with dwell times of 20 and 50 ms that resulted in a cycle time of 1.2303 s. The IntelliQuan algorithm included in the Analyst 1.4.5 software was used to calculate abundances based on the peak areas after integration.

To carry out this analysis we pooled the samples (four samples per pool). Correlations between proteins were evaluated by calculating Pearson product–moment correlation coefficient of every transition from every peptide analyzed for each protein with respect to every transition of the peptides of the other proteins. If the results of 1 out of the 3 assayed transitions were divergent from the other two, this transition was not considered. While evaluating correlations, when 2 or more peptides were measured, the most significant peptide was selected, while the second one was used as a qualifier for the correlation, which was rejected if this second peptide had divergent results. Pearson's correlation coefficient (r) was calculated as a mean of all coefficients from all transitions from the most significant peptide-to-peptide correlation. When any of the transition-to-transition correlations was not significant, but close to signification, the greater transition-to-transition p value of the most significant peptide-to-peptide correlation was considered.

RNA extraction

Heart samples were homogenized in TRIzol reagent in a TissueLysser LT (Qiagen, UK). All RNA extractions were performed using a PureLink Kit according to the manufacturer's instructions (Ambion Life Technologies, CA, USA). RNA was quantified using a NanoDrop1000 spectrophotometer (Thermo Fisher Scientific, UK), and the purity and integrity of the RNA samples were measured using an Agilent 2100 Bioanalyzer with an RNA 6000 Nano LabChip kit (Agilent Technologies, Spain). All samples showed a 260/280 ratio of >2.0 and an RNA integrity number of ≥9.

RNAseq

The RNA samples were isolated using a MicroPoly(A) Purist Kit (Ambion, USA). The total polyA-RNA samples were used to generate whole transcriptome libraries that were sequenced on a SOLiD 5500XL platform as per the manufacturer's recommendations (Life Technologies, CA). The amplified cDNA quality was analyzed using the Bioanalyzer 2100 DNA 1000 kit (Agilent Technologies, Spain), and the cDNA was quantified using the Qubit 2.0 Fluorometer (Invitrogen, UK). Whole transcriptome libraries were used to generate SOLiD templated beads by following the SOLiD Templated Bead Preparation guide. Bead quality was estimated based on WFA (workflow analysis) parameters. The samples were sequenced using the 50625 paired-end protocol, which generated 75 nt+35 nt (Paired-End) +5 nt (Barcode) sequences. Quality data were measured using the SETS software parameters (SOLiD Experimental Tracking System).

Computational analysis of RNAseq data

The initial whole transcriptome paired-end reads obtained from the sequencing were mapped against the latest version of the human genome (Version GRchr37/hg19) by using the Life Technologies mapping algorithm (http://www.lifetechnologies. com/). The aligned records were reported in the BAM/SAM format [25]. Bad quality reads (Phred score <10) were eliminated using the Picard Tools software [26].

The isoform and gene predictions were subsequently estimated using the cufflinks method [27], and the expression levels were calculated using the HTSeq software [28]. The Edge method was applied to analyze the differential expression between conditions [29]. This method relies on a Poisson model to estimate the RNAseq data variance for differential expression. We selected genes and isoforms that were calculated to exhibit p value<0.05 and fold-change >1.5.

Statistics

Data are presented as mean ± standard deviation (SD). The Kolmogorov–Smirnov test was used to analyze the normal distribution of the variables. Comparisons between 2 groups were performed using Student's t-test, while Pearson's correlation coefficient was calculated to analyze the association between variables. Analyses were considered significant when p value< 0.05. All statistical analyses were performed using SPSS software v. 20 for Windows (IBM SPSS Inc., Chicago, IL, USA).

Results

Patients' clinical characteristics

LV tissue samples were obtained from 17 patients with DCM (82% men; mean age, 54±11 years; ejection fraction, <40%). These patients had a New York Heart Association (NYHA) functional classification of III–IV and were previously diagnosed with significant comorbidities including hypertension, and diabetes mellitus. The patients' clinical and echocardiographic characteristics are summarized in Table 1. Eight non-diseased donor hearts were used as CNT samples (63% men; mean age, 55±8 years; ejection fraction, >50%).

Differentially expressed mitochondrial proteins in patients with HF of dilated etiology

The protein expressions of purified heart mitochondria from 8 DCM patients and 8 CNT were compared using two-dimensional differential gel electrophoresis (2D-DIGE). Each gel contained the mitochondrial proteome of DCM patients, CNT samples, and an internal standard. Gel images were imported into DeCyder Differential Analysis Software, which detected 1,171–1,418 protein spots. Reproducibility was tested by comparing the variation within the different gels in the same group. The t-test statistical analysis did not show significant differences. We focused on the identification of up- and downregulation of spot intensities where the fold change was ≥1.5 (p value<0.05). Considering these criteria, statistical analysis of the data in the DeCyder software revealed changes in the abundance of 19 protein spots corresponding to 17 mitochondrial proteins revealed by mass spectrometry. We encountered 16 significantly upregulated and 3

downregulated spots in DCM hearts (Figure_S1 in File S1). The tandem mass spectrometry (MS/MS) identification details are summarized in Table_S1 in File S1.

Table 2 shows that most of the proteins altered in the mitochondrial proteome of DCM patients are involved in cardiac energy metabolism, some implicated in substrate utilization such as ETFD or DLDH, while others are implicated in energy production such as ATPA. The remaining identified mitochondrial proteins were basically structural or implicated in the protein synthesis and stress response, such as coiled-coil-helix-coiled-coil-helix domain-containing protein 3, EFTU, and PRDX3, respectively. The majority of the altered proteins belong to the mitochondrial matrix or the inner membrane.

Validation of protein differential abundance and mRNA levels

A selection of representative proteins of each mitochondrial function and/or localization and involved in heart damage was validated using different techniques that compared its levels among DCM patients (n = 17) and CNT (n = 8). Therefore, using western blot techniques, we determined the levels of some relevant proteins involved in metabolism, including ODPA, ETFD, DLDH, AL4A1, ATPA; protein synthesis EFTU; and stress response PRDX3. As shown in Figures 1a-g, levels of all of the analyzed molecules were significantly increased in the pathological samples (ODPA, 152±19 vs. 100±18 au, p value<0.0001; ETFD, 157±35 vs. 100±37 au, p value<0.01; DLDH, 132±42 vs. 100±25 au, p value<0.05; AL4A1, 193±75 vs. 100±43 au, p value<0.01; ATPA, 149±36 vs. 100±25 au, p value<0.01; EFTU, 200±51 vs. 100±53 au, p value<0.0001; and PRDX3, 148±57 vs. 100±23 au, p value<0.01). These results coincided with those of the proteomic analysis.

We determined whether there was a relationship between these protein levels and the clinical characteristics shown in Table 1. The LV function parameters were completely available in 15 of 17 samples from DCM patients. We found significant relationships between PRDX3 levels and LV function (Fig. 1h). The results obtained showed that PRDX3 is significantly correlated with LV end systolic diameter, LV end diastolic diameter, fractional shortening, and ejection fraction (r = 0.73, p value<0.01; r = 0.71, p value<0.01; r = −0.61, p value<0.05; and r = −0.62, p value<0.05, respectively).

The immunofluorescence study findings were consistent with the increased levels observed by western blotting and proteomic analysis showing that the intensity of all validated proteins was higher in the dilated hearts than in the CNT samples. These proteins had a diffuse cytoplasmic distribution with a significantly higher percentage of fluorescence in the dilated group (ODPA, 21%, p value<0.01 [Fig. 2a]; ETFD, 44%, p value<0.01 [Fig. 2b]; DLDH, 24%, p value<0.05 [Fig. 2c]; AL4A1, 32%, p value<0.01 [Fig. 2d]; ATPA, 18%, p value<0.01 [Fig. 2e]; EFTU, 42%, p value<0.05 [Fig. 2f]; PRDX3, 29%, p value< 0.05 [Fig. 2g]).

Immunocytochemistry studies confirmed the previous results and the localization or distribution of these mitochondrial proteins. Figure 3 shows an increase in immunogold labeling in ODPA, ETFD, DLDH AL4A1, ATPA, EFTU, and PRDX3 in DCM hearts compared to CNT hearts. We observed similar localization of each protein in the pathological and CNT samples.

To validate the previous analyses and to evaluate the possible relationship between the altered proteins involved in the different mitochondrial processes, these molecules were monitorized by SRM (Table_S2 and S3 in File S1). Differential expression was confirmed in all three transitions per peptide (p value<0.05). We

Table 1. Clinical and echocardiographic characteristics of heart failure patients.

	DCM (n = 17)
Age (years)	54±11
Gender male (%)	82
BMI (kg/m²)	26±5
Prior hypertension (%)	29
Diabetes mellitus (%)	29
NYHA class	3.3±0.4
Hemoglobin (mg/dL)	13±2
Hematocrit (%)	40±7
Total cholesterol (mg/dL)	144±43
Duration of disease (months)	85±69
Echo-Doppler study	
Ejection fraction (%)	24±7
Fractional shortening (%)	13±4
Left ventricular end systolic diameter (mm)	66±13
Left ventricular end diastolic diameter (mm)	75±13
Left ventricle mass (g)	484±132
Left ventricle mass index (g/cm²)	255±80

Duration of disease from diagnosis of heart failure until heart transplant. BMI, body mass index; DCM, dilated cardiomyopathy; NYHA, New York Heart Association.

obtained a good relationship between the abundance of altered proteins involved in substrate utilization (ODPA vs. DLDH, r = 0.58, p value<0.05; ODPA vs. ETFD, r = 0.65, p value<0.01; DLDH vs. ETFD, r = 0.59, p value<0.05) and with other altered molecules implicated in the metabolic process (ODPA vs. AL4A1, r = 0.73, p value<0.001; DLDH vs. AL4A1, r = 0.58, p value< 0.05; ETFD vs. AL4A1, r = 0.76, p value<0.01) and protein synthesis (ODPA vs. EFTU, r = 0.66, p value<0.01; DLDH vs. EFTU, r = 0.74, p value<0.001; ETFD vs. EFTU, r = 0.67, p value<0.01). In addition, these molecules showed a relevant correlation with ATPA (ODPA vs. ATPA, r = 0.77, p value< 0.001; DLDH vs. ATPA, r = 0.74, p value<0.001; ETFD vs. ATPA, r = 0.61, p value<0.01; AL4A1 vs. ATPA, r = 0.78, p value<0.001; EFTU vs. ATPA, r = 0.89, p value<0.0001).

The mRNA differences between DCM patients and CNT were determined by RNAseq. The mRNA levels of the ODPA (*PDHA1*), ETFD (*ETFDH*), AL4A1 (*ALDH4A1*), ATPA (*ATP5A1*), and EFTU (*TUFM*) genes were increased in the DCM group (18-fold, p value<0.05; 62-fold, p value<0.01; 88-fold, p value<0.01; 79-fold, p value<0.01; and 25-fold, p value< 0.05, respectively) (Fig. 4a). These results also coincided with those of proteomic analysis demonstrating the same tendency between gene expression and protein levels. However, the DLDH (*DLD*) and PRDX3 (*PRDX3*) gene expressions did not reach statistical significance. Finally, we obtained a good relationship between the mRNA expression of altered molecules involved in substrate utilization (*PDHA1* vs. *ETFDH*, r = 0.60, p value<0.01) and with other altered genes implicated in the metabolic process (*PDHA1* vs. *ALDH4A1*, r = 0.50, p value<0.05; *ETFDH* vs. *ALDH4A1*, r = 0.62, p value<0.01) and protein synthesis (*PDHA1* vs. *TUFM*, r = 0.54, p value<0.05; *ETFDH* vs. *TUFM*, r = 0.70, p value< 0.01). In addition, these molecules showed a relevant correlation with *ATP5A1* (*ETFDH* vs. *ATP5A1*, r = 0.75, p value<0.001; *ALDH4A1* vs. *ATP5A1*, r = 0.62, p value<0.001; *TUFM* vs. *ATP5A1*, r = 0.70, p value<0.001) (Fig. 4b).

Discussion

Mitochondrial proteome of DCM human hearts

In the present study, we carried out a 2D-DIGE analysis of isolated cardiac mitochondria from the LV tissue of DCM patients to investigate mitochondrial cardiac protein expression changes. The mechanisms responsible for mitochondrial dysfunction in human hearts are poorly understood and the animal models used to elucidate it may not reflect the true pathophysiology of DCM. There are some studies based on proteomic approach in failing hearts [30–33], however this is the first study to analyze the mitochondrial proteome in pathological human hearts. As described so far, isolating and analyzing single organelles is a right approach to uncover essential information regarding their regulation and interplay [13].

Because the heart consumes more ATP than any other organ, it is particularly rich in mitochondria. During HF development, both energy demands and metabolism change show a remarkable decrease in oxidative phosphorylation and a shift toward glucose over fatty acid utilization [5]. Mitochondria are a major source and target of free radicals, and the collapse of the mitochondrial transmembrane potential can initiate the signaling cascades involved in apoptosis [34]. They also play an important role in antioxidant regeneration and are responsible for the majority of ATP production [35]. A recent study by Ahuja *et al.* showed that DCM is clearly associated with enhanced mitochondrial biogenesis and mtDNA deletion, making evident the essential role of mitochondria in this form of HF [36]. Thus, analysis of the mitochondrial proteome could provide new insights into cardiac dysfunction in DCM patients. Here, we identified 19 protein spots corresponding to 17 mitochondrial proteins altered in failing hearts (14 increased, 3 decreased). These changes comprise many aspects of mitochondrial function, including metabolism, transport, respiratory chain, stress response, protein synthesis, and cell death.

Table 2. Mitochondrial proteins differentially regulated in dilated cardiomyopathy *vs.* controls.

Spot	Accesion code	Protein name	Fold-change	*p*-value	Main Localization	Function
324	ETFD_HUMAN	Electron transfer flavoprotein-ubiquinone oxidoreductase	+1.74	0.047	Inner membrane	Metabolism/Transport
335	CH60_HUMAN	60 kDa heat shock protein	+1.95	0.023	Matrix	Stress response
344	ETFD_HUMAN	Electron transfer flavoprotein-ubiquinone oxidoreductase	+1.77	0.001	Matrix/Inner membrane	Metabolism/Transport
348	DLDH_HUMAN	Dihydrolipoyl dehydrogenase	+2.25	0.008	Matrix	Metabolism
361	AL4A1_HUMAN	Delta-1-pyrroline-5-carboxylate dehydrogenase	+1.79	0.002	Matrix	Metabolism
407	DLDH_HUMAN	Dihydrolipoyl dehydrogenase	+1.73	0.019	Matrix	Metabolism
472	ODO2_HUMAN	Dihydrolipoyllysine-residue succinyltransferase component of 2-oxoglutarate dehydrogenase complex	+1.51	0.0001	Matrix	Metabolism
601	ODPA_HUMAN	Pyruvate dehydrogenase E1 component subunit alpha, somatic form	+2.21	0.001	Matrix	Metabolism
604	ACADM_HUMAN	Medium-chain specific acyl-CoA dehydrogenase	+1.79	0.027	Matrix	Metabolism
	ODPA_HUMAN	Pyruvate dehydrogenase E1 component subunit alpha, somatic form			Matrix	Metabolism
614	KCRS_HUMAN	Creatine kinase S-type	−2.01	0.001	Inner membrane	Metabolism
689	EFTU_HUMAN	Elongation factor Tu	+2.41	0.008	Mitochondrion	Protein biosynthesis
732	MDHM_HUMAN	Malate dehydrogenase	−1.75	0.008	Matrix	Metabolism
861	CY1_HUMAN	Cytochrome c1, heme protein	+1.71	0.007	Inner membrane	Metabolism/Respiratory chain
	ECH1_HUMAN	Delta(3,5)-Delta(2,4)-dienoyl-CoA isomerase			Matrix	Metabolism
882	ATPA_HUMAN	ATP synthase subunit alpha	+2.23	0.01	Inner membrane	Metabolism/Respiratory chain
904	ATPA_HUMAN	ATP synthase subunit alpha	+1.93	0.049	Inner membrane	Metabolism/Respiratory chain
960	CHCH3_HUMAN	Coiled-coil-helix-coiled-coil-helix domain-containing protein 3	+1.79	0.006	Inner membrane	Structural
1019	PRDX3_HUMAN	Thioredoxin-dependent peroxide reductase	+1.73	0.031	Mitochondrion	Stress response
	NDUV2_HUMAN	NADH dehydrogenase [ubiquinone] flavoprotein 2			Inner membrane	Metabolism/Respiratory chain
1039	PRDX3_HUMAN	Thioredoxin-dependent peroxide reductase	+1.59	0.046	Mitochondrion	Stress response
1140	PRDX5_HUMAN	Peroxiredoxin-5	−1.89	0.023	Mitochondrion	Stress response

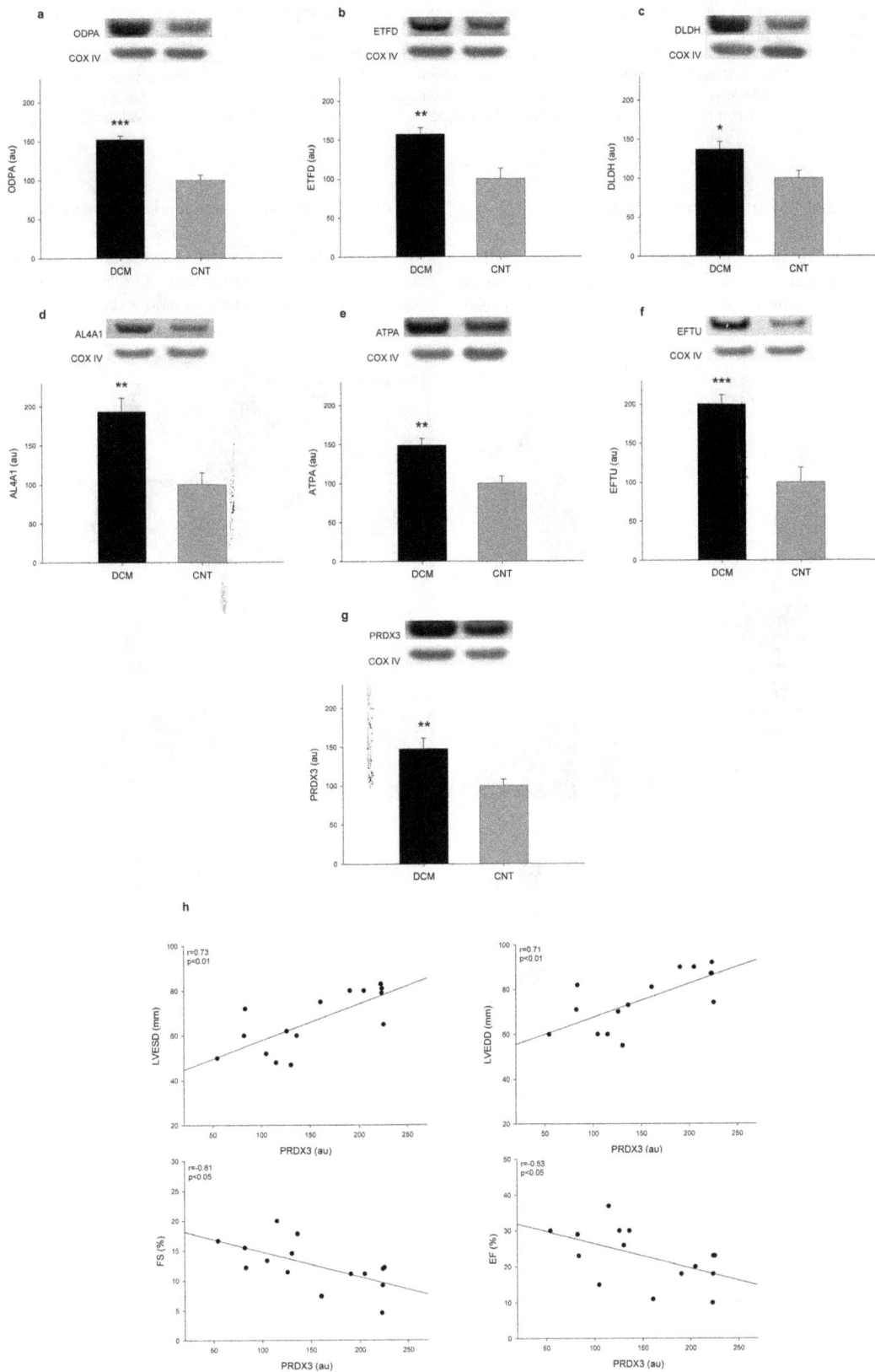

Figure 1. Mitochondrial protein overexpression in dilated human hearts and relationship between PRDX3 and left ventricular function. (a–g) The influence of dilated cardiomyopathy on the amount of each representative protein involved in cardiac energy metabolism (ODPA, ETFD, DLDH, AL4A1, and ATPA), protein biosynthesis (EFTU), and stress response (PRDX3) analyzed using western blotting techniques. As shown, all proteins were significantly increased in the DCM group (n = 17) compared with the CNT group (n = 8). The values from the controls were set to 100. Values were normalized to COX IV and finally to the CNT group. The data are expressed as mean+SEM in arbitrary units (optical density).

Images are representative of the results obtained for all of the patients with DCM and the CNT included in the study. (h) Scatter plots showing the relationship between PRDX3 protein levels and left ventricular function, specifically with fractional shortening, left ventricular end systolic diameter, and left ventricular end diastolic diameter. CNT, control; DCM, dilated cardiomyopathy; ODPA, pyruvate dehydrogenase E1 component subunit α, somatic form; ETFD, electron transfer flavoprotein-ubiquinone oxidoreductase; DLDH, dihydrolipoyl dehydrogenase; AL4A1, delta-1-pyrroline-5-carboxylate dehydrogenase; ATPA, ATP synthase subunit α; EFTU, elongation factor Tu; PRDX3, thioredoxin-dependent peroxide reductase; FA, fractional shortening; LVESD, left ventricular end systolic diameter; LVEDD, left ventricular end diastolic diameter. *p value<0.05, **p value<0.01, ***p value<0.0001.

Alterations in substrate utilization in DCM human hearts

Impaired cardiac energy metabolism is known to play a major role in HF [4,6]. Our results are in concordance with these studies, as we found that most of the proteins altered in the mitochondrial proteome of DCM patients are involved in cardiac energy metabolism, 12 of the 17 differentially regulated proteins identified (70%) are involved in this process. The metabolic machinery has three components: substrate utilization, energy production, and energy transfer and utilization. The main way that myocytes compensate for decreasing mitochondrial function is by increasing

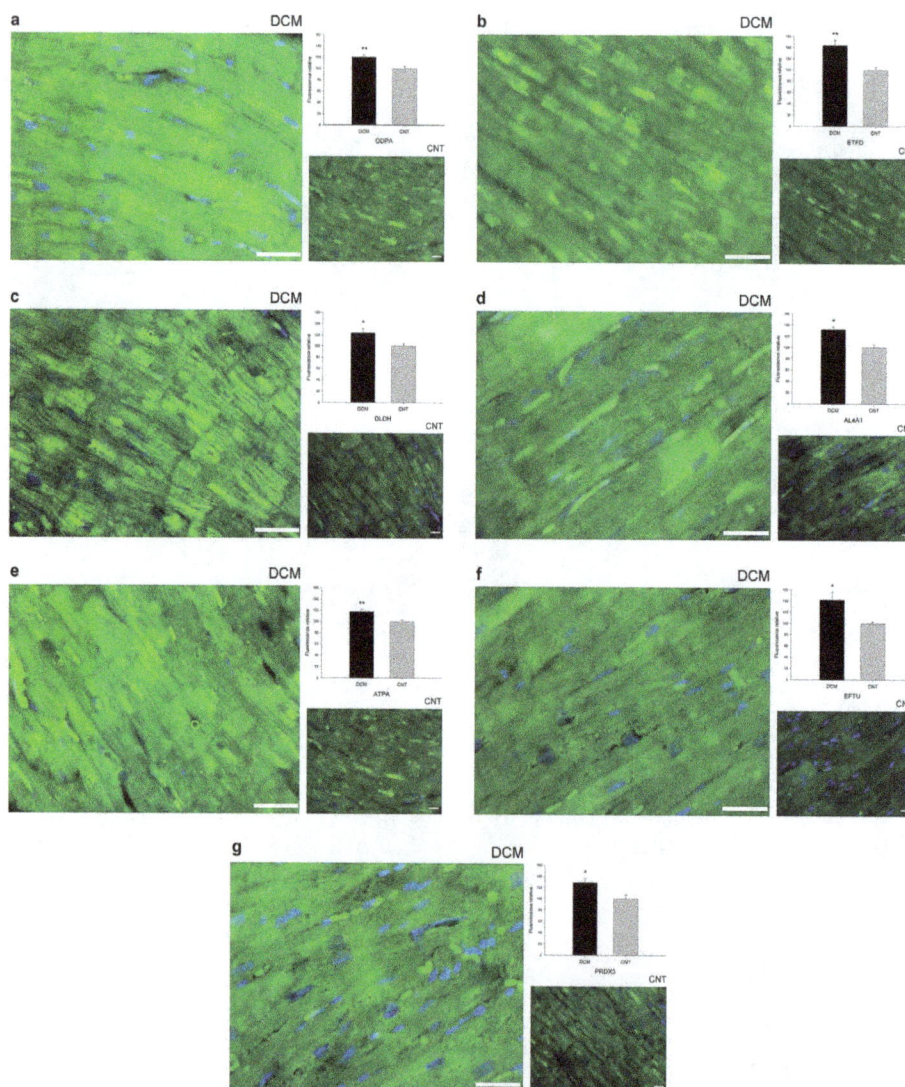

Figure 2. Mitochondrial protein overexpression in dilated human hearts according to immunofluorescence techniques. Influence of dilated cardiomyopathy on the amount of each representative protein involved in cardiac energy metabolism (ODPA, ETFD, DLDH, AL4A1, and ATPA), protein biosynthesis (EFTU), and the stress response (PRDX3). Immunofluorescence of (a) ODPA, (b) ETFD, (c) DLDH, (d) AL4A1, (e) ATPA, (f) EFTU, and (g) PRDX3 were significantly increased in patients with dilated cardiomyopathy compared with the control group. Here we show the nucleus co-stained with DAPI (blue). All of the micrographs are representative of the results obtained in four independent experiments for each group and protein studied, DCM (n = 4) and CNT (n = 4). The bar represents 100 μm. The bar graph shows the relative fluorescence intensity in dilated compared to control hearts. The data are expressed as mean ± SEM. CNT, control; DCM, dilated cardiomyopathy; ODPA, pyruvate dehydrogenase E1 component subunit α, somatic form; ETFD, electron transfer flavoprotein-ubiquinone oxidoreductase; DLDH, dihydrolipoyl dehydrogenase; AL4A1, delta-1-pyrroline-5-carboxylate dehydrogenase; ATPA, ATP synthase subunit α; EFTU, elongation factor Tu; PRDX3, thioredoxin-dependent peroxide reductase. *p value<0.05, **p value<0.01.

DCM   CNT

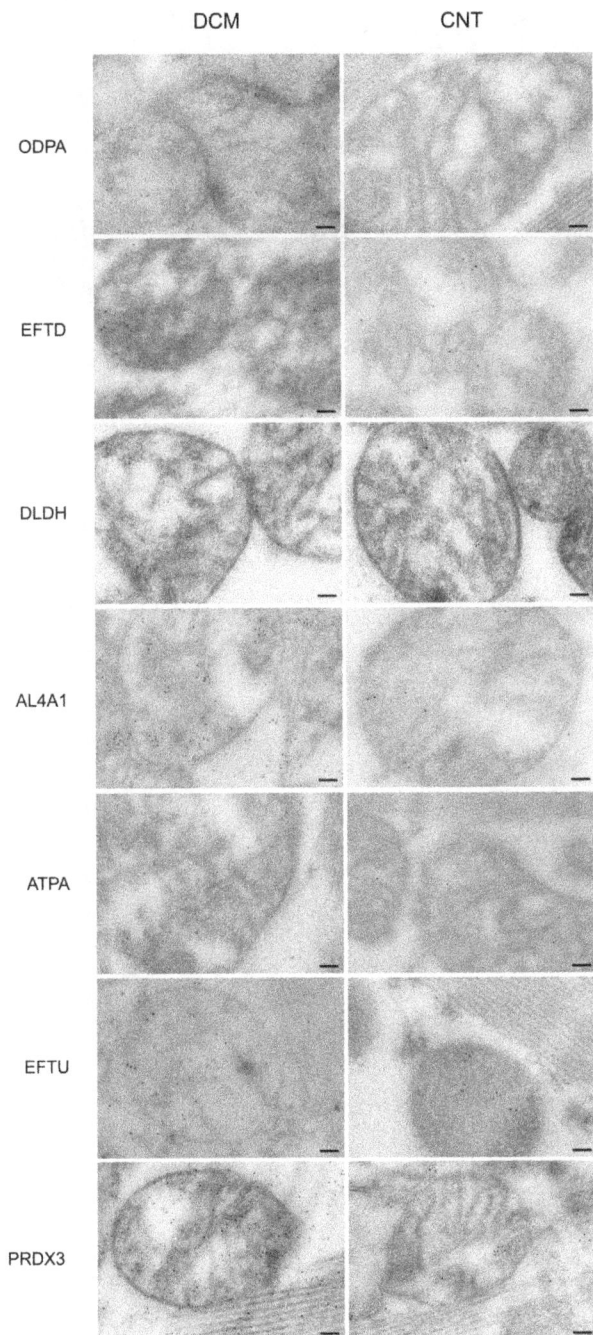

Figure 3. Mitochondrial protein localization and overexpression in dilated human hearts analyzed using transmission electron microscopy. Influence of dilated cardiomyopathy on the amount and localization of the representative proteins involved in cardiac energy metabolism (ODPA, ETFD, DLDH, AL4A1, and ATPA), protein biosynthesis (EFTU), and stress response (PRDX3). We observed an increase in immunogold labeling in all proteins studied in DCM hearts. We also confirmed the location of all proteins analyzed and observed a similar distribution of each protein upon comparing pathological with control samples. The bar represents 100 nm. CNT, control; DCM, dilated cardiomyopathy; ODPA, pyruvate dehydrogenase E1 component subunit α, somatic form; ETFD, electron transfer flavoprotein-ubiquinone oxidoreductase; DLDH, dihydrolipoyl dehydrogenase; AL4A1, delta-1-pyrroline-5-carboxylate dehydrogenase; ATPA, ATP synthase subunit α; EFTU, elongation factor Tu; PRDX3, thioredoxin-dependent peroxide reductase.

glycolysis-related protein expression [5]. Therefore, we focused on three overexpressed proteins involved in substrate utilization. The pyruvate dehydrogenase complex is a multi-enzyme system composed of multiple copies of 3 catalytic components, E_1 ($E_1\alpha$ and $E_1\beta$), E_2, and E_3. Specifically, ODPA catalyzes the overall conversion of pyruvate to acetyl-CoA and CO_2, and thereby links the glycolytic pathway to the tricarboxylic cycle [37,38]. We found that levels of this protein and its mRNA are significantly increased in DCM in concordance with the earlier study results of a canine HF model [39]. In addition, we found that ETFD was overexpressed. Electron transfer flavoproteins are heterodimeric proteins that transfer electrons between primary dehydrogenases and respiratory chains and link the oxidation of fatty acids and some amino acids to the mitochondrial respiratory system [40,41]. Finally, we also validated the overexpression of DLDH, a stable homodimer and essential component of the pyruvate dehydrogenase and glycine cleavage system as well as the α-ketoacid dehydrogenase complex [42]. This result is consistent with those published by our group and also by Li *et al.* in previous studies of total homogenate of LV tissue of DCM patients observing increased DLDH levels [43,44]. In addition, we found a good correlation between the protein levels and mRNA expression of these molecules and also with other altered proteins and its mRNA levels implicated in metabolic process and protein synthesis, specifically AL4A1 and EFTU, thereby interconnecting the alterations found in different processes.

Alterations in energy production in DCM human hearts

High myocardial energy production rates are required to maintain the constant demand of the working heart ATP and alterations in oxidative phosphorylation reduce cardiac function by providing an insufficient supply of ATP to cardiomyocytes [5]. Although the activity of electron transport chain complexes and ATP synthase (the complex responsible for ATP production) activity are known to be reduced in HF [45], reports on the individual levels of ATP synthase subunits in this syndrome are contradictory. While some authors observed that ATP synthase levels did not change [45] or diminish [46] in failing hearts, other studies revealed an increase in ATPA [47,48]. In the present work, we observed a significant overexpression of this protein. This lack of agreement between studies might be because of differences in the sample type or protocols and techniques used to detect these proteins. We also found a significant positive correlation of ATPA protein levels and mRNA expression with the overexpressed molecules involved in substrate utilization, highlighting the relationship between two principal components of the cardiac energy metabolism system. In other words, changes in some proteins involved in substrate utilization implicate modifications in specific components of oxidative phosphorylation. In addition, we found a good correlation between ATPA protein levels and mRNA expression with EFTU and *TUFM*, respectively. Thus, alterations in the energy production system are linked to higher activation of protein biosynthesis and oxidative damage in failing hearts since EFTU promotes the GTP-dependent binding of aminoacyl-tRNA to the A-site of ribosomes, and its downregulation increase reactive oxygen species [49].

Antioxidant PRDX3 overexpression is associated with impaired ventricular function

A large number of studies have reported that oxidative stress is important in the pathophysiology and development of HF via free radical production [50,51]. Reactive oxygen species play a key role in the onset and progression of coronary heart disease, tissue necrosis, and contractile dysfunction [52,53]. PRDX3 is a

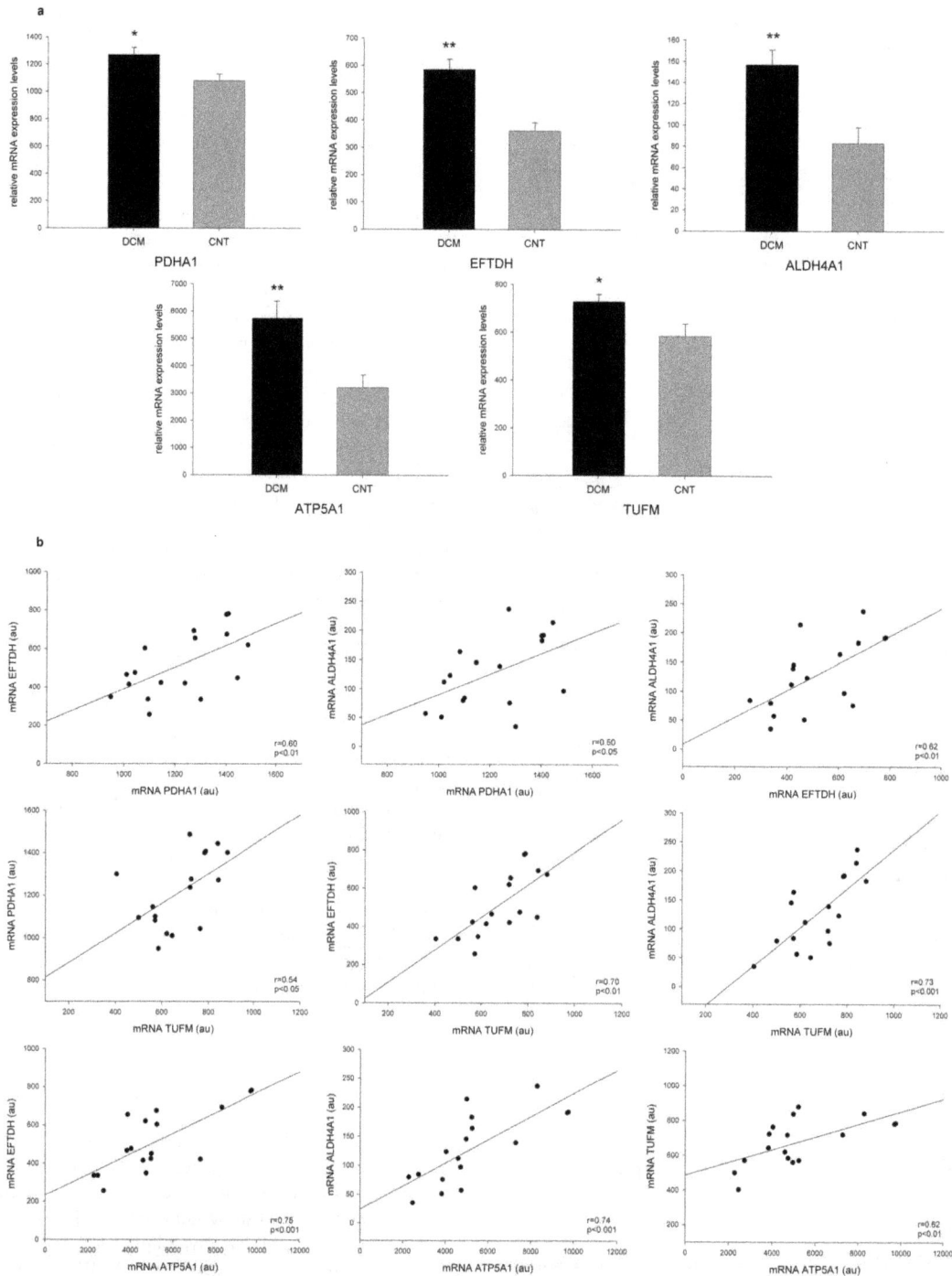

Figure 4. Levels of mRNA expression determined by RNAseq. (a) mRNA levels of the ODPA gene (*PDHA1*), ETFD gene (*ETFDH*), AL4A1 gene (*ALDH4A1*), ATPA gene (*ATP5A1*), and EFTU gene (*TUFM*) were increased in the dilated group compared to the control group. The data are expressed as mean ± SEM in mRNA relative expression. (b) The scatter plots show the relationship between the mRNA expression of molecules altered involved in substrate utilization (*PDHA1* and *ETFDH*) and also with other altered protein implicated in metabolic processes (*ALDH4A1*) and protein synthesis (*TUFM*). These genes showed a relevant correlation with *ATP5A1*, which is involved in energy production. CNT, control; DCM, dilated cardiomyopathy; ODPA, pyruvate dehydrogenase E1 component subunit α, somatic form; ETFD, electron transfer flavoprotein-ubiquinone oxidoreductase; AL4A1, delta-1-pyrroline-5-carboxylate dehydrogenase; ATPA, ATP synthase subunit α; EFTU, elongation factor Tu. *p value<0.05, **p value<0.01.

mitochondrial antioxidant protein that protects radical-sensitive enzymes against oxidative damage by a radical-generating system. Matsusshima *et al.* reported that PRDX3 overexpression protects the heart against post-myocardial infarct remodeling and failure in mice, reducing LV cavity dilation, dysfunction, fibrosis, and

apoptosis [54]. These results are consistent with our findings, since we found a significant increase in the protein levels of PRDX3 in the cardiac tissue of DCM patients probably due to its specific role in the attenuation mechanisms of these failing hearts. Although, we did not observe the same tendency between protein levels and gene expression, likely because of a different removal mechanism of this protein in failing hearts; however, the most remarkable finding was the good correlation between this protein level and LV function, indicating that an increased PRDX3 level is associated with impaired ventricular function. Thus, our findings demonstrated once again that mitochondrial oxidative stress is key player in the pathogenesis of cardiac failure and showed for the first time a direct relationship between the level of this antioxidant and LV dysfunction in human cardiac tissue, suggesting that it could be a primary line of defense against this disease process. With the objective to evaluate the causality of this significant relationship, further studies need to be done.

It is noteworthy that our results are consistent and have been validated by different established techniques and by novel and precise SRM analysis and RNAseq approach. Despite these significant data, more work is needed to fully understand the causal role and which of the alterations observed are adaptive or maladaptive. A promising way to obtain new therapeutic options is to explore strategies based on gene therapy to clarify the mechanism leading to energetic derangement and, moreover, to restore ventricular function in DCM patients. For example, silencing or inducing overexpression of *PRDX3* through gene therapy with the creation of a murine DCM model would allow us to investigate whether LV is aggravated or improved and to develop etiology-specific therapies in HF. In this way, we suggest new molecular targets and all these experiments could be providing a new therapeutic approach.

Study limitations

A common limitation of studies that use cardiac tissues from end-stage failing human hearts is the fact that there is high variability in disease etiology and treatment. To make our study population etiologically homogeneous, we chose DCM patients who did not report any family history of the disease.

Moreover, our tissue samples were taken from the transmural left ventricle apex, so our findings could not be generalized to all layers and regions of the left ventricle. However, we want to emphasize the importance of having carried out this study in a

significant number of samples from explanted human hearts from DCM patients undergoing cardiac transplantation. This has allowed us to extract the region of tissue that we wanted to analyze, which could not have been possible if the study had been made using biopsied tissues.

In summary, the present study is the first to analyze the mitochondrial proteome in cardiac tissue of DCM patients. We found significant and reproducible alterations in cardiac energy metabolism, especially in molecules involved in substrate utilization and energy production, obtaining a considerable relationship between the alterations detected in both processes. LV dysfunction is directly related with the antioxidant PRDX3 expression. This work provides new insight into the cellular mechanisms associated with the pathophysiology of DCM and could serve as a pivotal study to develop etiology-specific therapies in HF.

Supporting Information

File S1   Supporting figure and tables. Figure_S1: Representative two-dimensional DIGE gel. Table_S1: Additional data on MS protein identification of DCM spots with differential expression by MALDI-MS. Table_S2: Additional data on selected reaction monitoring (SRM) analysis. Table_S3: Additional data on selected reaction monitoring (SRM) analysis. Analyte peak area (counts) for each peptide in all samples.

Acknowledgments

The authors thank the Transplant Coordination Unit (Hospital Universitario La Fe, Valencia, Spain) for their help in obtaining the samples. Furthermore, we are grateful to Lorena Gómez and Guillermo Esteban for the assistance in microscopy procedures.

Author Contributions

Conceived and designed the experiments: ERL MR AS. Performed the experiments: ERL ET AO MO. Analyzed the data: ERL MGB MR MP JRGJ FL MO. Contributed reagents/materials/analysis tools: ERL MGB MMMN. Contributed to the writing of the manuscript: ERL MR.

References

1. Jefferies JL, Towbin JA (2010) Dilated cardiomyopathy. Lancet 375: 752–762.
2. Olson EN (2004) A decade of discoveries in cardiac biology. Nat Med 10: 467–474.
3. Bindoff L (2003) Mitochondria and the heart. Eur Heart J 24: 221–224.
4. Maloyan A, Sanbe A, Osinska H, Westfall M, Robinson D, et al. (2005) Mitochondrial dysfunction and apoptosis underlie the pathogenic process in alpha-B-crystallin desmin-related cardiomyopathy. Circulation 112: 3451–3461.
5. Neubauer S (2007) The failing heart–an engine out of fuel. N Engl J Med 356: 1140–1151.
6. Neubauer S, Horn M, Cramer M, Harre K, Newell JB, et al. (1997) Myocardial phosphocreatine-to-ATP ratio is a predictor of mortality in patients with dilated cardiomyopathy. Circulation 96: 2190–2196.
7. Rosca MG, Hoppel CL (2010) Mitochondria in heart failure. Cardiovasc Res 88: 40–50.
8. Sanbe A, Tanonaka K, Kobayasi R, Takeo S (1995) Effects of long-term therapy with ACE inhibitors, captopril, enalapril and trandolapril, on myocardial energy metabolism in rats with heart failure following myocardial infarction. J Mol Cell Cardiol 27: 2209–2222.
9. Bayeva M, Gheorghiade M, Ardehali H (2013) Mitochondria as a therapeutic target in heart failure. J Am Coll Cardiol 61: 599–610.
10. Horstkotte J, Perisic T, Schneider M, Lange P, Schroeder M, et al. (2011) Mitochondrial thioredoxin reductase is essential for early postischemic myocardial protection. Circulation 124: 2892–2902.
11. Lal H, Zhou J, Ahmad F, Zaka R, Vagnozzi RJ, et al. (2012) Glycogen synthase kinase-3alpha limits ischemic injury, cardiac rupture, post-myocardial infarction remodeling and death. Circulation 125: 65–75.
12. Liesa M, Luptak I, Qin F, Hyde BB, Sahin E, et al. (2011) Mitochondrial transporter ATP binding cassette mitochondrial erythroid is a novel gene required for cardiac recovery after ischemia/reperfusion. Circulation 124: 806–813.
13. Agnetti G, Husberg C, Van Eyk JE (2011) Divide and conquer: the application of organelle proteomics to heart failure. Circ Res 108: 512–526.
14. Deng WJ, Nie S, Dai J, Wu JR, Zeng R (2010) Proteome, phosphoproteome, and hydroxyproteome of liver mitochondria in diabetic rats at early pathogenic stages. Mol Cell Proteomics 9: 100–116.
15. Imahashi K, Schneider MD, Steenbergen C, Murphy E (2004) Transgenic expression of Bcl-2 modulates energy metabolism, prevents cytosolic acidification during ischemia, and reduces ischemia/reperfusion injury. Circ Res 95: 734–41.
16. Macrae DJ (2007) The Council for International Organizations and Medical Sciences (CIOMS) guidelines on ethics of clinical trials. Proc Am Thorac Soc 4: 176–178, discussion 178–179.
17. Swedberg K, Cleland J, Dargie H, Drexler H, Follath F, et al. (2005) Guidelines for the diagnosis and treatment of chronic heart failure: executive summary (update 2005): The Task Force for the Diagnosis and Treatment of Chronic Heart Failure of the European Society of Cardiology. Eur Heart J 26: 1115–1140.
18. Laemmli UK (1970) Cleavage of structural proteins during the assembly of the head of bacteriophage T4. Nature 227: 680–685.
19. Shevchenko A, Wilm M, Vorm O, Mann M (1996) Mass spectrometric sequencing of proteins silver-stained polyacrylamide gels. Anal Chem 68: 850–858.
20. Perkins DN, Pappin DJ, Creasy DM, Cottrell JS (1999) Probability-based protein identification by searching sequence databases using mass spectrometry data. Electrophoresis 20: 3551–3567.
21. Azorin I, Portoles M, Marin P, Lazaro-Dieguez F, Megias L, et al. (2004) Prenatal ethanol exposure alters the cytoskeleton and induces glycoprotein microheterogeneity in rat newborn hepatocytes. Alcohol Alcohol 39: 203–212.
22. Portoles M, Faura M, Renau-Piqueras J, Iborra FJ, Saez R, et al. (1994) Nuclear calmodulin/62 kDa calmodulin-binding protein complexes in interphasic and mitotic cells. J Cell Sci 107 (Pt 12): 3601–3614.

23. Reynolds WA (1963) The Effects of Thyroxine Upon the Initial Formation of the Lateral Motor Column and Differentiation of Motor Neurons in Rana Pipiens. J Exp Zool 153: 237–249.

24. Tomas M, Fornas E, Megias L, Duran JM, Portoles M, et al. (2002) Ethanol impairs monosaccharide uptake and glycosylation in cultured rat astrocytes. J Neurochem 83: 601–612.

25. Li H, Handsaker B, Wysoker A, Fennell T, Ruan J, et al. (2009) The Sequence Alignment/Map format and SAMtools. Bioinformatics 25: 2078–2079.

26. McKenna A, Hanna M, Banks E, Sivachenko A, Cibulskis K, et al. (2010) The Genome Analysis Toolkit: a MapReduce framework for analyzing next-generation DNA sequencing data. Genome Res 20: 1297–1303.

27. Trapnell C, Williams BA, Pertea G, Mortazavi A, Kwan G, et al. (2010) Transcript assembly and quantification by RNA-Seq reveals unannotated transcripts and isoform switching during cell differentiation. Nat Biotechnol 28: 511–515.

28. Anders S, Huber W (2010) Differential expression analysis for sequence count data. Genome Biol 11: R106.

29. Robinson MD, McCarthy DJ, Smyth GK (2010) edgeR: a Bioconductor package for differential expression analysis of digital gene expression data. Bioinformatics 26: 139–140.

30. de Weger RA, Schipper ME, Siera-de Koning E, van der Weide P, van Oosterhout MF, et al. (2011) Proteomic profiling of the human failing heart after left ventricular assist device support. J Heart Lung Transplant 30: 497–506.

31. Hammer E, Goritzka M, Ameling S, Darm K, Steil L, et al. (2011) Characterization of the human myocardial proteome in inflammatory dilated cardiomyopathy by label-free quantitative shotgun proteomics of heart biopsies. J Proteome Res 10: 2161–71.

32. Nishtala K, Phong TQ, Steil L, Sauter M, Salazar MG, et al. (2011) Virus-induced dilated cardiomyopathy is characterized by increased levels of fibrotic extracellular matrix proteins and reduced amounts of energy-producing enzymes. Proteomics 11: 4310–20.

33. Comunian C, Rusconi F, De Palma A, Brunetti P, Catalucci D, et al. (2011) A comparative MudPIT analysis identifies different expression profiles in heart compartments. Proteomics 11: 2320–8.

34. Green DR, Reed JC (1998) Mitochondria and apoptosis. Science 281: 1309–1312.

35. Kagan VE, Shvedova A, Serbinova E, Khan S, Swanson C, et al. (1992) Dihydrolipoic acid–a universal antioxidant both in the membrane and in the aqueous phase. Reduction of peroxyl, ascorbyl and chromanoxyl radicals. Biochem Pharmacol 44: 1637–1649.

36. Ahuja P, Wanagat J, Wang Z, Wang Y, Liem DA, et al. (2013) Divergent mitochondrial biogenesis responses in human cardiomyopathy. Circulation 127: 1957–1967.

37. Kato M, Wynn RM, Chuang JL, Tso SC, Machius M, et al. (2008) Structural basis for inactivation of the human pyruvate dehydrogenase complex by phosphorylation: role of disordered phosphorylation loops. Structure 16: 1849–1859.

38. Korotchkina LG, Patel MS (1995) Mutagenesis studies of the phosphorylation sites of recombinant human pyruvate dehydrogenase. Site-specific regulation. J Biol Chem 270: 14297–14304.

39. Heinke MY, Wheeler CH, Chang D, Einstein R, Drake-Holland A, et al. (1998) Protein changes observed in pacing-induced heart failure using two-dimensional electrophoresis. Electrophoresis 19: 2021–2030.

40. Gempel K, Topaloglu H, Talim B, Schneiderat P, Schoser BG, et al. (2007) The myopathic form of coenzyme Q10 deficiency is caused by mutations in the electron-transferring-flavoprotein dehydrogenase (ETFDH) gene. Brain 130: 2037–2044.

41. Goodman SI, Axtell KM, Bindoff LA, Beard SE, Gill RE, et al. (1994) Molecular cloning and expression of a cDNA encoding human electron transfer flavoprotein-ubiquinone oxidoreductase. Eur J Biochem 219: 277–286.

42. Brautigam CA, Chuang JL, Tomchick DR, Machius M, Chuang DT (2005) Crystal structure of human dihydrolipoamide dehydrogenase: NAD+/NADH binding and the structural basis of disease-causing mutations. J Mol Biol 350: 543–552.

43. Li W, Rong R, Zhao S, Zhu X, Zhang K, et al. (2012) Proteomic analysis of metabolic, cytoskeletal and stress response proteins in human heart failure. J Cell Mol Med 16: 59–71.

44. Rosello-Lleti E, Alonso J, Cortes R, Almenar L, Martinez-Dolz L, et al. (2012) Cardiac protein changes in ischaemic and dilated cardiomyopathy: a proteomic study of human left ventricular tissue. J Cell Mol Med 16: 2471–2486.

45. Rosca MG, Okere IA, Sharma N, Stanley WC, Recchia FA, et al. (2009) Altered expression of the adenine nucleotide translocase isoforms and decreased ATP synthase activity in skeletal muscle mitochondria in heart failure. J Mol Cell Cardiol 46: 927–935.

46. Agnetti G, Kaludercic N, Kane LA, Elliott ST, Guo Y, et al. (2010) Modulation of mitochondrial proteome and improved mitochondrial function by biventricular pacing of dyssynchronous failing hearts. Circ Cardiovasc Genet 3: 78–87.

47. Cieniewski-Bernard C, Mulder P, Henry JP, Drobecq H, Dubois E, et al. (2008) Proteomic analysis of left ventricular remodeling in an experimental model of heart failure. J Proteome Res 7: 5004–5016.

48. Yang J, Moravec CS, Sussman MA, DiPaola NR, Fu D, et al. (2000) Decreased SLIM1 expression and increased gelsolin expression in failing human hearts measured by high-density oligonucleotide arrays. Circulation 102: 3046–3052.

49. Zhang DX, Yan H, Hu JY, Zhang JP, Teng M, et al. (2012) Identification of mitochondria translation elongation factor Tu as a contributor to oxidative damage of postburn myocardium. J Proteomics 77: 469–479.

50. Keith M, Geranmayegan A, Sole MJ, Kurian R, Robinson A, et al. (1998) Increased oxidative stress in patients with congestive heart failure. J Am Coll Cardiol 31: 1352–1356.

51. Shiomi T, Tsutsui H, Matsusaka H, Murakami K, Hayashidani S, et al. (2004) Overexpression of glutathione peroxidase prevents left ventricular remodeling and failure after myocardial infarction in mice. Circulation 109: 544–549.

52. Brioschi M, Polvani G, Fratto P, Parolari A, Agostoni P, et al. (2012). Redox proteomics identification of oxidatively modified myocardial proteins in human heart failure: implications for protein function. PLoS One 7: e35841.

53. Ide T, Tsutsui H, Kinugawa S, Utsumi H, Kang D, et al. (1999) Mitochondrial electron transport complex I is a potential source of oxygen free radicals in the failing myocardium. Circ Res 85: 357–363.

54. Matsushima S, Ide T, Yamato M, Matsusaka H, Hattori F, et al. (2006) Overexpression of mitochondrial peroxiredoxin-3 prevents left ventricular remodeling and failure after myocardial infarction in mice. Circulation 113: 1779–1786.

14

A Bumpy Ride on the Diagnostic Bench of Massive Parallel Sequencing, the Case of the Mitochondrial Genome

Kim Vancampenhout[1], Ben Caljon[2], Claudia Spits[1], Katrien Stouffs[1,2], An Jonckheere[3], Linda De Meirleir[1,3], Willy Lissens[1,2], Arnaud Vanlander[4], Joél Smet[4], Boel De Paepe[4], Rudy Van Coster[4], Sara Seneca[1,2]*

1 Research Group Reproduction and Genetics (REGE), Vrije Universiteit Brussel (VUB), Brussels, Belgium, 2 Center for Medical Genetics, UZ Brussel, Vrije Universiteit Brussel (VUB), Brussels, Belgium, 3 Department of Pediatric Neurology, UZ Brussel, Vrije Universiteit Brussel (VUB), Brussels, Belgium, 4 Department of Pediatrics, Division of Pediatric Neurology and Metabolism, University Hospital Ghent, Ghent University, Ghent, Belgium

Abstract

The advent of massive parallel sequencing (MPS) has revolutionized the field of human molecular genetics, including the diagnostic study of mitochondrial (mt) DNA dysfunction. The analysis of the complete mitochondrial genome using MPS platforms is now common and will soon outrun conventional sequencing. However, the development of a robust and reliable protocol is rather challenging. A previous pilot study for the re-sequencing of human mtDNA revealed an uneven coverage, affecting predominantly part of the plus strand. In an attempt to address this problem, we undertook a comparative study of standard and modified protocols for the Ion Torrent PGM system. We could not improve strand representation by altering the recommended shearing methodology of the standard workflow or omitting the DNA polymerase amplification step from the library construction process. However, we were able to associate coverage bias of the plus strand with a specific sequence motif. Additionally, we compared coverage and variant calling across technologies. The same samples were also sequenced on a MiSeq device which showed that coverage and heteroplasmic variant calling were much improved.

Editor: Robert Lightowlers, Newcastle University, United Kingdom

Funding: This work was supported by Fonds voor Wetenschappelijk Onderzoek Vlaanderen (FWO; www.FWO.be) G.0.200; The 'Association Belge contre les Maladies Neuro-Musculaires (ABMM)' (http://www.hospichild.be/fr/associations/maladies-neuro-musculaires/association-belge-contre-les-maladies-neuro-musculaires-asbl-abmm-maladies-genetiques), and Vrije Universiteit Brussel (with reference OZR1928 and OZRMETH3). Authors who received funding: CS LDM SS. The funders had no role in study design, data collection and analysis, decision to publish, or preparation of the manuscript.

Competing Interests: The authors have declared that no competing interests exist.

* Email: sara.seneca@uzbrussel.be

Introduction

The human mitochondrial DNA (mtDNA) is a small circular double stranded molecule that comprises 16569 bp and codes for 13 protein genes, 22 tRNAs and 2 rRNAs. All these are essential elements to the correct function of the oxidative phosphorylation (OXPHOS) system, a fundamental process of the cellular role of mitochondria. For over 25 years, the pathogenicity of certain alterations of the mitochondrial genome has been clearly established in mtDNA disease. Despite the existence of mutation hotspot genes and regions, and the occurrence of recurrent mutations, these pathogenic aberrations are scattered over the entire mitochondrial genome. This makes it necessary to completely analyze this small genome to confirm or exclude pathogenic mtDNA changes. Molecular analysis often requires different and complementary methods, e.g. Southern blot, long range (LR)-PCR, Denaturing Gradient Gel Electrophoresis (DGGE), High Resolution Melting (HRM), quantitative (q)PCR and Sanger sequencing for the detection and quantification of mtDNA. The emergence of MPS technologies has provided the diagnostic bench with a new and highly valuable tool for the evaluation of human mtDNA integrity. However, these new sequencing platforms have pitfalls, and crucial biases might be created [1] such as the loss of coverage in regions with GC-extreme (high or low) content, or the limited ability to analyze homopolymeric stretches [2] [3]. As a result, heteroplasmic variant calling might be severely complicated or even erroneous, as the nucleotide representation can be too weak or unreliable in some of these regions. In a recent study by Seneca et al. [4], the mitochondrial genomes of 32 DNA samples were analyzed using an Ion Torrent PGM system after enrichment with LR-PCR amplification of the mtDNA. A major bias in read depth between the positive and negative strand was seen for almost 10% of the mitochondrial genome, despite the fact that the sequencing was carried out at an average coverage of 6000. Moreover, in some regions the data for the positive strand dropped severely, reaching a critically low coverage. This difference in read depth between both strands made it challenging to distinguish true low-level heteroplasmic variants from sequencing errors. Therefore, we tried to develop an improved MPS-based protocol for the analysis of the human mitochondrial genome. Several library preparation methods and sequencing technologies were tested in order to ameliorate the present sequencing protocol, and their outputs were compared. We were also able to identify the specific nature of the systematically undercovered nucleotide motifs. We are convinced that our findings are of interest to all laboratories working on MPS for the mtDNA, both in a research or clinical setting.

Materials and Methods

Ethics Statement

This study was approved by the ethics committee of the Institutional Review Board (IRB) of the University Hospital (UZ Brussel, Vrije Universiteit Brussel). For all control samples a written informed consent was obtained. The informed consent form was also reviewed and approved by the local ethics committee of the IRB. For the patient samples, during clinical consultation oral consent was given to study their genetic material by any methods relevant to diagnostically confirm or rule out mutations in their mtDNA. This procedure does not require a written consent by the patient, and oral consent is recorded in a protected medical patient file. This is a standard procedure that is approved within the Center for Medical Genetics and accepted by the ethics committee of the IRB of the hospital.

Sample collection and DNA

Six DNA samples, corresponding to three controls (samples 1, 2, 4 in [4]) and three patients (samples 9, 14, 21 in [4]), were randomly selected from the previous sample cohort [4]. Total DNA had been extracted from leukocytes using standard DNA isolation techniques (Chemagen, Perkin Elmer, Zaventem, Belgium). An overview of the samples and techniques used is given in Supporting Information S1.

Long range PCR

MPS data files, obtained from a previous study, were mainly generated by the sequencing of three overlapping LR-PCR fragments covering the whole mitochondrial genome (all six samples were amplified using the 'three overlapping' fragment approach, two were additionally generated with a 'single fragment' method) [4]. However, as was demonstrated in a previous study, one large single LR-PCR product allowed the detection of variants, indels and large deletions simultaneously, a situation that is advantageous due to time and cost constrains for clinical genetic testing. For this single LR-PCR a 16.2 kb fragment [5] was generated using the LongAmp *Taq* PCR kit (New England Biolabs, Bioke, Leiden, The Netherlands). The mitochondrial genome was amplified from 200 ng gDNA as template in a 50 μL PCR assay according to manufacturer's recommendations. The PCR protocol was adapted to an initial 30 s denaturation at 94°C, followed by 15 cycles with first a denaturation of 10 s at 92°C, annealing at 67°C for 30 s and an extension of 10 min at 68°C. This was followed by 18 cycles with a denaturation of 10 s at 92°C and an extension of 10 min +20 s every cycle at 68°C. A final extension step was performed at 68°C for 7 min. Successful PCR amplification was assessed using 0.8% agarose gel electrophoresis, and products were purified with AMpure beads (Analis, Champion, Belgium).

Ion Torrent PGM sequencing

Ion Torrent semi-conductor sequencing technology detects the incorporation of each of the four nucleotides as small changes in

pH that are provoked by the release of a proton. Library and template preparation include an amplification step. The latter is known as an emulsion PCR which takes place in aqueous droplets suspended in oil.

The data files of six samples, previously sequenced using the Ion Torrent PGM assay according to the manufacturer's instructions [4], were regarded as benchmark material for a comparative study of the new protocols described in the present study. We evaluated the following modifications to the standard protocol: different shearing methodologies and avoiding the amplification step in the library preparation of the Ion Torrent PGM protocol. To test the fragmentation methods, LR-PCR products were sheared using the Covaris M220 sonicator (Life Technologies Europe, Gent, Belgium) and the NEBNext dsDNA Fragmentase (Bioke). For the first fragmentation method, a dilution to 100 ng in 50 μL of LR-PCR products were subjected to sonication for 130 s with a duty factor of 20%, a peak incident power of 50W, a temperature of 20°C and 200 cycles per burst, to tailor the DNA molecules into fragments with a median size of 200 bp (Ion Xpress Plus gDNA Fragment Library Preparation, Appendix B). A standard procedure was followed for the NEBNext dsDNA Fragmentase assay. Briefly, 1 μg of PCR product was added to 2 μL 10x Fragmentase reaction buffer and 0.2 μL of 100x BSA. This mixture was placed on ice for 5 min prior to the addition of 2 μL of NEBNext dsDNA Fragmentase and an incubation at 37°C for 30 min. The reaction was stopped by adding 5 μL of 0.5 M EDTA solution to the DNA fragments. Sheared samples were purified using AMPure beads. The size distribution of the fragmented DNA was assessed on the Bioanalyzer (Agilent, Diegem, Belgium), using the High Sensitivity Assay (Agilent, Diegem, Belgium). All further downstream manipulations were performed according to the Ion Torrent PGM protocol's instructions (Ion Xpress Plus gDNA Fragment Library preparation, Life Technologies, Gent, Belgium). Briefly, samples were end repaired, ligated with adaptors, nick repaired and bead purified prior to amplification of size selected (E-gel system, Life Technologies) fragments around 330 bp long. Fragment sizes were assessed using the Bioanalyzer system and quantified with the Qubit 2.0 fluorimeter (Life Technologies, Gent, Belgium). Pooled libraries were used for emulsion PCR amplification. Sequencing reactions were run on the Ion Torrent PGM using Ion 316 version 2 chips and the Ion PGM 200 sequencing kit (Life Technologies, Gent, Belgium).

Illumina MiSeq sequencing

To obtain 350 bp fragments LR-PCR products were sheared with the Covaris M220 sonicator (Life Technologies Europe, Gent, Belgium) and the NEBNext dsDNA Fragmentase enzyme (Bioke, Leiden, The Netherlands), both starting with 1 μg LR-PCR product. Covaris sheared LR-PCR products were fragmented using custom instrument specifications (TruSeq DNA PCR-Free Sample Preparation Guide). The protocol described, before concerning the NEBNext dsDNA Fragmentase, was the same except for the incubation time that was adapted to 15 min to obtain 350 bp fragments. Next, samples were further processed using the TruSeq DNA PCR-Free Sample Preparation protocol as instructed by the supplier (Illumina, Eindhoven, The Netherlands). After fragmentation, end repair, adenylation, and indexed paired end adapter ligation, samples were pooled and processed on the MiSeq sequencer with the MiSeq Reagent Micro Kit, v2 (Illumina). Conversely, all six samples were also processed using the Nextera XT kit (Illumina). A single Nextera tagmentation enzymatic reaction was used where LR-PCR products were simultaneously fragmented and tagged with adaptors. Finally, a limited cycle PCR protocol (12 cycles) was applied, adding simultaneously sequencing indexes (Nextera XT DNA Sample Preparation Guide, Illumina).

Detection threshold determination for the MiSeq

The technical error rate of the MiSeq platform was determined with the methodology used for the Ion Torrent PGM system [4]. For the latter device, which unlike PhiX for the Illumina MiSeq lacks an endogenous control sample, a well typed pUC19 plasmid was used. The use of the same pUC19 DNA sample also allowed a comparison of sequencing results across platforms. One µg of pUC19 plasmid DNA (Thermo Fisher, Erembodegem-Aalst, Belgium) was sheared by the Covaris or NEBNext dsDNA Fragmentase. Subsequently, samples were processed using the TruSeq DNA PCR-Free Sample Preparation protocol, and sequenced on the MiSeq. The error rate of the sequencing process was computed by calculating the ratio of non-reference versus total bases per position. Taking the average of all ratios per position resulted in the average error rate of the pUC19 plasmid DNA.

Data analysis

FastQ files from all datasets, generated by either the Ion Torrent PGM or MiSeq platforms, were mapped to the mitochondrial revised Cambridge Reference Sequence (rCRS, NC 012920.1) using BWA-MEM (version 0.7.5) [6]. As a metric for coverage bias, the relative coverage was used. Applying the SAMtools software (version 0.1.18) [7] the number of reads mapping to each reference base was counted. The mean coverage was calculated by averaging this value across each base in the sequence. By computing the ratio of the coverage of a given reference base and the mean coverage of all reference bases, the relative coverage was obtained. This was calculated for the plus and minus strand separately, for the total coverage of both strands together, and was presented in graphical illustrations. To visualize the relative coverage resulting from all different protocols and methods tested, circular plots were generated with the freeware Circos-0.64 software [8]. The Circos plots demonstrated in this article are restricted to sample 1, as the coverage profiles were consistent across all samples. To compare different methodologies, datasets were down sampled to an average coverage of 3000 using Picard (http://picard.sourceforge.net). The average relative coverage was collected for all samples processed with the same protocol resulting in seven datasets (Ion Torrent standard, Ion Torrent without amplification step, Ion Torrent Covaris, Ion Torrent NEBNext dsDNA Fragmentase, TruSeq Covaris, TruSeq NEBNext dsDNA Fragmentase and Nextera XT). For each dataset the fraction with a relative coverage <0.50; <0.25; <0.10; <0.05 and <0.01 was determined. To identify the nucleotide composition of undercovered regions GC, AT along with CT, AG, AC and GT dinucleotide motif plots were created and correlated to the total relative coverage, as well as the relative coverage from each strand separately. Both the incidence (in percentages) of the dinucleotide motifs in the mtDNA molecule, and the relative coverage were calculated in bins of 150 nucleotides and illustrated as bias plots.

For variant calling, three different strategies were employed and compared. First, all data were analyzed using an in-house pipeline based on GATK. FastQ files were aligned to the rCRS using BWA-MEM and sorted. Next, GATK realignment around indels and recalibration was performed. The GATK Unified Genotyper was used for variant calling, without at random down sampling of reads to reduce coverage. Subsequently, all variants with a quality score <400 were filtered from the vcf data. Second, all data were also analyzed using the CLC Genomics Workbench (version 6.0.5)

against the rCRS. Only variants with an average quality score > 25 were selected. A third and last strategy was only implemented on the Ion Torrent data. PGM files were mapped and variants were called using the Torrent Suite 4.2.

For each sample analyzed with the Ion Torrent PGM or MiSeq device, the sequencing error was determined for each position of the genome sequence, with exception of the true variants (versus rCRS) detected in each sample. The average sequencing error and their standard deviations were determined for these six samples. Potential low heteroplasmic variant levels were compared to these values and utilized as a reliable baseline (index) to reduce the false positive rate of the data [4].

Results and Discussion

Assessment of different PGM protocols

We have recently studied the use of the Ion Torrent PGM sequencer system in a diagnostic setting for the nucleotide analysis of human mitochondrial genomes of patient and control samples. The results uncovered a rather poor performance for some of the mtDNA regions [4]. Although it is well known that the PGM sequencing technology has problems handling homopolymeric stretches, an additional limitation was revealed, as a major difference in read depth between both strands was exposed for about 10% of the mitochondrial genome regions. For these sequences, the relative coverage of the positive strand dropped below 0.1. These particular patterns were reproduced in replicates of the same and between different samples, but never observed for pUC19 plasmid samples (Figure 1A). The causes of this remained unknown. Previous experiments had already excluded primer, LR-PCR or sample dependence, and it was assumed that the discrepancy originated from the enzymatic shearing step included in the Ion Torrent assay [4]. Altering fragmentation in the original Ion Torrent PGM assay could thus promote a change of the coverage profile. Hence, the standard enzymatic shearing step was omitted and substituted with an enzymatic treatment with NEBNext dsDNA Fragmentase or with physical shearing with a Covaris M220 sonicator device, leaving all further downstream process steps unchanged. Nonetheless, MPS data demonstrated that none of the altered protocols induced an equilibrated strand representation. Neither did they show an improvement of the under-representation of the plus strand. Uneven coverage was still produced (Figure 1B). Both shearing methods resulted still in 7 to 7.8% of the 16.2 kb fragment to have a relative coverage of the plus strand <0.1. Moreover, 2% of the 16.2 kb region showed a relative coverage of the plus strand <0.01 (Table 1). Further experiments, such as omission of the first PCR amplification step in the PGM library preparation protocol were carried out and subjected to MPS. But also this intervention did not lead to a reduced bias (Figure 1C, Table 1). By exchanging the Platinum Taq DNA polymerase for Kapa HiFi in the nick translation and amplification step during library preparation, Quail et al. [9] had demonstrated a reduced bias in PGM data. Therefore, it was proposed that the DNA amplification treatment during the library preparation and/or the emulsion PCR mediated a bias interfering with all further analysis of the mitochondrial genome. In order to further characterize the underlying mechanisms of poor PGM results across parts of the mitochondrial genome, the depth of the relative coverage seen at each position was tabulated for both strands separately. Hence, a possible association with its nucleotide composition was investigated systematically. In-house Perl scripts were used to calculate the content of GC or AT rich motifs, as well as any other dinucleotide rich combination. This analysis did not disclose any relationship between GC or AT rich regions and poor strand representation (Figure 2). The findings of Quail et al. [9] about very low coverage from GC or AT rich motifs for *P.falciparum* were not confirmed by the analysis of mtDNA. In contrast, reduced coverage was detected for AC and CT rich motifs. Particularly, coverage of the plus strand was negatively influenced by these two motifs. Moreover, the relative coverage of the plus strand dropped almost to zero for 80% (and more) AC rich motifs (Figure 2). The sequencing bias is seen for a high AC-content (range 70–80%) which corresponds to the figures of 80% and more for the GC and AT motifs presented by Ross et al. [1]. It is already known for a long time that the nucleotide composition of both mtDNA strands is different. The plus strand or *light* strand is C-rich, while the minus strand or *heavy* strand is G-rich. The rCRS is based on the L-strand and corresponds to the underrepresented plus strand in our sequencing results. The analyses were also performed for the pUC19 plasmid DNA. As expected, no correlation between its nucleotide composition and coverage data was observed (Supporting Information S2). We therefore hypothesize that the troughs generated by the Ion Torrent PGM system rather originate from the proliferation of the sheared mtDNA sequences and not from the fragmentation method *per se*. In fact, it might be inherent to the combination of the DNA polymerases used in the PCR amplification steps included in the standard protocols, and the nature of the mitochondrial genome sequence.

Comparison PGM-MiSeq

We proceeded to study mitochondrial genome resequencing on a MiSeq platform, using two different strategies. The results of the PCR amplification free protocol of TruSeq were compared with those of the Nextera XT kit, a method including one PCR amplification step in the library preparation step. Experiments were carried out according to the manufacturer's instructions. The average read depth for the different datasets generated with the MiSeq were 3723, 4701 and 19418 for the TruSeq Covaris, TruSeq NEBNext dsDNA Fragmentase and the Nextera XT methods, respectively. The reads generated by MiSeq (paired end reads), had a 150 bp fixed length, while reads generated by Ion Torrent PGM showed a variable single-end read length with an average of 145 bp. To compare different methodologies, datasets were down sampled to an average coverage of 3000. Relative coverage analysis showed a major improvement in strand equilibration for the TruSeq data. Data from the TruSeq sheared with the Covaris protocol, and the TruSeq enzymatically digested with NEBNext dsDNA Fragmentase achieved an impressive relative coverage, with few areas (only 1.6% and 1.6%, respectively) of the plus strand <0.5. The Nextera XT data did not show strand bias as seen with the PGM data. However, a general unevenness of coverage of both strands was seen. Indeed, regions of both strands (9.2% of the plus strand and 9.6% of min strand) showed a relative coverage <0.5. (Figure 1D, Table 1) which were associated with CT rich motifs. Unlike for the PGM, where mainly the positive strand was involved, both strands were affected, however not as severe as for the Ion Torrent data (Figure 2).

Detection limit of the MiSeq

The detection threshold for the identification of base variants was set on 5% for the Ion Torrent chemistry. This value was based on the determination of the sequencing error and the sensitivity and specificity experiments previously performed [4]. To set the detection threshold for the MiSeq, the same pUC19 plasmid DNA sample was sheared with two different methods, once using the Covaris M220 sonicator and secondly using the NEBNext dsDNA

Fragmentase. Both differentially sheared samples were sequenced on the MiSeq following TruSeq PCR free library preparation and a 100% coverage was obtained with an average read depth of 30 440 and 30 966 respectively. Similar average sequencing error results were obtained with 0.27% and 0.19% for the Covaris sheared sample and the enzymatic sheared sample respectively. These values are in concordance with the error rate obtained by the PhiX, which presented with an error rate of 0.35%. These error rates in turn correspond to previously reported data for the MiSeq platform [9]. By applying these results to determine the variant threshold for the mitochondrial resequencing, a detection threshold level of 2% is possible. However as the PGM data were previously investigated with a detection threshold of 5%, these settings were also used for the MiSeq data.

Variant calling

Last, we assessed variant detection in all samples using the data panel of nucleotide alterations reflecting the Sanger sequencing previously performed. The majority of these variants were identified on both platforms (Table 2; Supporting Information S3). Results were collected for a PGM, TruSeq or Nextera XT dataset. Two variant calling pipelines, an in-house pipeline based on GATK and the Quality-based variant detection method (CLC Genomics Workbench) were applied to MiSeq datasets, and subsequently compared to the results of our previous study. The TS4.2 was only used with the PGM data. The first pipeline resulted in 99.5% of the variants detected in the TruSeq and Nextera XT dataset, while the PGM dataset showed a 92.4% concordance with the Sanger sequencing results. The CLC Genomics Workbench pipeline requires the variant to be present on both strands. 93.4%, 97.7% and 97.2% of the Sanger sequencing variants were called in the TruSeq Covaris, TruSeq NEBNext dsDNA Fragmentase and the Nextera XT dataset, respectively. Applying these terms to the PGM data resulted in 84.7% concordance with Sanger sequencing. However, omitting the strand parameter identified 95.2% of the variants for PGM data. These figures demonstrated clearly the effect of strand bias on variant calling for the PGM data. Indeed, 67 out of 98 false negative results were present on one strand only. An additional analysis with the TS4.2 software identified 96.6% of the variants. Three positions, m.294T>C, m.16183A>C and the polymorphic 302_316 region, presented as false negative results in the PGM data sets. An additional false negative variant, at position m.5899_5900insC escaped variant calling. All of these variants are situated near a homopolymeric stretch and, with the exception of m.5899_5900insC, are also located in regions with significant AC contents and its associated strand bias (relative coverage of the plus strand <0.2). It must be pointed out that, despite the well documented shortcoming in homopolymer calling, the propriety software is clearly well fitted for the PGM needs in variant calling. Comparing the various algorithms applied in this present and the previous study, the TS4.2 software was noticeably the better performer. Compared to the former TS3.6 version, a remarkable improvement was noticed for the false positive rate. Reanalyzing all PGM samples with the TS4.2 release showed a reduction in false positives from 13,4% to 8,9%, with a detection threshold level of 5%. The highest sensitivity for the MiSeq results (TruSeq and Nextera XT data) was obtained by our in-house pipeline based on GATK. Indeed, the only false negative result for these data was one specific variation in the polymorphic 302_316 region in sample 21. Two single nucleotide insertions were detected in this region with Sanger sequencing (m.309_310insC and m.315_316insC), but MiSeq identified them incorrectly as a heteroplasmic sequence mixture of molecules with an insertion of

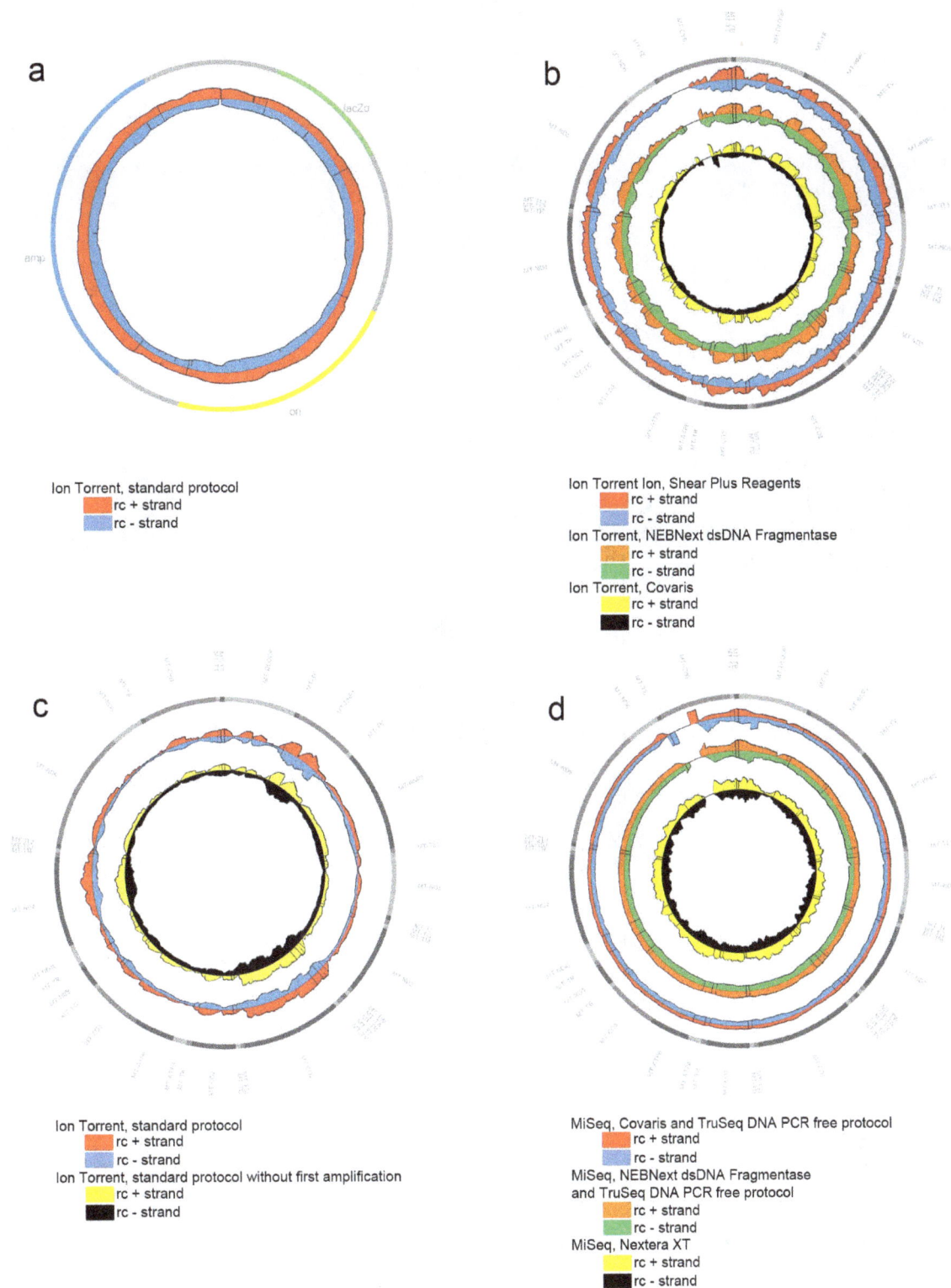

Figure 1. Genome Coverage plots. Representation of the MPS relative coverage of both strands (rc+: relative coverage of the plus strand, rc-: relative coverage of the negative strand) of the pUC19 plasmid, or mtDNA molecules obtained from the Ion Torrent PGM or MiSeq sequencing system. The outer circle symbolizes the pUC19 (A) or mtDNA (B, C, D) gene structure, respectively. **1A:** Use of the Ion Torrent PGM standard protocol on the pUC19 plasmid. **1B:** Use of three different fragmentation methods in combination with the Ion Torrent sequencing protocol on the mtDNA: Ion Shear Plus Reagents (enzymatic), NEBNext dsDNA Fragmentase (enzymatic) and Covaris (physical). **1C:** Use of an Ion Torrent PGM protocol without PCR amplification in the library construction on the mtDNA. **1D:** LR-PCR products of the mtDNA were Covaris (physical) or NEBNext dsDNA Fragmentase (enzymatic) sheared, followed by a TruSeq DNA PCR free protocol on a MiSeq instrument. The same six samples were processed with a Nextera XT kit (enzymatic shearing and PCR amplification in library preparation) prior to MiSeq analysis.

Table 1. Comparison between different methods and technologies based on relative coverage (RC) analysis of the data.

Ion Torrent PGM

RC	Standard			no library amplification			Covaris			NEBNext ds Fragmentase		
	Total	Plus	Min	Total	Plus	Min	Total	Plus	Min	Total	Plus	Min
<0.5	15.14	23.43	3.96	16.10	24.32	4.64	13.15	22.33	2.58	11.17	19.25	1.60
<0.25	1.36	13.12	0.02	1.10	13.73	0.09	0.88	12.70	0.35	0.27	11.52	0.29
<0.10	0.01	7.66	0.01	0.04	8.38	0.04	0.01	7.83	0.01	0.00	7.05	0.01
<0.05	0.00	5.47	0.01	0.00	6.02	0.01	0.00	5.81	0.01	0.00	4.95	0.00
<0.01	0.00	2.04	0.00	0.00	2.49	0.00	0.00	2.73	0.00	0.00	1.96	0.00

Illumina MiSeq

RC	TruSeq-Covaris			TruSeq-NEBNext ds Fragmentase			Nextera XT		
	Total	Plus	Min	Total	Plus	Min	Total	Plus	Min
<0.5	0.27	1.56	1.47	0.39	1.64	1.14	7.03	9.20	9.57
<0.25	0.01	0.81	0.83	0.01	0.73	0.42	1.64	2.61	2.48
<0.10	0.01	0.38	0.35	0.01	0.12	0.02	0.01	0.26	0.01
<0.05	0.00	0.23	0.20	0.00	0.01	0.00	0.01	0.06	0.01
<0.01	0.00	0.00	0.00	0.00	0.00	0.00	0.00	0.00	0.00

For all samples processed with a same protocol the average relative coverage was calculated and resulted in 7 different datasets. For each dataset, the fraction with a relative coverage <0.50, <0.25, <0.10, <0.05, <0.01 was determined. These analyses were performed for each strand separately (Plus, Min) and the total relative coverage (Total).

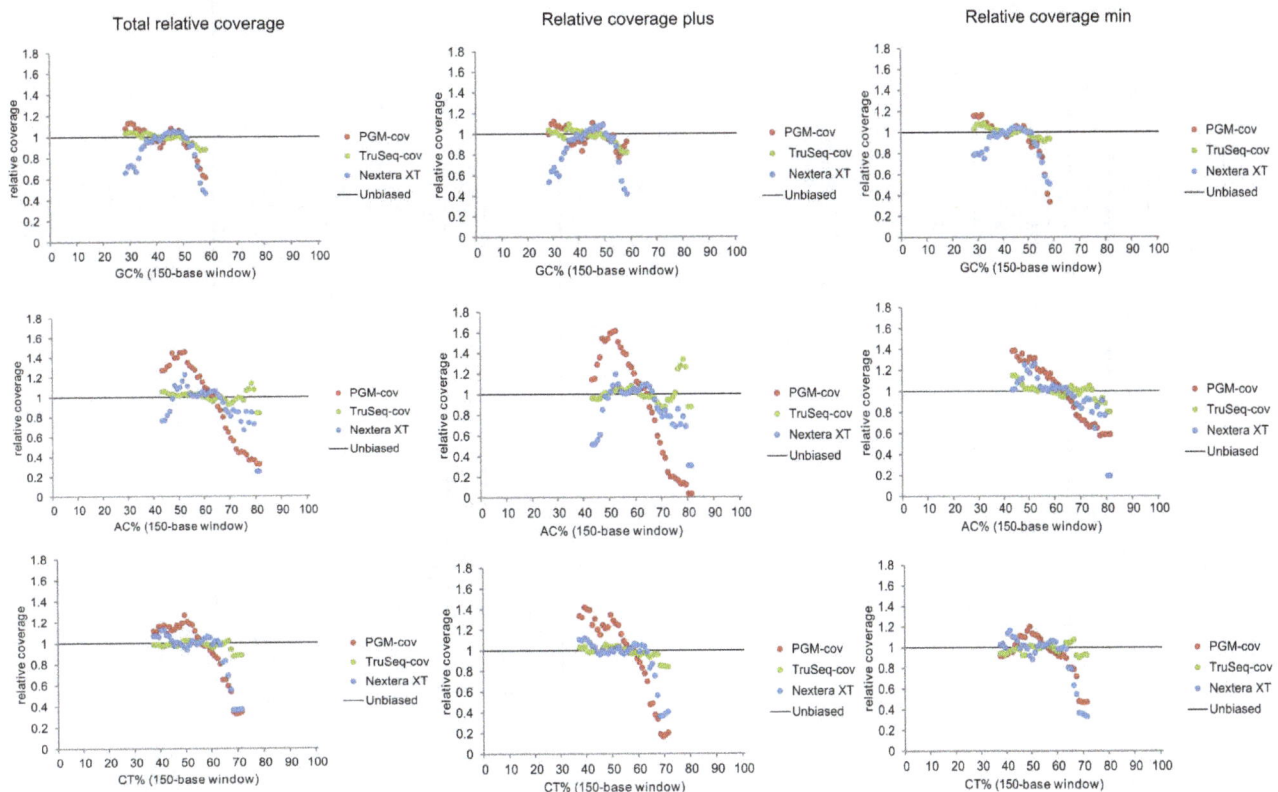

Figure 2. Nucleotide GC, AC and CT bias plots for the human mtDNA. The relative coverage as seen in this illustration is based on the average of the relative coverage of the six samples processed with the different protocols: Covaris shearing followed by the Ion Torrent protocol, Covaris shearing followed by the TruSeq procedure and the Nextera XT method. The average relative coverage was calculated for the total relative coverage and for both strand separately.

two or four C's at position 309. Analysis of the Ion Torrent PGM data previously had revealed four variants (m.7989T>C, m.9769T>C, m.10866T>C, m.12071T>C) hitherto not identified by Sanger sequencing. These same variants were identified by the Illumina system with analogous allele frequencies (Table 3). In this project, a detection threshold of 5% was used for all data analysis. However, it must be pointed out that a more stringent detection threshold of 2% is possible for both the PGM and MiSeq data. From a diagnostic perspective these low detection limits are not always relevant. Most pathogenic mutations have a disease threshold well >60%. Nonetheless, in the context of genetic counseling of asymptomatic female relatives for family planning, low detection limits might be indicated. Adjusting the detection limit to 2% in our sample cohort identified two additional heteroplasmic variants on the MiSeq platform. A novel heteroplasmic variant m.8207C>T (p.(Pro208Ser)) in the *MT-CO2* gene was revealed in the mtDNA of leukocytes of patient 9. Another heteroplasmic variant, m.5609T>C, was identified in leukocytes of patient 14 in the *MT-TA* gene. Both allele frequencies, 2% and 4% respectively, were below the applied detection limit of the PGM sequencer. Both nucleotide variants, however, were acknowledged by the PGM data as well, as was indicated by review of the BAM files in IGV and reanalysis of the data using a detection threshold of 0.8% (corresponding to the sequencing error rate of the PGM device). It must be pointed out that, although the accuracy of low level heteroplasmy determination is heavily dependent on the depth of coverage, it is also defined by the sequencing error of the system. The latter being related to

PCR, platform technologies, and the various algorithms implemented at the different steps of data processing.

Conclusion

MPS analysis is a powerful tool able to simultaneously detect and quantify sequencing variants. However, diagnostic settings have high demands regarding accuracy of test results. A high sensitivity is crucial to avoid a misdiagnosis, while a low false positive rate is necessary to minimize additional Sanger sequencing work for confirmation of pathogenic discoveries. Our current findings have illustrated that MPS protocols demand a thorough evaluation of their data, and validation of the result files before a possible implementation as a diagnostic test should be considered. In many laboratories MPS analysis is now part of daily diagnostic work. Selecting an appropriate methodology for MPS projects envisioned deserves the necessary attention. Assessment of the nucleotide content of DNA samples to be analyzed proved here to be an essential parameter, among others, for evaluation of the performance of a sequencing methodology or technology. In our hands, the current Ion Torrent PGM standard assay, even with modifications, suffered from lack of coverage consistency of the L-strand of the human mitochondrial genome, making an evaluation of heteroplasmy in these underrepresented regions cumbersome. Comparison of the PGM and MiSeq Nextera XT data results with the MiSeq PCR free sequencing method suggest that coverage bias might be generated by the enzymes involved in the amplification rounds of the MPS processes. Indeed, the Nextera XT method, which included a PCR amplification step, produced also more variation in coverage than the samples processed with

Table 2. Comparison between the number of variants detected with the MPS and Sanger sequencing technologies.

Sample	Sanger sequencing	PGM[a]									MiSeq[b]								
		Ion shear enzymes			Covaris			NEBNext dsDNA Fragmentase			Covaris			NEBNext dsDNA Fragmentase			Nextera XT		
		vs Sanger	FN	extra	vs Sanger	FN	extra	vs Sanger	FN	extra	vs Sanger	FN	extra	vs Sanger	FN	extra	vs Sanger	FN	extra
1	33	32	1	1	32	1	1	32	1	1	33	–	1	33	–	1	33	–	1
2	13	11	2	1	12	1	1	12	1	1	13	–	1	13	–	1	13	–	1
4	33	32	1	–	31	2	–	31	2	–	33	–	–	33	–	–	33	–	0
9	36	35	1	2	35	1	2	35	1	2	36	–	3	36	–	2	36	–	2
14	57	56	1	–	56	1	–	56	1	–	57	–	1	57	–	1	57	–	–
21	42	40	2	–	41	1	–	41	1	–	41	1	–	41	1	–	41	1	–
Total	**214**	**206**	**8**	**4**	**207**	**7**	**4**	**207**	**7**	**4**	**213**	**1**	**6**	**213**	**1**	**6**	**213**		

FN: false negative result, extra: additional low allele frequency variants identified compared to Sanger sequencing.
[a]Ion Torrent PGM results obtained with the TS4.2 software.
[b]MiSeq data obtained with the in-house GATK pipeline.

Table 3. Overview of the minor allele frequency (in %) of heteroplasmic variants detected in this study.

Sample	Variant	PGM					MiSeq				
		Ion shear	Covaris	NEBNext	average	stddev	Covaris	NEBNext	Nextera XT	average	stddev
1	m.12071T>C	12.2	12.2	12.1	12.2	0.1	12	12	23	15.7	6.4
2	m.7989T>C	17.1	14.7	14.1	15.3	1.6	19	19	13	17.0	3.5
9	m.9769T>C	9.7	9	8.5	9.1	0.6	8	8	8	8.0	0.0
9	m.10866T>C	7.3	6	6.4	6.6	0.7	7	7	7	7.7	1.2
9	m.8207C>T	1.5	1.3	1.5	1.4	0.1	2	1	2	1.7	0.6
14	m.5609T>C	4	8	4.8	5.6	2.1	4	4	4	4.0	0.0
14	m.7453G>A	52	55	53	53.3	1.5	53	54	56	54.3	1.5

the TruSeq DNA PCR-Free Sample Preparation protocol. Nextera XT certainly reduced, but did not resolve the coverage inconsistency. Further modifications, such as the use of another DNA polymerase in both amplification steps of the standard PGM workflow (one in the library preparation, and another in the emulsion PCR) may lead to further improvements. However, this might be a complex process and beyond the time management and financial scope of this project. Consequently, at this very moment the TruSeq DNA PCR-Free Sample Preparation protocol on the MiSeq system might be the most appropriate technology to address low copy number mtDNA heteroplasmy adequately.

Supporting Information

Supporting Information S1   Overview experiments.

Supporting Information S2   Nucleotide GC, AC and CT bias plots for the pUC19 plasmid. pUC19 DNA was processed with the Standard Ion Torrent protocol, Covaris

sheared followed by the TruSeq procedure and the Nextera XT method.

Supporting Information S3   Overview of the variant calling results obtained by the TS4.2 software for the Ion Torrent data and our in-house pipeline based on GATK for all MiSeq data and Ion Torrent PGM data. P: Ion Torrent PGM sequencing; T: TruSeq DNA PCR-Free Sample Preparation protocol; NXT: Nextera XT method; S: sample; Ion: Ion shear enzymes; C: Covaris; N: NEBNext dsDNA Fragmentase; TS4.2: analyzed with the Torrent Suite 4.2 software; GATK: analyzed with our in-house pipeline based on GATK.

Author Contributions

Conceived and designed the experiments: KV BC KS WL SS. Performed the experiments: KV BC. Analyzed the data: KV BC CS SS. Contributed reagents/materials/analysis tools: AJ LDM AV JS BDP RVC. Contributed to the writing of the manuscript: KV BC CS WL KS SS.

References

1. Ross MG, Russ C, Costello M, Hollinger A, Lennon NJ, et al. (2013) Characterizing and measuring bias in sequence data. Genome Biol 14: R51.
2. Dohm JC, Lottaz C, Borodina T, Himmelbauer H (2008) Substantial biases in ultra-short read data sets from high-throughput DNA sequencing. Nucleic Acids Res 36: e105.
3. Oyola SO, Otto TD, Gu Y, Maslen G, Manske M, et al. (2012) Optimizing illumina next-generation sequencing library preparation for extremely at-biased genomes. BMC Genomics 13: 1.
4. Seneca S, Vancampenhout K, Van Coster R, Smet J, Lissens W, et al. (2014) Analysis of the whole mitochondrial genome: translation of the Ion Torrent Personal Genome Machine system to the diagnostic bench? Eur. J. Hum. Genet. doi:10.1038/ejhg.2014.49.
5. Cheng S, Higuchi R, Stoneking M (1994) Complete mitochondrial genome amplification. Nat. Genet 7: 350–351.
6. Li H (2013) Aligning sequence reads, clone sequences and assembly contigs with BWA-MEM. Available: http://www.arxiv.org/abs/1303.3997. Accessed 29 April 2014.
7. Li H, Handsaker B, Wysoker A, Fennell T, Ruan J, et al. (2009) The Sequence Alignment/Map format and SAMtools. Bioinformatics 25: 2078–2079.
8. Krzywinski M, Schein J, Birol I, Connors J, Gascoyne R, et al. (2009) Circos: An information aesthetic for comparative genomics. Genome Res 19: 1639–1645.
9. Quail M, Smith ME, Coupland P, Otto TD, Harris SR, et al. (2012) A tale of three next generation sequencing platforms: comparison of Ion Torrent, pacific biosciences and illumina MiSeq sequencers. BMC Genomics 13: 341.

15

Mitochondrial Complex 1 Inhibition Increases 4-Repeat Isoform Tau by SRSF2 Upregulation

Julius Bruch[1,2◐], **Hong Xu**[1,2◐], **Anderson De Andrade**[1], **Günter Höglinger**[1,2]*

1 Department of Translational Neurodegeneration, German Centre for Neurodegenerative Diseases (DZNE), Munich, Germany, **2** Department of Neurology, Technische Universität München, Munich, Germany

Abstract

Progressive Supranuclear Palsy (PSP) is a neurodegenerative disorder characterised by intracellular aggregation of the microtubule-associated protein tau. The tau protein exists in 6 predominant isoforms. Depending on alternative splicing of exon 10, three of these isoforms have four microtubule-binding repeat domains (4R), whilst the others only have three (3R). In PSP there is an excess of the 4R tau isoforms, which are thought to contribute significantly to the pathological process. The cause of this 4R increase is so far unknown. Several lines of evidence link mitochondrial complex I inhibition to the pathogenesis of PSP. We demonstrate here for the first time that annonacin and MPP$^+$, two prototypical mitochondrial complex I inhibitors, increase the 4R isoforms of tau in human neurons. We show that the splicing factor SRSF2 is necessary to increase 4R tau with complex I inhibition. We also found SRSF2, as well as another tau splicing factor, TRA2B, to be increased in brains of PSP patients. Thereby, we provide new evidence that mitochondrial complex I inhibition may contribute as an upstream event to the pathogenesis of PSP and suggest that splicing factors may represent an attractive therapeutic target to intervene in the disease process.

Editor: Oscar Arias-Carrion, Hospital General Dr. Manuel Gea González, Mexico

Funding: J.B. was funded by the Bavarian Research Foundation (Bayerische Forschungsstiftung), H.X. by the DAAD (German Academic Exchange Service), G.U.H. by the DFG (Deutsche Forschungsgemeinschaft, HO2402/6-2). The funders had no role in study design, data collection and analysis, decision to publish, or preparation of the manuscript.

Competing Interests: The authors have declared that no competing interests exist.

* Email: hoeglinger@lrz.tum.de

◐ These authors contributed equally to this work.

Introduction

Tauopathies are a heterogeneous group of neurodegenerative diseases with the common feature of intracellular aggregation of the microtubule associated protein tau. They include, but are not limited to, Alzheimer's Disease, Progressive Supranuclear Palsy (PSP), Argyrophilic Grain Disease (AGD), Corticobasal Degeneration (CBD), Pick's Disease and some other forms of frontotemporal dementias. Different tauopathies vary significantly in their clinical and pathological phenotype [1].

In the human central nervous system there are six predominant splicing variants of the *MAPT* gene, encoding tau proteins. These depend on the exclusion or inclusion of exons 2, 3 and 10: 3R0N, 3R1N, 3R2N, 4R0N, 4R1N and 4R2N [2]. 0N signifies the inclusion of neither exon 2 or 3. 1N denotes the inclusion of exon 2 but not 3, whilst 2N denotes the inclusion of both exons 2 and 3. 3R denotes the absence of exon 10, 4R its presence. Exon 10 codes for an additional microtubule binding repeat, so that 4R isoforms have 4 binding repeats, whilst 3R isoforms have only 3.

Across different tauopathies the isoform constitution varies. A common classification of tauopathies, therefore, is between the 3R isoform and the 4R isoform tauopathies [3]. While in healthy adults and in Alzheimer's disease 3R and 4R isoforms are generally in balance, PSP, CBD and AGD feature a relative excess of 4R isoforms [4]. Pick's Disease, conversely, has a

relative excess of 3R isoforms. This imbalance is thought to play a major role in the pathogenesis of these tauopathies [5]. 4R isoforms are more prone to aggregation than 3R isoforms [5]. A single mutation in the *MAPT* gene affecting the inclusion of exon 10 to favour generation of 4R tau appears to be sufficient to trigger a tauopathy [6]. This has led to the hypothesis that an excess of 4R tau may be significantly pathogenic. Therefore, reducing the relative amount of 4R may be a strategy for therapy in 4R tauopathies [5,7].

Alternative splicing of exon 10 is regulated by a combination of *cis*-elements in exon 10 and intron 10, as well as by *trans*-acting factors [2]. It is through these *trans*-acting factors that alternative splicing can be modified and regulated by the cell. They are divided into heterogeneous nuclear ribonucleoproteins (hnRNPs) and serine/arginine-rich (SR) proteins or SR-like proteins. The SR proteins participate in the spliceosome and are thus involved in both constitutive splicing and the regulation of alternative splicing [8]. They are controlled through phosphorylation and acetylation and have been discussed as a potential drug target in the context of cancer treatment [9,10]. However, so far, the molecular mechanisms leading to preferential generation of 4R tau by alternative splicing of wild-type tau in sporadic 4R tauopathies are not understood.

There are several lines of evidence suggesting a role for dysfunction of the mitochondrial respiratory chain, particularly of mitochondrial complex I, in the pathogenesis of PSP. A study using transmitochondrial cytoplasmic hybrid (cybrid) cell lines expressing mitochondrial genes from persons with PSP found complex I activity to be reduced [11]. Dysfunctional complex I is a major emitter of reactive oxygen species [12] and evidence of oxidative stress has been found in autopsy material of PSP patients [13,14]. A study using combined phosphorus and proton magnetic resonance spectroscopy has identified evidence for cerebral depletion in high-energy phosphates and increased lactate levels in PSP, a pattern compatible with a primary failure of the mitochondrial respiratory chain [15]. Finally, there is also an epidemiological association between the consumption of soursop fruit containing the mitochondrial complex I inhibitor annonacin [16] and a PSP-like tauopathy on the island of Guadeloupe [17]. Annonacin has been shown to induce a tauopathy *in vitro* in cultured neurons [16,18], as well as *in vivo* [19]. So far described are four effects of annonacin that are typical features for tauopathies, namely increased tau protein levels, tau hyperphosphorylation, redistribution of tau from the axons to the somatodendritic compartment, and eventual cell death [18]. Here, we explore the effect of complex I inhibition on the alternative splicing of tau.

Materials and Methods

Cell Culture

Nunc Nunclon Delta 6-well (for protein and mRNA) or 48-well (for cell assays) plates (Thermo Fisher Scientific, Waltham, MA, USA) were coated with 100 µg/ml poly-L-lysine (Sigma-Aldrich, St. Louis, MO, USA) and 5 µg/ml fibronectin (Sigma-Aldrich). LUHMES (Lund Human Mesencephalic) cells, derived from female human embryonic ventral mesencephalic cells by conditional immortalization [20] (Tet-off v-myc over-expression) were seeded in a concentration of 130,000 cells/cm^2 to achieve a confluence of 50%. They were then differentiated for 8 days in a medium of DMEM/F12 (Sigma-Aldrich), 1 µg/ml Tetracycline, 2 mg/ml GDNF and 490 µg/ml dbcAMP into post-mitotic neurons [21] with a dopaminergic phenotype [20]. On day 8 post differentiation the cells were treated with 25 nM annonacin, 20 µM 6-OHDA or 10 µM MPP$^+$ for 48 h. For the intoxication period the medium was replaced with new medium containing glucose levels reduced to 250 µM, i.e. the physiological concentration in the human brain [22]. For the starving condition, cells were incubated for 24 hours in pure DMEM (Life Technologies, Grand Island, NY, USA) with no additives and no glucose.

Human Brain Tissue and Ethics Statement

Human fresh frozen brain sections of the *locus coeruleus* area were obtained from The Netherlands Brain Bank, Netherlands Institute for Neuroscience, Amsterdam (www.brainbank.nl). All Material has been collected from donors for or from whom written informed consent for a brain autopsy and the use of the material and clinical information for research purposes had been obtained by The Netherlands Brain Bank in accordance with the Declaration of Helsinki.

Quantitative Real-Time PCR

RNA from human tissue samples was extracted by grinding the tissue in liquid nitrogen to a powder and then dissolving it in the RA1 buffer supplied as part of the NucleoSpin RNA (Macherey Nagel, Düren, Germany) RNA extraction kit +1% (v/v) 2-Mercaptoethanol (Sigma-Aldrich). RNA from cells was extracted by scraping the cells from the culture plate with RA1

buffer +1% (v/v) 2-Mercaptoethanol. The remaining extraction procedure was according to the manufacturer's instructions for the NucleoSpin RNA kit. RNA concentrations were determined using the NanoDrop 2000c Spectrophotometer (Thermo Fisher Scientific). The RNA was then transcribed into cDNA with the iScript cDNA Synthesis Kit (BioRad, Berkeley, CA, USA) using the manufacturer's instructions. Real-Time PCR was performed on the Applied Biosystems StepOnePlus (Life Technologies) system using TaqMan Universal Master Mix II and TaqMan primers against total *MAPT, MAPT 0N, MAPT 1N, MAPT 2N, MAPT 3R, MAPT 4R, SRSF1, SRSF2, SRSF3, SRSF6, SRSF7, SRSF9, SRSF11* and *TRA2B. PSMC1* and *POL2A* were used as reference genes for relative quantification in all tau splicing factor experiments, while *PPIB* and *GAPDH* were used in all tau isoform experiments as they were determined to be the most stably expressed across the respective experimental conditions. All values are relative quantities compared to untreated (control) cells. Three biological repeats with three technical repeats each were analysed. Analysis was conducted with the Applied Biosystems StepOnePlus (Life Technologies) and Qbase+ (Biogazelle, Zwijnaarde, Belgium) software packages. Absolute quantification was performed by creating a standard curve with plasmids containing either the 2N3R or the 2N4R spliced variant of *MAPT* (obtained as a gift from Eva-Maria Mandelkow, DZNE Bonn, Germany). The absolute quantity was computed by deriving the relationship between CT values and absolute quantity with the StepOne Plus software.

Western Blotting

Protein was extracted from cells using the M-PER Mammalian Protein Extraction Reagent (Thermo Fisher Scientific). The protein solution was frozen at $-80°C$ immediately after retrieval and for a minimum of two hours. The solution was then thawed on ice, vortexed, centrifuged at 5000 g for 15 minutes at 4°C and the supernatant retrieved. Total protein concentrations were determined using the BCA kit (Thermo Fisher Scientific) by heating the samples at 60°C for 30 minutes and measuring the absorption on the NanoDrop 2000c Spectrophotometer (Thermo Fisher Scientific). 20 µg of total protein were then adjusted to equal concentrations between samples by dilution with M-PER and subsequently heated at 95°C for 5 minutes with 1× Roti-Load loading buffer (Carl Roth, Karlsruhe, Germany). SDS-PAGE was performed using precast Gels (anyKD, Bio-Rad) in a tris-glycine running buffer (14.4% glycine, 3% Tris, 1% SDS w/v, Carl Roth). The protein was blotted onto PVDF membrane (Bio-Rad) at 70 V for 65 minutes. The membrane was blocked with 1× Roti-Block solution (Carl Roth) for 1h and then incubated at 4°C overnight under gentle shaking with the primary antibody (table 1) in TBS with 5% BSA (Cell Signaling, Danvers, MA, USA) and 0.05% TWEEN (Sigma-Aldrich). The membranes were then washed and incubated with the appropriate secondary antibody at 1:2500 (v/v) in 1× Roti-Block solution for 2 h, followed by further washing and exposure to Clarity Western ECL Substrate (Bio-Rad) or, in the case of 4-repeat tau, to **ECL solution** (General Electric, Fairfield, CT, USA). Chemiluminescence was detected with the Gel image System (Bio-Rad) and analysed by background subtracted optical density analysis with ImageLab software (Bio-Rad).

siRNA Silencing

LUHMES cells were seeded out and differentiated as described above and allowed to adhere to the plate floor for 4 h. siRNA (Sigma-Aldrich) targeted against SRSF2 (final concentration 200 nM) and Lipofectamine RNAiMAX (Life Technologies) (final concentration 1.2 µl/ml) were dissolved in separate aliquots of

Table 1. Primary Antibody Concentrations Used.

Antigen	Clone	Species	Concentration (v/v)	Company
Human tau	HT7	Mouse	1:1000	Pierce Antibodies, Thermo
3-repeat tau	8E6/C11	Mouse	1:500	Millipore
4-repeat tau	1E1/A6	Mouse	1:300	Millipore
Actin (I-19)	Polyclonal	Goat	1:2500	Santa Cruz Biotechnologies

medium (OptiMEM, Life Technologies). The diluted siRNA was then added to the diluted Lipofectamine RNAiMAX. The combined solution was then allowed to incubate for 20 minutes before being added to the cells.

ATP Assay

ATP assays were conducted using the ATP test kit by Lonza according to the manufacturer's instructions. Luminescence was read with the FLUOstar Omega platereader (BMG Labtech). The data was analysed using the MARS Data Analysis Software (BMG Labtech).

MTT Assay

Thiazolyl Blue Tetrazolium Blue (MTT) (Sigma Aldrich) was dissolved in sterile PBS to a concentration of 5 mg/ml. This stock solution was added to the cells in culture medium to achieve a final concentration of 0.5 mg/ml. The 48-well culture plate was then incubated at 37°C for 1 h, the medium removed completely and frozen at −80°C for 1 h. The plate was then thawed, 300 μl DMSO (AppliChem, Darmstadt, Germany) was added per well and the plate was shaken to ensure complete dissolution of the violet crystals. 100 μl from each well were transferred to a new 96-well plate and the absorbance was read with the platereader at a wavelength of 590 nm (reference wave length 630 nm). The data was analysed using the MARS Data Analysis Software (BMG Labtech).

Statistics

Prism 6 (GraphPad Software, La Jolla, CA, USA) was used for statistical calculations and for the creation of line and bar graphs. Results were compared by 2-way ANOVA with Sidak post-hoc test, unless stated otherwise. Data are shown as mean ± SEM. P< 0.05 was considered significant.

Results

Annonacin Causes an Upregulation of the Tau 4-Repeat Isoform

We first characterized expression of tau isoforms in LUHMES cells, a cell culture line of human mesencephalic neurons, derived from female human embryonic ventral mesencephalic cells by conditional immortalization (Tet-off v-myc over-expression) [21]. These cells start expressing the 4-repeat (4R) isoform of tau from day 8 post differentiation into a neuronal phenotype. On day 10, 4R-spliced mRNA makes up 3.9% ±0.3 (n = 3) of the total MAPT mRNA. We used this human neuronal model for the present work as rodent cells express only 4R tau.

When treated with annonacin at a concentration of 25 nM for 48 h from days 8 to 10 post differentiation, LUHMES cells remain 60.7±0.4% viable (MTT assay) with an ATP concentration of 64±1% compared to that of untreated cells (figure 1A).

Under these conditions we observed the mRNA of the 4R isoforms of tau to be upregulated significantly (figure 1B) compared with untreated cells, as determined by quantitative PCR. There was no significant change in the relative quantity of 3R isoforms. The level of inclusion of exons 2 and 3 also did not change significantly, although there was a slight increase in the amount of 0N isoforms. This indicates that annonacin selectively increases inclusion of exon 10 with no or little relative effect on the alternative splicing of the other exons.

We also observed an upregulation of the 4R tau isoforms on the protein level by Western blot (figure 1C). 3R tau was again not significantly changed. The level of upregulation at the protein level is very similar to that on the mRNA level, suggesting a tight correlation between the regulation of alternative splicing and the isoform distribution seen at the protein level. There was no significant increase in the amount of total tau with annonacin, probably due to the greater proportion of 3R isoforms in LUHMES cells of this age.

The Splicing Factor SRSF2 Is Necessary for Annonacin-Mediated Alternative Splicing

We next explored the mechanism of how annonacin induces this isoform change. We tested 10 splicing factors known to influence the inclusion or exclusion of exon 10 in the MAPT gene [2] by quantitative PCR. An overview of the splicing factors tested is shown in table 2.

We found SRSF2 to be the only splicing factor significantly upregulated with annonacin treatment that is known to promote the inclusion of exon 10 (figure 2A). This prompted us to explore whether SRSF2 has a functional role to play in annonacin mediated 4R upregulation. We knocked down SRSF2 with siRNA starting from 6 hours post differentiation in LUHMES cells and treated these cells with annonacin from days 8–10, as in the previous experiments. At day 10, SRSF2 was reduced by half. In spite of this incomplete silencing efficiency, the 4R isoform of MAPT in annonacin treated cells was reduced dramatically compared to untreated cells (figure 2B). This suggests that SRSF2 plays a critical role in the upregulation of the 4R MAPT isoforms seen upon annonacin treatment.

More Splicing Factors Are Upregulated in Human PSP

We also tested splicing factor expression levels in human brain tissue of the *locus coeruleus* of 4 PSP patients and five control patients free of psychiatric or neurodegenerative diseases (table 3). This time, however, we limited our analysis to those splicing factors known to increase MAPT exon 10 inclusion. We confirmed the increase of the 4R isoform in the PSP patients compared to the controls (figure 2C). Expression of the splicing factors SRSF2 and TRA2B was also increased significantly. This suggests that the increase in 4R isoforms seen with annonacin treatment may

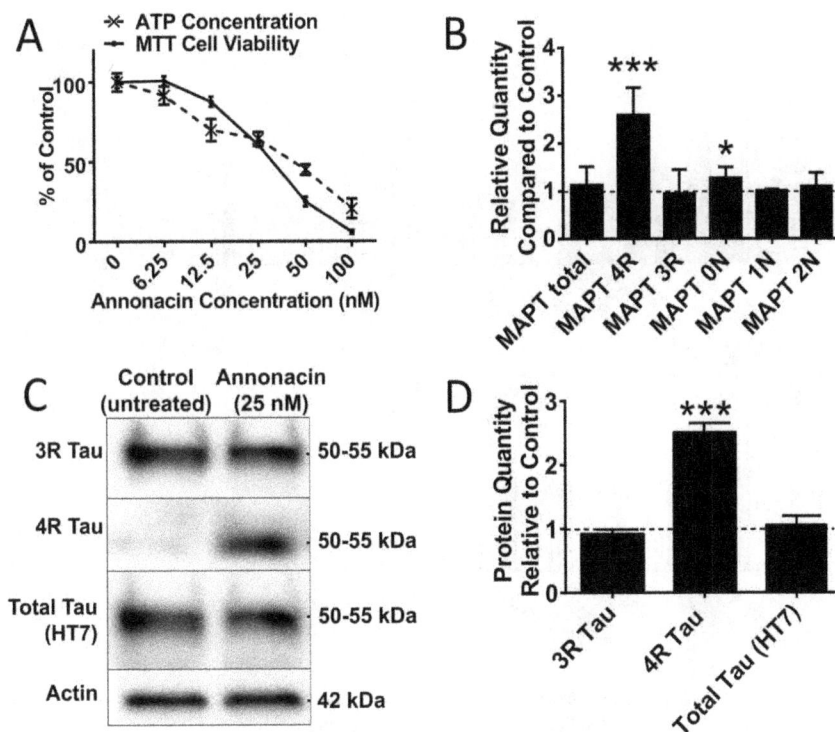

Figure 1. Annonacin Causes an Upregulation of the 4R Isoforms of Tau. A) LUHMES neurons were treated with different concentrations of annonacin for 48 h from day 8–10 post differentiation (n = 12). The MTT test, a measure for mitochondrial reducing function, and ATP concentration, are expressed as a relative percentage compared to untreated control cells. B) 4R isoform (exon 10) mRNA is upregulated with annonacin treatment. Quantitative PCR results showing the relative quantity of mRNA for different *MAPT* splicing variants in cells treated with 25 nM annonacin for 48 h from day 8–10 post differentiation compared to untreated cells (dotted line). 3 biological repeats with 3 technical repeats each. ***: p<0.001, *: p< 0.05 vs. untreated cells (2-way ANOVA with Sidak's post-hoc test). C) 4R isoform protein is upregulated with annonacin treatment. Western blot for 3R and 4R isoforms of tau protein, as well as total tau (detected with the HT7 antibody). LUHMES cells were either left untreated or treated with 25 nM annonacin. Actin was used as loading control. D) Quantification of figure 1C. Results show the relative quantity (fold-change) compared to untreated control cells (relative quantity = 1, represented by dotted line). 3 biological repeats. ***: p<0.001 vs. untreated cells (2-way ANOVA with Sidak's post-hoc test).

account partly for the mechanism by which 4R isoform tau is upregulated in PSP.

4R Tau Upregulation Occurs with Other Complex I Inhibitors but Not Oxidative Stress

We tested whether 4R isoform upregulation upon annonacin treatment is a non-specific consequence of neuronal injury,

specific to mitochondrial complex I inhibition or even more specific to annonacin. We therefore repeated the experiment with 1-methyl-4-phenylpyridinium (MPP+), another complex I inhibitor, 6-hydroxydopamine (6-OHDA), a neurotoxin known to be neurotoxic primarily through oxidative stress [23] and by starving the cells of glucose and nutrients. As shown in figures 3 A and B, a comparable level of toxicity and ATP reduction to

Table 2. Overview of the splicing factors known to influence MAPT exon 10 alternative splicing.

Splicing factor	Target *cis*-element	Effect on exon 10 splicing
SRSF1 (SRp30a, ASF)	PPE	Inclusion
SRSF2 (SRp30b, SC35)	SC35-like	Inclusion
SRSF3 (SRp20)	ND	Exclusion
SRSF4 (SRp75)	ND	Exclusion
SRSF6 (SRp55)	ND	Exclusion
SRSF7 (9G8)	ISS	Exclusion
SRSF9 (SRp30c)	ND	Inclusion
SRSF11 (SRp54)	PPE	Exclusion
TRA2B	PPE	Inclusion

Source: Adapted from [2].

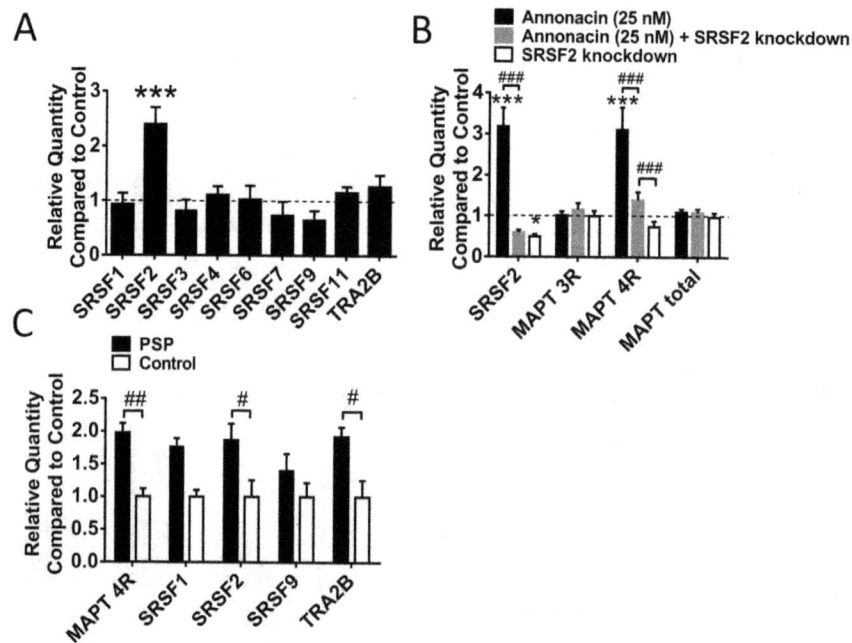

Figure 2. SRSF2 is a Critical Player in Annonacin Mediated Tau Alternative Splicing. A) Quantitative PCR results for 9 different splicing factors known to have an effect on exon 10 alternative splicing (table 2). Data shown are relative quantities compared to untreated cells (dotted line). Only SRSF2 was elevated significantly with annonacin treatment. All other splicing factors tested were not significantly elevated. 3 biological repeats with 3 technical repeats each. ***: p<0.001, *: p<0.05 vs. untreated control (2-way ANOVA with Sidak's post-hoc test). B) Quantitative PCR results for LUHMES cells on day 10 post differentiation treated with SRSF2 knockdown siRNA for 10 days and/or with annonacin for 48 h. 3 biological repeats with 3 technical repeats each. ***: p<0.001 vs. untreated control (dotted line); ###: p<0.001 (2-way ANOVA with Sidak's post-hoc test). C) Quantitative PCR results for the 4 splicing factors known to increase MAPT exon 10 inclusion in *locus coeruleus* tissue of four PSP patients and five controls without neurodegenerative diseases. 3 biological repeats with 3 technical repeats each. #: p<0.05, ##: p<0.01 (2-way ANOVA with Sidak's post-hoc test).

that of 25 nM annonacin is achieved by 6-OHDA at a concentration of 20 µM (22% ATP reduction) and by MPP+ at a concentration of 10 µM (37% ATP reduction). Therefore, we decided to use these concentrations to test the MAPT isoform changes with these toxins.

With MPP+ treatment we observed a significant increase in exon 10 inclusion on the mRNA level by qPCR (figure 3C) and in the levels of 4R tau isoforms by Western blot (figure 4A, B) compared to controls, as with annonacin. With 6-OHDA treatment and with starvation we only observed a slight reduction

in both 4R and 3R isoforms. In all three cases the inclusion of exons 2 and 3 did not increase (data not shown). This would suggest that complex I inhibition in general and not oxidative stress or neuronal suffering per se is responsible for the increased level of exon 10 inclusion observed with annonacin.

Finally, we explored the role of *SRSF2* in these observations. We found that MPP+ also acts via SRSF2 upregulation and that there is no *SRSF2* upregulation with 6-OHDA treatment or starvation.

Table 3. Overview of Human Tissue Used.

Case Number	Diagnosis	Cause of Death	Age at death	Braak Stage	Sex	Postmortem delay (hours: minutes)
P1	PSP	"Natural death"	73	2C	Male	4:20
P2	PSP	Acute heart failure	70	3	Male	6:50
P3	PSP	Aspiration pneumonia	73	2	Male	6:15
P4	PSP	Urinary tract infection	70	1A	Male	5:20
C1	Non-demented control	Pancreas carcinoma	70	0	Male	7:30
C2	Non-demented control	Prostate cancer	69	0	Male	5:55
C3	Non-demented control	Lung emboli (clinical suspicion)	73	0	Male	24:45
C4	Non-demented control	Sepsis	71	1	Male	7:40
C5	Non-demented control	Myocardial infarction	67	1B	Male	18:35

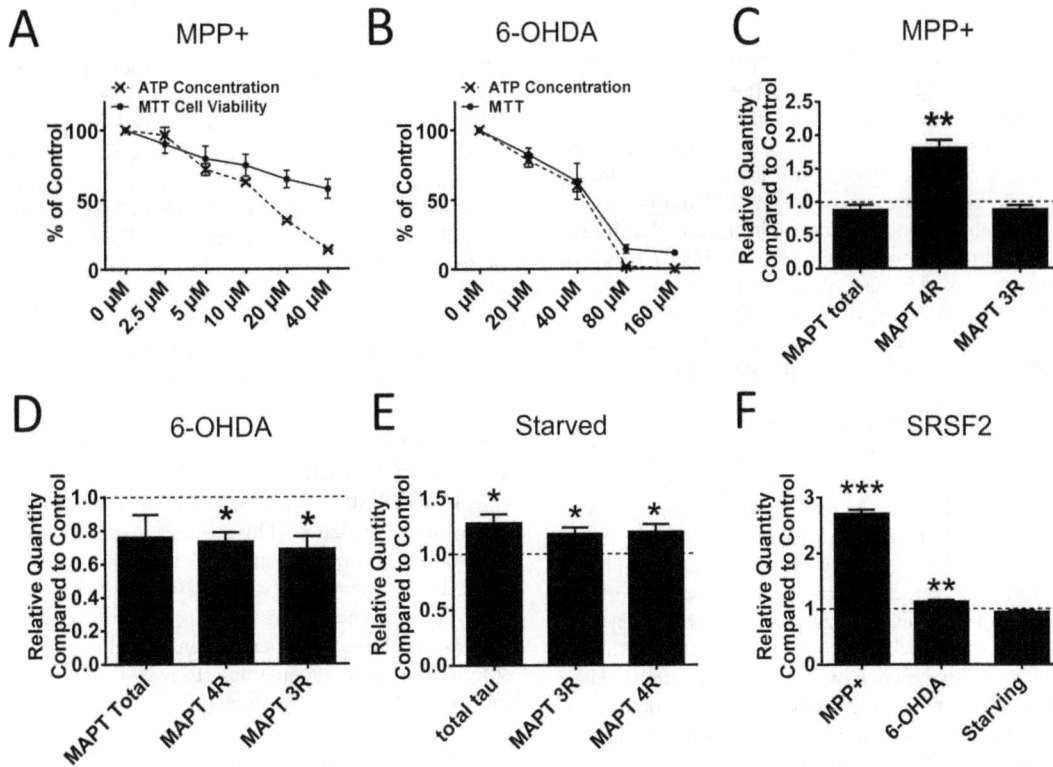

Figure 3. The 4R Isoform Shift Can Be Reproduced with Another Complex I Inhibitor but not with the Oxidative Stressor 6-OHDA. A) ATP concentration and MTT cell viability in LUHMES cells as measured by the MTT assay for different concentrations of 6-OHDA. Treatment was for 48 h from day 8–10 post differentiation. n = 12. B) ATP concentration and MTT cell viability in LUHMES cells as measured by the MTT assay for different concentrations of MPP+. Treatment was for 48 h from day 8–10 post differentiation. n = 12. C) Quantitative PCR results of MAPT splicing variants for LUHMES cells treated with 10 μM MPP$^+$ for 48 h from day 8–10 post differentiation. 3 biological repeats with 3 technical repeats each. **: p<0.01 vs. untreated cells (dotted line), (2-way ANOVA with Sidak's post-hoc test). D) Quantitative PCR results of MAPT splicing variants for LUHMES cells treated with 20 μM 6-OHDA for 48 h from day 8–10 post differentiation. 3 biological repeats with 3 technical repeats each. *: p<0.05 vs. untreated cells (dotted line), (2-way ANOVA with Sidak's post-hoc test). E) Quantitative PCR results of MAPT splicing variants for LUHMES cells starved of nutrients and glucose for 24 h from day 8–9 post differentiation. 3 biological repeats with 3 technical repeats each. *: p<0.05 vs. untreated cells (dotted line), (2-way ANOVA with Sidak's post-hoc test). F) Quantitative PCR results of SRSF2 for LUHMES cells treated with 10 μM MPP$^+$ or 20 μM 6-OHDA for 48 h from day 8–10 post differentiation or starved for 24 h from day 8–9 post differentiation. 3 biological repeats with 3 technical repeats each. *: p<0.05, **: p<0.01 vs. untreated cells (dotted line), (2-way ANOVA with Sidak's post-hoc test).

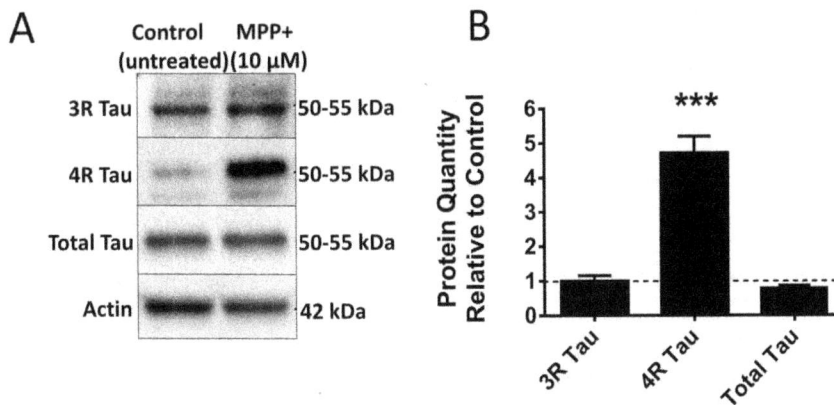

Figure 4. The 4R Isoform was upregulated in the protein level with MPP$^+$ treatment. A) 4R isoform protein is upregulated with MPP$^+$ treatment. Western blot for 3R and 4R isoforms of tau protein, as well as total tau (detected with the HT7 antibody). LUHMES cells were either left untreated or treated with 10 μM MPP$^+$. Actin was used as loading control. B) Quantification of figure 1G. Results show the relative quantity compared to the untreated control cells (dotted line). 3 biological repeats. ***: p<0.001 vs. untreated cells (2-way ANOVA with Sidak's post-hoc test).

Discussion

Mitochondrial Complex I Inhibition Reproduces the 4R Isoform Shift Seen in Several Tauopathies

In this paper we have been able to add the increase in 4R tau isoforms as an additional feature to the list of characteristics of PSP that annonacin treatment reproduces in cell culture. This makes annonacin treated neurons a good model for PSP and potentially other sporadic 4R tauopathies. It is unique in the fact that it does not rely on any genetic modification of the *MAPT* gene or artificial overexpression. The fact that it reliably produces an increase in the 4R tau isoforms would also allow it to be used to screen, test and develop candidate drugs targeting tau alternative splicing – something that would not be possible with overexpression-based models of tauopathy.

However, the effect on alternative splicing is not specific to annonacin. Rather, it seems to be related to mitochondrial complex I inhibition more generally. This is suggested by the fact that we have observed the same increase in 4R tau isoforms with MPP+, another complex I inhibitor. In fact, other features of tauopathy have also been reproduced by other complex I inhibitors [24,25]. However, due to the epidemiological evidence from Guadeloupe strongly linking annonacin consumption to a PSP-like tauopathy, annonacin makes a particularly convincing case as a cell culture based model for PSP. The only drawback of this model relying on immature human neurons is that despite the upregulation of 4R tau, after 10 days there still appears to be overall more 3R than 4R tau, whereas in adult human brain neurons, 3R and 4R are more or less balanced.

However, it is not yet fully understood to what extent the relative increase in the 4R tau isoform contributes to neurotoxicity or impairment of neural functioning. 4R tau isoform increases are only seen in a selection of tauopathies and are region specific. In Alzheimer's disease, there is no abnormal upregulation of 4R isoforms. In PSP, there is some evidence that the 4R isoform may not be upregulated in the frontal cortex, despite the existence of tau pathology in this region [26]. On the other hand, patients with FTDP-17 due to mutations that exclusively affect tau alternative splicing and result in an increase of 4R tau, are evidence that an upregulation of the 4R isoforms is sufficient to start tau aggregation [2,6].

SRSF2 Forms the Link Between Complex I Inhibitors and the Increase in 4R Isoforms

We have identified SRSF2 as a mediator essential for mitochondrial complex I inhibitor induced exon 10 inclusion. The fact that a knockdown of SRSF2 reverses the annonacin induced increase in 4R tau confirms that SRSF2 plays a necessary role for this isoform shift.

SRSF2 is controlled by several kinases including SRPK, AKT, topoisomerase I and CLK/STY family kinases, as well as lysine acetylation [27]. The histone deacetylase inhibitor sodium butyrate has already been demonstrated to increase SRSF2 levels [28], whilst the kinase activity of topoisomerase I can be inhibited with the antitumour drug NB-506 [9]. This suggests that, at least indirectly, SRSF2 is a potentially drugable target.

In our annonacin-treated cell cultures, which might be considered to be an acute model of a sporadic tauopathy, inhibition of SRS2 prevented the 4R isoform shift of tau but not the cell death induced by annonacin. This suggests that, in this model, the 4R tau is not necessary for cell death, since neurons might rather die from reduced energy production [18]. This does, however, not exclude that in a more chronic situation with even higher levels of 4R tau this isoform shift may become the predominant cause of neuronal dysfunction and death.

Complex I Inhibition Is Unlikely to Explain All of the Increase in 4R Isoforms in PSP

In human PSP patients both the SRSF2 and TRA2B splicing factors are upregulated. This suggests that the 4R upregulation is not exclusively due to complex I inhibition, as in that case we would have expected only SRSF2 to be upregulated. Therefore, exploring upstream events leading to TRA2B upregulation may lead to insights on further reasons for the increase in 4R tau isoforms in some tauopathies. It would also be interesting to compare the splicing factor expression levels in 3R tauopathies versus 4R tauopathies.

If SRSF2 is confirmed to be a key player in mediating the 4R isoform upregulation in PSP and other 4R tauopathies, this would make it a suitable drug target for reducing this isoform shift.

Conclusion

In summary, we can conclude that SRSF2 is a necessary mediator for mitochondrial complex I inhibitor induced tau 4R isoform upregulation. As SRSF2 is also increased in PSP patients this suggests mitochondrial complex I inhibition may play at least a partial role in the pathogenesis of 4R tauopathies such as PSP. However, other mechanisms are also likely to contribute.

Acknowledgments

Mr Robin Konhäuser and Ms Magda Baba were instrumental in maintaining the cell lines and cell culture. We would like to acknowledge the Netherlands Brain Bank for their generous contribution of the human brain tissue samples.

Author Contributions

Conceived and designed the experiments: JB HX GUH. Performed the experiments: JB HX. Analyzed the data: JB HX ADA. Contributed reagents/materials/analysis tools: ADA. Wrote the paper: JB HX GUH.

References

1. Williams DR (2006) Tauopathies: classification and clinical update on neurodegenerative diseases associated with microtubule-associated protein tau. Intern Med J 36: 652–660.
2. Liu F, Gong CX (2008) Tau exon 10 alternative splicing and tauopathies. Mol Neurodegener 3: 8.
3. Chen S, Townsend K, Goldberg TE, Davies P, Conejero-Goldberg C (2010) MAPT isoforms: differential transcriptional profiles related to 3R and 4R splice variants. J Alzheimers Dis 22: 1313–1329.
4. Buee L, Delacourte A (1999) Comparative biochemistry of tau in progressive supranuclear palsy, corticobasal degeneration, FTDP-17 and Pick's disease. Brain Pathol 9: 681–693.
5. Zhou J, Yu Q, Zou T (2008) Alternative splicing of exon 10 in the tau gene as a target for treatment of tauopathies. BMC Neurosci 9 Suppl 2: S10.
6. Spillantini MG, Murrell JR, Goedert M, Farlow MR, Klug A, et al. (1998) Mutation in the tau gene in familial multiple system tauopathy with presenile dementia. Proc Natl Acad Sci U S A 95: 7737–7741.
7. Avale ME, Rodriguez-Martin T, Gallo JM (2013) Trans-splicing correction of tau isoform imbalance in a mouse model of tau mis-splicing. Hum Mol Genet 22: 2603–2611.
8. Will CL, Luhrmann R (2011) Spliceosome structure and function. Cold Spring Harb Perspect Biol 3.
9. Pilch B, Allemand E, Facompre M, Bailly C, Riou JF, et al. (2001) Specific inhibition of serine- and arginine-rich splicing factors phosphorylation, spliceosome assembly, and splicing by the antitumor drug NB-506. Cancer Res 61: 6876–6884.

10. Zhong XY, Ding JH, Adams JA, Ghosh G, Fu XD (2009) Regulation of SR protein phosphorylation and alternative splicing by modulating kinetic interactions of SRPK1 with molecular chaperones. Genes Dev 23: 482–495.

11. Swerdlow RH, Golbe LI, Parks JK, Cassarino DS, Binder DR, et al. (2000) Mitochondrial dysfunction in cybrid lines expressing mitochondrial genes from patients with progressive supranuclear palsy. J Neurochem 75: 1681–1684.

12. Lenaz G, Baracca A, Fato R, Genova ML, Solaini G (2006) Mitochondrial Complex I: structure, function, and implications in neurodegeneration. Ital J Biochem 55: 232–253.

13. Stamelou M, de Silva R, Arias-Carrion O, Boura E, Hollerhage M, et al. (2010) Rational therapeutic approaches to progressive supranuclear palsy. Brain 133: 1578–1590.

14. Albers DS, Swerdlow RH, Manfredi G, Gajewski C, Yang L, et al. (2001) Further evidence for mitochondrial dysfunction in progressive supranuclear palsy. Exp Neurol 168: 196–198.

15. Stamelou M, Pilatus U, Reuss A, Magerkurth J, Eggert KM, et al. (2009) In vivo evidence for cerebral depletion in high-energy phosphates in progressive supranuclear palsy. J Cereb Blood Flow Metab 29: 861–870.

16. Lannuzel A, Michel PP, Hoglinger GU, Champy P, Jousset A, et al. (2003) The mitochondrial complex I inhibitor annonacin is toxic to mesencephalic dopaminergic neurons by impairment of energy metabolism. Neuroscience 121: 287–296.

17. Lannuzel A, Hoglinger GU, Verhaeghe S, Gire L, Belson S, et al. (2007) Atypical parkinsonism in Guadeloupe: a common risk factor for two closely related phenotypes? Brain 130: 816–827.

18. Escobar-Khondiker M, Hollerhage M, Muriel MP, Champy P, Bach A, et al. (2007) Annonacin, a natural mitochondrial complex I inhibitor, causes tau pathology in cultured neurons. J Neurosci 27: 7827–7837.

19. Yamada ES, Respondek G, Mussner S, de Andrade A, Hollerhage M, et al. (2014) Annonacin, a natural lipophilic mitochondrial complex I inhibitor, increases phosphorylation of tau in the brain of FTDP-17 transgenic mice. Exp Neurol 253C: 113–125.

20. Lotharius J, Falsig J, van Beek J, Payne S, Dringen R, et al. (2005) Progressive degeneration of human mesencephalic neuron-derived cells triggered by dopamine-dependent oxidative stress is dependent on the mixed-lineage kinase pathway. J Neurosci 25: 6329–6342.

21. Scholz D, Poltl D, Genewsky A, Weng M, Waldmann T, et al. (2011) Rapid, complete and large-scale generation of post-mitotic neurons from the human LUHMES cell line. J Neurochem 119: 957–971.

22. Silver IA, Erecinska M (1994) Extracellular glucose concentration in mammalian brain: continuous monitoring of changes during increased neuronal activity and upon limitation in oxygen supply in normo-, hypo-, and hyperglycemic animals. J Neurosci 14: 5068–5076.

23. Glinka Y, Gassen M, Youdim MB (1997) Mechanism of 6-hydroxydopamine neurotoxicity. J Neural Transm Suppl 50: 55–66.

24. Schapira AH (2010) Complex I: inhibitors, inhibition and neurodegeneration. Exp Neurol 224: 331–335.

25. Hoglinger GU, Lannuzel A, Khondiker ME, Michel PP, Duyckaerts C, et al. (2005) The mitochondrial complex I inhibitor rotenone triggers a cerebral tauopathy. J Neurochem 95: 930–939.

26. Chambers CB, Lee JM, Troncoso JC, Reich S, Muma NA (1999) Overexpression of four-repeat tau mRNA isoforms in progressive supranuclear palsy but not in Alzheimer's disease. Ann Neurol 46: 325–332.

27. Edmond V, Moysan E, Khochbin S, Matthias P, Brambilla C, et al. (2011) Acetylation and phosphorylation of SRSF2 control cell fate decision in response to cisplatin. EMBO J 30: 510–523.

28. Edmond V, Brambilla C, Brambilla E, Gazzeri S, Eymin B (2011) SRSF2 is required for sodium butyrate-mediated p21(WAF1) induction and premature senescence in human lung carcinoma cell lines. Cell Cycle 10: 1968–1977.

16

Exercise-Mediated Wall Shear Stress Increases Mitochondrial Biogenesis in Vascular Endothelium

Boa Kim[1,2], Hojun Lee[1], Keisuke Kawata[1], Joon-Young Park[1,2]*

1 Department of Kinesiology, Temple University, Philadelphia, Pennsylvania, United States of America, **2** Cardiovascular Research Center, Temple University, Philadelphia, Pennsylvania, United States of America

Abstract

Objective: Enhancing structural and functional integrity of mitochondria is an emerging therapeutic option against endothelial dysfunction. In this study, we sought to investigate the effect of fluid shear stress on mitochondrial biogenesis and mitochondrial respiratory function in endothelial cells (ECs) using *in vitro* and *in vivo* complementary studies.

Methods and Results: Human aortic- or umbilical vein-derived ECs were exposed to laminar shear stress (20 dyne/cm^2) for various durations using a cone-and-plate shear apparatus. We observed significant increases in the expression of key genes related to mitochondrial biogenesis and mitochondrial quality control as well as mtDNA content and mitochondrial mass under the shear stress conditions. Mitochondrial respiratory function was enhanced when cells were intermittently exposed to laminar shear stress for 72 hrs. Also, shear-exposed cells showed diminished glycolysis and decreased mitochondrial membrane potential ($\Delta\Psi$m). Likewise, in *in vivo* experiments, mice that were subjected to a voluntary wheel running exercise for 5 weeks showed significantly higher mitochondrial content determined by *en face* staining in the conduit (greater and lesser curvature of the aortic arch and thoracic aorta) and muscle feed (femoral artery) arteries compared to the sedentary control mice. Interestingly, however, the mitochondrial biogenesis was not observed in the mesenteric artery. This region-specific adaptation is likely due to the differential blood flow redistribution during exercise in the different vessel beds.

Conclusion: Taken together, our findings suggest that exercise enhances mitochondrial biogenesis in vascular endothelium through a shear stress-dependent mechanism. Our findings may suggest a novel mitochondrial pathway by which a chronic exercise may be beneficial for vascular function.

Editor: Feng Ling, RIKEN Advanced Science Institute, Japan

Funding: This work was supported by American Heart Association Grants 12SDG12070327 (to J-Y. Park) and Predoctoral Fellowship 12PRE11960049 (to B. Kim), Temple University's University Felowships (to B. Kim and H. Lee), and College of Health Professions and Social Work Research Seed Grant (J-Y. Park). The funders had no role in study design, data collection and analysis, decision to publish, or preparation of the manuscript.

Competing Interests: The authors have declared that no competing interests exist.

* Email: parkjy@temple.edu

Introduction

Mitochondria are multifunctional organelles. Not only are they metabolic hubs, but they are also involved in other vital cellular processes. In endothelial cells (ECs), the potential physiological role of mitochondria has been somewhat neglected because their energy supply is relatively independent of the mitochondrial respiration, although the accuracy of this notion as it relates to other mitochondrial functions in the cells is unknown. To this end, emerging evidence suggests that mitochondria are essential for maintaining various endothelial homeostasis such as ROS signaling, Ca^{2+} regulation, apoptosis and cell senescence [1–9]. Furthermore, mitochondrial dysfunction has appeared to be responsible for the range of cardiovascular diseases intimately related with endothelial dysfunction such as hypertension and atherosclerosis [1,3,4,8–13]. Thus, it is imperative to identify an effective intervention to manipulate mitochondrial networks in the endothelium.

The regular practice of physical activity is one of the most effective non-pharmacological interventions improving endothelial dysfunction. During the last two decades, the beneficial effects of exercise on the vascular endothelium have been extensively studied in various aspects of the endothelial function related to endothelium-dependent vasodilatory, anti-inflammatory, anti-thrombotic, and anti-apoptotic endothelial phenotypes [14–26]. Whilst exercise-induced uniaxial laminar flow has been thought to be the central signaling mechanism for the endothelial adaptations [27–33], a direct impact of this flow pattern on endothelial mitochondrial adaptations *in vivo* is unknown.

Mitochondrial biogenesis is a complex process involving the replication of mitochondrial DNA (mtDNA) and the expression of mitochondrial proteins encoded by both nuclear and mitochondrial genomes. Peroxisome proliferator-activated receptor-γ coactivator-1α (PGC-1α) transactivates nuclear respiratory factor 1 (NRF-1) which, in turn, activates mtDNA transcription factor A (TFAM) that regulates mtDNA transcription and replication. The activation of PGC-1α involves a dual-posttranslational modification involving AMP-activated protein kinase (AMPK) and NAD-dependent protein deacetylase, sirtuin 1 (SIRT1), but the specific regulatory mechanism in ECs remains controversial [34,35]. p53-inducible ribonucleotide reductase (p53R2) plays a crucial role in a salvage pathway to supply dNTPs for mtDNA synthesis [36]. In addition, upregulation of other mitochondrial contents including respiratory chain complexes and their assembly proteins (i.e., COX IV, SCO1 and SCO2) are also important for preventing dilution of the contents for a successful mitochondrial proliferation. Mitochondrial dynamics plays a crucial role in mitochondrial quality control. Mitochondrial fission is achieved through the action of a set of proteins, including dynamin-related protein, Drp1, and outer-membrane receptor-like protein, Fis1. Mitochondria fusion involves outer mitochondrial membrane proteins, mitofusins 1 and 2 (Mfn1 and Mfn2) and an inner membrane protein Opa1 [37]. Through proper fusion/fission dynamics coordinated with contents amplification, new daughter mitochondria are formed [38].

Recently, potential link between shear stress and mitochondrial biogenesis in ECs has been suggested [39–42]. Chen et al. reported that laminar flow upregulates the key mitochondrial biogenesis regulators including PGC-1α and SIRT1 as well as the MitoTracker Green signals in shear-exposed ECs [39]. In addition, a study reported that a short-term forced exercise on a motorized treadmill significantly altered mitochondrial dynamic protein profiles in the rat aortic tissues in a NO-dependent fashion [40]. Here, we report that laminar shear stress (LSS) increases mitochondrial biogenesis/dynamics and mtDNA content, and modulates their respiratory function and bioenergetics in human ECs. We also report that chronic voluntary running exercise increases mitochondrial density in the mouse endothelium in a shear stress-dependent manner. Findings from this study will help understand the effects of aerobic exercise-mediated increase in wall shear stress (WSS) on enhancing mitochondrial contents which might be a guide of therapeutic approach for improving cardiovascular health.

Materials and Methods

Cell culture and LSS protocol

Human aortic ECs (HAECs) and human umbilical vein ECs (HUVECs) (Lonza) were cultured in EGM-2 and M199 medium supplemented with 20% fetal bovine serum and endothelial cell growth supplement, respectively. Cells were exposed to the arterial levels of LSS for various time points by using a cone-and-plate shear system once they reach at 100% confluency. Overview of the LSS protocol is outlined in figure 1A. All experiments with HAECs and HUVECs were conducted between the 3–7 passages.

Immunoblotting

Cells were washed three times with cold DPBS and lysed in RIPA buffer (10 mM Tris-HCl, 5 mM EDTA, 150 mM NaCl, 1% Triton X-100, 0.1% SDS, 1% Deoxycholate, pH 7.5). Following precipitation of insoluble fraction of the RIPA samples by centrifugation (16,000 g for 15 min at 4°C), supernatants were collected and subjected to Bradford assay to quantify protein concentrations. The resulting protein samples underwent SDS-PAGE and were transferred to Immobilon-P membrane (Millipore). Subsequently, the membrane was blocked with 5% nonfat dry milk in TBST for 20 min at room temperature and incubated overnight with respective primary antibodies. Antibodies were purchased from the following sources: rabbit polyclonal anti-PGC-1α (Novus), mouse monoclonal anti-porin (anti-VDAC) (Invitro-gen), goat polyclonal anti-p53R2 (Santa Cruz), rabbit polyclonal anti-AMPKα (Cell signaling), rabbit polyclonal anti-phospho-AMPKα (Cell signaling), mouse monoclonal α-tubulin (Sigma-Aldrich). The membranes were then washed twice in TBST and incubated with HRP-conjugated secondary antibodies for an hour followed by washing three times with TBST. Then, membranes were subjected to standard enhanced chemiluminescence (Thermo Fisher Scientific) method for visualization.

mRNA isolation, cDNA synthesis, and real-time PCR

mRNAs were isolated using Dynabeads direct kit, and cDNA synthesis were performed on poly-dT magnetic beads by reverse transcription using superscript II (Invitrogen). mRNA expression levels were quantified by real-time PCR using SYBR green fluorescence. Cycle threshold (Ct) values were normalized to the housekeeping gene HPRT1. The primer sequences used are described in Table 1.

mtDNA content quantification

Total genomic DNAs were isolated by using the DNeasy kit (QIAGEN) and mtDNA contents were assessed by semi-quantitative PCR. The relative ratio between mitochondrial DNA (COX I; cytochrome c oxidase subunit I, COX II; cytochrome c oxidase subunit II, or ND II; NADH dehydrogenase subunit 2) compared to nuclear DNA (18s rRNA) amount was calculated. Primer sequences were as follows:

COXI (human)
Sense, 5′- CATAGGAGGCTTCATTCACTG – 3′
Antisense, 5′- CAGGTTTATGGAGGGTTCTTC – 3′
COXII (human)
Sense, 5′- CCATAGGGCACCAATGATACTG – 3′
Antisense, 5′- AGTCGGCCTGGGATGGCATC – 3′
NDII (mouse)
Sense, 5′- CCTATCACCCTTGCCATCAT – 3′
Antisense, 5′- GAGGCTGTTGCTTGTGTGAC – 3′
18s rRNA (human and mouse)
Sense, 5′-CTTAGAGGGACAAGTGGCGTTC-3′
Antisense, 5′-CGCTGAGCCAGTCAGTGTAG-3′

MitoTracker staining

Live HAECs exposed to either static (STT) or LSS were incubated with 200 nM pre-warmed MitoTracker Green FM or MitoTracker Red CMXRos (Molecular Probes) solution at 37°C for 30 min. After removal of the incubation solution, cells were washed three times with pre-warmed PBS and then mounted in Hank's balanced salt solution. For quantitative analyses, more than 100 images per each group were acquired using an epi-fluorescence upright microscope with a 63x objective oil lens. For MitoTracker Green FM staining, excitation/emission wavelengths were set at 470/525 nm (FL filter Set 38, Zeiss), and for MitoTracker Red CMXRos staining, excitation/emission wavelengths were set at 587/647 nm (FL filter Set 64HE). Images were initially acquired using an AxioCam MRm and AxioVision image processing system (Zeiss), and the fluorescence intensities were assessed using Image J software (NIH).

Mitochondrial respiration

HUVECs were subjected to intermittent LSS at 20 dynes/cm^2 for up to 72 hours while the STT control group was maintained in the absence of LSS. Cells were subcultured as needed to avoid becoming over-confluent for the duration of experiments. Cells were then harvested and the oxygen consumption was measured using a Clark-type oxygen electrode in complete media. Final

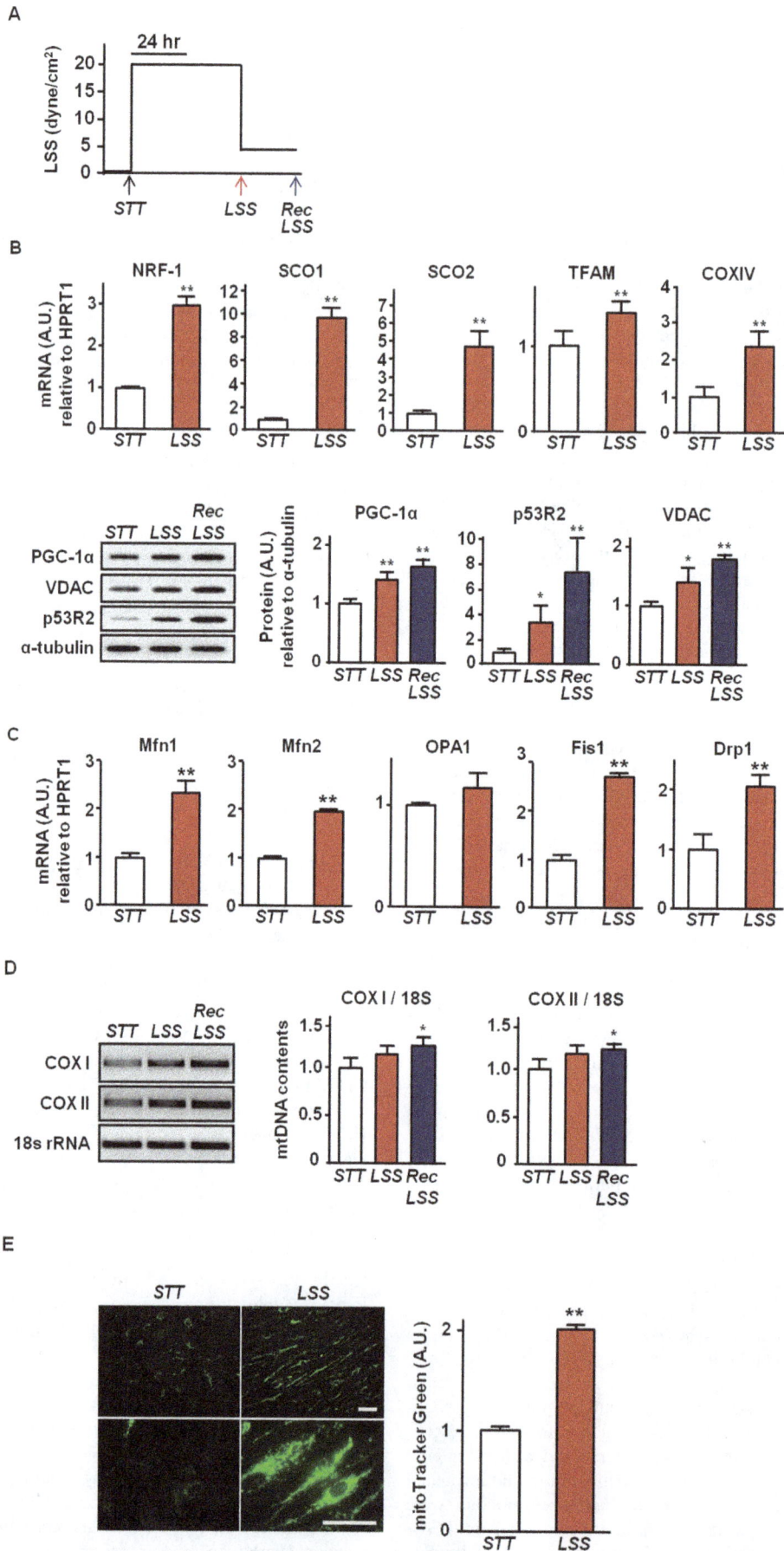

Figure 1. Increased mitochondrial biogenesis markers by LSS in HAECs. (a) An overview of LSS protocol used. HAECs were exposed to exercise-mimicking LSS at 20 dyne/cm^2 for 48 hrs, and then, recovery (Rec) LSS at 5 dyne/cm^2 was followed for another 24 hrs. (b) Effect of LSS on the mRNA and protein expression of mitochondrial biogenesis markers. mRNA expression of NRF-1, SCO1, SCO2, TFAM, and COX IV were assessed by real-time PCR and protein contents of PGC-1α, VDAC, and p53R2 were analyzed by western blot. (c) Effect of LSS on the mRNA expression of mitochondrial dynamics markers. mRNA expression of Mfn1, Mfn2, OPA1, Fis1, and Drp1 were assessed by real-time PCR. (d) Effect of LSS on mtDNA contents. Relative mtDNA contents are expressed as a ratio of COX I and II to 18s rRNA. (e) Effect of LSS on mitochondrial mass. Mitochondria were labeled with MitoTracker Green in live HAECs. Representative fluorescence micrographs under STT (left panel) and after 48 hrs of LSS at 20 dyne/cm^2 (right panel) are shown. Bar = 50 μm. The MitoTracker Green fluorescence intensities were analyzed using the Image J (NIH) software. All densitometry analyses values are shown as mean ± SE; * $P<0.05$ vs. STT; ** $P<0.01$ vs. STT.

oxygen consumption was normalized to the number of cells (nmol O_2/min/10^8 cells).

Lactate production measurement

Lactate concentration in cell culture medium was measured by a colorimetric enzymatic assay according to the manufacturer's instructions (Sigma). Briefly, when cells were grown at ≈80% confluency, cell culture medium was replaced with fresh basal M199 medium. Then, media samples were collected at 12, 24, and 36 hours after incubation and filtered through 10 kDa molecular weight cut-off spin columns (Milipore) before being subjected to lactate assays. Lactate concentration was normalized to corresponding viable cell numbers determined by trypan blue exclusion quantification.

Microarray analysis

To gain insight into global patterns of metabolic gene expression, microarray analysis was performed. RNA was isolated by using RNeasy kit (QIAGEN). Microarray analysis were performed from STT (n = 4) and LSS (n = 6) exposed HUVECs by using Affymetrix whole-genome arrays containing 45,101 probe sets corresponding to ≈34,000 genes. Heat map was created with Gene-E ver. 3.0.214 (Broad Institute, Inc).

Ethics statement

This study was carried out in strict accordance with the recommendations and the Guide for the Care and Use of Laboratory Animals of the National Institutes of Health. The protocol was approved by the Temple University Institutional Animal Care and Use Committee (Permit Number: 4159). All sacrifices were performed under isoflurane anesthesia, and all efforts were made to minimize suffering.

Experimental animals and voluntary wheel exercise

After three days of acclimation period, twenty inbred C57Bl/6J mice were randomly assigned to either sedentary (SED) (n = 10) or voluntary wheel (VW) running exercise (n = 10) group. VW group animals were individually housed in a rat-sized cage with a metal wheel with a diameter of 11.5 cm (Prevue) fitted with digital magnetic counter. SED group animals were singly housed in the same sized cage without the running wheel. All animals were given water and food (Purina chow) ad libitum. VW running exercise began at an age of 8 to10-week-old and continued for 5 weeks.

Blood vessel isolation

Mice were euthanized two days after the end of 5-weeks of VW exercise period. For the preparation of RNA, protein, and DNA,

Table 1. Primer Sequences for Real-Time PCR.

Species	Genes	Primer sequences (5′ – 3′)	
		Sense	**Antisense**
human	NRF-1	CCAAGTGAATTATTCTGCCG	TGACTGCGCTGTCTGATATCC
	SCO1	GGCACAGCCAGTGCATTCCTGCCTG	GCATCACACTCGTGATCAATATCCTC
	SCO2	GCAGCCTGTCTTCATCACTGTGGACC	CCGCACACTGTCTGAGATCTGCTC
	TFAM	AGCTAAGGGTGATTCACCGC	GCAGAAGTCCATGAGCTGAA
	COX IV	ACGAGCTCATGAAAGTGTTGTG	AATGCGATACAACTCGACTTTCTC
	HPRT1	GACACTGGCAAAACAATGCAG	AGTCTATAGGCTCATAGTGC
	MFN1	AGTAACAGGATTGGCGTCCG	CGTTTCCTCCTATCATGGTCACC
	MFN2	ATGCATCCCCACTTAAGCAC	CCAGAGGGCAGAACTTTGTC
	OPA1	GGCTCTGCAGGCTCGTCTCAAGG	TTCCGCCAGTTGAACGCGTTTACC
	DRP1	CACAGGAGGAGGTGGACAGC	CGCCTCCTTCAGTGCGTGGT
	FIS1	ATGGAGGCCGTGCTGAAC	TCAGGATTTGGACTTGGA
mouse	PGC-1α	ACGGTTTACATGAACACAGCTGC	CTTGTTCGTTCTGTTCAGGTGC
	NRF-1	GAACGCCACCGATTTCACTGTC	CCCTACCACCCACGAATCTGG
	TFAM	CTGATGGGTATGGAGAAGGAGG	CCAACTTCAGCCATCTGCTCTTC
	p53R2	CCAGGTTACCATGGTTGTGG	CCAGTGCACTCAGTAGCTGTG
	SCO1	CTAGCTTAGCACAATAGCAAGGGCAGGCTAC	CCCAGGAATGCAGTTATGACATGACAGCAAAGGCAG
	SCO2	CAGCCTGTCTTCATCACTGTGGA	GACACTGTGGAAGGCAGCTATGTGCC
	TIF	CTGAGGATGTGCTGTCTGGGAA	CCTTTGCCTCCACTTCGGTC

abdominal aorta was isolated after whole body perfusion with ice-cold PBS at a pressure of approximately 100 mmHg. For *en face* staining, several different regions of blood vessels including aortic arch, thoracic aorta, femoral artery, and mesenteric artery were isolated after the perfusion with ice-cold PBS and a fixative, 2% paraformaldehyde.

En face immunostaining

Isolated blood vessels were post-fixed at 0.4% paraformaldehyde overnight at room temperature. The vessels were then washed five times with PBS and permeabilized by using 0.3% Triton-X in 2% BSA/PBS. Mitochondrial contents were assessed by using anti-VDAC (1:100) (Abcam) antibody and Alexafluor488-conjugated anti-rabbit secondary antibody (Invitrogen). EC were identified by co-staining using anti-CD31 (1:100) (Millipore) antibody conjugated to the Alexafluor647-conjugated anti-hamster secondary antibody (Jackson ImmunoResearch). Primary antibodies were incubated overnight at 4°C with gentle agitation. After rinsing in 2% BSA/PBS, secondary antibodies were incubated for 2 hours at room temperature. Immunostained vessels were placed on slide glass and cut longitudinally and mounted in ProlongGold with DAPI solution (Invitrogen). The fluorescence was analyzed under fluorescence microscope (Axioimager, Zeiss) with 64x oil objective lens.

Statistics

The results are presented as mean ± SE for a minimum of three independent experiments in triplicate. Depending on how many conditions were compared, either two tailed t-test analysis or one-way ANOVA with the Fisher's least significant difference test was conducted. $P<0.05$ was considered statistically significant for all analyses.

Results

LSS enhances mitochondrial biogenesis in human ECs

As shown in figure 1B, we observed that LSS upregulates mRNA and protein expression of key genes that are related to mitochondrial biogenesis in HAECs. mRNA expressions of NRF-1, TFAM, COX IV, SCO1 and SCO2 were significantly increased in the ECs exposed to LSS. As well, protein expressions of PGC1α, p53R2, and VDAC were increased when cells were exposed to LSS. To confirm the LSS-induced increase in mitochondrial biogenesis, we stained HAECs with MitoTracker Green FM, a fluorescence dye which stains mitochondria in a mass-dependent fashion, and observed two-fold increase in mitochondrial mass in LSS-exposed HAECs (Fig. 1E). As shown in figure 1D, mtDNA contents were also significantly increased by LSS. In addition, expression of both profusion (Mfn1 and Mfn2) and profission (Drp1 and Fis1) factors were significantly increased after LSS exposure (Fig. 1C).

Next, we sought to examine whether LSS-induced mitochondrial biogenesis was functionally relevant to the mitochondrial bioenergetic properties. As shown in figure 2A, the rate of oxygen consumption was significantly enhanced in HUVECs after being exposed to LSS for 72 hours. To evaluate a potential occurrence of metabolic shift from glycolytic to aerobic metabolism in these cells, we evaluated cellular lactate production and performed gene expression array experiments on a number of genes related to the glycolytic pathways. Cellular lactate production was significantly suppressed in the LSS-exposed ECs compared to the STT-exposed ECs (Fig. 2D). Moreover, among the twenty-one genes related to glycolysis pathway, the vast majority of genes were down-regulated under LSS (Fig. 2C and Table S1). Notably, these

genes include key rate-limiting enzymes for glycolysis such as hexokinase II (HK2) and phospohofructokinase (PFK)-related genes (i.e., PFKFB1, PFKFB2, and PFKP). Mitochondrial membrane potential (ΔΨm), which was determined by Mito-Tracker Red CMXRos, was significantly decreased in LSS-exposed ECs compared to STT-exposed ECs (Fig. 2B).

Five weeks of VW running induces mitochondrial biogenesis in blood vessel and it is mediated by exercise-induced increase in WSS on vascular endothelium

Given our observation that LSS is positively related to mitochondrial biogenesis *in vitro*, we hypothesized that exercise-mediated increase in WSS would enhance mitochondrial biogenesis in mouse endothelium. As shown in figure 3A, expressions of genes that are related to mitochondrial biogenesis were analyzed in abdominal aorta isolated from SED and VW group mice. Elevated mRNA expressions of mitochondrial biogenesis markers which include PGC-1α, NRF1, TFAM, p53R2, and SCO1 were observed in VW group mice compared to SED. Also, western blot analysis revealed that phosphorylated AMPKα and VDAC were increased by three-fold in VW group compared to SED (Fig. 3B). Furthermore, greater mtDNA content was found in VW group compared to SED (Fig. 3C). We also hypothesized that differential hemodynamic flow in different vessel beds may lead to distinct responses depending on their geometrical location in the vascular tree. *En face* staining experiment revealed that the level of VDAC protein in greater curvature, lesser curvature, thoracic aorta, and femoral artery was higher in VW group compared to the SED group (Fig 4). VW running elicited greater mitochondrial adaptation in lesser curvature compared to greater curvature. The greatest increase in mitochondrial content was observed in femoral artery. In mesenteric artery, decreased level of mitochondrial content was observed in VW compared to SED group.

Discussion

Here, we report that LSS enhances mitochondrial biogenesis, mitochondrial dynamics, and mtDNA copy number in primary cultured human ECs. Consistent with these findings, we also demonstrate that voluntary aerobic exercise training increases mitochondrial content in the endothelium in a region-specific fashion. In addition, we found that a long-term shear-exposure is sufficient to improve mitochondrial respiration and to alter substrates metabolism from anaerobic glycolysis to oxidative phosphorylation-dependent mechanisms in ECs. These findings are particularly important because potential metabolic contributions of the endothelial mitochondria have been widely neglected as they are highly glycolytic cells containing relatively small number of mitochondria (only 2-5% of the entire cytoplasmic volume) compared to other energy demanding tissues [43]. Furthermore, studies have demonstrated that, under stress conditions, fatty acids are the major substrate for ATP generation in ECs suggesting an important contribution of mitochondria-dependent metabolism for endothelial homeostasis [44]. To this end, our data suggest that LSS-induced mitochondrial biogenesis may have important implications for preventing endothelial dysfunction although future researches are needed to investigate the effect of LSS (or aerobic exercise training) on the energy metabolism and the substrate utilization in ECs *in vivo*.

In this study, we also demonstrated that a long-term LSS at a physiological level decreased ΔΨm. This result is consistent with a previous report showing that shear stress induces a decrease in ΔΨm and an increase in the endogenous ATP [45]. In contrast, a

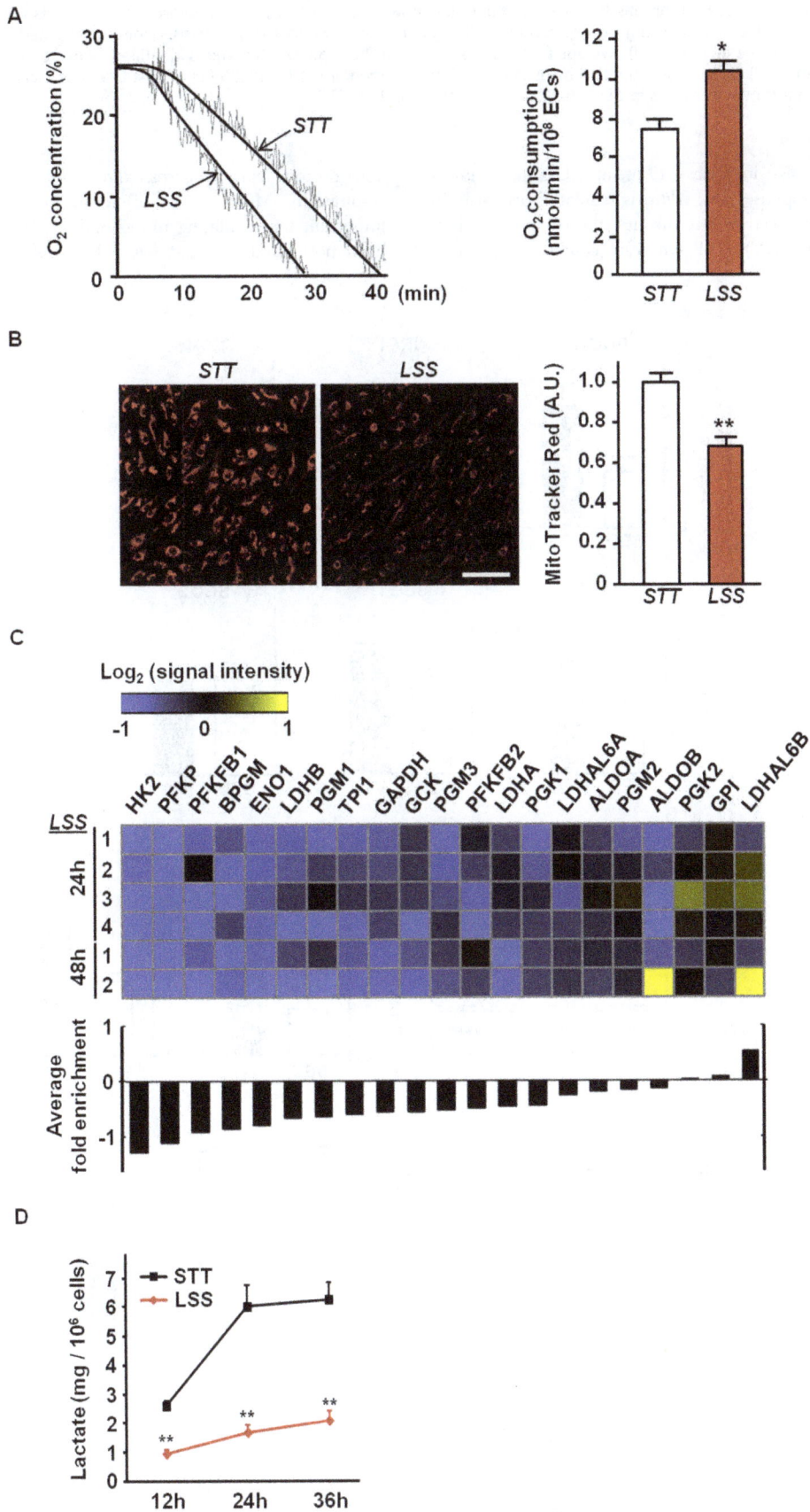

Figure 2. Effect of LSS on endothelial metabolism. (a) Enhanced mitochondrial respiration in LSS-exposed HUVECs. Oxygen consumption of HUVECs was measured after the intermittent LSS exposure for up to 72 hours. Representative strips of the oxygen consumption measured (left panel). Normalized values to the number of cells (right panel). (b) Effect of LSS on $\Delta\Psi$m in ECs. $\Delta\Psi$m was estimated by using MitoTracker Red CMX

Ros. Representative fluorescence micrographs for each condition are shown. Bar = 100 μm. The fluorescence intensities were analyzed using the Image J (NIH) software. (c) Heat map showing the expression of glycolysis markers by microarray analysis. Genes upregulated are presented in yellow and downregulated are in blue (upper panel). Average fold change of each of those glycolysis markers identified by microarray analysis are shown in a bar graph (lower panel). (d) Lactate concentration measured in cell culture medium at 12, 24, 36 hrs of post LSS or STT. Values were normalized to viable cell number. Data shown as means ± SE; * $P<0.05$ vs. STT; ** $P<0.01$ vs. STT.

short-term shear stress increases $\Delta\Psi$m in ECs suggesting a biphasic temporal response [46]. $\Delta\Psi$m is regulated primarily by the balance between electron flux through the respiratory chain (Complexes I, III, and IV), ATP synthesis (coupled respiration),

and proton leakage across the inner membrane (uncoupled respiration). Maintenance of $\Delta\Psi$m at physiological range is important for regulating mitochondrial ROS production. It has been postulated that there is a U-shaped curve describing the

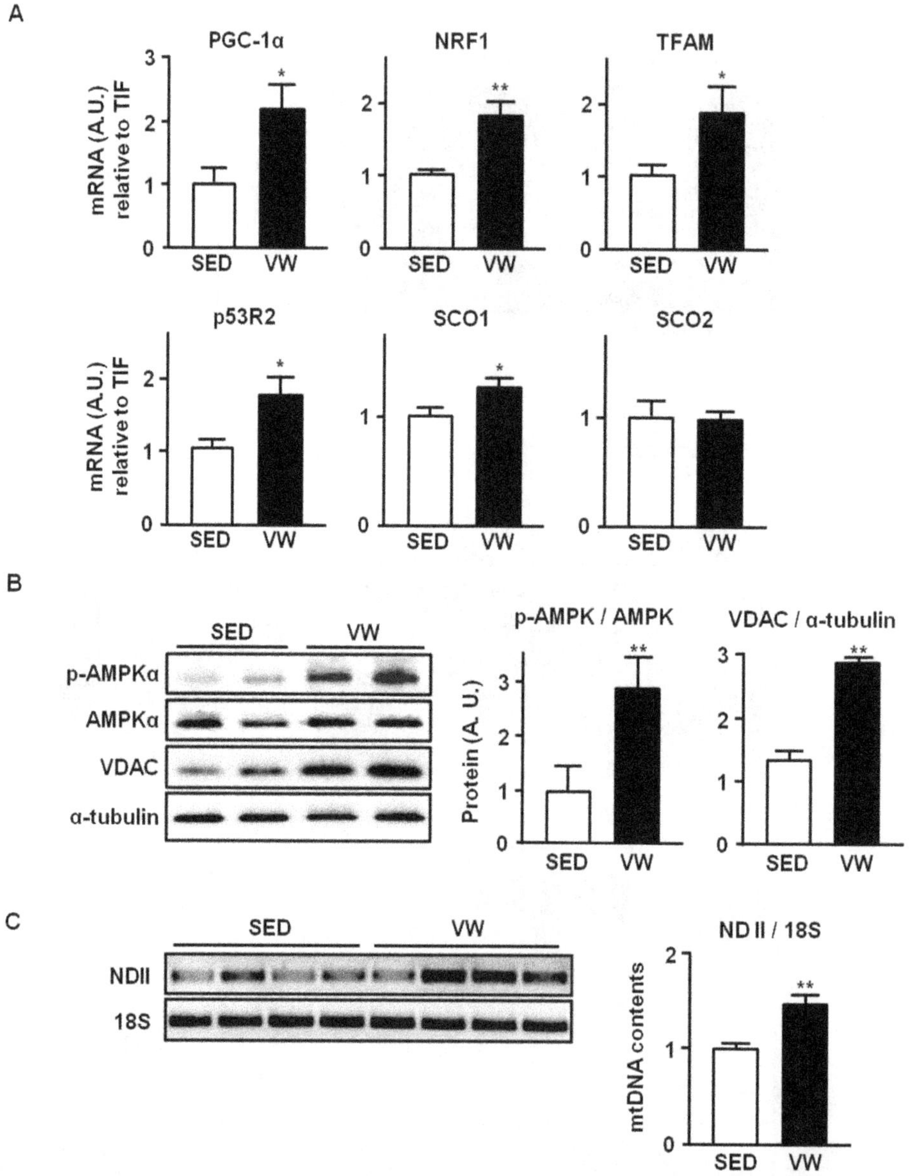

Figure 3. Effect of five weeks of voluntary wheel (VW) exercise on mitochondrial biogenesis markers in mice abdominal aorta (AA).
(a) Effect of VW running on mRNA expression of mitochondrial biogenesis markers in AA. mRNA expressions of PGC-1α, NRF1, TFAM, p53R2, SCO1, and SCO2 were examined by real-time PCR. Values were normalized to the level of housekeeping gene, TIF. (b) Effect of VW running on protein expression of mitochondrial biogenesis markers in AA. Tissue extracts of the AA from SED and VW group mice were subjected to western blot. The amount of phosphorylated- AMPKα was normalized by the amount of AMPKα protein. Protein content of mitochondrial biogenesis marker VDAC was also measured. The loading volume was normalized by the expression level of α-tubulin. (c) Effect of VW running on mtDNA content in AA. mtDNA contents were compared in between SED and VW run mice. Relative mtDNA content are expressed as a ratio of NADH dehydrogenase subunit 2 (ND II) to 18s rRNA. All densitometry analyses values are shown as means ± SE. Data shown represent results from a total of 5 mice per group; * $P<0.05$ vs. SED; ** $P<0.01$ vs. SED.

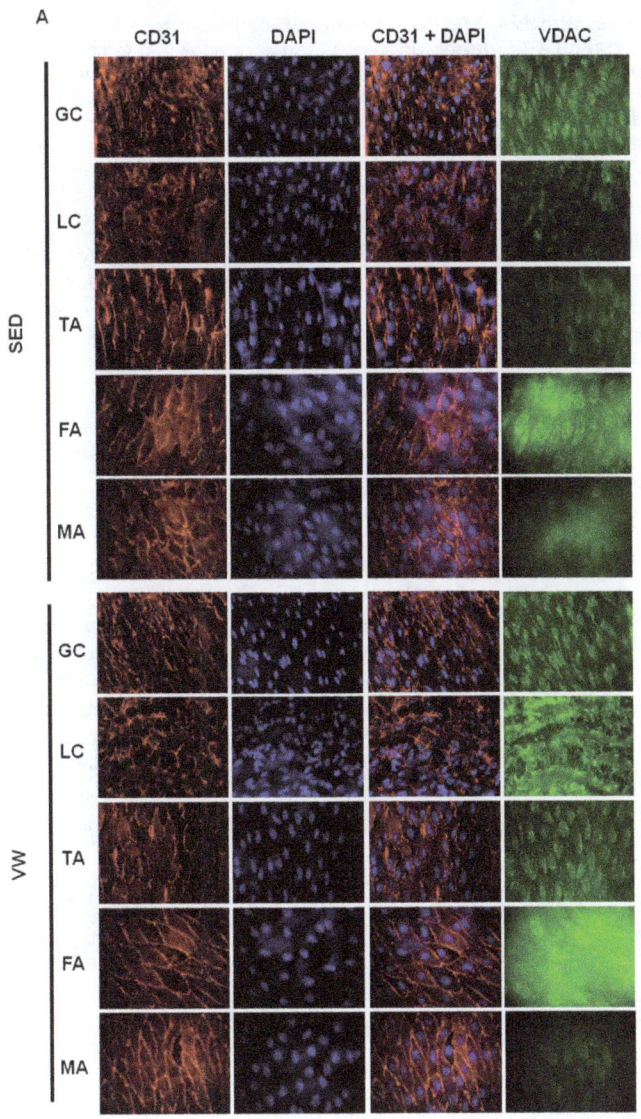

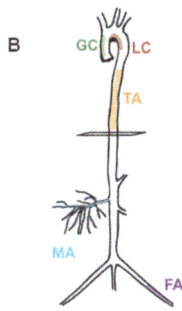

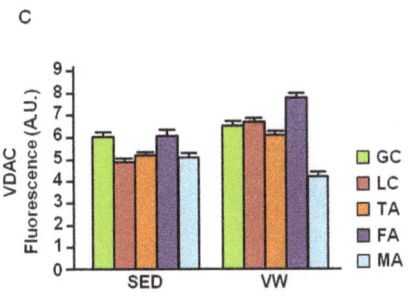

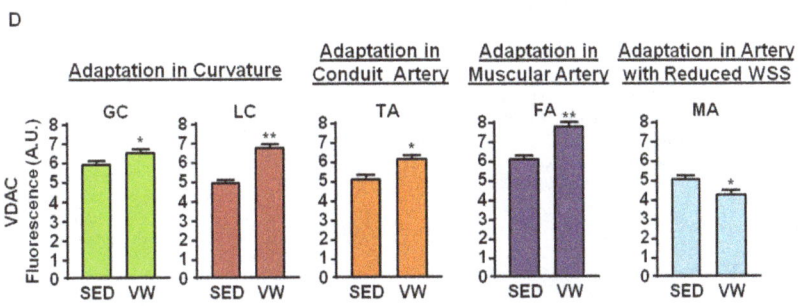

Figure 4. Effect of five weeks of VW running on mitochondrial in mouse endothelium. (A) Representative fluorescence micrographs of *en face* immunostaining. Endothelium of the greater curvature (GC), lesser curvature (LC), thoracic aorta (TA), femoral artery (FA), and mesenteric artery (MA) were stained in the sedentary (SED) and voluntary wheel (VW) run C57BL6 mice. The green fluorescent staining indicates mitochondrial density stained by VDAC, and the red color represents ECs stained by CD31 (an endothelial cell specific marker). Nuclei were counterstained with DAPI. Shown are representative images of *en face* staining labeled. (B) Illustration of mouse arterial tree. (C) Summary of densitometry analysis. Green fluorescence intensities by VDAC staining were analyzed using the Image J (NIH) software. Data shown as mean ± SE; Data shown represent results from a total of 10 mice per group * $P<0.05$ vs. SED. ** $P<0.01$ vs. SED.

relationship between $\Delta\Psi$m and ROS formation [47]. Furthermore, numerous studies have shown that hyperpolarization of the mitochondria (above ~-140 mV) triggers release of superoxide predominantly at complex III [48]. We observed that UCP2 expression is dramatically elevated under the same shear paradigm used in this study (unpublished data). Combined with evidence that UCP2 inhibits formation of ROS [49], it is plausible that the depolarization of the mitochondria would prevent ROS release. Together, shear stress may improve cellular redox state, at least in part, by modulating $\Delta\Psi$m in favor of reduced mitochondrial ROS production which compliment other shear-mediated mechanisms such as a down-regulation of NAD(P)H oxidase activity [50] and an increase in antioxidant system [51,52].

Different vascular beds are exposed to distinct flow patterns depending on their structural and functional properties. For example, in the aortic arch, greater curvature is exposed to a high-grade unidirectional shear stress where lesser curvature is exposed to a low-grade oscillatory shear stress [53]. Lesser curvature has been shown to be predisposed to atherosclerotic plaque formation. In sedentary mice, we observed that mitochondrial content is higher in the greater curvature compared to the lesser curvature, suggesting a direct correlation between flow pattern and mitochondrial content in the endothelium.

It is well known that a process termed 'blood redistribution' occurs during exercise [54,55]. At rest, only 15–20% of cardiac output is redirected to skeletal muscle and the majority of it goes to the other organs. Once exercise commence, however, 87% of blood is redirected to exercising muscles. Muscle blood flow has been shown to be increased up to 80-fold [56,57]. Corresponding to this concept, amount of blood extracted by the celiac, mesenteric, and renal arteries is decreased during exercise [27,54,55,58,59]. Interestingly, we observed the greatest adaptation in muscle feeding (femoral) artery (Fig. 4D) whereas the endothelial mitochondrial content in the mesenteric artery was found even lower in VW than SED.

During exercise, the magnitude of WSS is increased to higher levels ranged from 15 to 30 dynes/cm² in human arteries [29,30,33]. As an attempt to investigate underlying mechanisms of EC response to shear stress, and to better understand the effect of hemodynamics in endothelial/vascular health *in vivo*, several *in vitro* shear systems have been developed. Effects of the enhanced shear stress have been tested in numerous studies using an *in vitro* flow system, and these findings are consistent with those determined by *in vivo* studies [60]. In this study, we used 20 dyne/cm² of high LSS as an exercise-mimicking flow condition, as it is within the range of arterial level shear stress [61].

In conclusion, our data support an idea that aerobic exercise enhances mitochondrial integrity in vascular endothelium which is essential for endothelial function. Shear stress seems to modulate signal transduction pathways towards mitochondrial biogenesis. Therefore, regulation on mitochondrial remodeling may represent one of the mechanisms whereby exercise-mediated increase in WSS confers a vasculoprotective effect. Future research is warranted to investigate the downstream and upstream of the shear-sensing mechanism and clinical implications of the shear stress-induced mitochondrial remodeling in preventing endothelial dysfunction.

Author Contributions

Conceived and designed the experiments: BK JP. Performed the experiments: BK HL KK. Analyzed the data: BK JP. Contributed reagents/materials/analysis tools: BK JP. Wrote the paper: BK JP.

References

1. Krzywanski DM, Moellering DR, Fetterman JL, Dunham-Snary KJ, Sammy MJ, et al. (2011) The mitochondrial paradigm for cardiovascular disease susceptibility and cellular function: a complementary concept to Mendelian genetics. Lab Invest 91: 1122–1135.
2. Addabbo F, Ratliff B, Park HC, Kuo MC, Ungvari Z, et al. (2009) The Krebs cycle and mitochondrial mass are early victims of endothelial dysfunction: proteomic approach. Am J Pathol 174: 34–43.
3. Ballinger SW (2005) Mitochondrial dysfunction in cardiovascular disease. Free Radic Biol Med 38: 1278–1295.
4. Dikalova AE, Bikineyeva AT, Budzyn K, Nazarewicz RR, McCann L, et al. (2010) Therapeutic targeting of mitochondrial superoxide in hypertension. Circ Res 107: 106–116.
5. Doughan AK, Harrison DG, Dikalov SI (2008) Molecular mechanisms of angiotensin II-mediated mitochondrial dysfunction: linking mitochondrial oxidative damage and vascular endothelial dysfunction. Circ Res 102: 488–496.
6. Groschner LN, Waldeck-Weiermair M, Malli R, Graier WF (2012) Endothelial mitochondria–less respiration, more integration. Pflugers Arch 464: 63–76.
7. Kluge MA, Fetterman JL, Vita JA (2013) Mitochondria and endothelial function. Circ Res 112: 1171–1188.
8. Madamanchi NR, Vendrov A, Runge MS (2005) Oxidative stress and vascular disease. Arterioscler Thromb Vasc Biol 25: 29–38.
9. Yu E, Mercer J, Bennett M (2012) Mitochondria in vascular disease. Cardiovasc Res 95: 173–182.

10. Ren J, Pulakat L, Whaley-Connell A, Sowers JR (2010) Mitochondrial biogenesis in the metabolic syndrome and cardiovascular disease. J Mol Med (Berl) 88: 993–1001.
11. Chistiakov DA, Sobenin IA, Bobryshev YV, Orekhov AN (2012) Mitochondrial dysfunction and mitochondrial DNA mutations in atherosclerotic complications in diabetes. World J Cardiol 4: 148–156.
12. Ong SB, Hall AR, Hausenloy DJ (2013) Mitochondrial dynamics in cardiovascular health and disease. Antioxid Redox Signal 19: 400–414.
13. Sobenin IA, Sazonova MA, Postnov AY, Bobryshev YV, Orekhov AN (2013) Changes of mitochondria in atherosclerosis: possible determinant in the pathogenesis of the disease. Atherosclerosis 227: 283–288.
14. Spence AL, Carter HH, Naylor LH, Green DJ (2013) A prospective randomized longitudinal study involving 6 months of endurance or resistance exercise. Conduit artery adaptation in humans. J Physiol 591: 1265–1275.
15. van Duijnhoven NT, Green DJ, Felsenberg D, Belavy DL, Hopman MT, et al. (2010) Impact of bed rest on conduit artery remodeling: effect of exercise countermeasures. Hypertension 56: 240–246.
16. Durrant JR, Seals DR, Connell ML, Russell MJ, Lawson BR, et al. (2009) Voluntary wheel running restores endothelial function in conduit arteries of old mice: direct evidence for reduced oxidative stress, increased superoxide dismutase activity and down-regulation of NADPH oxidase. J Physiol 587: 3271–3285.
17. DeSouza CA, Shapiro LF, Clevenger CM, Dinenno FA, Monahan KD, et al. (2000) Regular aerobic exercise prevents and restores age-related declines in

endothelium-dependent vasodilation in healthy men. Circulation 102: 1351–1357.

18. Eskurza I, Monahan KD, Robinson JA, Seals DR (2004) Effect of acute and chronic ascorbic acid on flow-mediated dilatation with sedentary and physically active human ageing. J Physiol 556: 315–324.

19. Seals DR, Desouza CA, Donato AJ, Tanaka H (2008) Habitual exercise and arterial aging. J Appl Physiol 105: 1323–1332.

20. Taddei S, Virdis A, Ghiadoni L, Salvetti G, Bernini G, et al. (2001) Age-related reduction of NO availability and oxidative stress in humans. Hypertension 38: 274–279.

21. Denvir MA, Gray GA (2009) Run for your life: exercise, oxidative stress and the ageing endothelium. J Physiol 587: 4137–4138.

22. Adams V, Linke A, Krankel N, Erbs S, Gielen S, et al. (2005) Impact of regular physical activity on the NAD(P)H oxidase and angiotensin receptor system in patients with coronary artery disease. Circulation 111: 555–562.

23. Sessa WC, Pritchard K, Seyedi N, Wang J, Hintze TH (1994) Chronic exercise in dogs increases coronary vascular nitric oxide production and endothelial cell nitric oxide synthase gene expression. Circ Res 74: 349–353.

24. Green DJ, Maiorana A, O'Driscoll G, Taylor R (2004) Effect of exercise training on endothelium-derived nitric oxide function in humans. J Physiol 561: 1–25.

25. Green DJ, Spence A, Rowley N, Thijssen DH, Naylor LH (2012) Vascular adaptation in athletes: is there an 'athlete's artery'? Exp Physiol 97: 295–304.

26. Kasikcioglu E, Oflaz H, Kasikcioglu HA, Kayserilioglu A, Umman S, et al. (2005) Endothelial flow-mediated dilatation and exercise capacity in highly trained endurance athletes. Tohoku J Exp Med 205: 45–51.

27. Taylor CA, Hughes TJ, Zarins CK (1999) Effect of exercise on hemodynamic conditions in the abdominal aorta. J Vasc Surg 29: 1077–1089.

28. Cheng C (2003) Abdominal aortic hemodynamic conditions in healthy subjects aged 50–70 at rest and during lower limb exercise: in vivo quantification using MRI. Atherosclerosis 168: 323–331.

29. Suh GY, Les AS, Tenforde AS, Shadden SC, Spilker RL, et al. (2011) Hemodynamic changes quantified in abdominal aortic aneurysms with increasing exercise intensity using mr exercise imaging and image-based computational fluid dynamics. Ann Biomed Eng 39: 2186–2202.

30. Schlager O, Giurgea A, Margeta C, Seidinger D, Steiner-Boeker S, et al. (2011) Wall shear stress in the superficial femoral artery of healthy adults and its response to postural changes and exercise. Eur J Vasc Endovasc Surg 41: 821–827.

31. Cheng CP, Herfkens RJ, Lightner AL, Taylor CA, Feinstein JA (2004) Blood flow conditions in the proximal pulmonary arteries and vena cavae: healthy children during upright cycling exercise. Am J Physiol Heart Circ Physiol 287: H921–926.

32. Hjortdal VE, Emmertsen K, Stenbog E, Frund T, Schmidt MR, et al. (2003) Effects of exercise and respiration on blood flow in total cavopulmonary connection: a real-time magnetic resonance flow study. Circulation 108: 1227–1231.

33. Tang BT, Cheng CP, Draney MT, Wilson NM, Tsao PS, et al. (2006) Abdominal aortic hemodynamics in young healthy adults at rest and during lower limb exercise: quantification using image-based computer modeling. Am J Physiol Heart Circ Physiol 291: H668–676.

34. Lagouge M, Argmann C, Gerhart-Hines Z, Meziane H, Lerin C, et al. (2006) Resveratrol improves mitochondrial function and protects against metabolic disease by activating SIRT1 and PGC-1alpha. Cell 127: 1109–1122.

35. Higashida K, Kim SH, Jung SR, Asaka M, Holloszy JO, et al. (2013) Effects of resveratrol and SIRT1 on PGC-1alpha activity and mitochondrial biogenesis: a reevaluation. PLoS Biol 11: e1001603.

36. Bourdon A, Minai L, Serre V, Jais JP, Sarzi E, et al. (2007) Mutation of RRM2B, encoding p53-controlled ribonucleotide reductase (p53R2), causes severe mitochondrial DNA depletion. Nat Genet 39: 776–780.

37. Chan DC (2006) Mitochondrial fusion and fission in mammals. Annu Rev Cell Dev Biol 22: 79–99.

38. Ryan MT, Hoogenraad NJ (2007) Mitochondrial-nuclear communications. Annu Rev Biochem 76: 701–722.

39. Chen Z, Peng IC, Cui X, Li YS, Chien S, et al. (2010) Shear stress, SIRT1, and vascular homeostasis. Proc Natl Acad Sci U S A 107: 10268–10273.

40. Miller MW, Knaub LA, Olivera-Fragoso LF, Keller AC, Balasubramaniam V, et al. (2013) Nitric oxide regulates vascular adaptive mitochondrial dynamics. Am J Physiol Heart Circ Physiol 304: H1624–1633.

41. Knaub LA, McCune S, Chicco AJ, Miller M, Moore RL, et al. (2013) Impaired response to exercise intervention in the vasculature in metabolic syndrome. Diab Vasc Dis Res 10: 222–238.

42. Al-Mehdi A-B (2007) Mechanotransduction of Shear-Stress at the Mitochondria. Mitochondria: Springer New York. pp. 169–181.

43. Oldendorf WH, Cornford ME, Brown WJ (1977) The large apparent work capability of the blood-brain barrier: a study of the mitochondrial content of capillary endothelial cells in brain and other tissues of the rat. Ann Neurol 1: 409–417.

44. Dagher Z, Ruderman N, Tornheim K, Ido Y (2001) Acute regulation of fatty acid oxidation and amp-activated protein kinase in human umbilical vein endothelial cells. Circ Res 88: 1276–1282.

45. Kudo S, Morigaki R, Saito J, Ikeda M, Oka K, et al. (2000) Shear-stress effect on mitochondrial membrane potential and albumin uptake in cultured endothelial cells. Biochem Biophys Res Commun 270: 616–621.

46. Li R, Beebe T, Cui J, Rouhanizadeh M, Ai L, et al. (2009) Pulsatile shear stress increased mitochondrial membrane potential: implication of Mn-SOD. Biochem Biophys Res Commun 388: 406–412.

47. Daiber A (2010) Redox signaling (cross-talk) from and to mitochondria involves mitochondrial pores and reactive oxygen species. Biochim Biophys Acta 1797: 897–906.

48. Zamzami N, Marchetti P, Castedo M, Decaudin D, Macho A, et al. (1995) Sequential reduction of mitochondrial transmembrane potential and generation of reactive oxygen species in early programmed cell death. J Exp Med 182: 367–377.

49. Shimasaki Y, Pan N, Messina LM, Li C, Chen K, et al. (2013) Uncoupling protein 2 impacts endothelial phenotype via p53-mediated control of mitochondrial dynamics. Circ Res 113: 891–901.

50. Duerrschmidt N, Stielow C, Muller G, Pagano PJ, Morawietz H (2006) NO-mediated regulation of NAD(P)H oxidase by laminar shear stress in human endothelial cells. J Physiol 576: 557–567.

51. Harrison DG, Widder J, Grumbach I, Chen W, Weber M, et al. (2006) Endothelial mechanotransduction, nitric oxide and vascular inflammation. J Intern Med 259: 351–363.

52. Wang J, Pan S, Berk BC (2007) Glutaredoxin mediates Akt and eNOS activation by flow in a glutathione reductase-dependent manner. Arterioscler Thromb Vasc Biol 27: 1283–1288.

53. Suo J, Ferrara DE, Sorescu D, Guldberg RE, Taylor WR, et al. (2007) Hemodynamic shear stresses in mouse aortas: implications for atherogenesis. Arterioscler Thromb Vasc Biol 27: 346–351.

54. Flamm SD, Taki J, Moore R, Lewis SF, Keech F, et al. (1990) Redistribution of regional and organ blood volume and effect on cardiac function in relation to upright exercise intensity in healthy human subjects. Circulation 81: 1550–1559.

55. Jorfeldt L, Wahren J (1971) Leg blood flow during exercise in man. Clin Sci 41: 459–473.

56. Boushel R, Langberg H, Green S, Skovgaard D, Bulow J, et al. (2000) Blood flow and oxygenation in peritendinous tissue and calf muscle during dynamic exercise in humans. J Physiol 524 Pt 1: 305–313.

57. Laughlin MH (1996) Section 12: Control of blood flow to cardiac and skeletal muscle during exercise. Handbook of pathophysiology Regulation and integration of multiple systems.

58. Rowell LB (1974) Human cardiovascular adjustments to exercise and thermal stress. Physiol Rev 54: 75–159.

59. Bradley SE, Childs AW, Combes B, Cournand A, Wade OL, et al. (1956) The effect of exercise on the splanchnic blood flow and splanchnic blood volume in normal man. Clin Sci (Lond) 15: 457–463.

60. Chiu JJ, Chien S (2011) Effects of disturbed flow on vascular endothelium: pathophysiological basis and clinical perspectives. Physiol Rev 91: 327–387.

61. Traub O, Berk BC (1998) Laminar shear stress: mechanisms by which endothelial cells transduce an atheroprotective force. Arterioscler Thromb Vasc Biol 18: 677–685.

17

Bacterial Fucose-Rich Polysaccharide Stabilizes MAPK-Mediated Nrf2/Keap1 Signaling by Directly Scavenging Reactive Oxygen Species during Hydrogen Peroxide-Induced Apoptosis of Human Lung Fibroblast Cells

Sougata Roy Chowdhury[1,2], Suman Sengupta[2☾], Subir Biswas[2☾], Tridib Kumar Sinha[1], Ramkrishna Sen[3]*, Ratan Kumar Basak[1], Basudam Adhikari[1], Arindam Bhattacharyya[2]*

1 Materials Science Centre, Indian Institute of Technology Kharagpur, West Bengal, India, **2** Immunology lab, Department of Zoology, University of Calcutta, West Bengal, India, **3** Department of Biotechnology, Indian Institute of Technology Kharagpur, West Bengal, India

Abstract

Continuous free radical assault upsets cellular homeostasis and dysregulates associated signaling pathways to promote stress-induced cell death. In spite of the continuous development and implementation of effective therapeutic strategies, limitations in treatments for stress-induced toxicities remain. The purpose of the present study was to determine the potential therapeutic efficacy of bacterial fucose polysaccharides against hydrogen peroxide (H_2O_2)-induced stress in human lung fibroblast (WI38) cells and to understand the associated molecular mechanisms. In two different fermentation processes, *Bacillus megaterium* RB-05 biosynthesized two non-identical fucose polysaccharides; of these, the polysaccharide having a high-fucose content (~42%) conferred the maximum free radical scavenging efficiency *in vitro*. Structural characterizations of the purified polysaccharides were performed using HPLC, GC-MS, and ^1H/^{13}C/2D-COSY NMR. H_2O_2 (300 µM) insult to WI38 cells showed anti-proliferative effects by inducing intracellular reactive oxygen species (ROS) and by disrupting mitochondrial membrane permeability, followed by apoptosis. The polysaccharide (250 µg/mL) attenuated the cell death process by directly scavenging intracellular ROS rather than activating endogenous antioxidant enzymes. This process encompasses inhibition of caspase-9/3/7, a decrease in the ratio of Bax/Bcl2, relocalization of translocated Bax and cytochrome c, upregulation of anti-apoptotic members of the Bcl2 family and a decrease in the phosphorylation of MAPKs (mitogen activated protein kinases). Furthermore, cellular homeostasis was re-established via stabilization of MAPK-mediated Nrf2/Keap1 signaling and transcription of downstream cytoprotective genes. This molecular study uniquely introduces a fucose-rich bacterial polysaccharide as a potential inhibitor of H_2O_2-induced stress and toxicities.

Editor: Andreas Villunger, Innsbruck Medical University, Austria

Funding: All the financial support of this paper is from the institutional fund of Indian Institute of Technology Kharagpur, Kharagpur, India. The funders had no role in study design, data collection and analysis, decision to publish, or preparation of the manuscript.

Competing Interests: The authors have declared that no competing interests exist.

* Email: rksen@yahoo.com (RS); arindam19@yahoo.com (AB)

☾ These authors contributed equally to this work.

Introduction

Reactive oxygen species (ROS), which are derivatives of cellular metabolic reactions, modulate the fundamental physiological functions of aerobic life. Normally, aerobic organisms use oxygen as the terminal electron receptor during oxidative phosphorylation, which is likely the greatest source of free oxygen radicals for cells under normal circumstances. Oxidative stress is a specific cellular stress in which the physiological ratio between oxidants and reductants stands in favor of the oxidant, creating species such as free oxygen radicals. At times, excessive generation of these radicals due to a high specific-growth rate, a surge in the level of respiration or from exogenous factors may trigger degenerative cellular disorders in eukaryotes. There is increasing evidence indicating that ROS and other oxygen-derived free radicals may contribute to a variety of pathological effects and induce diseases, including aging, cancer, neurodegenerative disorders, atherosclerosis, lung damage, diabetes and rheumatoid arthritis [1,2].

The mechanism by which oxygen exerts toxicity has been extensively studied. Free radicals, which are formed by a one-electron reduction of molecular oxygen (O_2), tend to stabilize themselves by retracting electrons from biological macromolecules such as proteins, lipids and DNA. ROS accelerate the peroxidation of membrane lipids (phospholipids and lipoproteins), thus damaging the cell membrane in a chain reaction [3–5]. In aerobic life, endogenous antioxidants shield against the ill effects of oxidative stress. However, various pathological processes inhibit these protective mechanisms. Therefore, supplementation of antioxidants becomes necessary at times. Cellular oxidative damage is challenged by both synthetic and natural antioxidants. Different antioxidants of synthetic origin are commercially available, although in recent studies, many were found to have deleterious side effects [2,6,7]. Hence, natural antioxidants with the fewest side effects and acceptable biocompatibility remain in demand. Literature review reveals that many polysaccharides of plant and microbial origin stimulate a range of biological effects [8,9], including free radical scavenging and antioxidant activity [10].

Fucose is found only in its L-conformation in vertebrate glycoconjugates. Phylogenetically, fucose first appeared in algal and fungal polysaccharides. Later, fucose also appeared in bacterial and plant glycoconjugates. L-fucose, as well as some other monosaccharides, exhibits several interesting biological properties independent from its metabolic fate. Recently, researchers have demonstrated that L-fucose and fucose-rich oligosaccharides are potent free radical scavengers during ascorbate- and ROS-induced toxicity [11,12].

Apoptosis, unlike necrosis, is a tightly regulated type of cell death in which a cell effectively executes its own demise in a programmed manner. Recent evidence confirmed that ROS and the resulting oxidative stress play a critical role in apoptosis [13,14]. Apoptotic cell death is usually associated with the activation of the extrinsic or intrinsic pathway and occasionally with both. The extrinsic pathway is initiated by the activation of death receptors, which leads to cleavage of caspase-8. The intrinsic pathway involves mitochondrial dysfunction, release of cytochrome c and subsequent activation of caspase-9. Mitochondria unarguably play a pivotal role in the induction and control of stress-mediated apoptosis. The pro- and anti-apoptotic proteins of the Bcl-2 (B-cell lymphoma 2) family are reported to be the crucial regulators of cell signaling in the mitochondria-dependent intrinsic pathway of apoptosis [15–19]. The Bcl-2 family, together with downstream proteins, maintains a dynamic balance between cell death and cell survival [19]. Several synthetic and natural antioxidants are reported to influence Bcl-2 family proteins. N-acetyl cysteine and Chaga extract effectively upregulated Bcl-2 against glucose/glucose oxidase (G/GO) and hydrogen peroxide treatments, respectively [20,21].

In redox-sensitive signaling pathways, the Nrf2/Keap1 (nuclear factor [erythroid-derived 2]-like 2/Kelch-like ECH-associated protein 1) axis plays a crucial role as a regulator of cellular response against endogenous or exogenous electrophilic assaults. Under normal conditions without any predominant electrophilic stress, Nrf2 remains associated with Keap1 primarily in the cytoplasm and is later subjected to proteasomal degradation [22–24]. The Nrf2/Keap1 pathway effectively regulates the downstream transcription of genes encoding phase II detoxifying enzymes such as heme oxygenase-1 (OH-1/HMOX1), glutathione S-transferase a 2 (GASTA2), nicotinamide adenine dinucleotide phosphate quinone oxidoreductase 1 (NQO1), and glutathione peroxidase 1 (GPX1), among others [25]. However, when their components interact with the Nrf2/Keap1 complex, some major signaling pathways may affect the regulation of antioxidant response element (ARE)-responsive genes. In some previous reports, the mitogen-activated protein kinases (MAPKs), which are important mediators of stress-induced apoptosis, were also reported to be the mediators of the Nrf2/Keap1/ARE signaling pathway [26,27]. The MAPK family primarily consists of three relatively well-studied kinases: the extracellular signal-regulated protein kinases (ERKs), the c-Jun N-terminal kinases (JNKs), and the p38 kinases, all of which trigger the signaling cascade by phosphorylation on either serine or threonine residues flanking a proline residue [16,28]. ERK2 and p38 MAPK positively regulate Nrf2 activity to initiate the transcription of antioxidant genes [29].

In this study, two non-identical, bacterially synthesized fucose polysaccharides were identified and compared as potent free radical scavengers in $vitro$. Furthermore, the polysaccharide showing greater antioxidant potential was challenged with hydrogen peroxide-induced stress in a human embryonic lung fibroblast cell line (WI38). In addition to initial structural characterization, experiments were performed to identify the underlying cell signaling mechanisms during H_2O_2-induced stress and polysaccharide treatment. This study uniquely highlights composition-based antioxidant efficiency of bacterial fucose polysaccharides and the mechanism of combating stress to attain cellular homeostasis via direct ROS scavenging rather than activating endogenous antioxidant enzymes.

Results

Bacterial exopolysaccharides (EPSs)

An EPS-producing bacterial strain, $Bacillus$ $megaterium$ RB05, was isolated from riverine sediments and identified by 16S rRNA gene sequencing. Two EPSs were biosynthesized, one in glucose mineral salts medium (GMSM) and one in GMSM-supplemented jute culture (JC). The EPSs were duly purified as described previously [30,31]. The apparent average molecular weight of the JC-derived EPS was 1.28×10^5 Da, whereas an apparent molecular weight of 1.7×10^5 Da was observed for the EPS synthesized in GMSM. The carbohydrate and protein contents of the EPS produced in GMSM were 95.7 ± 3.5 and 2.5 ± 0.4 wt.%, respectively, whereas the corresponding values for the JC-derived EPS were 93.9 ± 3.1 and 3.5 ± 0.8 wt.%, respectively. After hydrolysis and anthranilic acid derivatization, the purified EPSs were analyzed for sugar composition by HPLC. Galactose (37.6%) was found to be the major monosaccharide by weight for the GMSM-derived EPS, followed by arabinose (20.2%), mannose (19.3%) and glucose (14.0%); fucose (4.9%) and N-acetyl glucosamine (4.0%) were present as minor fractions. For the JC-derived EPS, fucose was identified as the primary functional monomer (41.9%), followed by glucose (26.6%), mannose (15.8%), galactose (12.2%), and N-acetyl glucosamine (3.5%). Based on the weight percentage of fucose content, the two EPSs were classified as a low-fucose-content (LFC) polysaccharide and a high-fucose-content (HFC) polysaccharide [30–33].

Free radical scavenging potential of the polysaccharides in vitro

The free radical scavenging potential of the polysaccharides was tested chemically before introduction into human cell lines. Free radicals and electrophiles were generated chemically in a system and scavenged by the added polysaccharides. The in $vitro$ free radical scavenging potential of the polysaccharides in terms of their IC_{50} values are shown in Table S1. In most cases, the values were found to be similar to the positive control values. Figures 1A and B show the ability of the polysaccharides to scavenge hydroxyl radicals in site-specific and non-specific reactions. In the non-specific cases, the LFC and HFC polysaccharides displayed similar scavenging performances at 1 mg/mL (95% and 93%), whereas for the site-specific reactions, the activity of the HFC polysaccharide was statistically superior ($p<0.05$) at the same concentration. Both the LFC and HFC polysaccharides scavenged H_2O_2 almost as well as the standard control sodium pyruvate (Figure 1C), although the IC_{50} values were comparatively higher; this might be due to the concentration of H_2O_2 used during the experiment. The polysaccharides showed a moderate dose-dependent scavenging effect against the singlet oxygen species, with IC_{50} values of 0.425±0.024 mg/mL and 0.175±0.028 mg/mL, respectively, for the LFC and HFC polysaccharides (Figure 1D). Figures 1E and F show the quenching of superoxide radicals in non-enzymatic and enzymatic reactions. The scavenging potential of the polysaccharides was found to be superior to that of the reference compound. In the enzymatic reaction, administration of these polysaccharides inhibited the process of free radical generation throughout their lifespan. As depicted in Table S1 and Figure 1G, the HFC (69%) and LFC (59%) polysaccharides successfully scavenged DPPH radical, but to a lower degree than the standard ascorbic acid (92%). The total antioxidant activity of these polysaccharides was evaluated in two different systems, one based on scavenging performance against $ABTS^{\cdot+}$ and the other via bleaching of β-carotene in the β-carotene-linoleate model system. In the first model, the total antioxidant capacity of the polysaccharides was calculated from the decolorization of $ABTS^{\cdot+}$. The polysaccharides were found to be more effective than the standard Trolox at suppressing $ABTS^{\cdot+}$ at the same concentrations. The results, which are expressed as percentage inhibition of absorbance, are shown in Figure 1H. In the β-carotene-linoleate model system (Figure 1I), the antioxidant performance of the LFC and HFC polysaccharides was found to be 82% and 83.5%, respectively, at 0.1 mg/mL. The antioxidant behavior of the reference compound butylated hydroxyanisole (BHA) increased in a concentration-independent manner in the present system. From the above results, one could easily consider these polysaccharides to be potent free radical scavengers in vitro. However, in most cases, the HFC polysaccharide was found to perform better than the LFC polysaccharide. The HFC polysaccharide was therefore subjected to further experiments to understand the structure and the cellular biology of the antioxidant during oxidative stress.

Structural elucidation of the HFC polysaccharide

The sugar linkages in alditol acetates of the methylated sugars derived from the HFC polysaccharides were elucidated using GC-MS (Table S2). The linkages between the monomers were analyzed according to the description by Bjorndal et al. [34]. The ^1H NMR spectrum (Figure 2A) of the HFC polysaccharide at 40°C showed three signals at δ 5.11, 5.24, and 5.23 ppm for three anomeric protons corresponding to α-linked glycopyranose residues and five signals at δ 4.62, 4.54, 4.64, 4.55, and 4.68 ppm for five anomeric protons corresponding to β-linked residues. The sugar residues were designated as A–H according to their decreasing anomeric proton chemical shifts (Table S3). In the ^{13}C NMR spectrum (Figure 2B) at 40°C, three signals appeared at δ 100.94, 102.02, and 100.19 ppm for three anomeric carbons corresponding to α-linked glycopyranose residues and five signals at δ 102.59, 105.17, 101.82, 103.60, and 95.75 ppm for five anomeric carbons corresponded to β-linked residues. The sugar residues were designated as A–H according to their decreasing anomeric carbon chemical shifts (Table S3). The anomeric proton chemical shift for residue A at δ 5.11 ppm and a carbon chemical shift of 100.94 ppm indicated that it is an α-linked anomer. The downfield shift of C-1(δ 100.94 ppm), C-2 (δ 77.40 ppm) and C-4 (δ 74.76 ppm) with respect to the standard values [35,36] and the characteristic $J_{H=1, H=2}$ coupling constant value (3.4 Hz) indicated that residue A is a (1 → 2,4)-linked-α-D-mannopyranosyl moiety. The anomeric proton chemical shift for residue B at δ 5.24 ppm and a carbon chemical shift of 102.02 ppm indicated that it is an α-linked anomer. The downfield shift of C-1(δ 102.02 ppm), C-4 (δ 74.84 ppm) and C-6 (δ 69.79 ppm) and the characteristic $J_{H=1, H=2}$ coupling constant value (3.5 Hz) indicated that residue B is a (1 → 4,6)-linked-α-D-mannopyr-

anosyl moiety. Residue C displayed an anomeric proton signal at δ 5.23 ppm and a carbon chemical shift at δ 100.19 ppm. The downfield shift of C-1 (δ 100.19 ppm) and C-3 (δ 74.24 ppm) along with a characteristic $J_{H=1, H=2}$ coupling constant value of 3.7 Hz signified residue C as (1 → 3)-linked-α-D-fucopyranosyl moiety. Additionally, a slight downfield shift of C-4 (δ 74.07 ppm) also suggested the presence of a sulfate group (SO_3^-) within this moiety. Residue D was found to be a (1 → 2,4)-linked-β-D-galactopyranosyl moiety showing anomeric proton and carbon signals at δ 4.62 and 4.54 ppm, respectively, and residue E was identified as a (1 → 4)-linked-β-D-galactopyranosyl moiety with anomeric proton and carbon signals at δ 102.59 and 105.17 ppm, respectively. Downfield shifts were observed at C-1(δ 102.59 ppm), C-2 (δ 79.57 ppm), and C-4 (δ 78.61 ppm) for residue D and at C-1 (δ 105.17 ppm) and C-4 (δ 78.63 ppm) for residue E with a $J_{H=1, H=2}$ coupling constant value of ~8.2 Hz. Residue F was found to be a (1 → 2,4)-linked-β-D-glucopyranosyl moiety showing anomeric proton and carbon signals at δ 4.64 and 4.55 ppm, respectively, and residue G was identified as a (1 → 4)-linked-β-D-glucopyranosyl moiety with anomeric proton and carbon signals at δ101.82 and 103.60 ppm, respectively. Downfield shifts were observed at C-1(δ 101.82 ppm), C-2 (δ 81.14 ppm), and C-4 (δ 79.41 ppm) for residue F and at C-1 (δ 103.60 ppm) and C-4 (δ 79.66 ppm) for residue G with a characteristic $J_{H=1, H=2}$ coupling constant value of ~8 Hz. Residue H displayed an anomeric proton signal at δ 4.68 ppm and a carbon chemical shift at δ 95.75 ppm. The downfield shift of only C-4 (δ 74.24 ppm) along with the characteristic $J_{H=1, H=2}$ coupling constant value of ~9 Hz signified residue H as a terminal β-D-GlcNAcp. Figure 3 shows the postulated structure of the HFC polysaccharide.

Effect of H_2O_2 and polysaccharides on the growth of human lung fibroblast cells

MTT (3-(4,5-dimethyl-2-thiazolyl)-2,5-diphenyl-2H-tetrazolium bromide) assays were performed for primary dose selection. WI38 cells were treated with H_2O_2 at different doses (100–500 μM) for the indicated period of time (0–24 h). As shown in Figure S1, cell growth was inhibited by 76% after treatment with 300 μM H_2O_2 for 24 h. The cells were pre-treated with the LFC and HFC polysaccharides for 2 h prior to H_2O_2 administration. After 24 h of treatment, the inhibition of cell growth by H_2O_2 in the presence of the LFC and HFC polysaccharides was found to be reduced to 27.3 and 13.4%, respectively, of the inhibition of cell growth by H_2O_2 alone at 250 μg/mL. However, at higher concentrations (500 μg/mL), the polysaccharides further inhibited growth compared with H_2O_2 treatment alone. Henceforth, selection of the optimal dose is critical to achieve the maximum response. Collectively, these data indicated that both polysaccharides displayed protective effects against the growth inhibition caused by oxidative stress over a range of concentrations, although the performance index of the HFC polysaccharide was comparatively superior. Thus, the HFC polysaccharide was preferentially adopted for treatments in the future experiments.

The viability of the cells during both the stress and polysaccharide treatment period was also determined via optical imaging by phase-contrast microscopy (Figure 4). During the stress period (300 μM H_2O_2), fewer live cells were found, and the cells appeared to be morphologically misshapen and distorted as time progressed. The number of viable cells increased with polysaccharide treatment in a time- and dose-dependent manner, and the normal cell shape and size was observed after a treatment of at least 12 h.

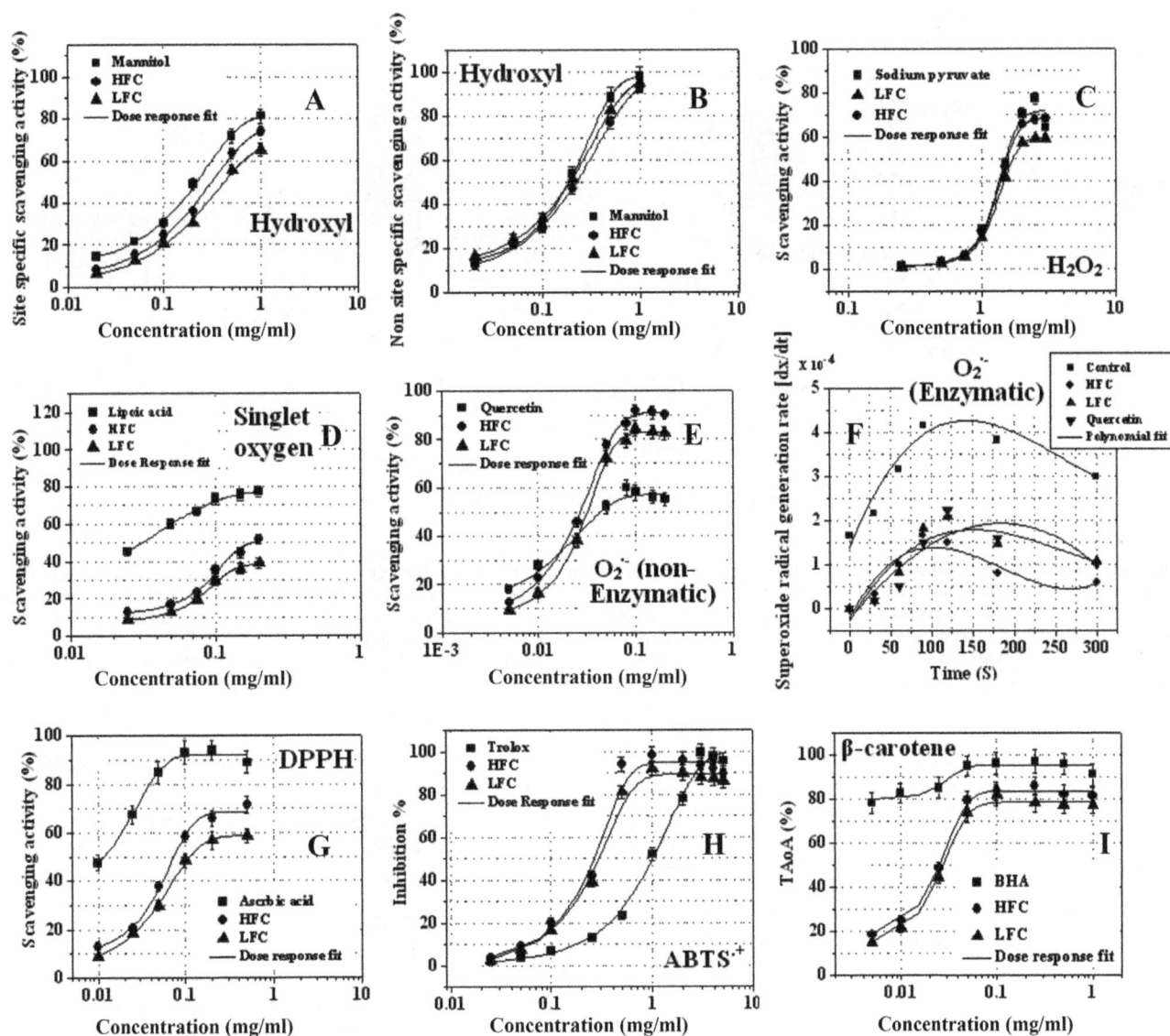

Figure 1. Scavenging performance of purified LFC and HFC polysaccharides against chemically-generated free radicals *in vitro*. A) hydroxyl radical (site specific); **B)** hydroxyl radical (non- specific); **C)** hydrogen peroxide; **D)** singlet oxygen; **E)** superoxide radical anion (non-enzymatic); **F)** superoxide radical anion (enzymatic); **G)** DPPH radical, **H)** total antioxidant capacity (ABTS^{+} scavenging); and **I)** total antioxidant capacity (β-carotene linoleate model). Results are representative of three independent experiments performed in triplicate and are represented as mean ± SD. Inhibition concentration (IC$_{50}$) values were calculated from pharmacological dose-response curve fit (sigmoidal) the equation: Y = Bottom + (Top-Bottom)/[1+10^{(LogIC50-X)* Hill slope}].

Polysaccharide inhibited H$_2$O$_2$-induced apoptosis

To determine whether H$_2$O$_2$-induced toxicity involves alteration in the cell cycle, the DNA content of the cells was analyzed by flow cytometry. Figure 5 shows that H$_2$O$_2$ (300 μM) significantly increased (p<0.05) the sub-G$_0$/G$_1$cell population with time. The cell population in this phase was found to be 60% of the total cells after 12 h exposure to H$_2$O$_2$. The HFC polysaccharide efficiently reduced the accumulation of cells in the sub-G$_0$/G$_1$ phase in a dose- and time-dependent manner. A polysaccharide dose of 250 μg/mL induced a significant decrease (p<0.05) in the percentage of cells in sub-G$_0$/G$_1$ phase after 24 h. This value decreased to 7.4%, which is similar to the value of the control cells (3.1%). However, H$_2$O$_2$ stress also disrupted the arrangement of phospholipids in the cell membrane [37]. The externalization of phosphatidylserine on the plasma membrane was detected by

annexin V-FITC staining in WI38 cells. Annexin V staining of the cells (Figure 6) indicated that H$_2$O$_2$-induced cell death was apoptotic and significantly increased (p<0.05) with time. During the polysaccharide treatment, the number of annexin V-positive cells was found to be significantly decreased (p<0.05). The HFC polysaccharide (250 μg/mL) reduced the percentage of annexin V-positive cells from 42.7% to 6.8% after 24 h. Similarly, a considerable reduction in the necrotic population was also observed during the polysaccharide treatment period.

Effect of ROS and polysaccharide treatment on H$_2$O$_2$-induced apoptosis

A fluorescent probe, H$_2$DCF-DA (2',7'-dichlorodihydrofluorescein diacetate), was used to measure the generation of intracellular ROS within WI38 cells during H$_2$O$_2$ stress. As

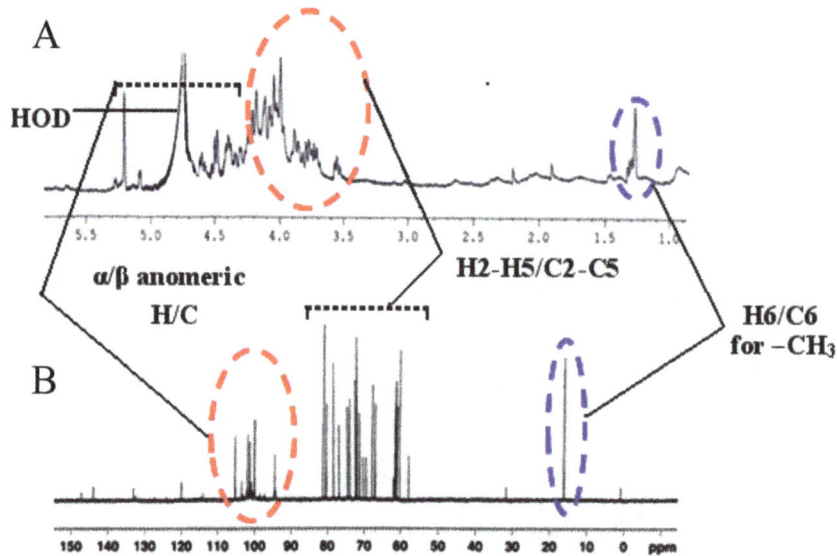

Figure 2. Partial structure elucidation of HFC polysaccharide by NMR spectroscopy. The purified samples were exchanged with deuterium by lyophilizing several times with D_2O. The polysaccharide was dissolved in 0.7 mL of D_2O (99.96%) at concentrations of 15 mg/mL (for ^1H NMR) and 30 mg/mL (for ^{13}C NMR). Spectra were run at a probe temperature of 40°C. **A)** in the ^1H NMR spectrum, black dotted line indicating α and β proton anomers, and dotted circles displaying H2–H5 and H6 (–CH_3 group) of the polysaccharide directed from downfield to upfield. **B)** in the ^{13}C NMR spectrum, downfield dotted circle stands for α and β carbon anomers, black dotted line indicating C2–C5, and upfield dotted circle signifying C6 (–CH_3 group) of the polysaccharide.

shown in Figure 7, the intensity of the DCF-liberated fluorescent signal from the stress-treated cells was found to increase gradually with time (maximum after 3 h), indicating an elevated level of intracellular ROS compared with the control basal level of ROS. Pre-treatment with the HFC polysaccharide significantly decreased (p<0.05) the cell fluorescence with time. This decrease in the fluorescence signal was dose specific. Treatment with the polysaccharide inhibited fluorescence to a greater degree than treatment with N-acetyl cysteine (NAC), which was used as positive control.

Regulation of mitochondrial function and expression of Bcl-2 family proteins

Mitochondrial membrane potential ($\Delta\Psi_m$) regulates mitochondrial permeability, which may be critical for inducing or arresting the stress-induced apoptotic pathways [37,38]. The effect of H_2O_2-induced stress on $\Delta\Psi_m$ of WI38 cells was examined using flow cytometry following the DiOC6 (3,3'-dihexyloxacarbocyanine iodide) staining method. As shown in Figure 8A, when WI38 cells were exposed to 300 μM H_2O_2, $\Delta\Psi_m$ decreased, resulting in a significant drop in potential (p<0.05). Although the membrane regained potential with time, this change was not significant (p> 0.05). The HFC polysaccharide (250 μg/mL) induced a gradual

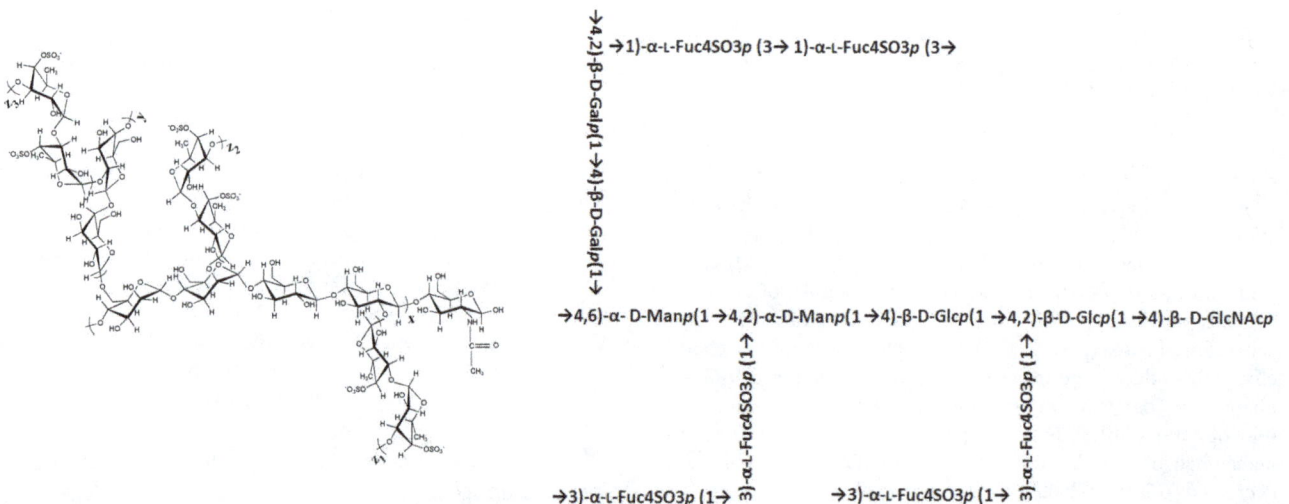

Figure 3. The possible structure interpreted for HFC polysaccharide.

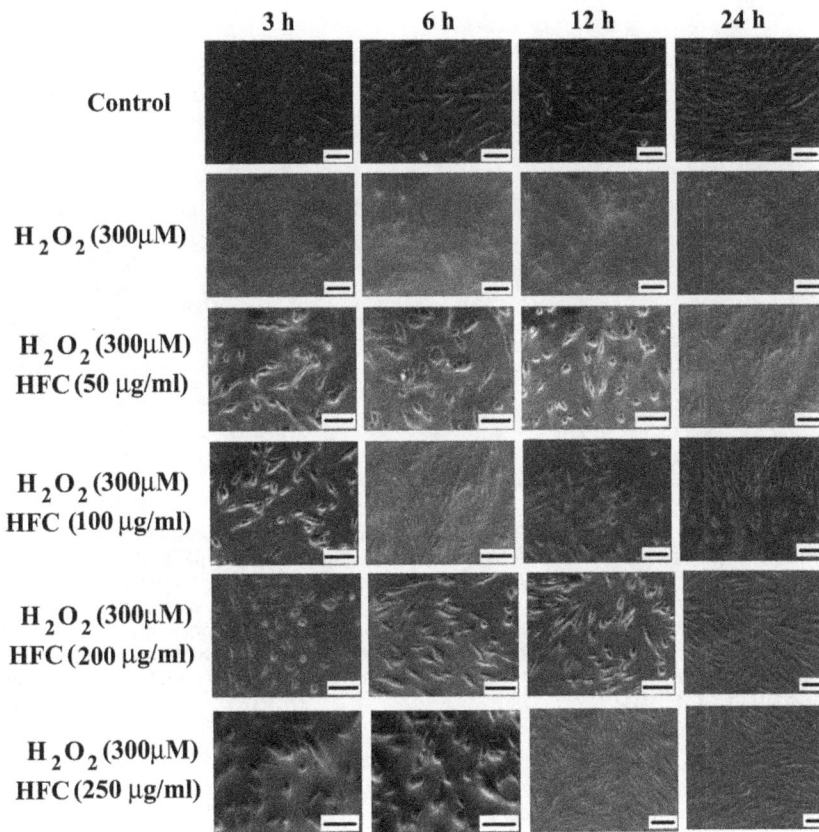

Figure 4. The effect of HFC polysaccharide on H$_2$O$_2$-induced morphological changes in WI38 cells. The cells were incubated in presence (100, 200 and 250 µg/mL) or in absence of HFC polysaccharide for 1 h followed by the treatment with 300 µM H$_2$O$_2$ in both the cases for varying periods of time (0–24 h). Cell morphology was observed under microscope in phase contrast mode. Results are representative of three independent experiments performed in triplicate. Indicated *scale bars* signify 50 µm distance and photographs were taken at 10× zoom.

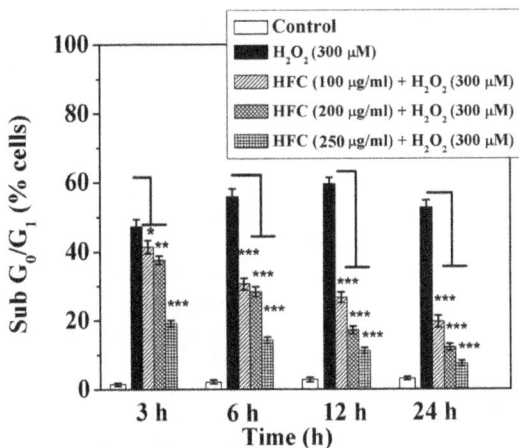

Figure 5. The protective effect of HFC polysaccharide on H$_2$O$_2$-induced DNA damage during cell cycle progression. WI38 cells were incubated in presence (100, 200 and 250 µg/mL) or in absence of HFC polysaccharide for 1 h followed by the treatment with 300 µM H$_2$O$_2$ in both the cases for varying periods of time (0–24 h). The cell cycle progression was assayed by PI staining with flow-cytometry. The corresponding data are shown as bar graphs. Results are representative of three independent experiments performed in triplicate and are represented as mean ± SD. A one-way analysis of variance (ANOVA, Bonferroni corrections for multiple comparisons) was performed, where significant level stands for *p<0.05, **p<0.001, ***p<0.0001.

increase in mitochondrial membrane potential in a time- and dose- dependent manner. No significant loss or gain in potential was observed after the administration of the polysaccharide without H$_2$O$_2$ exposure. The Bcl-2 family proteins have been reported to regulate $\Delta\Psi_m$ [19]. Therefore, the expression of Bcl-2 family proteins and their corresponding genes was measured in H$_2$O$_2$- and HFC polysaccharide-treated WI38 cells.

Immunoblotting (Figure 8B and Figure S2) revealed that the expression of anti-apoptotic proteins such as Bcl-2 and Bcl-xl was found to be suppressed in the cells where stress was induced (the levels were 44% and 48%, respectively, of those of the control after 24 h). Bad was the least responsive against the applied stress (p> 0.05). The HFC polysaccharide relieved the stress by arresting the apoptotic pathway in a time-dependent manner. The ratio between Bax (Bcl2-associated × protein) and Bcl-2 was measured using band-densitometry. As shown in Figure 8C, H$_2$O$_2$-induced stress increased the ratio (4.8-fold increase compared with the control after 24 h), whereas polysaccharide treatment for 24 h decreased the ratio to a level that was not significantly different from that of the control cells (p>0.05).

The expression of the Bcl-2 family proteins was quantified using real-time reverse transcriptase (RT-PCR) (Figure 8D). Even at the mRNA level, significant alterations (p<0.05) in expression pattern, in terms of fold change ($2^{\Delta\Delta Ct}$), were observed for Bcl-2, Bcl-xl, and Bax during the stress and treatment period. The mRNA expression of Bcl-2 and Bcl-xl was downregulated (5.9- and 5.2-fold decreases, respectively, compared with the control after 24 h),

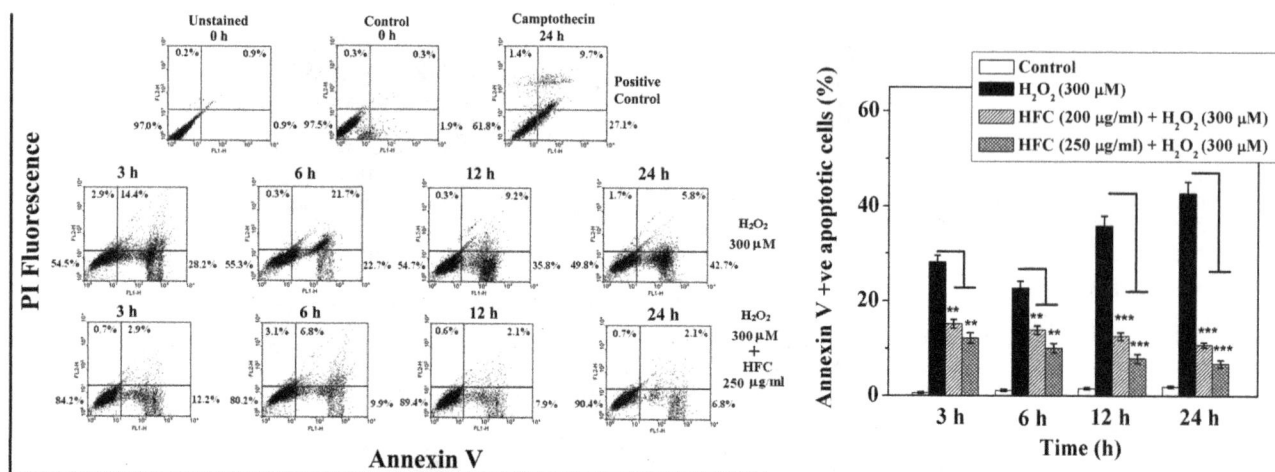

Figure 6. Inhibitory effects of HFC polysaccharide on H₂O₂-induced apoptosis of WI38 cells. WI38 cells were incubated in presence (200 and 250 µg/mL) or in absence of HFC polysaccharide for 1 h followed by the treatment with 300 µM H₂O₂ in both the cases for varying periods of time (0–24 h). Cellular apoptosis was assayed by annexin V-FITC and PI counterstaining and analyzed with flow cytometry. Camptothecin was used as positive control. Dual parameter dot plot of FITC fluorescence (x-axis) versus PI fluorescence (y-axis) is represented as logarithmic fluorescence intensity. Quadrants: upper left necrotic cells, lower left live cells, lower right apoptotic cells, and upper right necrotic or late phase of apoptotic cells. The corresponding data are shown as bar graphs. Results are representative of three independent experiments performed in triplicate and are represented as mean ± SD. A one-way analysis of variance (ANOVA, Bonferroni corrections for multiple comparisons) was performed, where significant level stands for ** p<0.001, *** p<0.0001.

whereas the expression of Bax was upregulated (5.7-fold increase compared with the control after 24 h) with H₂O₂ treatment. For Bad and cytochrome c, the fold changes were moderate (2.8- and 1.9-fold increases, respectively, compared with the control after 24 h). The HFC polysaccharide actively regulated the aforementioned genes. After 24 h of treatment, a significant decrease (p<

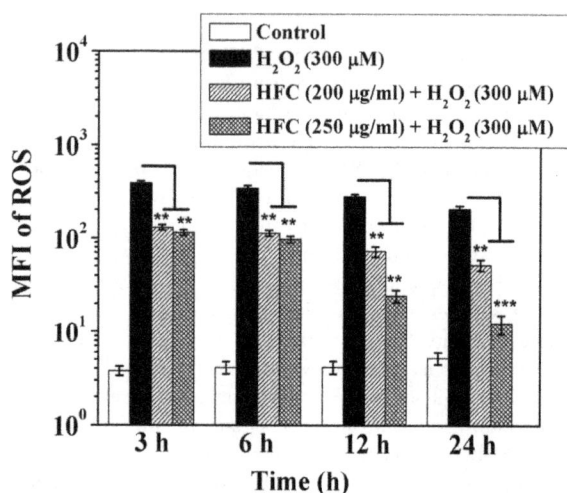

Figure 7. Stabilization of H₂O₂-induced intracellular ROS by HFC polysaccharide. WI38 cells were incubated in presence (200 and 250 µg/mL) or in absence of HFC polysaccharide for 1 h followed by the treatment with 300 µM H₂O₂ in both the cases for varying periods of time (0–24 h). ROS levels were monitored by flow cytometry using H₂DCF-DA. N-acetyl cysteine (NAC) was used as positive control. The mean fluorescence indices (MFI) are shown as bar graphs. Results are representative of three independent experiments performed in triplicate and are represented as mean ± SD. A one-way analysis of variance (ANOVA, Bonferroni corrections for multiple comparisons) was performed, where significant level stands for ** p<0.001, *** p<0.0001.

0.05) in the mRNA expression was observed, and the values were observed to be similar to those of untreated cells.

Translocation of Bax and cytochrome c

Mitochondrial damage as a result of stress-induced apoptosis is often accompanied by the release of apoptotic factors into the cytosol [19]. Evidence indicates that the release of cytochrome c into the cytosol induces cleavage of caspase-9. This, in turn, activates caspase-3, which plays an important role in the execution of stress-induced apoptosis in different cell types [39]. Therefore, the translocation of cytochrome c and Bax was studied using western blotting (Figure 9A and Figure S2). The localization of proteins during the time-course of stress and polysaccharide treatment was also determined by immunofluorescence using specific antibodies. In the untreated control cells, cytochrome c immunoreactivity (green fluorescence) colocalized with the Mito-Tracker red fluorescence, indicating the mitochondrial association of cytochrome c (Figure 9B). After H₂O₂ exposure, the stressed cells exhibited diffusion of the green fluorescence from the mitochondria to the cytoplasm. A visual decrease in the fluorescence intensity was observed for the mitochondria-associated cytochrome c, thus indicating a release of cytochrome c from the mitochondria to the cytoplasm. However, with HFC polysaccharide treatment, cytochrome c successfully translocated back to and re-localized in the mitochondria in a time-dependent manner.

The Bax protein was immunostained using a Bax-specific antibody and a green fluorescent secondary antibody (Figure 9B). In the untreated control cells, Bax was found to be localized throughout the cytosol, but H₂O₂ treatment triggered Bax localization in the mitochondria. This mitochondrial association of Bax decreased with time as the mitochondrial membrane regained potential after polysaccharide administration. Western blots (Figure 9A) demonstrated an increase in cytosolic and mitochondrial pro-apoptotic Bax (4.2- and 3.4-fold increases, respectively, compared with the control after 24 h) during the

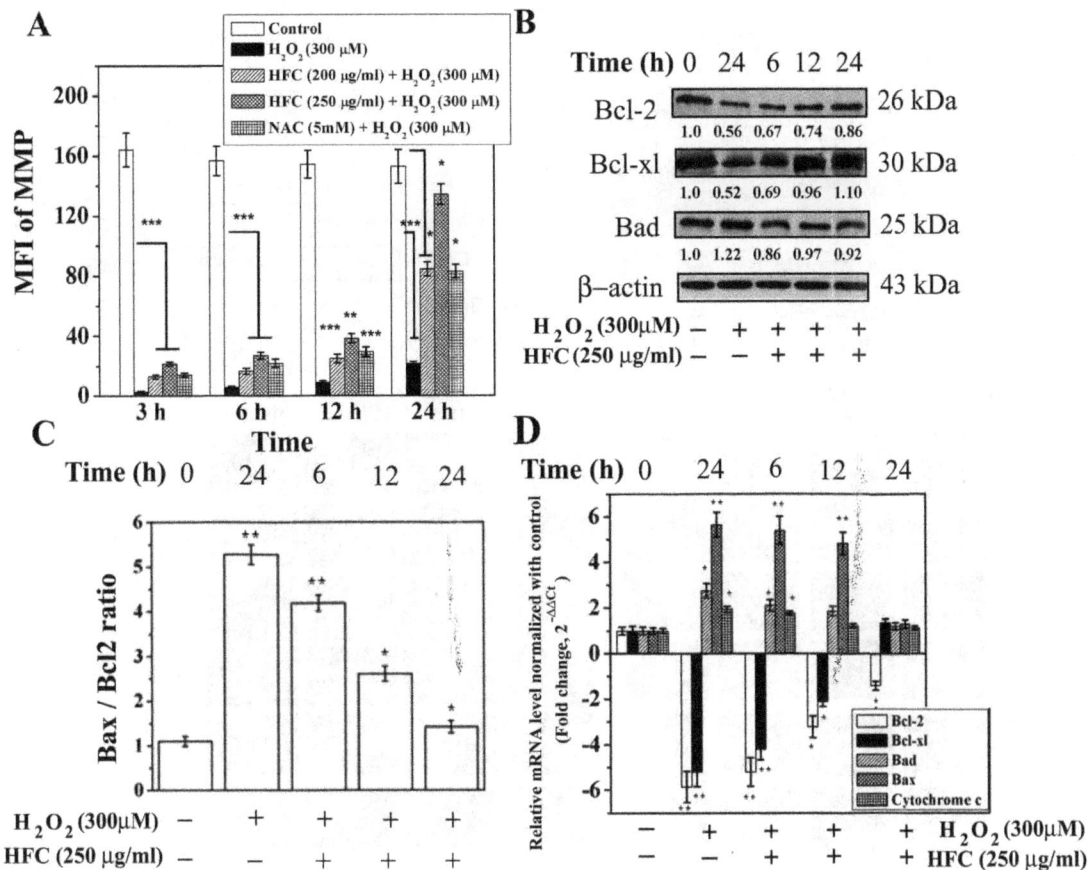

Figure 8. The effects of HFC polysaccharide treatment on H₂O₂-induced regulation of mitochondrial functions. The polysaccharide prevented H₂O₂-induced changes in the expression of Bcl2 family at both mRNA and protein level. WI38 cells were incubated in presence (200 and 250 µg/mL) or in absence of HFC polysaccharide for 1 h followed by the treatment with 300 µM H₂O₂ in both the cases for varying periods of time (0–24 h). **A)** Mitochondrial membrane potential (MMP) was monitored by DiOC6 staining with flow cytometry. The mean fluorescence indices (MFI) are shown as bar graphs. **B)** Protein level expression of Bcl2, Bcl-xl, and Bad was evaluated by immunoblotting. β-actin was used as loading control. Fold changes are represented as relative values of band densitometries normalized to control and are shown as numbers below the immunoblots. Results are representative of three independent experiments performed in triplicate and are represented as mean value. **C)** The ratio between Bax and Bcl2 were calculated from band densitometries of corresponding protein level expressions and are shown as bar graphs. **D)** Fold changes of Bcl2, Bcl-xl, Bad, Bax, and cytochrome c at mRNA level were calculated using real-time RT-PCR (SYBR green method). Fold changes are represented as relative values normalized to control and quantified in the terms of $2^{-\Delta\Delta Ct}$. GAPDH was used as internal control. Results are representative of three independent experiments performed in triplicate and are represented as mean ± SD. A one-way analysis of variance (ANOVA, Bonferroni corrections for multiple comparisons) was performed, where significant level stands for * $p<0.05$, ** $p<0.001$, *** $p<0.0001$.

stress period. At the same time, cytochrome c level in the cytosol was significantly increased (2.1-fold increase compared with the control after 24 h), although the mitochondrial level of cytochrome c was found to be significantly decreased (68% decrease compared with the control after 24 h). Treating the cells with the HFC polysaccharide reversed the translocation process over time. As a result, the mitochondrial level of cytochrome c increased with time (12% decrease compared with the control after 24 h), with an effective post-treatment increase in $\Delta\Psi_m$. The Bax level was decreased in both the cytosol and the mitochondria as a part of the polysaccharide treatment against stress (24% and 12% decreases, respectively, compared with the control after 24 h).

Caspase-dependent signaling pathway

The involvement of caspase proteins during the stress and the polysaccharide treatment periods was investigated by western blotting analysis. The ratio of pro and active caspases was calculated from band-densitometry. As depicted in Figure 10 and Figure S2, H₂O₂ downregulated the expression of procaspase-9

and significantly increased active caspase-9 (5.4-fold decrease in the ratio of pro/active caspase-9 compared with the control after 24 h). Similarly, a significant increase ($p<0.05$) in the levels of active effector caspase-3 and -7 (7.4- and 5.3-fold decreases in the ratio of pro/active caspase-3 and -7, respectively, compared with the control after 24 h), as well as cleavage of PARP (5.4-fold increase compared with the control after 24 h), was observed. In contrast, treatment with the HFC polysaccharide markedly decreased ($p<0.05$) the expression of active caspases, effector caspases, and cleavage of PARP in a time-dependent fashion (1.5-, 1.7-, and 1.3-fold decreases in the ratio of pro/active caspase-9, -3 and -7, respectively, and 1.4-fold increase in PARP cleavage compared with the control after 24 h).

In caspase inhibition assays, WI38 cells were pre-treated with the pan-caspase inhibitor Z-VAD-fmk (50 µM) for 1 h and then treated with the polysaccharide and H₂O₂ for 24 h. As shown in Figure 11A and B, Z-VAD-fmk effectively inhibited the H₂O₂-induced apoptosis of WI-38 cells (81.5% decrease in the percentage of apoptotic cells) and significantly reduced the level

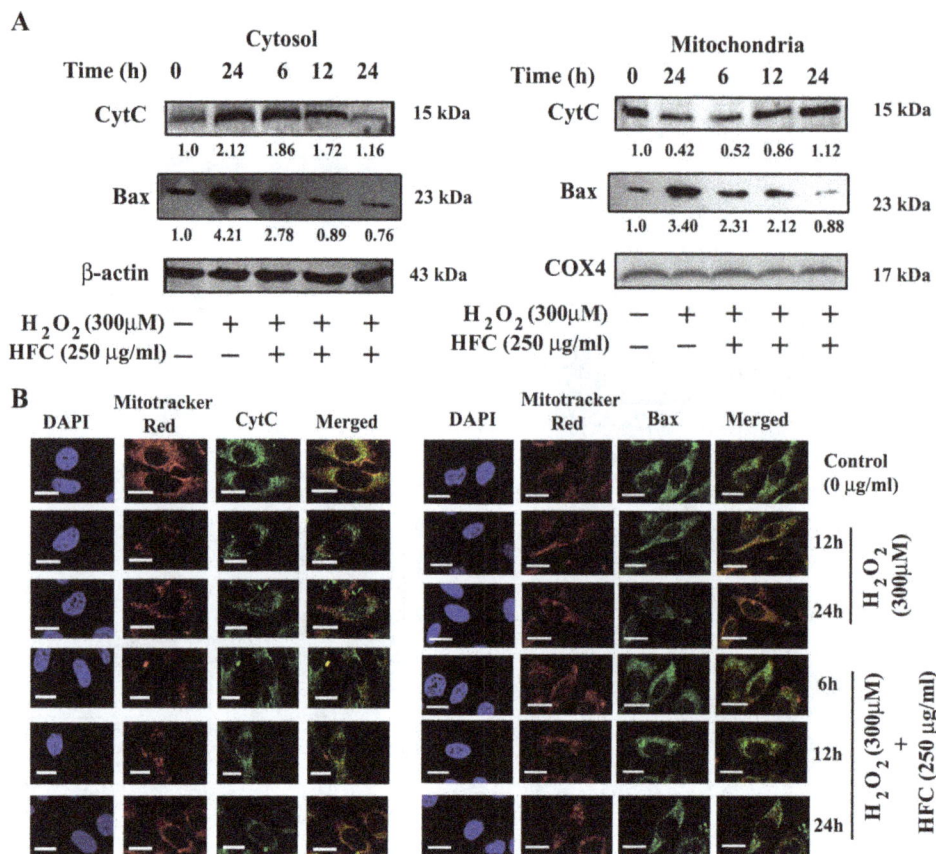

Figure 9. Translocation of Bax and cytochrome c induced by H$_2$O$_2$ and HFC polysaccharide during the stress and polysaccharide treatment period, respectively. WI38 cells were incubated in presence (250 µg/mL) or in absence of HFC polysaccharide for 1 h followed by the treatment with 300 µM H$_2$O$_2$ in both the cases for varying periods of time (0–24 h). **A)** Protein level expression of Bax and cytochrome c in both cytosolic and mitochondrial fractions was observed by immunoblotting. β-actin and COX4 were used as loading control. Fold changes are represented as relative values of band densitometries normalized to control and are shown as numbers below the immunoblots. Results are representative of three independent experiments performed in triplicate and are represented as mean value. A one-way analysis of variance (ANOVA, Bonferroni corrections for multiple comparisons) was performed, where significant level stands for * p<0.05, ** p<0.001. **B)** H$_2$O$_2$-induced release of cytochrome c from mitochondria to cytosol and re-localization into mitochondria again during the polysaccharide treatment were monitored under fluorescence microscope using fluorescence-tagged (green florescence) specific antibodies. Similarly, mitochondrial translocation of Bax and their cytosolic re-localization was also tracked following the same procedure. The cells were treated with 100 nM MitoTracker Red (red florescence) for 30 min before cell-fixation for mitochondrial staining. Each image shown is representative of 20 random fields observed. Indicated *scale bars* signify 10 µm distance and photographs were taken at 100× zoom.

of active caspase-9 (p<0.05). For further confirmation, the effects of the caspase-9-specific inhibitor Z-LEHD-fmk and the caspase 3/7-specific inhibitor Ac-DEVD-CHO were evaluated using flow cytometry (annexin V-PI staining). The inhibitors attenuated H$_2$O$_2$-induced apoptotic cell death (82.2% and 78.4% decreases for Z-LEHD-fmk and Ac-DEVD-CHO, respectively, compared with the cells treated with H$_2$O$_2$ only). Additionally, together with the HFC polysaccharide, the inhibitors further reduced the apoptotic cell numbers (a 90.4% decrease compared with the cells treated with H$_2$O$_2$ only).

The HFC polysaccharide downregulates H$_2$O$_2$-mediated phosphorylation of MAP kinases

Similar to caspase activation, the MAPK family is involved in a process that induces stress-mediated cell death [19]. In view of this evidence, the regulation of the MAPK signaling pathway (JNK, p38, and ERK) was studied during the stress and polysaccharide treatment periods. As shown in Figure S3, the phosphorylation level of all three kinases was found to be maximal after 6 h of

H$_2$O$_2$ treatment (83%, 114%, and 112% increases for JNK, p38, and ERK, respectively, compared with the control). The phosphorylation level significantly (p<0.05) decreased in response to the polysaccharide treatment after 6 h (25%, 33.6% and 28.3% decreases for JNK, p38, and ERK, respectively, compared with the cells treated with only H$_2$O$_2$ after 6 h). But at the later time periods, the corresponding rates of decrease was found not to be statistically significant (p>0.05) [data not shown].

Involvement of MAPK-mediated Nrf2/Keap1 signaling and transcriptional regulation of downstream genes

Nrf2 has been demonstrated to be a transcription factor that is usually sequestered in the cytoplasm by Keap1, but during oxidative stress, Nrf2 translocates to the nucleus, resulting in transcriptional activation of cytoprotective genes encoding phase II detoxifying enzymes. Therefore, in the current effort, the protein levels of Nrf2 and Keap1 in the cytoplasm and the nucleus were determined using western blotting, and the mRNA expression was examined using real-time RT-PCR. The translo-

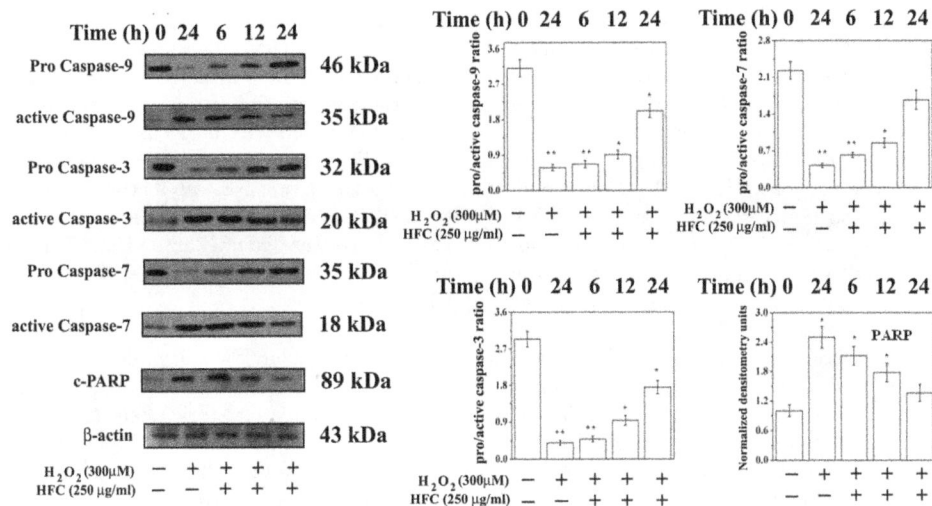

Figure 10. Polysaccharide imposed protection against H$_2$O$_2$-induced intrinsic apoptosis of WI38 cells via caspase inhibition at protein level. WI38 cells were incubated in presence (250 µg/mL) or in absence of HFC polysaccharide for 1 h followed by the treatment with 300 µM H$_2$O$_2$ in both the cases for varying periods of time (0–24 h). Protein level expression of various pro- and active forms of caspases (caspase-9, -3, and -7) and cleaved PARP was evaluated by immunoblotting. β-actin was used as loading control. Band densitometries are represented as a ratio between pro and active caspases in the form of bar graphs, Results are representative of three independent experiments performed in triplicate and are represented as mean ± SD. A one-way analysis of variance (ANOVA, Bonferroni corrections for multiple comparisons) was performed, where significant level stands for * p<0.05, ** p<0.001.

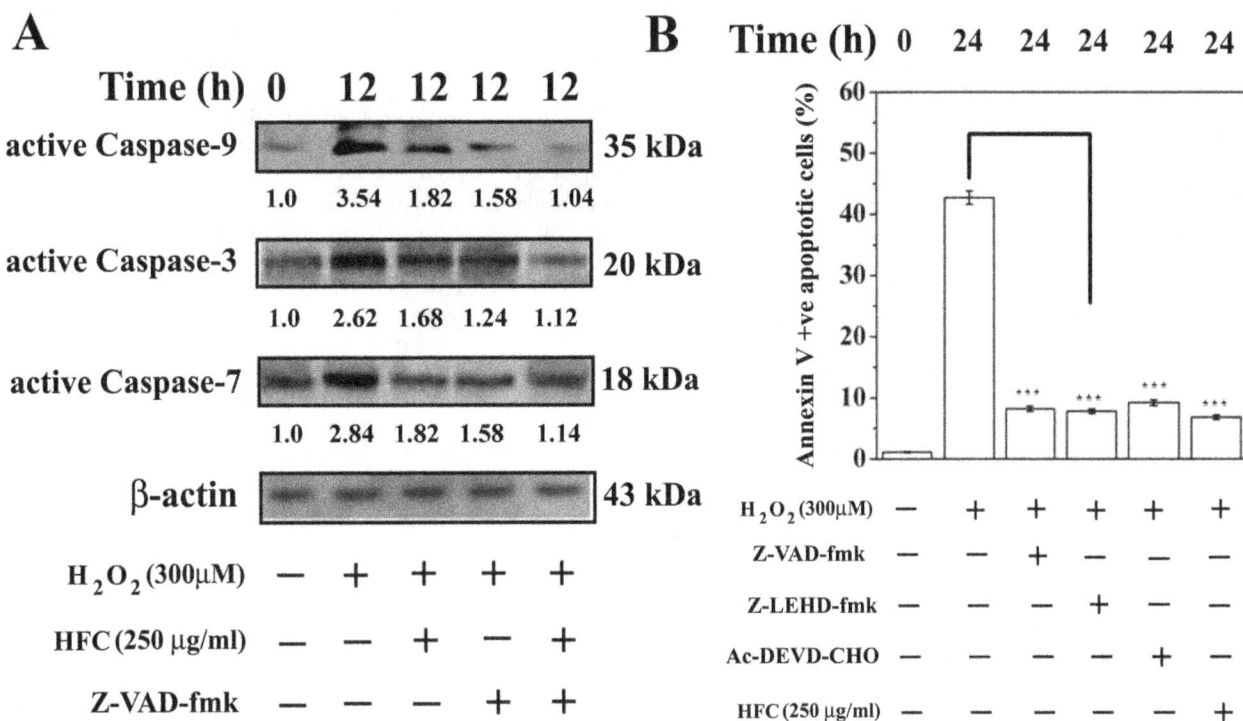

Figure 11. Caspase inhibition assay using pan- and specific caspase inhibitors for intrinsic pathway. The cells were pre-incubated for 1 h individually with 100 µM pan caspase-inhibitor Z-VAD-fmk, caspase-9-specific inhibitor Z-LEHD-fmk (50 µM), caspase-3 and -7-specific inhibitor Ac-DEVD-CHO (50 µM). Then the cells were incubated in presence (250 µg/mL) or in absence of HFC polysaccharide for 1 h followed by the treatment with 300 µM H$_2$O$_2$ in both the cases for varying periods of time (0–24 h). **A)** The expression of active forms of caspases (caspase-9, -3, and -7) in the presence of the inhibitors was evaluated by immunoblotting. **B)** Apoptosis was quantified by flow cytometry as described earlier. β-actin was used as loading control. Fold changes are represented as relative values of band densitometries normalized to control and are shown as numbers below the immunoblots. Results are representative of three independent experiments performed in triplicate and are represented as mean value. A one-way analysis of variance (ANOVA, Bonferroni corrections for multiple comparisons) was performed, where significant level stands for * p<0.05, *** p<0.0001.

cation was viewed microscopically using immunofluorescence. As depicted in Figure 12A and Figure S2, H_2O_2 induced a significant decrease ($p<0.05$) in the levels of Nrf2 and Keap1 in the cytosol (66% and 61% decreases, respectively, compared with the control after 24 h) and simultaneously facilitated a significant increase ($p<0.05$) in their nuclear levels (2.3- and 2.6-fold increases, respectively, compared with the control after 24 h). Immunofluorescence (Figure 12B) also supported significant nuclear translocation of Nrf2 and Keap1 after H_2O_2 treatment. In accordance with protein expression, mRNA expression of Nrf2 and Keap1 was also upregulated (8.6- and 7.5-fold increases, respectively, compared with the control after 24 h) with the applied stress in a time-dependent manner (Figure 13B). The nuclear translocation of Nrf2 subsequently triggered the transcription of a downstream battery of genes. As shown in Figure 13C, H_2O_2 significantly upregulated ($p<0.05$) the mRNA expression of HMOX1, NQO1, GSTA2, SOD1, and GPX1 (11.5-, 12.8-, 9.8-, 11.8-, and 10.6-fold increases, respectively, compared with the control after 24 h). However, the polysaccharide treatment successfully decreased the intracellular ROS level, which in turn inhibited the nuclear translocation of Nrf2 (Figure 12B), resulting in a time-dependent stabilization of the Nrf2/Keap1 signaling pathway. Therefore, the increase in the nuclear Nrf2 and Keap1 was found to be insignificant ($p>0.05$) as compared with the control (1.5- and 1.2-fold increases, respectively, after 24 h), which occurred in parallel to the significant increase ($p<0.05$) in cytosolic levels. As shown in Figure 13B, the mRNA expression of Nrf2 and Keap1 decreased gradually with time back to a level similar to that of the control values after 24 h of polysaccharide treatment. As a result, the post-treatment increase in the mRNA level expression of cytoprotective genes (Figure 13C) was found to be statistically insignificant ($p>0.05$) (1.4-, 2.2-, 1.9-, 1.9-, and 1.7-fold increases for HMOX1, NQO1, GSTA2, SOD1, and GPX 1, respectively, compared with the control after 24 h), thus passively indicating a decrease in the intracellular level of ROS.

At times, ROS-generating oxidants may confer a relative cooperation between major death signals and cellular defense mechanisms. Hence, to detect whether the nuclear translocation of Nrf2 was mediated by MAPK, the cells were pre-treated with MAPK-specific inhibitors 1 h before any other treatment. As shown in Figure 13A, SP600125 (a JNK-specific inhibitor) and SB203580 (a p38-specific inhibitor) effectively attenuated the phosphorylation and nuclear translocation of Nrf2. The cytosolic level of Nrf2 increased significantly ($p<0.05$) compared with the cells treated with only H_2O_2 (increases of 74% and 62% for SP600125 and SB203580, respectively, after 6 h). The combined treatment with inhibitor and polysaccharide further reduced the phosphorylation level of JNK and p38 and increased the level of Nrf2 in the cytosol (increases of 144% and 126% for SP600125 and SB203580, respectively, after 12 h). The results suggest that phosphorylation of the MAPKs JNK and p38 may be partially responsible for transducing signals involved in Nrf2 translocation in H_2O_2-treated WI38 cells.

Discussion

The present study was primarily performed to identify the utility of functional bacterial polysaccharides as therapeutic agents against cellular injury during oxidative stress. Experiments were also performed to understand the underlying mechanism of cellular response and signal coordination within human lung fibroblasts during hydrogen peroxide-induced stress and subsequent polysaccharide treatment against stress. *Bacillus megaterium* RB-05, a fresh-water bacterial isolate, biosynthesized two non-

identical EPSs, one in GMSM and one in GMSM-supplemented JC. Structural elucidation revealed the differences in the compositions of the polysaccharides, especially in fucose content. In earlier reports, fucose-rich polysaccharides from algae, fungi and bacteria were effectively used in therapeutics. The role of fucose-rich polysaccharides as antioxidants has been widely accepted and has consistently been employed for healing critical diseases. Both polysaccharides, the HFC and LFC, were found to have antioxidant potential and effectively scavenged free radicals *in vitro*, although performance of the HFC polysaccharide was found to be superior.

The oxidative stress induced by an overproduction of ROS in the lung causes many clinical conditions including cancer, asthma, cystic fibrosis, ischemia-reperfusion injury, and aging [40]. During this investigation, a fucose-rich bacterial polysaccharide exhibited cytoprotective effects against H_2O_2-induced oxidative stress and the apoptosis of WI38 cells by scavenging intracellular ROS and regulating mitochondrial membrane potential, the Bcl-2 family proteins, the caspase cascade, and the phosphorylation status of MAPKs. The effective decrease in intracellular ROS resulted in the stabilization of Nrf2/Keap1-mediated redox signaling, leading to cellular homeostasis.

Dosage selection remains a critical issue when supplementing natural antioxidants. These exogenous materials may also act as pro-oxidants depending on concentration and the buffering capacity of the cell [13]. In some previous studies, it was shown that fucoidans induce ROS-mediated apoptosis of MCF-7 cells at concentrations of 1 mg/mL [19,41] and 820 µg/mL [19] via caspase-dependent and caspase-independent pathways, respectively, and exhibit antitumor activity in Huh7 cells by downregulating chemokine expression [19,42]. Considering these results, the present investigation optimized the IC_{50} values to determine the free radical scavenging potential and the cellular cytotoxicity of the polysaccharides *in vitro*. Both the polysaccharides displayed a significant cytoprotective effect during H_2O_2 exposure in WI38 cells at the dose of 250 µg/mL. Flow cytometry provided evidence for the apoptotic death of WI38 cells in response to H_2O_2 administration, with significant changes in the percentages of annexin V-positive and sub-G0/G1 cells; these characteristics were considerably decreased upon HFC polysaccharide treatment. Interestingly, at a similar dose, the polysaccharide had minimal effects on the cell cycle progression of untreated WI38 cells. Usually, during the treatment period, ROS is capable of degrading and depolymerizing polysaccharides, presumably leading to their dysfunction. Prolonged treatment with polysaccharides such as dextran has been shown to limit exogenous and endogenous ROS levels, resulting in smaller fragmentation products of polysaccharides (~10 kDa) [43–46]. Hence, it can be concluded that the external application of these biomacromolecules may passively shield cellular components from oxidative assaults triggered by ROS.

Mitochondria play a pivotal role in the ROS-mediated apoptotic process [20]. It has been found that H_2O_2 induces mitochondrial dysfunction followed by a rapid efflux of intracellular ROS, which further triggers apoptosis of the cells. Literature reports also suggest that the exposure to low doses of H_2O_2 promotes apoptosis rather than necrosis [13]. During an early event of apoptosis, ROS increase the permeabilization and depolarization of the mitochondrial membrane to facilitate a rapid loss of membrane potential [28]. Similarly, in this study, the intracellular ROS led to mitochondrial dysfunction in WI38 cells after H_2O_2 exposure. The HFC polysaccharide successfully suppressed the formation of intracellular ROS, leading to a substantial regaining of mitochondrial membrane potential.

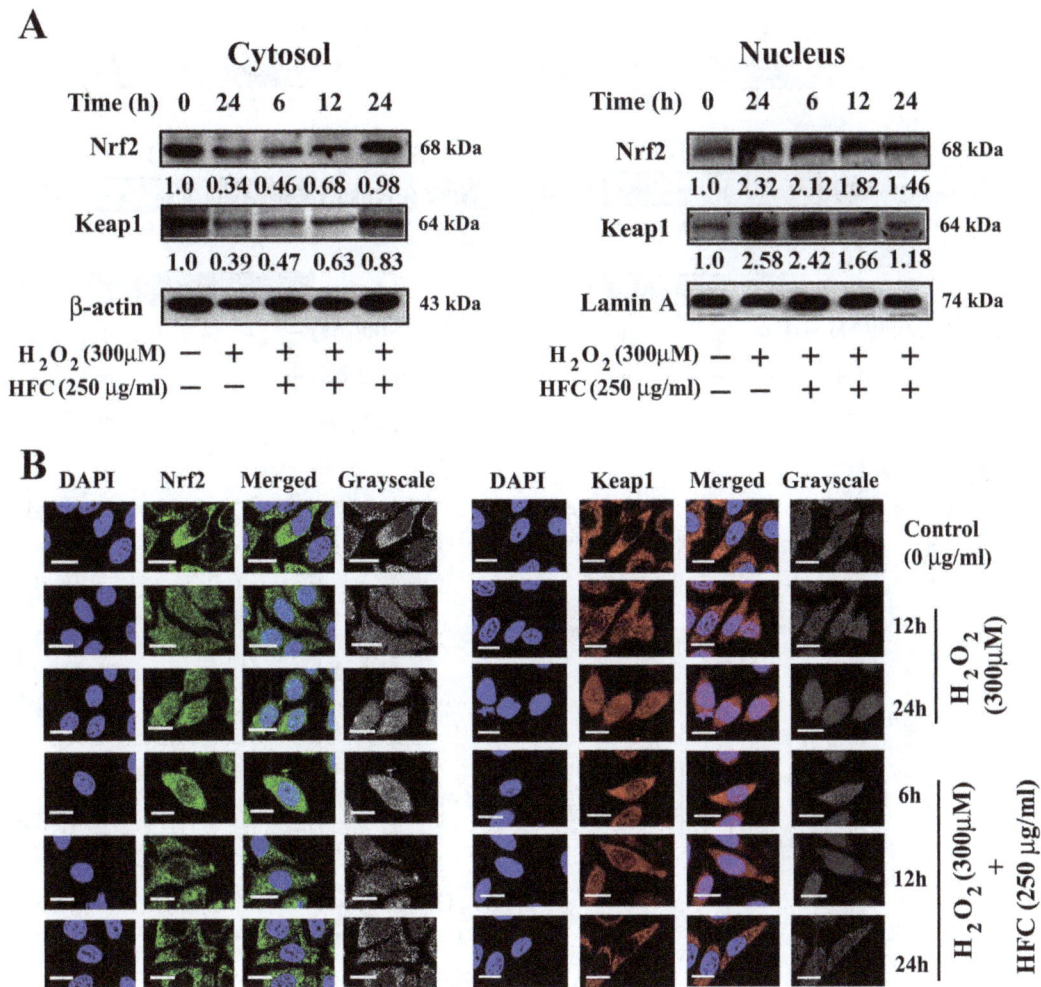

Figure 12. Stabilization of Nrf2/Keap1 signaling and their nuclear translocation induced by HFC polysaccharide. WI38 cells were incubated in presence (250 μg/mL) or in absence of HFC polysaccharide for 1 h followed by the treatment with 300 μM H_2O_2 in both the cases for varying periods of time (0–24 h). **A)** Protein level expression of Nrf2 and Keap1 in both cytosolic and nuclear fractions was observed by immunoblotting. β-actin and Lamin A were used as loading control. Fold changes are represented as relative values of band densitometries normalized to control and are shown as numbers below the immunoblots. Results are representative of three independent experiments performed in triplicate and are represented as mean value. A one-way analysis of variance (ANOVA, Bonferroni corrections for multiple comparisons) was performed, where significant level stands for *$p < 0.05$. **B)** H_2O_2-induced translocation of Nrf2 and Keap1 from cytosol to nucleus and re-localization into cytosol once again during the polysaccharide treatment were monitored under fluorescence microscope using fluorescence-tagged (red and green florescence, respectively, for Nrf2 and Keap1) specific antibodies. The cells were counterstained with DAPI (blue fluorescence) to visualize nuclear morphology. Each image shown is representative of 20 random fields observed. Indicated *scale bars* signify 10 μm distance and photographs were taken at 100× zoom.

Thus far, the best-studied regulator of apoptosis is the Bcl-2 family, which contains the integral outer-membrane oncoproteins of the mitochondria. The HFC polysaccharide significantly upregulated the expression of anti-apoptotic proteins and down-regulated the apoptotic proteins, which further attenuated ROS-induced cellular injury and likely inhibited the peroxidation of lung lipids, including pulmonary surfactants, and also reduced membrane permeability and surfactant activity [47]. The Bax/Bcl-2 ratio, a determinant of cell susceptibility to death signals, was found to be decreased in a time-dependent manner with the polysaccharide treatment. ROS-induced imbalances in protein expression during apoptosis form mitochondrial permeability transition pores (PTPs), through which cytochrome c is released to the cytosol from the intermembrane space. However, controversy remains regarding the timing of this phenomenon. As per one opinion, cytochrome c is released before any detectable loss of

$\Delta\Psi_m$, whereas other researchers have postulated that Bax plays a direct role in membrane permeability and promotes loss of $\Delta\Psi_m$ after translocating to the mitochondria, which is followed by the release of cytochrome c. The third group has proposed that the loss of $\Delta\Psi_m$ precedes the mitochondrial translocation of Bax and cytochrome c is released concurrently with membrane collapse [28,48–51]. In the present work, it was shown that the ROS-mediated translocation of cytochrome c and Bax occurred simultaneously and concurrently with $\Delta\Psi_m$ loss. The release of cytochrome c facilitates the formation of the apoptosome complex, thus activating the binding of procaspase-9 and Apaf-1 in the presence of ATP [52]. HFC polysaccharide treatment in stressed cells inhibited the formation of the apoptosome complex and the cleavage of procaspase-9, which is known to be a mediator of caspase-3 activation. This further suppressed the signal that

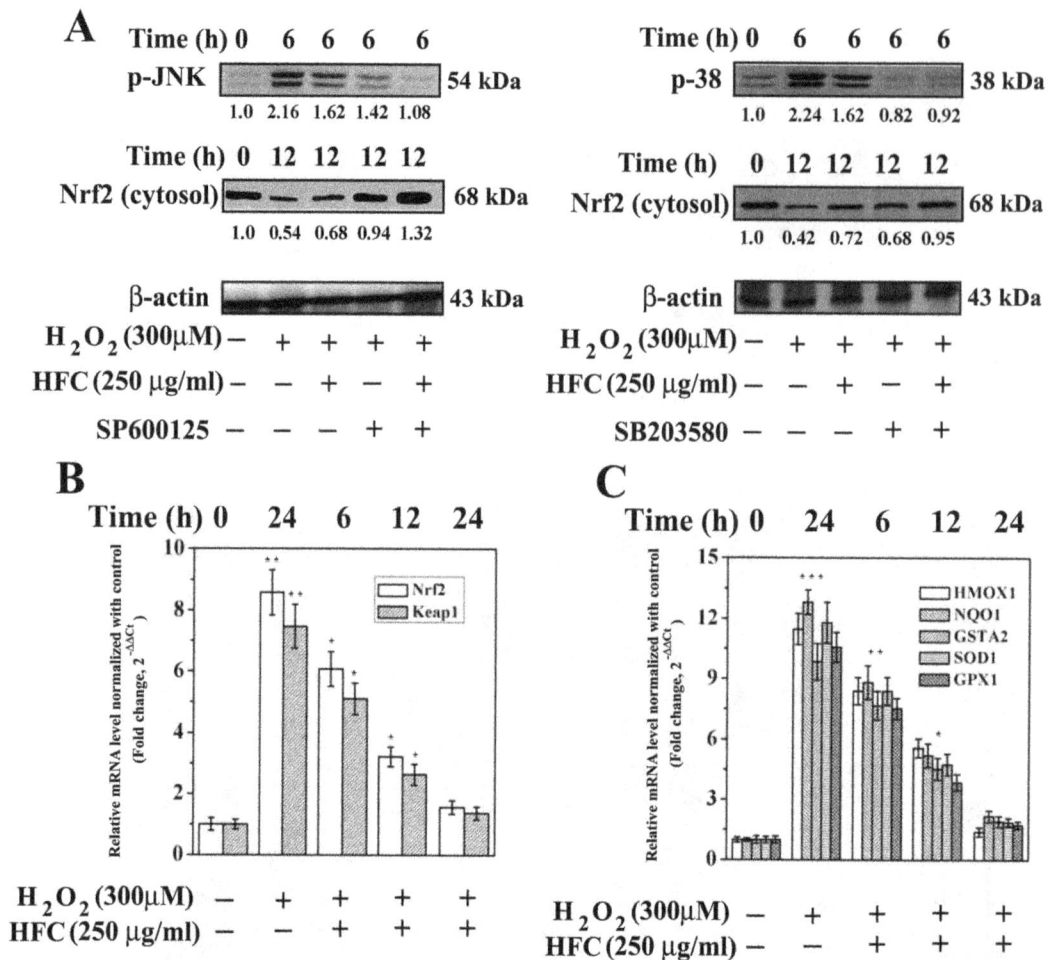

Figure 13. MAPK-mediated activation of Nrf2/Keap1 signaling during H$_2$O$_2$-induced apoptosis of WI38 cells. The cells were incubated in presence (250 µg/mL) or in absence of HFC polysaccharide for 1 h followed by the treatment with 300 µM H$_2$O$_2$ in both the cases for varying periods of time (0–24 h). **A)** Protein level expression of phosphorylated JNK and p38, and cytosolic Nrf2 in the presence of JNK- and p38- specific inhibitors, SP600125 and SB203580, respectively. Before any other treatment, WI38 cells were pre-incubated with 10 µM inhibitors for 1 h separately. β-actin was used as loading control. Fold changes are represented as relative values of band densitometries normalized to control and are shown as numbers below the immunoblots. Results are representative of three independent experiments performed in triplicate and are represented as mean value. **B)** The mRNA level expression of Nrf2 and Keap1 was quantified using real-time RT-PCR. RNA was extracted from the treated and untreated WI38 cells and after enzymatic reverse transcription, the cDNA content was analyzed by electrophoresis in 2% agarose gel containing 0.1% ethidium bromide. Fold changes were calculated using real-time RT-PCR (SYBR green method). **C)** Similarly, the expression of HMOX1, NQO1, SOD1, GPX1, and GASTA2 at mRNA level was quantified. Fold changes are represented as relative values normalized to control and quantified in the terms of $2^{-\Delta\Delta Ct}$. GAPDH was used as internal control. Results are representative of three independent experiments performed in triplicate and are represented as mean ± SD. A one-way analysis of variance (ANOVA, Bonferroni corrections for multiple comparisons) was performed, where significant level stands for * $p<0.05$, ** $p<0.001$, *** $p<0.0001$.

induces cleavage of polyadenosine diphosphate ribose polymerase (PARP), which may inhibit apoptosis [53].

Nrf2, a transcription factor that is part of the redox homeostatic gene regulatory network, is activated under any oxidative or electrophilic stress to enhance the expression of the phase II detoxifying enzymes [54]. Under basal conditions, Nrf2 remains an integral part of an E3 ubiquitin ligase complex (the Nrf2-Keap1-Cul3-Rbx1E3 ubiquitin ligase complex) and destined for proteasomal degradation in the cytosol. During any electrophilic assault, a conformational change may take place in the cysteine-rich domain of Keap1 that dissociates Nrf2 from the E3 ubiquitin ligase complex, thus leading to nuclear translocation of Nrf2. Accumulating Nrf2 in the nucleus binds to its transcriptional associate, Maf, forming a heterodimer that binds to the ARE binding site on DNA and activates the transcription of a battery of cytoprotective genes. Therefore, the expression status of genes such as HMOX1, NQO1, GSTA2, SOD1 and GPX1 can provide an indication of the intracellular ROS level [54–56]. As evident from the present study, the upregulation of the above genes provides the primary signal for increased ROS level after H$_2$O$_2$ exposure, whereas after polysaccharide administration, mRNA expression of the cytoprotective genes decreased to baseline with time. Here, it could be hypothesized that the HFC polysaccharide directly scavenged H$_2$O$_2$-generated ROS rather than implementing phase II enzymes for the detoxification of ROS, resulting in cellular homeostasis via stabilization of Nrf2/Keap1 signaling. Keap1 plays a post-inductive repressive role in the stabilization of the Nrf2/Keap1 signaling system. A nuclear export sequence (NES) present in Keap1 has been reported to terminate Nrf2/Keap1/ARE signaling. During stress recovery, Keap1 also translocates into the nucleus to dissociate Nrf2 from the ARE. The Nrf2/Keap1 complex then traffics out of the nucleus and reassociates with the Cul3-E3 ubiquitin ligase in the cytosol. Simultaneously, supplemented antioxidants lower the intracellular ROS and free radical levels and eventually create a reduced intracellular environment that maintains Keap1 in a reduced configuration. With a reduction in the amount of oxidized Keap1, Nrf2 undergoes regular ubiquitin-mediated degradation, which inhibits the nuclear translocation of Nrf2. This subsequently reduces the transcription of cytoprotective genes, and the levels of endogenous phase II enzymes lead to cellular homeostasis [25,54–56].

Earlier reports suggest that oxidative stress triggers the phosphorylation of MAPKs. The most studied kinases, such as p38, ERK1/2 and JNK1/2, play important roles in regulating the pathways involved in H_2O_2-induced cell death [28,39,57–59]. Therefore, the expression status of MAPKs and crosstalk between the MAPK pathway and other signaling pathways were investigated during the stress and polysaccharide treatment periods. An earlier study described the importance of the duration and intensity of JNK activation during cell death [19]. In alveolar epithelial cells, the phosphorylation of JNK begins within a few minutes of H_2O_2 exposure and persists for at least 1–6 h. Levels of both the p-JNK1 and p-JNK 2 isoforms were reportedly increased [60]. In the present study, WI38 cells showed a marked difference ($p<0.05$) in the phosphorylation status of MAPKs between the stress and treatment periods.

MAPK signaling has been implicated in the induction of the Nrf2/Keap1 pathway by many previous reports; this may occur via phosphorylation of Nrf2 or through the nuclear translocation of Nrf2 [26]. The overexpression of JNK2 enhances site-specific phosphorylation of Nrf2 *in vivo*, although this phosphorylation has only a limited role in regulating Nrf2. In another study, the administration of chemical stimulants only slightly altered the ARE-dependent transcription and the protein levels of Nrf2 [26]. In the present investigation, the JNK and p38 MAPK inhibitors SP 600125 and SB 203580 significantly inhibited the nuclear translocation of Nrf2 after H_2O_2 exposure. This suggests that JNK and p38 MAPKs are important mediators of the Nrf2 signaling network in H_2O_2-treated WI38 cells. Simultaneous treatment with the HFC polysaccharide and JNK and p38 MAPKs inhibitors attenuated the phosphorylation of JNK and p38 and relocalized Nrf2 in the cytosol. Figure 14 highlights the evident crosstalk between the death-regulatory signaling system and cellular defense mechanisms.

Conclusions

In the eukaryotic system, antioxidants function by directly by scavenging free radicals and/or inducing the activity of antioxidant enzymes. The present study most likely displayed the first. Both the HFC and LFC polysaccharides showed efficient free radical scavenging activities *in vitro*, although the activity of the HFC polysaccharide was comparatively better. The HFC polysaccharide readily scavenged H_2O_2-induced intracellular ROS, which decreased the oxidative stress in WI38 cells through stabilization of the Nrf2/Keap1 signaling system and regulation of cytoprotective genes involving the MAPK and mitochondria-mediated pathways. In future studies, this molecular mechanism needs to be validated *in vivo*; a positive result, if confirmed, would provide priceless information to the development of new approaches for effective stress-responsive antioxidants.

Materials and Methods

Materials

Human embryonic fibroblast lung cell line (WI38) was purchased from the National Centre for Cell Science (Pune, India). Dulbecco's modified Eagle medium (DMEM), fetal bovine serum (FBS) and antibiotic/antimycotic solution and gentamycin were purchased from HyClone, Thermo Fisher Scientific (Waltham, MA). All primary and secondary antibodies were procured from Santa Cruz Biotechnology (Santa Cruz, CA) and Cell Signaling Technology (Beverly, MA). Alexa fluor 660-conjugated anti-goat secondary antibody and FITC-conjugated anti-rabbit, anti-mouse, and anti-goat secondary antibodies were from Invitrogen (Carlsbad, CA) and Millipore (Billerica, MA, USA). The pan caspase inhibitor (Z-VAD-fmk) and the caspase-3/7 inhibitor (Ac-DEVD-CHO) were purchased from Promega (Madison, WI). The caspase-8 inhibitor (Ac-IETD-CHO) and the caspase-9 inhibitor (Z-LEHD-fmk) were purchased from BD Biosciences (San Diego, CA). The JNK inhibitor SP600125 and the p38 inhibitor SB203580 were purchased from Enzo Life Sciences International, Inc. (Plymouth, PA). DiOC6, H_2DCF-DA, MitoTracker Red, SYBR Green Master Mix, and TRIzol reagent were procured from Invitrogen, Carlsbad, CA. All other chemicals were from Merck (Mumbai, India and Darmstadt, Germany), Sigma Aldrich (St. Louis, MO), and Himedia (Mumbai, India) unless indicated otherwise.

Extracellular polysaccharide (EPS) production in different fermentation media

Bacillus megaterium RB-05 biosynthesized EPSs in glucose mineral salts medium (GMSM) and jute culture (JC). The isolation of the crude EPSs from the cultures and their subsequent purification were performed as described previously [30,31]. Carbohydrate content was spectrophotometrically determined at 490 nm (750 Lambda Double Beam UV–Vis Spectrophotometer, Perkin Elmer) using the phenol-sulfuric acid method [61], and protein content was analyzed at 595 nm according to the Lowry method [62]. Molecular weights of the purified EPSs were derived from the standard plot of the reference dextran.

Analysis of monosaccharide composition

Lyophilized polysaccharide (10 mg) was hydrolyzed with 2 mol/L trifluoroacetic acid (TFA) for 2 h at 121°C. TFA was removed using a rotary vacuum evaporator. The monosaccharide composition was identified by HPLC separation of their anthranilic acid derivatives, which were obtained as described by Anumula [63].

Structural elucidation

A solution of 5 mg of purified EPS in 0.5 mL of DMSO (dimethyl sulfoxide) was methylated by adding finely powdered NaOH (20 mg) and methyl iodide (0.1 mL) prior to sonication for 15 min. The methylated products were extracted with $CHCl_3$ and H_2O (5:2, v/v). The $CHCl_3$ phase was separated and dried under N_2, hydrolyzed in 2 M TFA at 100°C for 6 h. The hydrolyzed EPS was reduced with 50 mM sodium borohydride at room temperature for 4 h, evaporated three times from a mixture of acetic acid/methyl alcohol (1:1), and acetylated in 50:50 acetic anhydride/pyridine at 100°C for 90 min. Alditol acetates of the methylated sugars were analyzed using a Shimadzu GCMS-QP2010. NMR spectra were obtained using a Bruker Avance DPX-500 MHz NMR Spectrometer (Bruker Co., Billerica, MA).

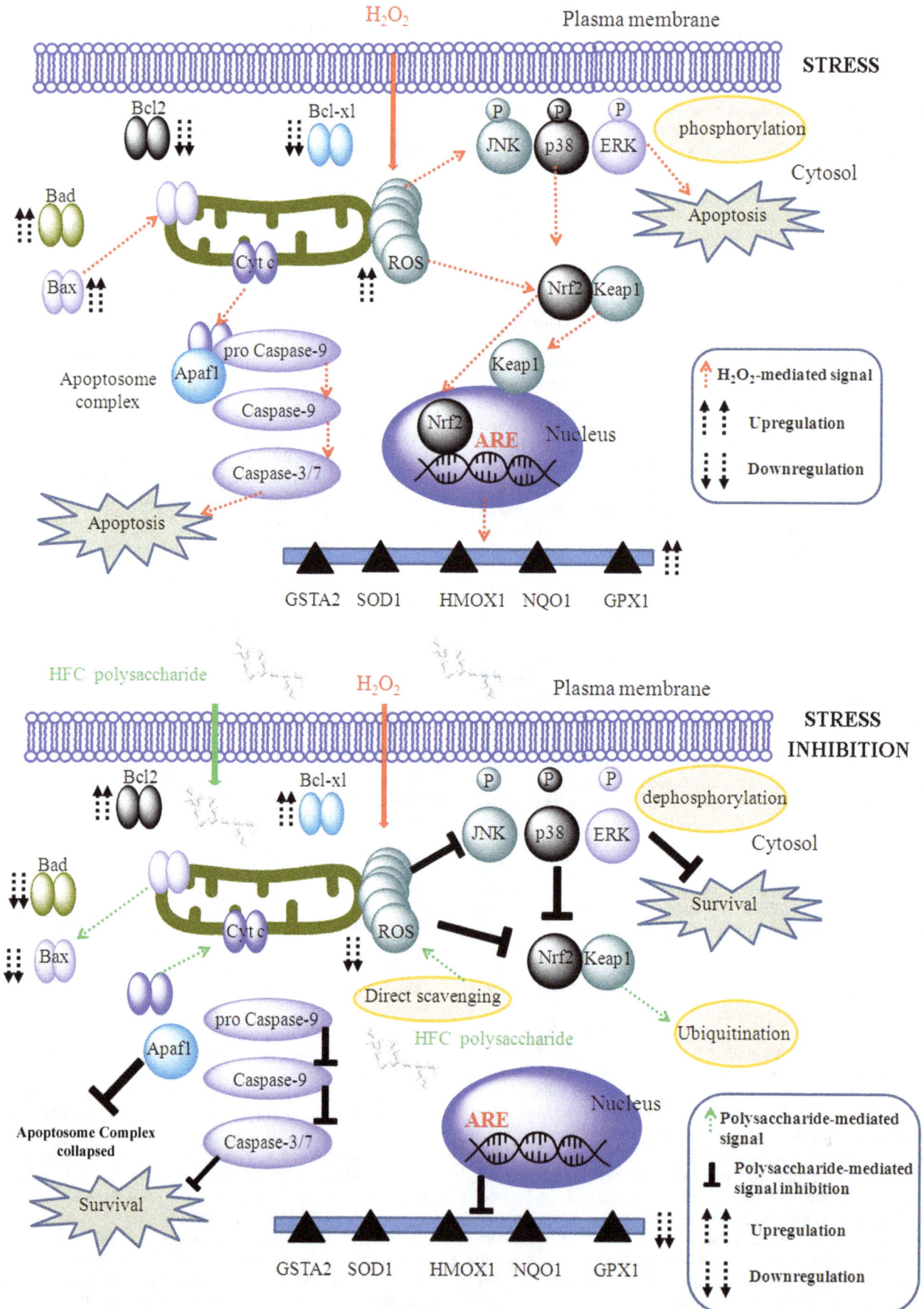

Figure 14. Plausible signaling cross-talk involved in the HFC polysaccharide treatment against H₂O₂-induced apoptosis of WI38 cells.

The proton and ^{13}C spectra were run at a probe temperature of $40°C$. The purified samples were dried in a vacuum over P_2O_5 and then exchanged with deuterium by lyophilizing several times with D_2O. The EPS was dissolved in 0.7 mL of D_2O (99.96%) at concentrations of 15 mg/mL (^1H NMR) and 30 mg/mL (^{13}C NMR). The 2D COSY experiment was recorded at a mixing time of 150 ms.

Free radical scavenging activity of the EPSs

Non-specific hydroxyl radical scavenging activity was assayed as described by Elizabeth and Rao [64] with slight modifications. The procedure for measuring site-specific hydroxyl radical scavenging activity was identical to that detailed above, but EDTA was replaced by an equal volume of buffer as reported previously [65]. Hydrogen peroxide scavenging activity was determined according to Hazra et al. (2008) [2]. The production of singlet oxygen ($^1O_2$) was determined by monitoring N, N-dimethyl-4-nitrosoaniline (RNO) bleaching following Hazra et al. (2008) [2]. Superoxide radical scavenging activity was measured by the reduction of nitro blue tetrazolium (NBT) [66,67]. In enzymatic assays, superoxide radicals were generated by the xanthine/xanthine oxidase (XO) system [66,67]. The effect of the tested samples on XO activity was evaluated by measuring the formation of uric acid from xanthine at 295 nm. The DPPH (1,1-diphenyl-2-picrylhydrazyl) radical scavenging was performed following Hazra et al. [2] Antioxidant capacity was measured based on the scavenging of $ABTS^{.+}$ by the test samples in comparison to a Trolox standard [2]. The antioxidant activity of the polysaccharides was also evaluated using the β-carotene linoleate model system.

Cell culture

The WI38 cell line was cultured in DMEM supplemented with 10% FBS and antibiotic/antimycotic solution (100 units) and gentamycin (50 μg/mL) with Na pyruvate (1 mM) in a humid incubator at $37°C$ in 5% CO_2. After 24 h, the medium was replaced with fresh medium containing various concentrations of GMSM- and JC-derived fucose polysaccharides for the initial screening. Cells grown in medium containing phosphate-buffered saline (PBS) without polysaccharide served as control.

MTT assays

WI38 cells were seeded in 96-well plates at a density of 3000 cells/well in 200 μl medium. The cells were incubated in presence (50–500 μg/mL) or in absence of HFC polysaccharide for 1 h followed by the treatment with H_2O_2 (0–500 μM) in both the cases for varying periods of time (0–24 h). After completion of the treatments, 10 μl of MTT (5 mg/mL) was added to each well. The formazan complex was dissolved in 100 μl dimethyl sulfoxide (DMSO) after 3 h of incubation. The absorbance of each well was measured at 570 nm with a microplate reader.

Phase contrast microscopy

WI38 cells were seeded onto 96-well plates (5×10^3 cells/well). The cells were incubated in presence (50, 100, 200 and 250 μg/mL) or in absence of HFC polysaccharide for 1 h followed by the treatment with 300 μM H_2O_2 in both the cases for varying periods of time (0–24 h). Cell growth and morphology were captured throughout the treatment period using a U-TVO 63× C microscope fit with an Olympus CAMEDIA digital wide-zoom camera, model C-7070 (Olympus Corp., Tokyo, Japan).

Cell cycle analysis and determination of apoptotic cells by flow cytometry

WI38 cells were seeded in a 6-well culture plate at a density of 5×10^4 cells/well and incubated in DMEM containing 10% FBS. The cells were incubated in presence (50, 100, 200 and 250 μg/mL) or in absence of HFC polysaccharide for 1 h followed by the treatment with 300 μM H_2O_2 in both the cases for varying periods of time (0–24 h). Evaluation of the cell cycle distribution and sub-G_0/G_1 peaks by flow cytometry was performed by measuring PI-fluorescence with a BD FACS Calibur flow cytometer (Becton Dickinson, San Jose, CA, USA) through an FL-2 filter (585 nm). For each sample, 1×10^4 events were recorded. Flow cytometry data were analyzed using Cell Quest.

The presence of apoptotic cells was determined by the ability of cells in suspension to bind to annexin V. Control and treated WI38 cells were adjusted to 5×10^5 cells/mL in binding buffer (10 mM HEPES [(4-(2-hydroxyethyl)-1-piperazineethanesulfonic acid] [pH 7.4], 140 mM NaCl, 2.5 mM $CaCl_2$). Then, 10 μl of FITC-annexin V was added to 190 μl of cell suspension. The mixture was incubated for 10 min at room temperature. After centrifugation, the cells were resuspended in 190 μl binding buffer, and 10 μl PI (1 mg/mL) solution was added. The cells were then analyzed in a FACS Calibur flow cytometer; for each sample, 1×10^4 events were acquired. The cells positive for both PI and annexin V were considered to be necrotic cells and thus excluded from the analysis.

Determination of intracellular ROS and mitochondrial membrane potential (MMP)

The generation of intracellular ROS was analyzed with the oxidation-sensitive fluorescent probe H_2DCF-DA using a FACS Calibur flow cytometer at excitation and emission wavelengths of 488 nm and 544 nm, respectively. The loss of mitochondrial transmembrane potential was verified by flow cytometry at the single-cell level. Control and WI38 cells treated for different periods of time were incubated with DiOC6 (50 nM) for 15 min at $37°C$ in the dark. Loss of DiOC6 fluorescence indicates disruption of the mitochondrial inner transmembrane potential. The probe was excited at 488 nm, and emission was measured through a 530 nm band pass filter with the FACS Calibur.

Preparation of sub-cellular fractions

For determination of Nrf2 and Keap1 protein levels in cytosolic and nuclear compartments, control and treated WI38 cells of different time periods were lysed in a hypotonic buffer (10 mM HEPES [pH 7.9], 1.5 mM $MgCl_2$, 10 mM KCl, 0.5 mM DTT, and 1× Complete Protease Inhibitor Cocktail [Roche, Molecular Biochemicals, Indianapolis, IN]). After centrifugation ($20,000 \times g$), the cytosolic proteins in the supernatant were collected, and the nuclear pellets were extracted with a high salt buffer (20 mM HEPES, pH 7.9, 1.5 mM $MgCl_2$, 20% glycerol, 0.2 mM EDTA, 300 mM KCl, 0.5 mM DTT, 1× Complete Protease Inhibitor Cocktail) by placing it into the ice for 30 min, which was followed by centrifugation ($20,000 \times g$) to collect the nuclear extracts.

Mitochondrial fraction was isolated following Frezza et al., [68] to determine the expression levels of cytochrome c and Bax protein in the control and treated cells of different time periods. PBS-washed cells (120×10^6) were suspended in ice-cold isolation buffer (0.1 M Tris–MOPS [3-(N-morpholino) propanesulfonic acid], 0.1 M EGTA/Tris, and 1 M sucrose, pH 7.4) for 30 min, homogenized, and centrifuged at $600 \times g$ for 10 min at $4°C$. Collected supernatant was centrifuged at $7000 \times g$ for 10 min at $4°C$. The pellet was washed with isolation buffer, resuspended in

200 µl of ice-cold isolation buffer, and then again centrifuged at 7000×g for 10 min at 4°C. The supernatant was discarded to obtain the pellet containing mitochondria.

To measure the expression levels of Bcl-2, caspase and MAPK family proteins, PBS-washed cells were lysed in ice-cold RIPA lysis buffer (150 mM sodium chloride, 1.0% TritonX-100, 50 mM Tris pH 8.0, 0.01% SDS, 0.5% sodium deoxycholate) containing 1 mM phenylmethylsulfonyl fluoride (PMSF) (SRL, Mumbai, India), 1 µg/mL of aprotinin, leupeptin and pepstatin. The samples were incubated on ice for 30 min and centrifuged at 20,000×g for 15 min at 4°C and protein containing supernatant was collected. Protein concentration of all the fractions was determined using the Bradford reagent (Sigma-Aldrich, St. Louis, MO) and subsequent measurement of absorbance was done at 595 nm in a UV-1700 PharmaSpec, Shimadzu spectrophotometer (Shimadzu Scientific Instruments, Columbia, MD). The remaining supernatant was stored at −20°C.

Western blot analysis

Cell lysates were diluted in sample buffer (0.312 mM Tris-HCl (pH 6.8), 50% glycerol, 10% SDS, 25% β-mercaptoethanol, and 0.25% bromophenol blue) at a final protein concentration of 5 µg/µl, and were then boiled at 100°C for 5 min. Aliquots of each sample (10 µl containing 50 µg protein) were loaded into dedicated wells of 9–12% polyacrylamide gels and separated by electrophoresis for 3 h at 100 V. Proteins were transferred to polyvinylidene difluoride membrane (Amersham Biosciences, Piscataway, NJ) for 1.5 h at 300 mA. After blocking of nonspecific binding with 5% nonfat dry milk in TBST (Tris-buffered saline-Tween), the membranes were then probed with primary antibodies (Text S1) and incubated overnight at 4°C. The membranes were washed with Tris-buffered saline-0.01% (v/v) containing Tween-20 at room temperature for 15 min and then incubated with alkaline phosphatase (AP)- or horseradish perox-idase (HRP)-conjugated secondary antibodies (anti-rabbit, anti-goat and anti-mouse IgG; 1:1000 dilution) in TBST for 2 h at room temperature. The membranes were developed according to the laboratory protocols. The band intensity of the detected protein was measured by densitometry (Gel Doc XR+ System, Bio-Rad Laboratories, Berkeley, CA). β-actin, Lamin A, and COX 4 were chosen as loading controls for constitutive expression in cytosolic, nuclear and mitochondrial fraction, respectively.

Inhibitors

WI38 cells were treated with different caspase inhibitors (Z-VAD-fmk, Ac-DEVD-CHO, Z-IETD-fmk and Z-LEHD-fmk), MAPK inhibitors (SP600125, JNK specific inhibitor and SB203580, p38 specific inhibitor) for 1–2 h prior to polysaccharide and H_2O_2 treatment.

RNA isolation, reverse transcription (RT), and real-time PCR

RNA was extracted using the TRIzol reagent and first-strand cDNA was synthesized from 1 µg total RNA using Random Hexamer (Promega, Madison, WI) and MMLV high performance reverse transcriptase enzyme (Epicentre Biotechnologies, Madison, WI) according to the manufacturer's instructions. The reverse transcriptase PCR was performed using the 2720 Thermal Cycler from Applied Biosystems (Foster City, MA). The PCR program involved subsequent incubations at 25°C for 10 min, 37°C for 60 min, 85°C for 5 min, and 4°C for 1 min. The cDNA content was analyzed by electrophoresis in 2% agarose gel containing 0.1% ethidium bromide. Real-time PCR amplifications were performed in triplicate using the SYBR Green I assay and were carried out using ABI PRISM 7000 Sequence Detection System (Applied Biosystems, Foster City, MA). The reactions were carried out in a 96-well plate in 20- µl reactions containing 2× SYBR Green Master Mix, 2 pmol each of forward and reverse primer, and a cDNA template corresponding to 25 ng total RNA. The real time PCR program included activation at 50°C for 2 min, initial denaturation at 95°C for 10 min, followed by 40 cycles of denaturation at 95°C for 15 s and annealing at 60°C for 1 min. Presence of <200 bp amplicons was checked in 2% agarose gel. The threshold cycle (C_t) was obtained from the PCR curves and expression levels of the target mRNA were quantified in terms of the C_t values of the untreated and treated samples and were normalized with the C_t values of GAPDH (internal control). The mRNA expression was quantified in terms of $2^{-\Delta\Delta Ct}$. Primer sequences (Table S4) were designed using the NCBI-Primer BLAST online tool and synthesized commercially.

Immunofluorescence analysis

WI38 cells were grown in poly-L-lysine (0.1 mg/mL) coated sterile cover slip. Control and treated WI38 cells of different time periods were incubated with 100 nM MitoTracker Red for 30 min for mitochondrial staining. Then the cells were rinsed three times with PBS, fixed for 10 min in 4% p-formaldehyde, and permeabilized for 10 min with 0.1% Triton X-100, followed by rinsing with PBS containing 1.0% bovine serum albumin for three times. The permeabilized cells were incubated with primary antibodies for 1 h at 37°C in a moist chamber. After PBS wash, the cells were incubated with Alexa fluor 660- or FITC-conjugated secondary antibodies for 1 h at 37°C in a moist chamber. DAPI (4′,6-diamidino-2-phenylindole, dihydrochloride) was used as a nuclear stain. The cells were visualized using U-TVO 63×C microscope fit with Olympus CAMEDIA digital wide zoom camera, Model C-7070 (Olympus Corp., Tokyo, Japan).

Statistical analysis

The experiments were performed in triplicates. Data shown are representative of at least three experiments and are represented as mean ± SD. Inhibition concentration (IC_{50}) values were calculated from pharmacological dose-response curve fit (sigmoidal) following the equation:

$$Y = \text{Bottom} + (\text{Top-Bottom})/$$
$$[1 + 10^{\wedge}\{(\text{LogIC50-X}) * \text{Hill slope}\}]$$

ANOVA (Bonferroni corrections for multiple comparisons) was employed to assess the statistical significances of differences among pair of data sets with a $p < 0.05$ considered to be significant. The statistical analysis was performed using Origin 8 software.

GenBank

16 rRNA gene sequence of *Bacillus megaterium* RB-05: GenBank Accession Number HM371417

Supporting Information

Figure S1 Protective effect of LFC and HFC polysaccharides on H_2O_2-induced cytotoxicity. Initially, WI38 cells were treated with various concentrations of H_2O_2 alone for 24 h and cell viability was determined by MTT assay for initial dose optimization. The cells were incubated in presence (50–500 µg/mL) or in absence of HFC polysaccharide for 1 h followed by the treatment with H_2O_2 (0–500 μM) in both the cases for varying periods of time (0–24 h). Cell viability was determined by MTT assay. Results are representative of three independent experiments performed in triplicate and are represented as mean ± SD.

Figure S2 The protective effect of HFC polysaccharide on H_2O_2-induced apoptosis of WI38 cells by regulating protein level expression of Bcl-2 family, caspase family and translocation of Bax, cytochrome c, Nrf2 and Keap1. The cells were incubated in presence (250 μg/mL) or in absence of HFC polysaccharide for 1 h followed by the treatment with 300 μM H_2O_2 in both the cases for varying periods of time (0–24 h). Band densitometries of the immunoblots were compared between only H_2O_2 treatment and combining treatment of H_2O_2 and HFC polysaccharide. β-actin, COX4 and Lamin A were used as loading control.

Figure S3 The protective effect of HFC polysaccharide on H_2O_2-induced apoptosis of WI38 cells by regulating phosphorylation of MAPKs. The cells were incubated in presence (250 μg/mL) or in absence of HFC polysaccharide for 1 h followed by the treatment with 300 μM H_2O_2 in both the cases for varying periods of time (0–24 h). Protein level expression of total and phosphorylated JNK, p38, and ERK was evaluated by immunoblotting. β-actin was used as loading control. Fold changes are represented as relative values of band densitometries normalized to control and are shown as numbers below the immunoblots. Results are representative of three independent experiments performed in triplicate and are represented as mean value. A one-way analysis of variance (ANOVA, Bonferroni corrections for multiple comparisons) was performed, where significant level stands for * $p<0.05$, ** $p<0.001$.

Table S1 Antioxidant and free radical scavenging potential of HFC and LFC polysaccharides with respect

to standard materials. Inhibition concentration (IC_{50}) values were determined from the fitted sigmoidal dose-response curves as derived from the equation mentioned in the 'materials and methods'.

Table S2 Partially methylated alditol acetate derivatives of HFC polysaccharide. Calculated from peak areas and response factors obtained using a flame ionization detector.

Table S3 ^{13}C NMR^d and ^{1}H NMR^c chemical shifts for HFC polysaccharide recorded in D_2O at 40°C.

Table S4 Designed primers for real-time PCR. The primers were designed using online database from National Center for Biotechnology Information (NCBI), NCBI-BLAST, and online PrimerQuest tool of Integrated DNA Technologies (IDT) as discussed in the 'Materials and methods' section.

Text S1 List of used primary antibodies and the sources.

Acknowledgments

All the authors thankfully acknowledge Indian Institute of Technology Kharagpur and University of Calcutta for providing the experimental facilities. SRC would like to thankfully acknowledge Council for Scientific and Industrial Research, Govt. of India and Dr. DS Kothari PDF program, University Grant Commission, Govt. of India for his fellowship.

Author Contributions

Conceived and designed the experiments: SRC SS RKB RS BA AB. Performed the experiments: SRC SS SB RKB TKS. Analyzed the data: SRC SS SB RKB TKS. Contributed reagents/materials/analysis tools: RS BA AB. Wrote the paper: SRC SS SB RKB RS BA AB.

References

1. Stadtman ER (1992) Protein oxidation and aging. Science 257: 1220–1224.
2. Hazra B, Biswas S, Mandal N (2008) Antioxidant and free radical scavenging activity of *Pondias pinnata*. BMC Complement Altern Med 8: 63–72.
3. Pryor WA, Houk KN, Foote CS, Fukuto JM, Ignarro LJ, et al. (2006) Free radical biology and medicine: it's a gas, man! Am J Physiol Regul Integr Comp Physiol 29: 491–511.
4. Hazra B, Sarkar R, Biswas S, Mandal N (2010) Comparative study of the antioxidant and reactive oxygen species scavenging properties in the extracts of the fruits of *Terminalia chebula*, *Terminalia belerica* and *Emblica officinalis*. BMC Complement Altern Med 10: 20–34.
5. Braca A, Sortino C, Politi M, Morelli I, Mendez JJ (2002) Antioxidant activity of flavonoids from *Licania licaniaeflora*. J Ethnopharmacol 79: 379–381.
6. Yazdanparast R, Ardestani A (2007) *In vitro* antioxidant and free radical scavenging activity of *Cyperus rotundus*. J Med Food 10: 667–674.
7. Yazdanparast R, Bahramikias S, Ardestani A (2008) *Nasturtium oficinale* reduces oxidative stress and enhances antioxidant capacity in hypercholesterolaemic rats. Chem Biol Interact 172: 176–184.
8. Nergard CS, Kiyohara H, Reynolds JC, Thomas-Oates JE, Matsumoto T, et al. (2006) Structures and structure-activity relationships of three mitogenic and complement fixing pectic arabinogalactans from the malian antiulcer plants *Cochlospermum tinctorium* A. Rich and *Vernonia kotschyana* Sch. Bip. ex Walp. Biomacromolecules 7: 71–79.
9. Inngjerdingen KT, Coulibaly A, Diallo D, Michaelsen TE, Paulsen BS (2006) A complement fixing polysaccharide from *Biophytum petersianum* Klotzsch, a medicinal plant from Mali, West Africa. Biomacromolecules 7: 48–53.
10. Liu J, Luo J, Ye H, Sun Y, Lu Z, et al. (2009) Production, characterization and antioxidant activities *in vitro* of exopolysaccharides from endophytic bacterium *Paenibacillus polymyxa* EJS-3. Carbohydr Polym 78: 275–281.
11. Péterszegi G, Fodil-Bourahla I, Robert AM, Robert L (2003) Pharmacological properties of fucose. Applications in age-related modifications of connective tissues. Biomed Pharmacother 57: 240–245.
12. Péterszegi G, Robert AM, Robert L (2003) Protection by L-fucose and fucose-rich polysaccharides against ROS-produced cell death in presence of ascorbate. Biomed Pharmacother 57: 130–133.
13. Clutton S (1997) The importance of oxidative stress in apoptosis. Br Med Bull 33: 662–668.
14. Curtin JF, Donovan M, Cotter TG (2002) Regulation and measurement of oxidative stress in apoptosis. J Immunol Methods 265: 49–72.
15. Hengartner MO (2000) The biochemistry of apoptosis. Nature 407: 770–776.
16. Luo X, Budihardjo I, Zou H, Slaughter C, Wang X (1998) Bid, a Bcl2 interacting protein, mediates cytochrome c release from mitochondria in response to activation of cell surface death receptors. Cell 94: 481–490.
17. Delivani P, Martin SJ (2006) Mitochondrial membrane remodeling in apoptosis: an inside story. Cell Death Differ 13: 2007–2010.
18. Alaimo A, Gorojod RM, Kotler ML (2011) The extrinsic and intrinsic apoptotic pathways are involved in manganese toxicity in rat astrocytoma C6 cells. Neurochem Int 59: 297–308.
19. Zhang Z, Teruya K, Eto H, Shirahata S (2011) Fucoidan extract induces apoptosis in MCF-7 cells via a mechanism involving the ROS-dependent JNK activation and mitochondria-mediated pathways. PLoS ONE 6: e27441.
20. Kumar S, Sitasawad SL (2009) N-acetylcysteine prevents glucose/glucose oxidase-induced oxidative stress, mitochondrial damage and apoptosis in H9c2 cells. Life Sci 84: 328–336.
21. Nakajima Y, Nishida H, Nakamura Y, Konishi T (2009) Prevention of hydrogen peroxide-induced oxidative stress in PC12 cells by 3,4-dihydroxybenzalacetone isolated from Chaga (*Inonotus obliquus* (persoon) Pilat). Free Radic Biol Med 47: 1154–1161.
22. Ade N, Leon F, Pallardy M, Peiffer JL, Kerdine-Romer S, et al. (2009) HMOX1 and NQO1 genes are upregulated in response to contact sensitizers in dendritic cells and THP-1 cell line: Role of the Keap1/Nrf2 pathway. Toxicol Sci 107: 451–460.
23. Stępkowski TM, Kruszewski MK (2011) Molecular cross-talk between the NRF2/KEAP1 signaling pathway, autophagy, and apoptosis. Free Radic Biol Med 50: 1186–1195.

24. Furukawa M, Xiong Y (2005) BTB Protein Keap1 Targets Antioxidant Transcription Factor Nrf2 for Ubiquitination by the Cullin 3-Roc1 Ligase. Mol Cell Biol 25: 162–171.

25. Nguyen T, Nioi P, Pickett CB (2009) The Nrf2-antioxidant response element signaling pathway and its activation by oxidative stress. J Biol Chem 28:, 13291–13295.

26. Sun Z, Huang Z, Zhang DD (2009) Phosphorylation of Nrf2 at multiple sites by MAP Kinases has a limited contribution in modulating the Nrf2-dependent antioxidant response. PLoS ONE 4: e6588.

27. Shi X, Zhou B (2010) The role of Nrf2 and MAPK pathways in PFOS-induced oxidative stress in Zebrafish embryos. Toxicol Sci 115: 391–400.

28. Raman M, Chen W, Cobb MH (2007) Differential regulation and properties of MAPKs. Oncogene 26: 3100–3112.

29. Zipper LM, Mulcahy RT (2003) Erk activation is required for Nrf2 nuclear localization during pyrrolidine dithiocarbamate induction of glutamate cysteine ligase modulatory gene expression in HepG2 cells. Toxicol Sci 73: 124–134.

30. Chowdhury SR, Basak RK, Sen R, Adhikari B (2011) Production of extracellular polysaccharide by *Bacillus megaterium* RB-05 using jute as substrate. Bioresour Technol 102: 6629–6632.

31. Chowdhury SR, Basak RK, Sen R, Adhikari B (2012) Utilization of lignocellulosic natural fiber (jute) components during a microbial polymer production. Mater Lett 66: 216–218.

32. Chowdhury SR, Manna S, Saha P, Basak RK, Roy D, et al. (2011) Composition analysis and material characterization of an emulsifying extracellular polysaccharide (EPS) produced by *Bacillus megaterium RB-05*: A hydrodynamic sediment-attached isolate of fresh water origin. J Appl Microbiol 111: 1381–1393.

33. Chowdhury SR, Basak RK, Sen R, Adhikari B (2011) Optimization, dynamics, and enhanced production of a free radical scavenging extracellular polysaccharide (EPS) from hydrodynamic sediment attached *Bacillus megaterium* RB-05. Carbohydr Polym 86: 1327–1335.

34. Björndal H, Hellerqvist CG, Lindberg B, Svensson S (1970) Gas-liquid chromatography and mass spectrometry in methylation analysis of polysaccharides. Angew Chem Int Ed Engl 9: 610–619.

35. Agrawal PK (1992) NMR spectroscopy in the structural elucidation of oligosaccharides and glycosides. Phytochemistry 31: 3307–3330.

36. Rinaudo M, Vincendon M (1982) 13C NMR structural investigation of scleroglucan. Carbohydr Polym 2: 135–144.

37. Liu CL, Xie LX, Li M, Durairajan SSK, Goto S (2007) Salvianolic acid B inhibits hydrogen peroxide-induced endothelial cell apoptosis through regulating PI3K/Akt signaling. PLoS ONE 2: e1321.

38. Smaili SS, Hsu YT, Sanders KM, Russell JT, Youle RJ (2001) Bax translocation to mitochondria subsequent to a rapid loss of mitochondrial membrane potential. Cell Death Differ 8: 909–920.

39. Liu B, Jian Z, Li Q, Li K, Wang Z, et al. (2012) Baicalein protects human melanocytes from H₂O₂-induced apoptosis via inhibiting mitochondria-dependent caspase activation and the p38 MAPK pathway. Free Radic Biol Med 53: 183–193.

40. Dalle-Donne I, Rossi R, Colombo R, Giustarini D, Milzani A (2006) Biomarkers of oxidative damage in human disease. Clin Chem 52: 601–623.

41. Yamasaki-Miyamoto Y, Yamasaki M, Tachibana H, Yamada K (2009) Fucoidan induces apoptosis through activation of caspase-8 on human breast cancer MCF-7 cells. J Agric Food Chem 57: 8677–8682.

42. Nagamine T, Hayakawa K, Kusakabe T, Takada H, Nakazato K (2009) Inhibitory effect of fucoidan on Huh7 Hepatoma Cells through downregulation of CXCL12. Nutr Cancer 61: 340–347.

43. Duan J, Kaspe DL (2011) Oxidative depolymerization of polysaccharides by reactive oxygen/nitrogen species. Glycobiology 21: 401–409.

44. Hammel KE, Kapich AN Jr, Jensen KA, Ryan ZC (2002) Reactive oxygen species as agents of wood decay by fungi. Enzyme Microb Technol 30: 445–453.

45. Kirk TK, Ibach R, Mozuch MD, Conner AH, Highley TL (1991) Characteristics of cotton cellulose depolymerized by a brown-rot fungus, by acid, or by chemical oxidants. Holzforschung 45: 239–244.

46. Miller JG, Fry SC (2001) Characteristics of xyloglucan after attack by hydroxyl radicals. Carbohydr Res 332: 389–403.

47. Bowler RP, Arcaroli J, Crapo JD, Ross A, Slot JW, et al. (2001) Extracellular superoxide dismutase attenuates lung injury after hemorrhage. Am J Respir Crit Care Med 164: 290–294.

48. Pastorino JG, Chen ST, Tafani M, Snyder JW, Farber JL (1998) The overexpression of Bax produces cell death upon induction of the mitochondrial permeability transition. J Biol Chem 273: 7770–7775.

49. Marzo I, Brenner C, Zamzami N, Susin SA, Beutneur G, et al. (1998) The permeability transition pore complex: A target for apoptosis regulation by caspases and Bcl-2-related proteins. J Exp Med 187: 1261–1271.

50. Bossy-Wetzel E, Newmeyer DD, Green DR (1998) Mitochondrial cytochrome release in apoptosis occurs upstream of DEVD-specific caspase activation independently of mitochondrial transmembrane depolarization. EMBO J 17: 37–49.

51. Goldstein JC, Waterhouse NJ, Juin P, Evan GI, Green DR (2000) The coordinate release of cytochrome c during apoptosis is rapid, complete and kinetically invariant. Nat Cell Biol 2: 156–162.

52. Zou H, Li Y, Liu X, Wang X (1999) An APAF-1 cytochrome c multimeric complex is a functional apoptosome that activates procaspase-9. J Biol Chem 274: 11549–11556.

53. Khan N, Afaq F, Mukhtar H (2007) Apoptosis by dietary factors: the suicide solution for delaying cancer growth. Carcinogenesis 28: 233–239.

54. Kensler TW, Wakabayashi N, Biswal S (2007) Cell survival responses to environmental stresses via the Keap1-Nrf2-ARE pathway. Annu Rev Pharmacol Toxicol 47: 89–116.

55. Lee JM, Johnson JA (2004) An important role of Nrf2-ARE pathway in the cellular defense mechanism. J Biochem Mol Biol 37: 139–143.

56. Reichard JF, Motz GT, Puga A (2007) Heme oxygenase-1 induction by NRF2 requires inactivation of the transcriptional repressor BACH1. Nucleic Acids Res 35: 7074–7086.

57. Lee HJ, Noh YH, Lee DY, Kim YS, Kim KY, et al. (2005) Baicalein attenuates 6-hydroxy dopamine-induced neurotoxicity in SH-SY5Y cells. Eur J Cell Biol 84: 897–905.

58. Chen YC, Chow JM, Lin CW, Wu CY, Shen SC (2006) Baicalein inhibition of oxidative-stress-induced apoptosis via modulation of ERKs activation and induction of HO-1 gene expression in rat glioma cells C6. Toxicol Appl Pharmacol 216: 263–273.

59. Lin HY, Shen SC, Lin CW, Yang LY, Chen YC (2007) Baicalein inhibition of hydrogen peroxide-induced apoptosis via ROS-dependent hemeoxygenase 1 gene expression. Biochim Biophys Acta 1773: 1073–1086.

60. Carvalho H, Evelson P, Sigaud S, Gonza'lez-Flecha B (2004) Mitogen-Activated Protein Kinases modulate H(2)O(2)-induced apoptosis in primary rat alveolar epithelial cells. J Cell Biochem 92: 502–513.

61. Dubois M, Gilles KA, Hamilton J, Rebers PA, Smith F (1956) Colorimetric method for determination of sugar and relative substances. Anal Chem 28: 350–366.

62. Lowry OH, Rosebrough NJ, Farr AL, Randall RJ (1951) Protein measurement with the folin phenol reagent. J Biol Chem 193: 265–275.

63. Anumula KR (1994) Quantitative determination of monosaccharides in glycoproteins by high-performance liquid chromatography with highly sensitive fluorescence detection. Anal Biochem 220: 275–283.

64. Elizabeth K, Rao MWA (1990) Oxygen radical scavenging activity of Curcumin. Int J Pharmaceu 58: 237–240.

65. Kitts DD, Wijewichreme AN, Hu C (2000) Antioxidant properties of a North American ginseng extract. Mol Cell Biochem 203: 1–10.

66. Fernandes E, Borges F, Milhazes N, Carvalho FD, Bastos ML (1999) Evaluation of superoxide radical scavenging activity of gallic acid and its alkyl esters using an enzymatic and a non-enzymatic system (abstract). Toxicol Lett 109: 42 (P078).

67. Valentão P, Fernandes E, Carvalho F, Andrade PB, Seabra RM, et al. (2002) Studies on the antioxidant activity of *Lippia citriodora* infusion: Scavenging effect on superoxide radical, hydroxyl radical and hypochlorous acid. Biol Pharm Bull 25: 1324–1327.

68. Frezza C, Cipolat S, Scorrano L (2007) Organelle isolation: functional mitochondria from mouse liver, muscle and cultured fibroblasts. Nat Protoc 2: 287–295.

18

Evaluation of the Efficacy & Biochemical Mechanism of Cell Death Induction by *Piper longum* Extract Selectively in *In-Vitro* and *In-Vivo* Models of Human Cancer Cells

Pamela Ovadje[1], Dennis Ma[1], Phillip Tremblay[1], Alessia Roma[1], Matthew Steckle[1], Jose-Antonio Guerrero[2], John Thor Arnason[2], Siyaram Pandey[1]*

1 Department of Chemistry & Biochemistry, University of Windsor, Windsor, ON, Canada, 2 Department of Biology, University of Ottawa, Ottawa, ON, Canada

Abstract

Background: Currently chemotherapy is limited mostly to genotoxic drugs that are associated with severe side effects due to non-selective targeting of normal tissue. Natural products play a significant role in the development of most chemotherapeutic agents, with 74.8% of all available chemotherapy being derived from natural products.

Objective: To scientifically assess and validate the anticancer potential of an ethanolic extract of the fruit of the Long pepper (PLX), a plant of the *piperaceae* family that has been used in traditional medicine, especially Ayurveda and investigate the anticancer mechanism of action of PLX against cancer cells.

Materials & Methods: Following treatment with ethanolic long pepper extract, cell viability was assessed using a water-soluble tetrazolium salt; apoptosis induction was observed following nuclear staining by Hoechst, binding of annexin V to the externalized phosphatidyl serine and phase contrast microscopy. Image-based cytometry was used to detect the effect of long pepper extract on the production of reactive oxygen species and the dissipation of the mitochondrial membrane potential following Tetramethylrhodamine or 5,5,6,6'-tetrachloro-1,1',3,3'-tetraethylbenzimidazolylcarbocyanine chloride staining (JC-1). Assessment of PLX *in-vivo* was carried out using Balb/C mice (toxicity) and CD-1 nu/nu immunocompromised mice (efficacy). HPLC analysis enabled detection of some primary compounds present within our long pepper extract.

Results: Our results indicated that an ethanolic long pepper extract selectively induces caspase-independent apoptosis in cancer cells, without affecting non-cancerous cells, by targeting the mitochondria, leading to dissipation of the mitochondrial membrane potential and increase in ROS production. Release of the AIF and endonuclease G from isolated mitochondria confirms the mitochondria as a potential target of long pepper. The efficacy of PLX in *in-vivo* studies indicates that oral administration is able to halt the growth of colon cancer tumors in immunocompromised mice, with no associated toxicity. These results demonstrate the potentially safe and non-toxic alternative that is long pepper extract for cancer therapy.

Editor: Stephanie Filleur, Texas Tech University Health Sciences Center, United States of America

Funding: This study was funded by Windsor & Essex County Cancer Centre Foundation by Seeds4Hope Grant (URL: http://windsorcancerfoundation.org/). The funders had no role in study design, data collection and analysis, decision to publish, or preparation of the manuscript.

* Email: spandey@uwindsor.ca

Introduction

The continuing increase in the incidence of cancer signifies a need for further research into more effective and less toxic alternatives to current treatments. In Canada alone, it was estimated that 267,700 new cases of cancer will arise, with 76,020 deaths occurring in 2012 alone. The global statistics are even more dire, with 12.7 million cancer cases and 7.6 million cancer deaths arising in 2008 [1,2]. The hallmarks of cancer cells uncover the difficulty in targeting cancer cells selectively. Cancer cells are notorious for sustaining proliferative signaling, evading growth suppression, activating invasion and metastasis and

resisting cell death among other characteristics [3]. These characteristics pose various challenges in the development of successful anticancer therapies. The ability of cancer cells to evade cell death events has been the center of attention of much research, with focus centered on targeting the various vulnerable aspects of cancer cells to induce different forms of Programmed Cell Death (PCD) in cancer cells, with no associated toxicities to non-cancerous cells.

Apoptosis (PCD type I) has been studied for decades, the understanding of which will enhance the possible development of more effective cancer therapies. This is a form of cell death that is required for regular cell development and homeostasis, as well as a defense mechanism to get rid of damaged cells; cells undergoing apoptosis invest energy in their own demise so as not to become a nuisance [2]. Cancer cells evade apoptosis in order to confer added growth advantage and sustenance, therefore current anticancer therapies endeavour to exploit the various vulnerabilities of cancer cells in order to trigger the activation of apoptosis through either the extrinsic or intrinsic pathways [4,5]. The challenges facing some of the available cancer therapies are their abilities to induce apoptosis in cancer cells by inducing genomic DNA damage. Although this is initially effective, as they target rapidly dividing cells [6], they are usually accompanied by severe side effects caused by the non-selective targeting of normal non-cancerous cells, suggesting a need for other non-common targets for apoptosis induction without the associated toxicities.

Natural health products (NHPs) have shown great promise in the field of cancer research. The past 70 years have introduced various natural products as the source of many drugs in cancer therapy. Approximately 75% of the approved anticancer therapies have been derived from natural products, an expected statistic considering that more than 80% of the developing world's population is dependent on the natural products for therapy [7]. Plant products especially contain many bioactive chemicals that are able to play specific roles in the treatment of various diseases. Considering the complex mixtures and pharmacological properties of many natural products, it becomes difficult to establish a specific target and mechanism of action of many NHPs. With NHPs gaining momentum, especially in the field of cancer research, there is a lot of new studies on the mechanistic efficacy and safety of NHPs as potential anticancer agents [8].

Long pepper, from the Piperaceae family, has been used for centuries for the treatment of various diseases. Several species of long pepper have been identified, including *Piper longum* (the extract of which is being used in this study), *Piper betle*, *Piper retrofactum*, extracts of which have been used for years in the treatment of various diseases. A long list of uses and benefits are associated with extracts of different *Piper spp*, with reports indicating their effectiveness as good digestive agents and pain and inflammatory suppressants [9]. However, there is little to no scientific validation, only anecdotal evidence, for the benefits associated with the use of long pepper extracts. There are scientific studies have been carried out on several compounds present in extracts of long pepper, including piperines, which has been shown to inhibit many enzymatic drug bio-transforming reactions and plays specific roles in metabolic activation of carcinogens and mitochondrial energy production [10–13], and various piperidine alkaloids, with fungicidal activity [9,14]. Some of these compounds have shown potent anticancer activity [15], suggesting that Long pepper extracts could represent a new NHP, with better selective efficacy against cancer cells.

In this study, we examine the efficacy of an ethanolic extract of Long pepper fruit (PLX) against various cancer cells, as well as attempt to elucidate the mechanism of action, following treatment. Results from this study demonstrate that PLX reduced the viability of various cancer cell types in a dose and time dependent manner, where apoptosis induction was observed, following mitochondrial targeting and the release of pro-apoptotic factors. Due to the low doses of PLX required to induce apoptosis in cancer cell, it was easy to find the therapeutic window of this extract. The induction of apoptosis was found to be caspase-independent, although there was activation of both the extrinsic and intrinsic pathways and the production of ROS was not essential to the mechanism of cell death induction by PLX. The complex polychemical extract of the fruit of the long pepper plant, as a natural health product with unprecedented anticancer activity, provides a way to target multiple vulnerabilities of cancer cells. Even in the presence of certain inhibitors, PLX was efficacious in inducing apoptosis suggesting the potential application of developing PLX as a safe and efficacious cancer therapy.

Materials and Methods

Animal studies were carried out according to the animal care committee protocol approved by the University of Windsor Animal Care Committee; This protocol and project was approved by the animal care committee – Protocol number: AUPP 10–17), in accordance with the Canadian Council of Animal Care (CCAC) guidelines.

Cell Culture

The malignant melanoma cell line G-361, human colorectal cancer cell lines HT-29 and HCT116 (American Type Culture Collection, Manassas, VA, USA Cat. No. CRL-1687, CCL-218 & CCL-247, respectively) were cultured with McCoy's Medium 5a (Gibco BRL, VWR, Mississauga, ON, Canada) supplemented with 10% (v/v) FBS (Thermo Scientific, Waltham, MA, USA) and 40 mg/ml gentamicin (Gibco, BRL, VWR). The ovarian adeno-carcinoma cell line OVCAR-3 (American Type Culture Collection, Cat. No. HTB-161) was cultured in RPMI-1640 media (Sigma-Aldrich Canada, Mississauga, ON, Canada) supplemented with 0.01 mg/mL bovine insulin, 20% (v/v) fetal bovine serum (FBS) standard (Thermo Scientific, Waltham, MA, USA) and 10 mg/mL gentamicin. The pancreatic adenocarcinoma cell line BxPC-3 (American Type Culture Collection, Cat. No. CRL-1424) was cultured in RPMI-1640 medium, supplemented with 10% (v/v) fetal bovine serum (FBS) standard and 40 mg/mL gentamicin. Normal-derived colon mucosa NCM460 cell line (INCELL Corporation, LLC., San Antonio, TX, USA) was grown in INCELL's M3Base medium (INCELL Corporation, LLC., Cat. No. M300A500) supplemented with 10% (v/v) FBS and 10 mg/mL gentamicin.

All cells were grown in optimal growth conditions of 37°C and 5% CO2. Furthermore, all cells were passaged for ≤6 months.

Long Pepper Extraction

Ripe and dried Indian long pepper fruits were obtained from Quality Natural Foods limited, Toronto Ontario. The plant material was ground up and extracted in anhydrous ethanol (100%) in a ratio of 1:10 (1 g plant material to 10 ml ethanol). The extraction was carried out overnight on a shaker at room temperature. The extract was passed through a P8 coarse filter, followed by a 0.45 μm filter. The solvent was evaporated using a RotorVap at 40°C and reconstituted in dimethylsulfoxide (Me$_2$SO) at a final stock concentration of 450 mg/ml.

Cell Treatment

Cells were plated and grown to 60–70% confluence, before being treated with Long Pepper Extracts (PLX), N-Acetyl-L-cysteine (NAC) (Sigma-Aldrich Canada, Cat. No. A7250), and broad-spectrum caspase inhibitor, Z-VAD-FMK (EMD Chemicals, Gibbstown, NJ, USA) at the indicated doses and durations. NAC was dissolved in sterile water. Z-VAD-FMK was dissolved in dimethylsulfoxide (Me$_2$SO). PLX was extracted as previously described, reconstituted in Me$_2$SO. Before treatment, a dilute working concentration of 10 mg/ml in PBS was prepared. Cells were treated with the 10 mg/ml to obtained the final concentrations indicated in the results section.

Table 1. Analysis of five well-known piperamides and crude long pepper extract at a flow rate of 1.0 mL/min with a mobile phase constituted of H_2O and methanol.

Time (mins)	H$_2$0 (%)	MeOH (%)
0.0	37.5	62.5
15.0	35.0	65.0
35.0	0.0	100.0
45.0	0.0	100.0
46.0	37.5	62.5

Assessing the Efficacy of Long Pepper Extract (PLX) In Cancer Cells

WST-1 Assay for Cell Viability. To assess the effect of PLX on cancer cells, a water-soluble tetrazolium salt (WST-1) based colorimetric assay was carried out as per manufacturer's protocol (Roche Applied Science, Indianapolis, IN, USA), to quantify cell viability as a function of cellular metabolism. Equal number of cells were seeded onto 96-well clear bottom tissue culture plates then treated with the indicated treatments at the indicated concentrations and durations. Following treatment, cells were incubated with the WST-1 reagent for 4 hours at 37°C with 5% CO2. The WST-1 reagent is cleaved to formazan by cellular enzymes in actively metabolizing cells. The formazan product was quantified by taking absorbance readings at 450 nm on a Wallac Victor[3] 1420 Multilabel Counter (PerkinElmer, Woodbridge, ON,

Canada). Cellular viability as a measure of metabolic activity was expressed as percentages of the solvent control groups.

Nuclear Staining. Subsequent to treatment, the nuclei of cells were stained with 10 μM Hoechst 33342 dye (Molecular Probes, Eugene, OR, USA) or 1 mg/ml propidium iodide (PI) (Sigma Aldrich, Mississauga, ON. Canada), to monitor nuclear morphology for apoptosis induction at designated time points and overall cell death. Cells were incubated with 10 μM Hoechst dye and 1 mg/ml PI for 10 minutes and micrographs were taken with a Leica DM IRB inverted fluorescence microscope (Wetzlar, Germany) at 400× magnification. Image-based cytometry was used to quantify the amount of cell death occurring with PI staining.

Annexin V Binding Assay. To confirm the induction of apoptosis, the binding of Annexin V to externalized phosphatidylserine on the outer cellular surface, was assessed. Following

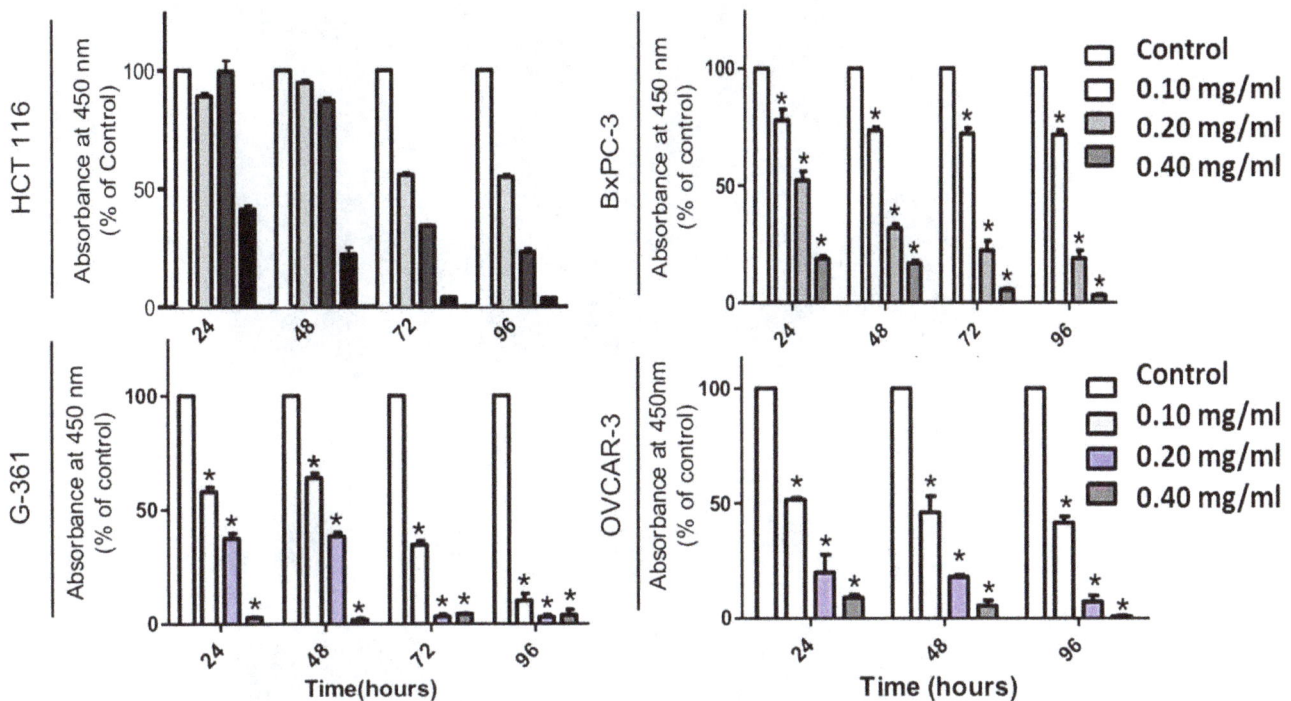

Figure 1. Crude Ethanolic Extract of Long Pepper (PLX) Effectively Reduces the Percentage of Viable Cancer cells in a Dose & Time Dependent Manner. Colon (HCT116), Ovarian (OVCAR-3), Pancreatic (BxPC-3) cancer and Melanoma (G-361) cells were treated with a crude ethanolic extract of long pepper (PLX), following which they were incubated with WST-1 cell viability dye for 4 hours. Absorbance was read at 450 nm and expressed as a percent of the control. Values are expressed as mean ± SD from quadruplicates of 3 independent experiments. **P< 0.0001.

A

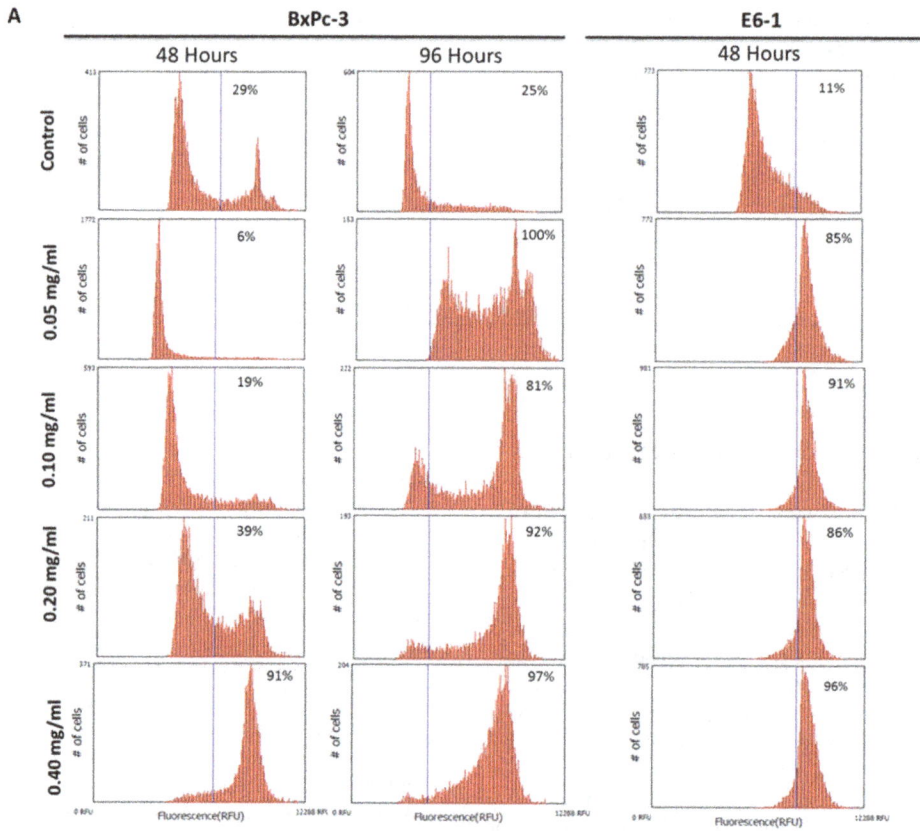

B

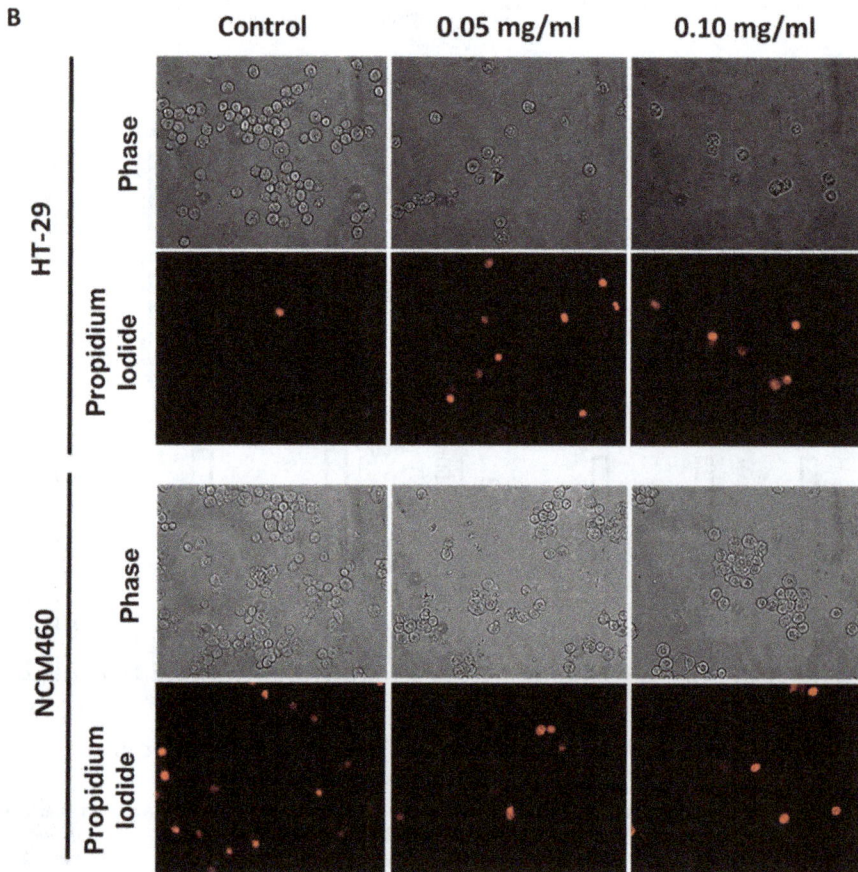

Figure 2. PLX Selectively Induces Cell Death in Human Cancer Cells in a Dose & Time Dependent Manner. (A) Following treatment of Human pancreatic (BxPc-3) cancer and T cell leukemia cells with PLX, at indicated time points, cells were incubated with propidium iodide and assessed for the induction of cell death by image-based cytometry. (B) Similar experiments were carried out in human colon cancer cells (HT-29) and normal colon epithelial cells (NCM460). Fluorescence microscopy was used to assess the induction of cell death as characterized by presence of propidium iodide positive cells. Images were taken at 400× magnification on a fluorescent microscope. Scale bar = 15 μm.

treatment with PLX, cells were washed twice in phosphate buffer saline (PBS). Subsequently, cells were resuspended and incubated in Annexin V binding buffer (10 mM HEPES, 10 mM NaOH, 140 mM NaCl, 1 mM CaCl2, pH 7.6) with Annexin V Alexa-Fluor-488 (1:50) (Invitrogen, Canada, Cat No. A13201) for 15 minutes. In the final 10 minutes of incubation, 10 μM Hoechst and 1 mg/ml propidium iodide were added to the microcentrifuge tube and incubated for the final 10 minutes in the dark. Micrographs were taken at 400× magnification on a Leica DM IRB inverted microscope (Wetzlar, Germany) and image-based cytometry was used to quantify the percentage of programmed cell death (annexin V positive cells) occurring after treatment.

TUNEL Staining to Detect DNA Damage and Quantify Apoptosis. Following PLX treatment, HT-29 cells were labeled with the Terminal deoxynucleotidyl transferase dUTP nick end labeling (TUNEL) assay. The assay was performed according to the manufacturer's protocol (Molecular Probes, Eugene, OR), in order to detect DNA damage. Cells were treated with PLX or VP-16 (as a positive control) at indicated concentrations and time points and analyzed for the fragmentation of DNA. Following treatment, cells were fixed by suspending them in 70% (v/v) ethanol and stored at −20°C overnight. The sample was then incubated with a DNA labeling solution (10 μL reaction buffer, 0.75 μL TdT enzyme, 8 μL BrdUTP, 31.25 μL of dH2O) for 1 hour at 25°C. Each sample was exposed to an antibody solution (5 μL Alexa Fluor 488 labeled anti-BrdU antibody and 95 μL rinse solution). The cells were incubated with the antibody solution for 20 minutes and TUNEL positive cells were quantified by image-based cytometry.

Whole Cell ROS Generation. Following treatment with PLX, cells were incubated with 2′,7′-Dichlorofluorescin diacetate H₂DCFDA (Catalog No. D6883, Sigma Aldrich, Mississauga ON. Canada) for 45 minutes. Cells were collected, washed twice in PBS and green fluorescence was observed using a TALI image-based cytometer (Invitrogen, Canada). NAC was used to assess the dependence of PLX on ROS generation and viability.

Assessment of Mitochondrial Function Following PLX Treatment

Tetramethylrhodamine Methyl Ester (TMRM) Staining. To monitor mitochondrial membrane potential (MMP), tetramethylrhodamine methyl ester (TMRM) (Gibco BRL, VWR, Mississauga, ON, Canada) or 5,5,6,6′-tetrachloro-1,1′,3,3′-tetraethylbenzimidazolylcarbocyanine chloride (JC-1) (Invitrogen, Canada) were used. Cells were grown on coverslips, treated with the indicated concentrations of treatments at the indicated time points, and incubated with 200 nM TMRM for 45 minutes at 37°C. Micrographs were obtained at 400× magnification on a Leica DM IRB inverted fluorescence microscope (Wetzlar, Germany). To confirm the results obtained by fluorescence microscopy, image-based cytometry was used to detect red fluorescence. Cells were seeded in 6-well plates and following treatment, cells were incubated with TMRM for 45 minutes, washed twice in PBS and placed in TALI slides. Red fluorescence was obtained using a TALI image-based cytometer (Invitrogen, Canada).

Mitochondrial Isolation to Assess Mitochondrial Targeting. Cells were collected by trypsin, washed once in cold PBS, resuspended in cold hypotonic buffer (1 mM EDTA, 5 mM Tris–HCl, 210 mM mannitol, 70 mM sucrose, 10 μM Leu-pep and Pep-A, 100 μM PMSF), and manually homogenized. The homogenized cell solution was centrifuged at 3000 rpm for 5 minutes at 4°C. The supernatant was centrifuged at 12,000 rpm for 15 minutes at 4°C and the mitochondrial pellet was resuspended in cold reaction buffer (2.5 mM malate, 10 mM succinate, 10 μM Leu-pep and Pep-A, 100 μM PMSF in PBS). The isolated mitochondria were treated with PLX at the indicated concentrations and incubated for 2 hours in cold reaction buffer. The control group was treated with solvent (ethanol). Following 2 hour incubation with extract, mitochondrial samples were vortexed and centrifuged at 12,000 rpm for 15 minutes at 4°C. The resulting supernatant and mitochondrial pellets (resuspended in cold reaction buffer) were subjected to Western Blot analysis to assess for the mitochondrial release/retention of pro-apoptotic factors.

Western Blot Analyses. Protein samples were subjected to SDS-PAGE, transferred onto a nitrocellulose membrane, and blocked with 5% w/v milk TBST (Tris-Buffered Saline Tween-20) solution for 1 hour. Membranes were incubated overnight at 4°C with an anti-endonuclease G (EndoG) antibody (1:1000) raised in rabbits (Abcam, Cat. No. ab9647, Cambridge, MA, USA), an anti-succinate dehydrogenase subunit A (SDHA) antibody (1:1000) raised in mice (Santa Cruz Biotechnology, Inc., sc-59687, Paso Robles, CA, USA), or an anti-apoptosis inducing factor (AIF) antibody raised in rabbits (1:1000) (Abcam, Cat. No. ab1998, Cambridge, MA, USA). After primary antibody incubation, the membrane was washed once for 15 minutes and twice for 5 minutes in TBST. Membranes were incubated for 1 hour at room temperature with an anti-mouse or an anti-rabbit horseradish peroxidase-conjugated secondary antibody (1:2000) (Abcam, ab6728, ab6802, Cambridge, MA, USA) followed by three 5-minute washes in TBST. Chemiluminescence reagent (Sigma-Aldrich, CPS160, Mississauga, ON, Canada) was used to visualize protein bands and densitometry analysis was performed using ImageJ software.

In-Vivo Assessment of Long Pepper Extract

Toxicity Assessment. Six week old Balb/C mice were obtained from Charles River Laboratories and housed in constant laboratory conditions of a 12-hour light/dark cycle, in accordance with the animal protocols outlined in the University of Windsor Research Ethics Board- AUPP 10–17. Following acclimatization, mice were divided into three groups (3 animals/control (untreated), 3 animals/gavage control (vehicle treatment) and 4 animals/ treatment group). The control untreated group was given plain filtered water, while the second and third group was given 50 mg/ kg/day vehicle (Me₂SO) or PLX, respectively for 75 days. During the period of study, toxicity was measured by weighing mice twice a week and urine was collected for protein urinalysis by urine dipstick and Bradford assays. Following the duration of study, mice were sacrificed and their organs (livers, kidneys and hearts) were obtained for immunohistochemical and toxicological analysis by Dr. Brooke at the University of Guelph.

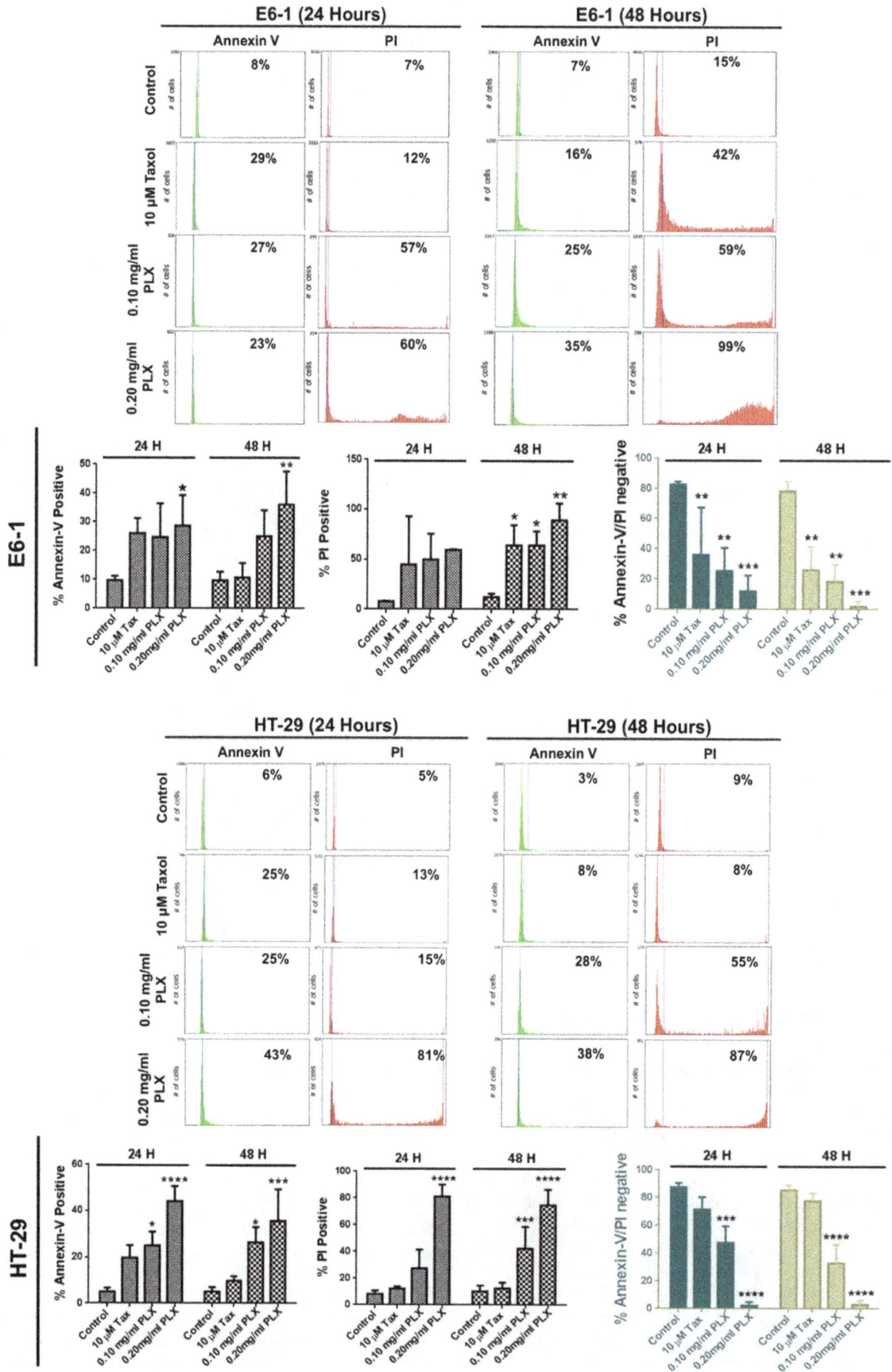

Figure 3. Quantification of Cell Death Induction Following PLX Treatment. Image-Based cytometry was used to quantify apoptotic induction (% Annexin V positive), followed by necrosis (% PI positive) in E6-1 and HT-29 cells following PLX treatment. The lack of annexin V or PI staining was used as an indication of live cells following treatment (%Annexin V/PI negative cells) (*P<0.05, ** P<0.003, ***P<0.0001). (E) To further confirm the induction of apoptosis.

Efficacy of PLX in Tumor Xenograft Models of Immunocompromised Mice. Six week old male CD-1 nu/nu mice were obtained from Charles River Laboratories and housed in constant laboratory conditions of a 12-hour light/dark cycle, in accordance with the animal protocols outlined in the University of Windsor Research Ethics Board- AUPP 10–17). Following acclimatization, the mice were injected subcutaneously in the right and left hind flanks with a colon cancer cell suspension (in Phosphate buffered saline) at a concentration of $2 * 10^6$ cells/mouse (HT-29, $p53^{-/-}$, in the left flank and HCT116, $p53^{+/+}$, in the right flank). Tumors were allowed to develop (approximately a week), following which the animals were randomized into treatment groups of 4 mice per group, a control group, a gavage control group given plain filtered sterile water, as well as gavage regimen of the vehicle (5 µL Me_2SO in PBS) twice a week. The final group was given filtered water supplemented with long pepper extract at a concentration of 100 µg/mL, as well as gavage regimen of long pepper extract (5 µL extract in PBS), twice a week, corresponding to 50 mg/kg/day. The tumors were assessed every other day by measuring the length, width and height, using a standard caliper and the tumor volume was calculated according to the formula $\pi/6$*length*width. The mice were also assessed for any weight loss every other day for the duration of the study, which lasted 75 days, following which the animals were sacrificed and their organs and tissues (liver, kidneys, heart and tumors) were

obtained and stored in 10% formaldehyde for immunohistochemical and toxicological analysis.

Hematoxylin & Eosin (H & E) Staining. Mice organs were fixed in 10% formaldehyde, following which they were cryosectioned into 10 µm sections and placed on a superfrost/Plus microscope slides (Fisherbrand, Fisher Scientific). Sections of organs were stained according to a standardized H & E protocol [16].

Phytochemical Analysis of Long Pepper Extract by HPLC

HPLC analysis of the long pepper crude extract was carried out at University of Ottawa in the Arnason lab. A total of five well-known piperamides were analyzed and compared to the crude long pepper extract. The extracts and piperamide standards were analyzed on a Luna C18-5u-250×4.6 mm column at 45°C at a flow rate of 1.0 mL/min with a mobile phase constituted of H_2O and methanol as outlined in Table 1. Chromatogram profiles were used to detect the any differences between a sample standard of known piperamides in the crude long pepper extracts.

Statistical Analysis

All experiments were repeated at least three independent times. Representative fluorescence images were shown, where appropriate. Statistical analysis was performed using GraphPad Prism 6.0

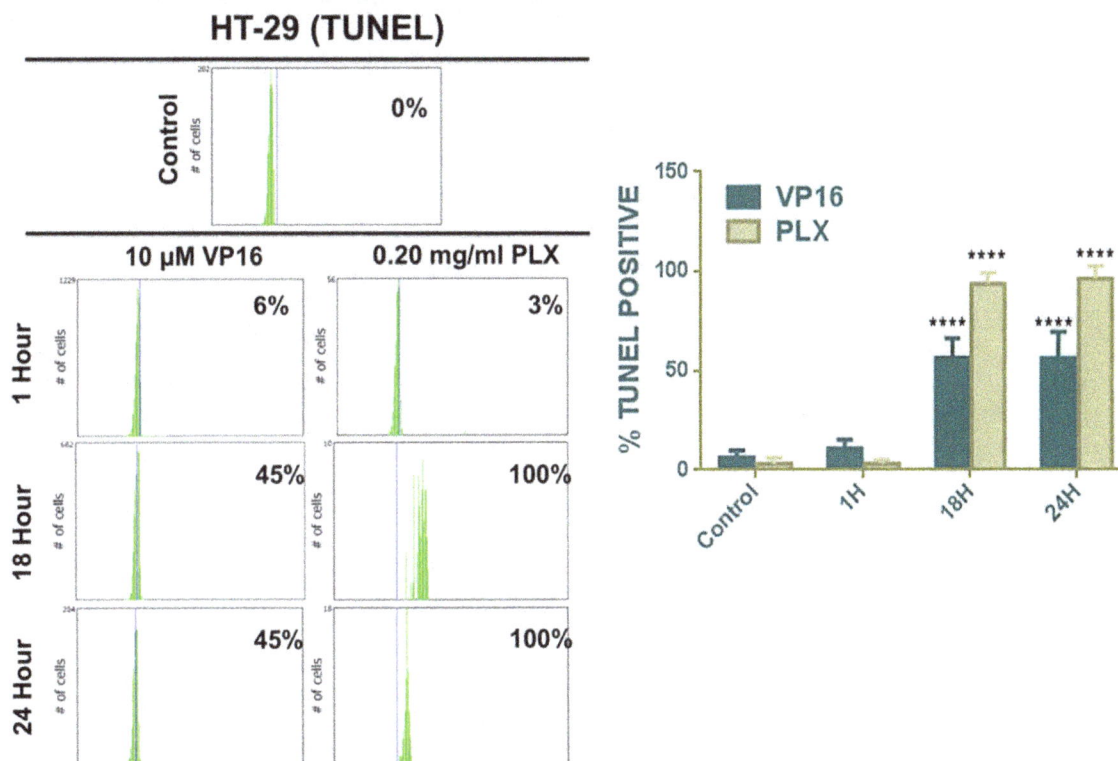

Figure 4. PLX Induces Double-stranded DNA Breaks in Cancer Cells. TUNEL labeling was used to detect DNA fragmentation. Following PLX and VP16 (as a positive control for DNA damage) treatment, cells were labelled with DNA staining solution and quantified by image-based cytometry. Treated cells were compared to the control untreated cell sample. (****P<0.0001).

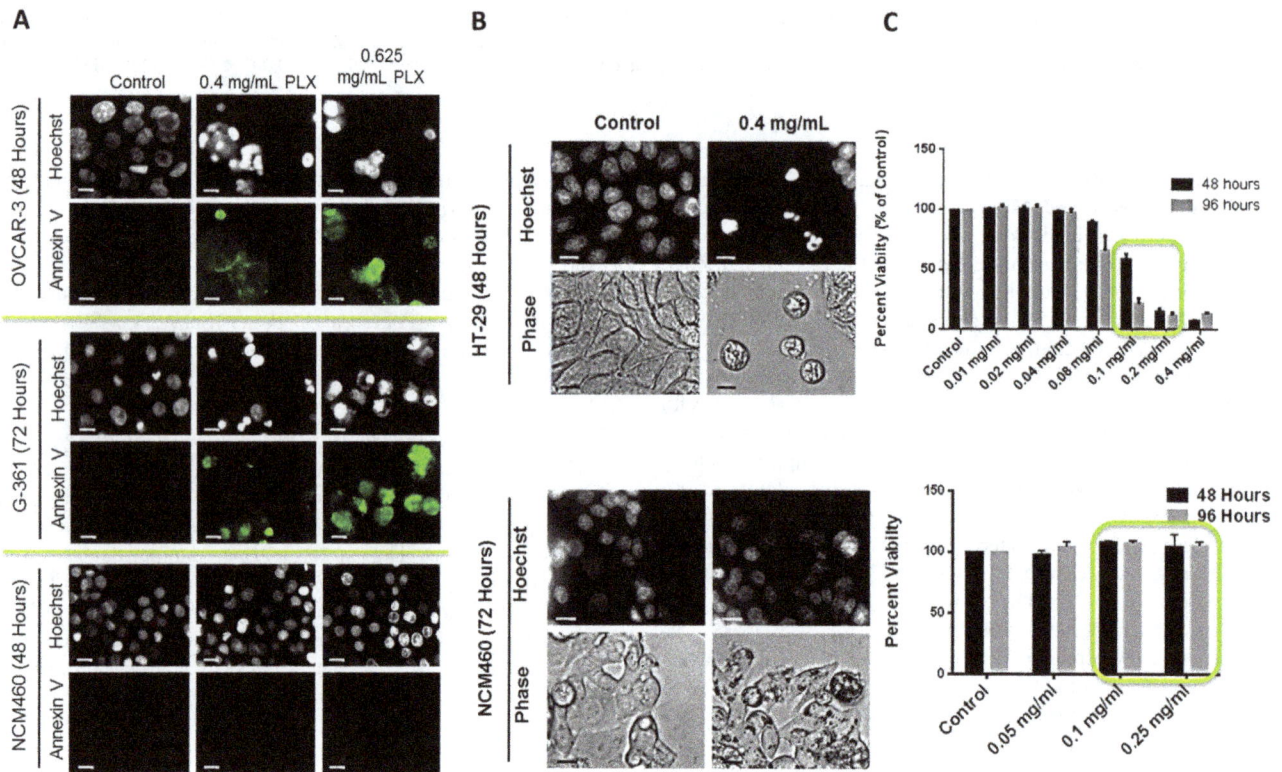

Figure 5. PLX Selectively Targets Cancer Cells for Apoptosis Induction. Subsequent to treatment with PLX, cells (Ovarian; OVCAR-3, Melanoma; G-361 and Normal Colon Epithelia cells (NCM460) were stained with Hoechst to characterize nuclear morphology and Annexin-V to detect apoptotic cells (A) and cellular morphology by phase contrast microscopy (B); Images were taken at 400× magnification on a fluorescent microscope. Scale bar = 15 μm. (C) Following PLX treatment, HT-29 colorectal cancer cells and non-cancerous NCM460 cells were incubated with WST-1 cell viability dye for 4 hours and absorbance was read at 450 nm and expressed as a percent of the control. Values are expressed as mean ± SD of 3 independent experiments. **P<0.0001.

288 software. The mean and standard error of three independent experiments were analyzed for the quantification data. The Student's T-test and two-way Anova were used for statistical analysis.

Results

Ethanolic Extract of Long Pepper (PLX) Effectively and Selectively Reduces the Viability of & Induces Apoptosis in Cancer cells in a Dose & Time Dependent Manner

The first step in understanding the effect of long pepper extract in this study was to assess the effect of PLX on the viability of cancer cells. Following treatment with increasing concentration of PLX at increasing time points, cells were incubated with a water soluble tetrazolium salt, which gets metabolized to a red formazan product by viable cells with active metabolism. This product can then be quantified by absorbance spectrometry. We observed the efficacy of crude PLX in reducing the viability of cancer cells, including colon (HCT116), pancreatic (BxPC-3), ovarian cancer (OVCAR-3) and melanoma cells. This effect was dose and time dependent (Figure 1). To further evaluate the anticancer activity of PLX, we wanted to assess its role in cell death and its selectivity to cancer cells. Our results demonstrate that PLX is able to selectively induce cell death in cancer cells (colon, pancreatic and leukemia) in a dose and time dependent manner, as characterized by the increase in propidium iodide positive cells in cancer cells treated with PLX (Figure 2). Furthermore, this effect was selective, as normal colon epithelial cells remained unaffected by this

treatment, at the same concentrations and time-points (Figure 2B). These results were quantified using image-based cytometry to determine the percentage of cells undergoing apoptosis and total cell death. We observed a 30–40% increase in annexin V positive cells, following PLX treatment and an 80–100% PI positive increase in the same cell samples, confirming the induction of apoptosis, following by necrosis in cultured cancer cells (Figure 3).

DNA fragmentation is a key biochemical feature of apoptosis. To further confirm this induction of apoptosis, TUNEL labelling to detect DNA fragmentation was employed. Quantification results from image-based cytometry show the efficacy of PLX in inducing apoptosis, following DNA fragmentation in HT-29 colon cancer cells in a time dependent manner. VP16, a known chemotherapeutic agent with DNA damaging capabilities, was used as a positive control (Figure 4).

Additionally, apoptosis induction in various cancer cells, melanoma (G-361), ovarian and colon cancer (HT-29) cells, was confirmed by Annexin-V binding assay. This induction of apoptosis was confirmed to be selective to cancer cells, as normal colon cells (NCM460) remained unaffected by PLX treatment. This was indicated by nuclear condensation, cell morphology and externalization of phosphatidyl serine to the outer leaflet of the cell membrane, as indicated by Hoechst staining, phase contrast images and binding of annexin V dye respectively (Figure 5A and B). The selectivity of PLX to cancer cells was further confirmed by the WST-1 cell viability assay that showed that PLX was highly effective at such low doses, a therapeutic window was easily observed (Figure 5C). Treatment of HT-29 with 0.20 mg/ml

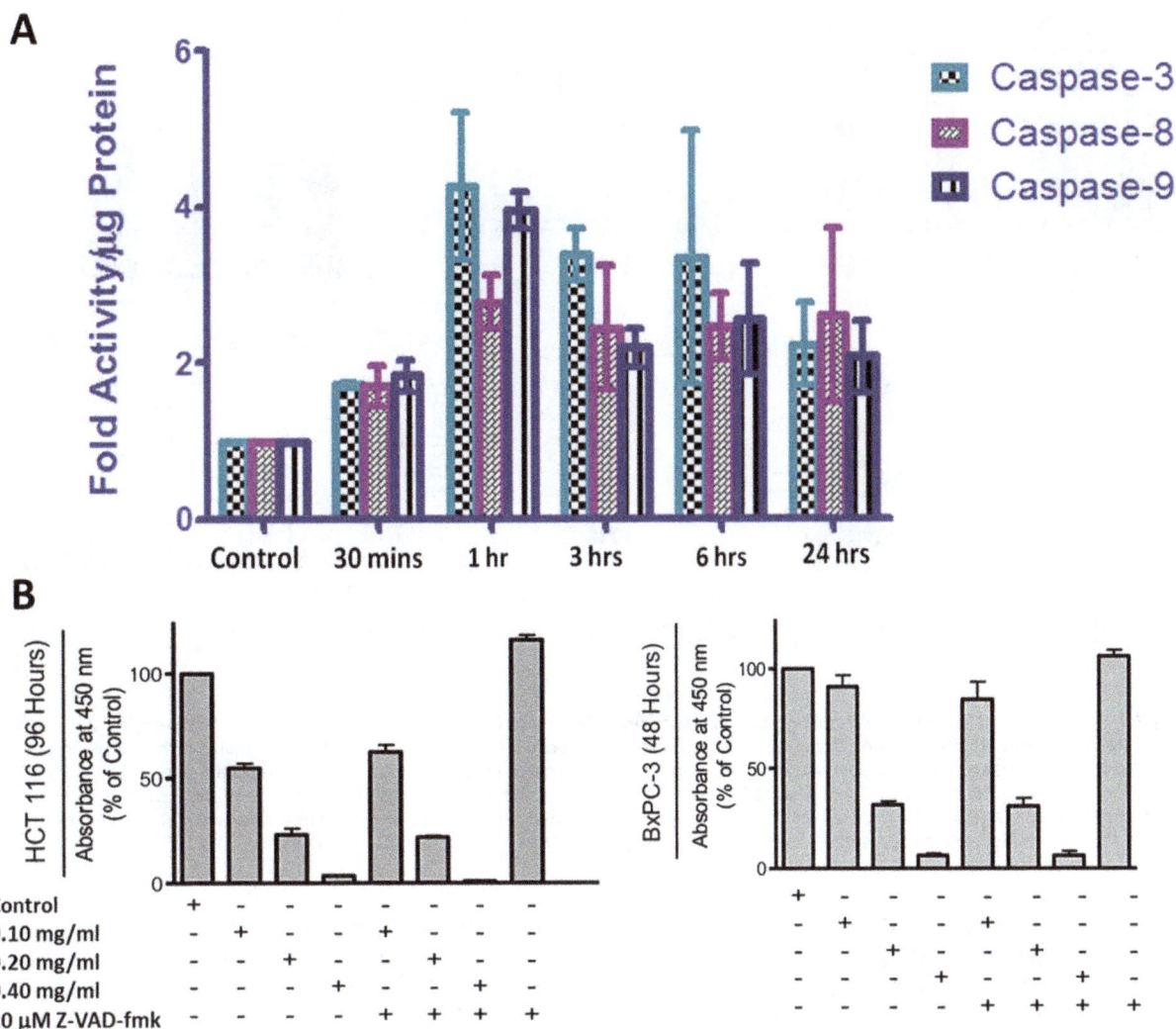

Figure 6. Long Pepper Extract (PLX) Activates the Extrinsic & Intrinsic Pathways of Apoptosis. Following treatment with 0.10 mg/ml PLX, at indicated time points, BxPc-3 cells were collected, washed and incubated with lysis buffer to obtain cell lysate. The cell lysate was incubated with caspase substrates, specific to each caspase (3, 8 and 9) and incubated for an hour. Fluorescence readings were obtained using a spectrofluorometer. An average of 6 readings per well and a minimum of three wells were run per experiment. The results here are reported as activity per μg of protein (in fold) and the average of three independent experiments is shown. **(B)** The reduction in viability was caspase independent, as a pan-caspase inhibitor, Z-VAD-fmk could not prevent the loss of viability induced by PLX treatment in colon and pancreatic cancer cells. Absorbance was read at 450 nm and expressed as a percent of the control. Values are expressed as mean ± SD from quadruplicates of 3 independent experiments. **P< 0.0001.

effectively reduced the viability by approximately 90%, while NCM460 cells remained at 100% viability at the same dose. This indicates that PLX can be more effective at very low doses, further reducing the chances of toxicity associated with treatment.

PLX Induces Caspase-Independent Apoptosis in Human Cancer Cells

Caspases are cysteine aspartic proteases that play a predominant role as death proteases [17]. Their roles in various cell death processes remains controversial, as their activation or inhibition could be essential to the progression of inhibition of cell death pathways [18,19]. To assess the role of caspases in our study, following treatment with 0.10 mg/ml PLX, at indicated time points, BxPc-3 cells were collected, washed and incubated with lysis buffer to obtain cell lysate. The cell lysate was incubated with caspase substrates, specific to each caspase (3, 8 and 9) and

incubated for an hour. Fluorescence readings were obtained using a spectrofluorometer. Our results indicate that PLX is able to activate both pathways (extrinsic and intrinsic apoptosis) in a time dependent manner. This was observed as rapid activation of caspases-3, 8 and 9 were observed as early as an hour, following treatment (Figure 6A).

To determine the importance of these activated caspases to the apoptosis-inducing effect of PLX, colon (HCT116) and pancreatic (BxPc-3) cancer cells were pre-treated with a pan-caspase inhibitor, Z-VAD-fmk (20 μM), for an hour before treatment with PLX. Following treatments, the WST-1 cell viability assay was used to assess for viability and efficacy of PLX. Our results indicate that the inhibition of caspases could not prevent the reduction of viability (Figure 6B), signifying that the effect of PLX in cancer cells is independent of caspase activation.

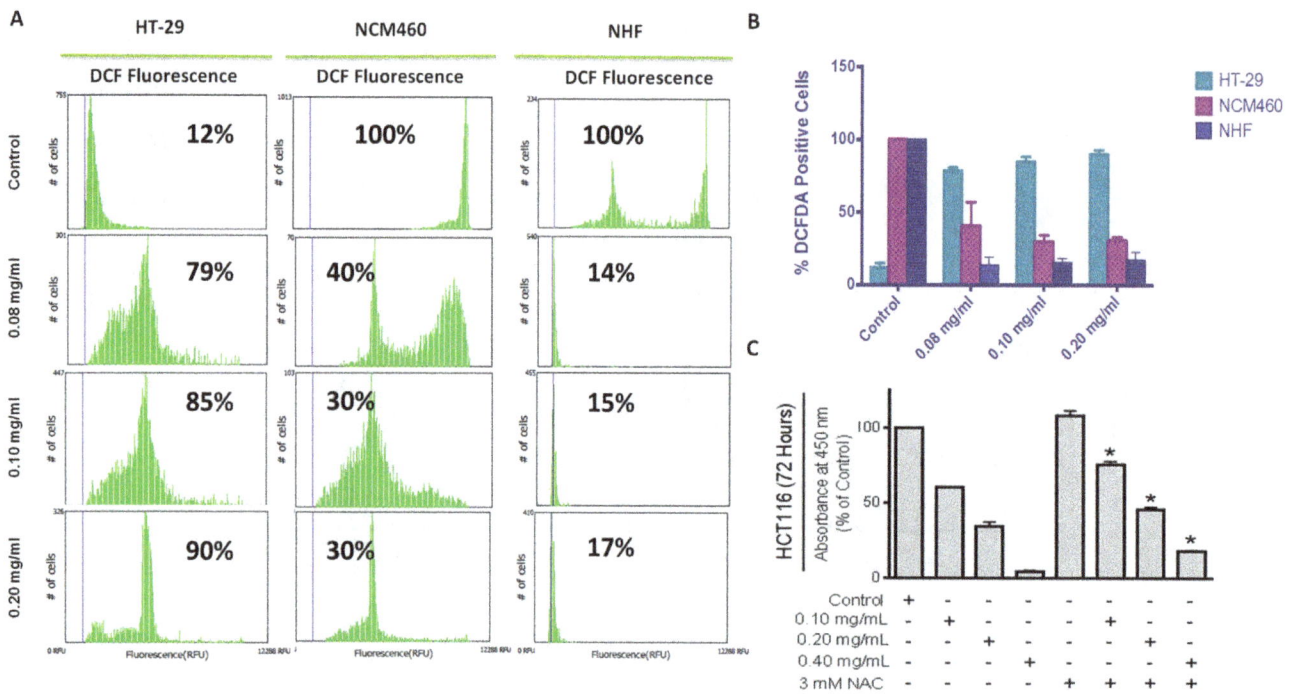

Figure 7. PLX Causes but is Not Dependent on the Production of Reactive Oxygen Species (ROS). (A) Colon cancer (HT-29), Normal Colon Epithelial (NCM460) and Normal Human Fibroblast (NHF) cells were treated with PLX for 48 hours, following which, they were incubated with H_2DCFDA and fluorescence results were obtained using an image based cytometer. Results were quantified using Graphpad prism 6.0 (B). (C) HCT116 colon cancer cells were treated with 3 mM N-acetylcysteine for an hour prior to PLX treatment. Cells were then treated PLX at indicated concentrations for 72 hours, following which the WST-1 assay was performed. Absorbance readings were taken at 450 nm and expressed as a percent of the control. Values are expressed as mean ± SD from quadruplicates of 3 independent experiments. *$p < 0.05$.

Long Pepper Extract Induces Oxidative Stress and Targets the Mitochondria of Cancer Cells

Generation of oxidative stress has been well established as a major player in the induction of several cell death processes, especially apoptosis [20,21]. The next part of our study focused on the role of oxidative stress in PLX induced apoptosis. Following treatment with PLX for 48 hours, cells were incubated with 2′,7′-Dichlorofluorescin diacetate H_2DCFDA for 45 minutes. The resulting green fluorescence histograms were obtained using a TALI image-based cytometer. From the results, it was observed that PLX induced extensive generation of whole cell reactive oxygen species (ROS) in HT-29 colon cancer cells, while acting to suppress any ROS present in the non-cancerous cell lines, NCM460 and normal human fibroblasts (NHF) (Figure 7A & B). This confirms our results of selectivity and indicates that PLX might act as a pro-oxidant in cancer cells in order to induce apoptosis.

To determine if this oxidative stress was essential to PLX activity, HCT116 colon cancer cells were pre-treated with N-acetyl-L-cysteine (NAC), a well-established anti-oxidant, used extensively in vitro studies [22,23], before treatment with PLX. Subsequent to PLX treatment, cells were analyzed for effect of PLX on viability, using the WST-1 viability assay. The results suggest that although PLX acts to induce oxidative stress to cause apoptosis, this oxidative stress is not essential to its activity. Both the cells treated with PLX alone and NAC followed by PLX showed a reduction in their viability (Figure 7C).

The mitochondria have also been shown to play a major role in the progression and execution of apoptosis. The permeabilization of the mitochondrial membrane usually leads to the release of pro-apoptotic factors, including cytochrome c, apoptosis inducing

factor (AIF) and endonuclease G (EndoG) [5,24]. These factors cause a caspase-independent pathway for apoptosis to pass through and could bypass the antioxidant effects of NAC observed in figure 7C.

To assess the efficacy of PLX on the mitochondria of cancer cells, OVCAR-3, HT-29 and NCM460 cells were stained with TMRM, a cationic dye that accumulates in healthy mitochondria. Mitochondrial membrane potential (MMP) dissipation was only observed in OVCAR-3 and HT-29 cells as seen with the dissipation of red TMRM fluorescence, by fluorescence microscopy and image-based cytometry (Figure 8A, B & C). Following mitochondrial membrane collapse, we wanted to determine if there was release of some pro-apoptotic factors. Western blot analysis was used to monitor for the release of AIF and EndoG from isolated OVCAR-3 mitochondria. Results demonstrate that PLX directly caused the release of both AIF and EndoG from the mitochondria of OVCAR-3 cells (Figure 8D). These results provide an insight to the mechanism of PLX action, where the mitochondria appear to be a direct target of PLX for the reduction of viability and the induction of apoptosis.

Long Pepper Extract is Well-Tolerated in Animal Models

Long pepper extracts (mainly water extracts) have been used for centuries and have been associated with various benefits [9]. With all these anecdotal reports of benefits, there have been no reports of toxicities associated with its use. To further scientifically evaluate and validate the safety of PLX, balb/c mice were orally gavaged with 50 mg/kg/day vehicle (DMSO) or PLX for 75 days and the mice were observed for signs of toxicity. To assess for toxicity, mice were weighed twice a week, urine was collected for protein urinalysis studies and following period of treatment, mice

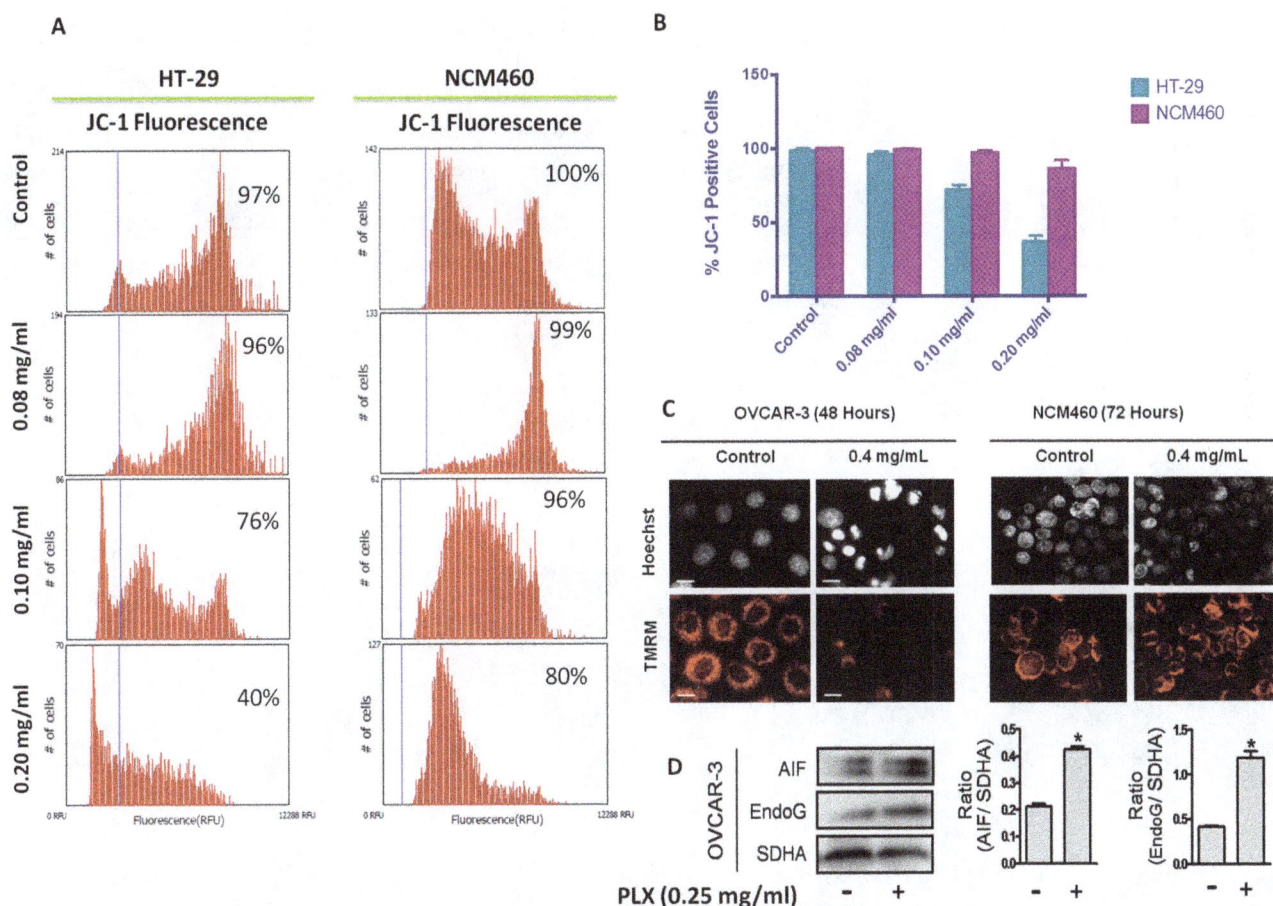

Figure 8. PLX Destabilizes the Mitochondrial Membrane of Cancer Cells. Colon cancer (HT-29), Ovarian cancer (OVCAR-3) and Normal Colon Epithelial (NCM460) cells were treated for 48 hours with PLX, following which, they were incubated with JC-1 (A) or TMRM (C) cationic mitochondrial membrane permeable dyes. Fluorescence readings were obtained using image based cytometry (A) and fluorescence microscopy; corresponding Hoechst dye images are also shown (C). Images were taken at 400× magnification on a fluorescent microscope. Scale bar = 15 μm. (D) Isolated mitochondria of OVCAR-3 cells were treated directly with PLX or solvent control (ethanol) for 2 hours. Following treatment, samples were centrifuged, to obtain mitochondrial supernatants, which were examined for the release of pro-apoptotic factors, AIF and EndoG via western blot analyses, and mitochondrial pellets which were probed for SDHA to serve as loading controls. Image is representative of 3 independent experiments demonstrating similar trends. Values are expressed as mean ± SD of quadruplicates of 1 independent experiment; *p<0.01 versus solvent control (ethanol).

were sacrificed and their organs were obtained for pathological analysis by a certified pathologist at the University of Guelph (Dr. Brooke). Results from this part of the study demonstrate that there was no weight loss overall in mice that were given PLX supplemented water (Figure 9).

To further assess toxicity, urine was collected from mice once a week and protein urinalysis was performed using a urine dipstick and a Bradford protein concentration assay. Protein urinalysis results indicate that there were trace amounts of protein in the urine of mice both from the control and the PLX group, with trace readings corresponding to protein concentrations between 5 and 20 mg/dL. Bradford assays confirm the results obtained by dipstick urinalysis (Figure 9A). There was no major difference between the control group and PLX group, confirming the lack of toxicity associated with oral administration of PLX in drinking water. Furthermore, the hearts, livers and kidneys were obtained following the toxicity study, sliced and stained with hematoxylin and eosin. Results show no gross morphologic difference between the control and the treatment group, confirming the lack of toxicity associated with PLX treatment. Results from the pathologist, indicate that the presence of any lesions in the tissues

are minimal or mild and interpreted as either background or incidental lesions and the lack of lesion type and frequency was enough to conclude no toxicological effect of PLX to the balb/c mice (Table 2).

Oral Administration of Long Pepper Extract Halts the Growth of Human Colon Cancer Xenografts in Immunocompromised Mice

Following efficacy studies, we wanted to further study the efficacy of PLX. For this study, CD-1nu/nu immunocompromised mice were subcutaneously injected with HT-29 cells (left) and HCT116 cells (right). Following the establishment of tumors, mice were separated into three groups, a control group, a vehicle (Me$_2$SO) group and a PLX treated group. Mice were observed for 75 days, with weights and tumor volumes measured twice a week. Results demonstrate that oral administration of PLX could suppress the growth of both p53 WT (HCT116) and p53 mutant (HT-29) tumors *in-vivo*. There were no signs of toxicity, as indicated by increasing weights during the study (Figure 10A & B). Furthermore, H & E staining revealed less nuclei in the PLX

A

	17-May	21-May	30-May	4-Jun	11-Jun	14-Jun
C1	13.697	21.080	27.584		0.977	27.139
C2	21.899	15.543	22.694	31.353		31.478
C3	15.854	17.717	25.700		12.692	24.968
C4	14.027		22.918	32.830	-3.509	33.179
C5		30.405	44.646		6.810	46.738
P1	9.843		22.560		2.273	16.173
P2	30.405	31.913			4.965	34.703
P3	40.077	20.840			4.766	19.105
P4	15.458	34.154				32.123

B

C

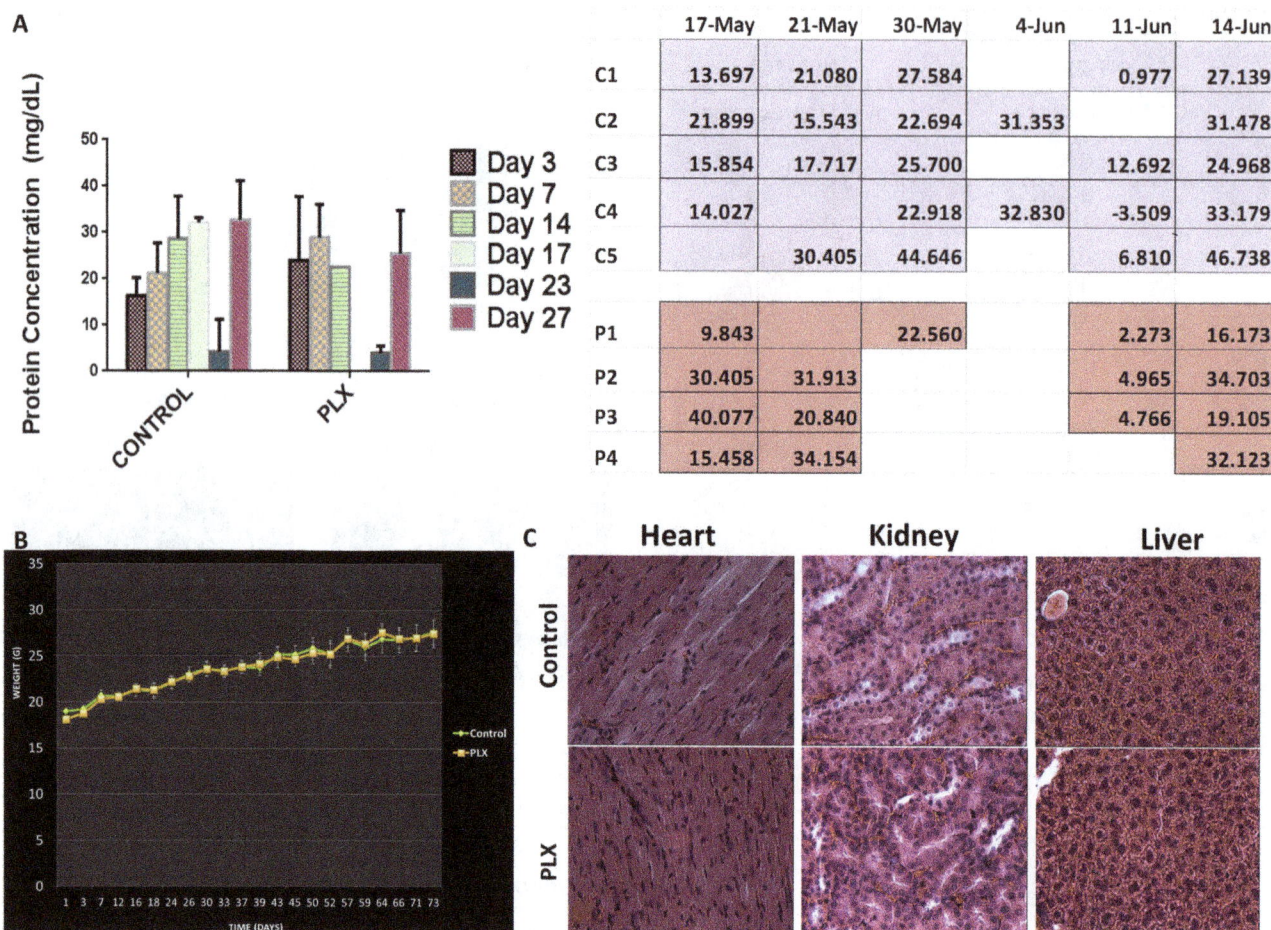

Figure 9. PLX is Well-Tolerated in Mice Models. Balb/C mice were divided into three groups (3 animals/control (untreated), 3 animals/gavage control (vehicle treatment) and 4 animals/treatment group). The control untreated group was given plain filtered water, while the second and third group was given 50 mg/kg/day vehicle (DMSO) or PLX, respectively. Mice were assessed for toxicity with protein urinalysis by Bradford Assay and dipstick analysis (A) and weight changes (B). (C) Hematoxylin and Eosin stained tissue sections of the liver, heart and kidney of control versus PLX treated group. Images were obtained on a bright field microscope at 63× objective.

treated group, compared to the control group, however, as observed in the toxicity studies, there were no gross morphological differences in the livers, kidneys and hearts of the control and PLX groups (Figure 10C).

Analysis of Long Pepper Extract

The availability of several species of long pepper and the host of compounds present within them make it essential to characterize the long pepper extract that has shown potent anticancer activity, both in *in-vitro* and *in-vivo* studies. We ran an HPLC profile study on the crude ethanolic extracts, compared with a piperamide standard mix. The chromatogram profile show that our PLX extract contained several classes of compounds known to be present in piper species, including piperines, piperlongumine and dihydropiperlongumine (Figure 11A & B), suggesting that our extract is a member of the *Piper longum* species.

Discussion

In this report we demonstrate for the first time, the selective anticancer potential of an ethanolic extract of the fruit of long pepper (PLX) in several cancer cell lines. PLX effectively reduced the viability of cancer cells and induced apoptosis in a dose- and

time-dependent manner, at low doses, allowing for a greater therapeutic window in *in-vitro* studies (Figure 1–5). This apoptosis inducing effect was found to be independent of caspases, cysteine aspartic proteases that play a role in the progression and execution of apoptosis (Figure 6B). These results suggest that PLX is not toxic to non-cancerous cells at such low doses, as was observed in the cancer cells. Selectivity and lack of toxicity was confirmed with *in-vivo* toxicological studies.

Damage to the kidneys is a common occurrence during various types to toxic therapies. This damage to the kidney results in large amounts of protein (>3.5 g/day) leaking into the urine [25,26], and this can be measured by various assays. Lack of toxicity was confirmed by the lack of increased protein concentration in the urine samples collected from both the control group and PLX treated group, by two different assays. The urine dipstick method indicated that all urine samples from the control and PLX groups had trace amounts of protein, corresponding to concentrations between 5 mg/dL and 20 mg/dL, well within the acceptable concentration range. Bradford protein assay showed a concentration of approximately 30 mg/dL most days urine was collected (Figure 9A). This is still within the acceptable range of protein concentration in urine. These results confirm anecdotal studies that suggest no associated toxicity or side effects observed with take

Table 2. Summary of Histological Lesions in Balb/C Mice on PLX regimen.

	No Treatment		Vehicle (Gavage Control)			Long Pepper Extract (Treatment group)			
	M1	M2	M1	M2	M3	M1	M2	M3	M4
Liver:									
-Infiltration, leukocyte, predominantly mononuclear, minimal		X	X		X	X			X
-Focal mineralization, minimal									X
-Hepatocyte necrosis, minimal									
-Focus of cellular alteration, eosinophilic, minimal			X	X				X	
-Hepatocyte vacuolation, lipid type, minimal			X	X			X		
- Hepatocyte vacuolation, lipid type, mild	X			X				X	X
Fibrin thrombus			X						
Heart:									
-Infiltration, leukocyte, predominantly mononuclear, minimal		X				X			X
Myofiber separation and vaculation, minimal (suspect artifact)		X	X						X
Kidney:									
- Infiltration, leukocyte, predominantly mononuclear, minimal	X	X		X		X		X	
Tubule vacuolation, minimal					X				X
Fibrin or other extracellular matrix, glomerulus								X	

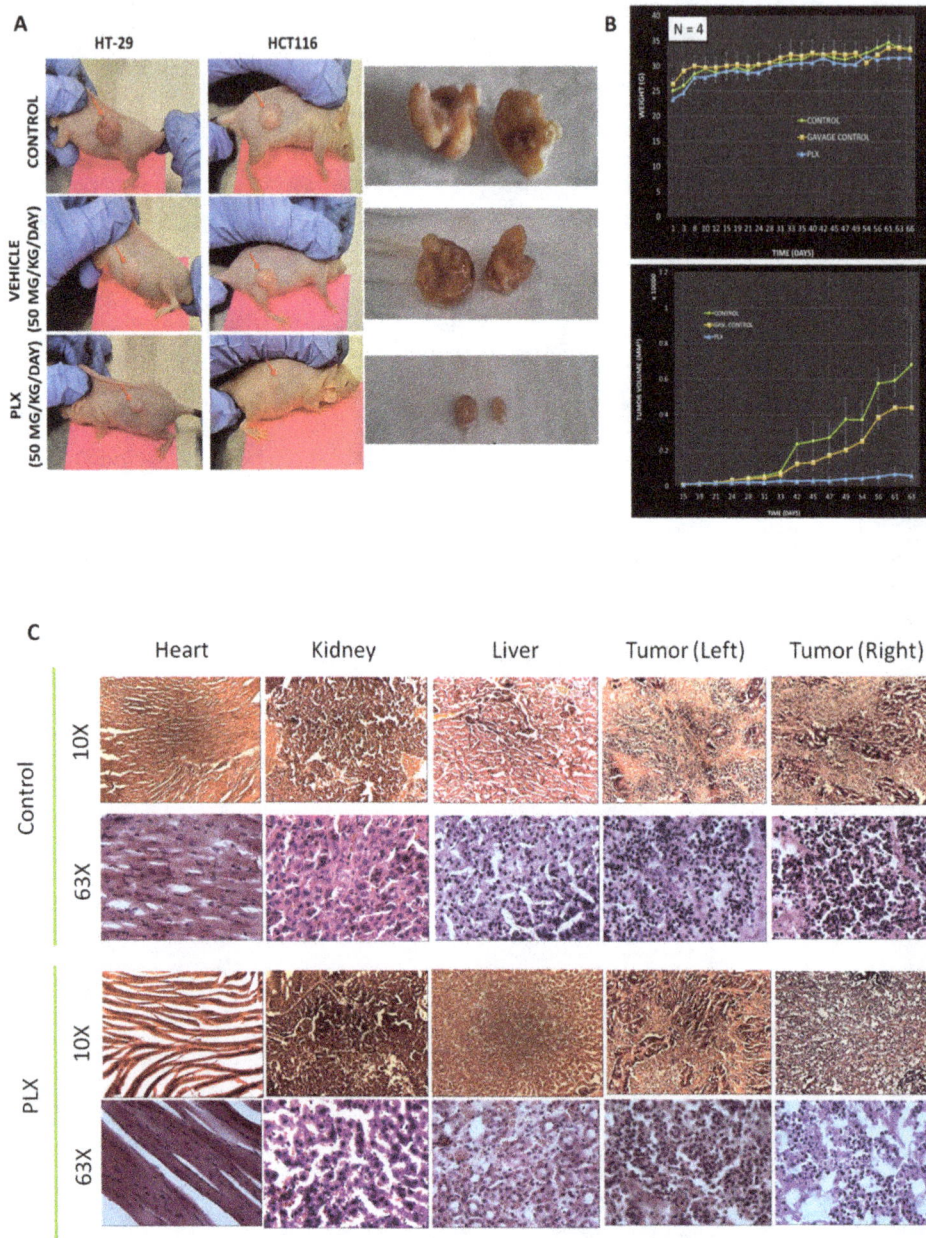

Figure 10. PLX Halts Growth of Colon Tumors in Xenograft Models. CD-1 nu/nu mice were subcutaneously injected with colon cancer cells; HT-29 (p53$^{-/-}$) on the left flank and HCT116 (p53$^{+/+}$) on the right flank. (A) Representative tumor size control mice and 50 mg/kg/day vehicle or PLX treated mice, respectively. PLX halted the growth of both HT-29 and HCT116 tumors *in-vivo*. (B) Average body weights of control and PLX treated mice. The body weights did not vary significantly during the study. Tumor volumes were measured and tumor curve shows the efficacy of 50 mg/kg/day oral administration of PLX. (C) Histopathological analysis of tissue samples obtained from control and PLX-treated animals. Hematoxylin and Eosin stained tissue sections of the livers, hearts, kidneys and tumors. Images were obtained on a bright field microscope at 10× and 63× objective.

long pepper extracts. The efficacy of PLX in *in-vivo* models also showed that not only was PLX well-tolerated, it was also effective at halting the growth of human tumor xenografts of colon cancer in nude mice (Figure 9A and B).

The next step in understanding the effect of PLX on cell death induction in cancer cells was to identify the mechanism of apoptosis induction observed following PLX treatment. The role of oxidative stress in cell death processes has been well characterized. It is well established the reactive oxygen species (ROS) could be the cause or effect of apoptosis induction in cells [20]. Some studies have suggested cancer cells to be more

dependent on cellular response mechanisms against oxidative stress and have exploited this feature to selectively target cancer cells [10]. The role of ROS generation in PLX-induced apoptosis was assessed following treatment. In this study, we found that PLX induced whole cell ROS production in a dose dependent manner, as indicated by the increase in green fluorescence of H$_2$DCFDA dye, cleaved by intracellular esterases and oxidized by ROS present (Figure 7A & B). However, we observed that ROS generation was not completely essential to PLX activity, as the presence of N-acetylcysteine could not entirely hamper the ability of PLX to reduce the viability of colon cancer cells (Figure 7C).

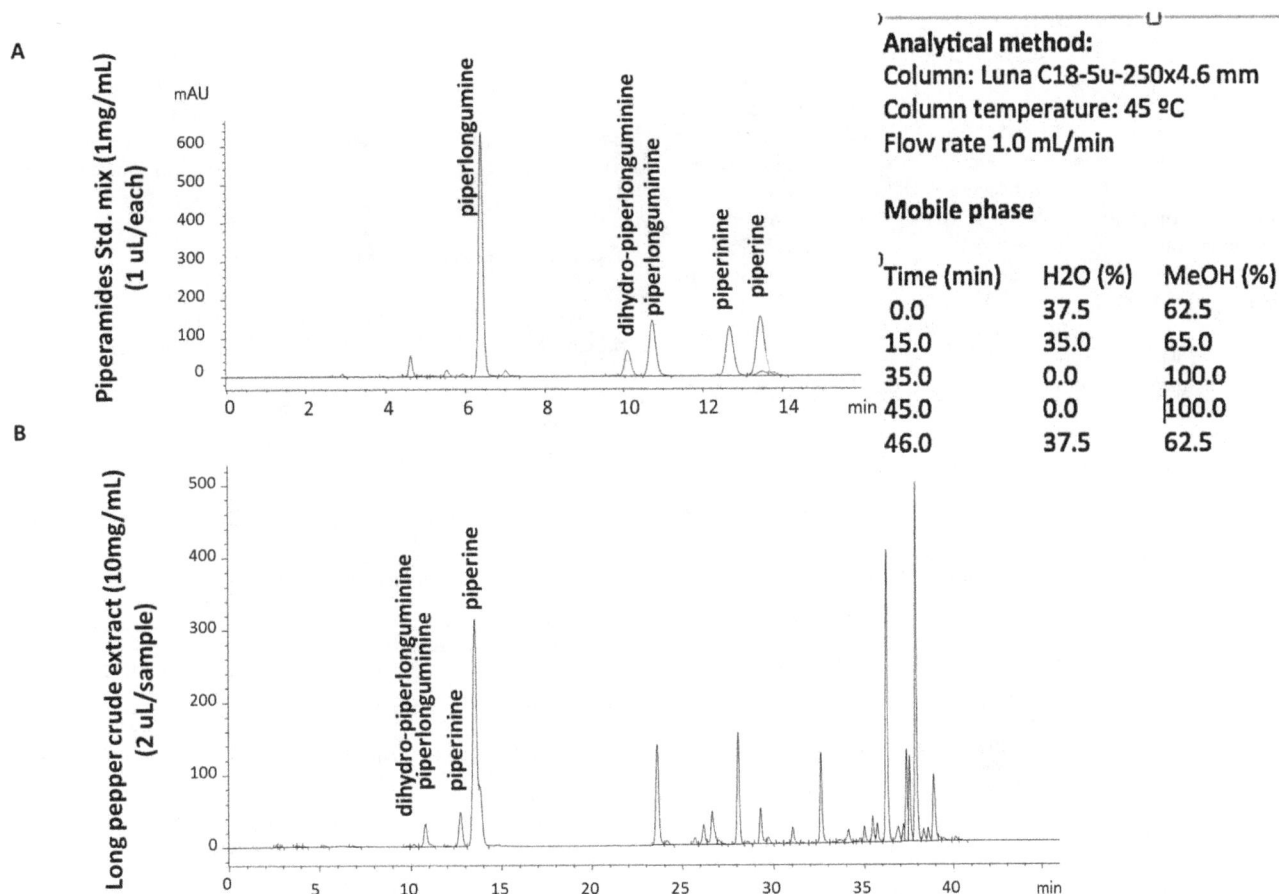

Figure 11. HPLC Analysis of PLX. Chromatograms of *Piper longum* extract (PLX) used for this study (10 mg/mL at 2 µL/Sample) (B) compared to Piperamides Standard mix (1 mg/mL at 1 µL/standard) (A).

The caspase-independence observed in figure 6B, suggest that PLX is acting through pro-apoptotic factors other than caspases. The mitochondria play a major role in the progression and execution of apoptosis. The permeabilization of the mitochondrial membrane usually leads to the release of pro-apoptotic factors, including cytochrome c, apoptosis inducing factor (AIF) and endonuclease G (EndoG) [5,17]. AIF and EndoG execute apoptosis in a caspase-independent possibly leading to the caspase- and partial ROS-independence observed. We show here that PLX caused MMP dissipation in cancer cells, while non-cancerous NCM460 cell mitochondria remained intact following treatment (Figure 8A–C). The dissipation of the mitochondrial membrane led to the release of AIF and EndoG (Figure 8D), allowing for the progression and execution of apoptosis in the absence of caspases and oxidative stress, providing insight to the mechanism of PLX action in cancer cells. Cancer cells differ from non-cancerous cells in variety of ways, which could enhance the selectivity of PLX to cancer cells. The Warburg effect is characterized by the high dependence of cancer cells on glycolysis and low dependence on mitochondria for energy production in cancer cells, therefore creating a more vulnerable target in cancer cell mitochondria [27]. Moreover, various anti-apoptotic proteins associated to the mitochondria have been reported to be highly expressed in cancer cells. Such proteins could serve as targets for selective cancer [28–30].

Unlike isolated natural compounds, there are usually more benefits to using a whole plant extract, with multiple pharmaco-logically active phytochemicals, than a single isolated compound. Multiple components within extracts could have many different intracellular targets, which may act in a synergistic way to enhance specific activities (including anticancer activities), while inhibiting any toxic effects of one compound alone. Additionally, the presence of multiple components may possibly decrease the chances of developing chemoresistance [31]. Moreover, natural extracts can be administered orally to patients, as a safe mode of administration. Some known compounds of the long pepper plants have been isolated and studied for their various activities [9–14].

We now report that the botanical identification of long pepper that was used for this study is *Piper longum* L. (Piperaceae), obtained from India. The phytochemical analysis of our material confirmed that this is the *Piper longum* species. We used the extract of the fruit in this study, since it is the usual part used medicinally and can be harvested sustainably. Although the fruit contains related piperamides, it does not contain piperlongumine, which is mainly present in the root of this plant [32]. The other piperamides present in the fruit such as dihydropiperlongumine likely have similar bioactivity. The small peak of piperlongumine observed in the HPLC chromatogram in Figure 11, as piperlon-gumine may be due to the reduction of piperlongumine to the larger dihyropiperlongumine peak that we observe. The analysis is consistent with *P. longum* fruits, and this is very important and points to the novel findings regarding the anticancer activity of this composition of components.

In a previous study that showed the efficacy of piperlongumine, high concentrations of 10 μM were required for significant cell death induction in cancer cells [10]. In this present study, we report the use of low concentrations of the complex mixture of the ethanolic long pepper extract (containing many bioactive and pharmacologically active compounds) was sufficient in inducing apoptosis in cancer cells selectively. This indicates that the individual bioactive compounds (present in sub micromolar concentrations within the extract) could act synergistically to induce apoptosis in cancer cells at very low concentrations, unlike a single identified compound. These findings highlights that the *Piper spp.* contain novel compounds with potent anticancer activity, in addition to piperlongumine.

In conclusion, our results demonstrate that long pepper extract (PLX), with a long historical use in traditional medicine, is selective in inducing apoptotic cell death in cancer cells by targeting non-genomic targets (e.g. the mitochondria). It is well tolerated in mice models and effective in reducing the growth of human tumor xenotransplants in animal models, when delivered orally. This could open a window of opportunity to develop a novel, safer cancer treatment, using complex natural health products from the Long Pepper.

Author Contributions

Conceived and designed the experiments: PO DM JAG JTA SP. Performed the experiments: PO DM PT AR MS JAG JTA SP. Analyzed the data: PO DM PT AR MS JAG JTA SP. Wrote the paper: PO DM SP.

References

1. Jemal A, Bray F, Ferlay J (2011) Global Cancer Statistics, 61(2): 69–90. doi:10.3322/caac.20107
2. Canadian Cancer Society's Steering Committee on Cancer Statistics (2012) Canadian Cancer Statistics 2012. Toronto, ON: Canadian Cancer Society; 2012.
3. Hanahan D, Weinberg RA (2011) Hallmarks of cancer: the next generation. Cell 144(5): 646–74. doi:10.1016/j.cell.2011.02.013
4. Fadeel B, Orrenius S (2005) Apoptosis: a basic biological phenomenon with wide-ranging implications in human disease. Journal of internal medicine 258(6): 479–517. doi:10.1111/j.1365-2796.2005.01570.x
5. Elmore S (2007) Apoptosis: A review of programmed cell death. Toxicologic pathology 35(4): 495–516. doi:10.1080/01926230701320337
6. Fulda S, Debatin KM (2006) Extrinsic versus intrinsic apoptosis pathways in anticancer chemotherapy. Oncogene 25(34): 4798–811. doi:10.1038/sj.onc.1209608
7. Davidson D, Amrein L, Panasci L, Aloyz R (2013) Small Molecules, Inhibitors of DNA-PK, Targeting DNA Repair, and Beyond. Frontiers in pharmacology 4(January): 5. doi:10.3389/fphar.2013.00005
8. Newman DJ, Cragg GM (2012) Natural products as sources of new drugs over the 30 years from 1981 to 2010. Journal of natural products 75(3): 311–35. doi:10.1021/np200906s
9. Bao N, Ochir S, Sun Z, Borjihan G, Yamagishi T (2013) Occurrence of piperidine alkaloids in Piper species collected in different areas. Journal of natural medicines. doi:10.1007/s11418-013-0773-0
10. Raj L, Ide T, Gurkar AU, Foley M, Schenone M, et al. (2012) Selective killing of cancer cells by a small molecule targeting the stress response to ROS. Nature 481(7382): 534–534. doi:10.1038/nature10789
11. Golovine KV, Makhov PB, Teper E, Kutikov A, Canter D, et al. (2013) Piperlongumine induces rapid depletion of the androgen receptor in human prostate cancer cells. The Prostate 73(1): 23–30. doi:10.1002/pros.22535
12. Jarvius M, Fryknäs M, D'Arcy P, Sun C, Rickardson L, et al. (2013) Piperlongumine induces inhibition of the ubiquitin-proteasome system in cancer cells. Biochemical and biophysical research communications 431(2): 117–23. doi:10.1016/j.bbrc.2013.01.017
13. Meghwal M, Goswami TK (2013) Piper nigrum and Piperine: An Update. Phytotherapy research: PTR, (February). doi:10.1002/ptr.4972
14. Lee SE, Park BS, Kim MK, Choi WS, Kim HT, et al. (2001) Fungicidal activity of pipernonaline, a piperidine alkaloid derived from long pepper, Piper longum L., against phytopathogenic fungi. Crop Protection 20(6): 523–528. doi:10.1016/S0261-2194(00)00172-1
15. Bezerra DP, Militão CG, de Castro FO, et al. (2007) Piplartine induces inhibition of leukemia cell proliferation triggering both apoptosis and necrosis pathways. Toxicol In Vitro vol. 21, no. 1, pp. 1–8
16. Fischer AH, Jacobson KA, Rose J, Zeller R (2008) Hematoxylin and eosin staining of tissue and cell sections. Cold Spring Harbor Protocols 2008(5): pdb-prot4986.
17. Earnshaw WC, Martins LM, Kaufmann SH (1999) Mammalian caspases: structure, activation, substrates, and functions during apoptosis. Annual review of biochemistry 68: 383–424. doi:10.1146/annurev.biochem.68.1.383
18. Thorburn A (2008) Apoptosis and autophagy: regulatory connections between two supposedly different processes. Apoptosis: an international journal on programmed cell death 13(1): 1–9. doi:10.1007/s10495-007-0154-9
19. Zhivotovsky B, Orrenius S (2010) Cell death mechanisms: cross-talk and role in disease. Experimental cell research 316(8): 1374–83. doi:10.1016/j.yexcr.2010.02.037
20. Simon HU, Haj-Yehia A, Levi-Schaffer F (2000) Role of reactive oxygen species (ROS) in apoptosis induction. Apoptosis vol. 5, no. 5, pp. 415–8
21. Madesh G, Hajnóczky G (2001) VDAC-dependent permeabilization of the outer mitochondrial membrane by superoxide induces rapid and massive cytochrome c release. J Cell Biol vol. 155, no. 6, pp. 1003–15.
22. Dekhuijzen P NR (2004) Antioxidant properties of N-acetylcysteine: their relevance in relation to chronic obstructive pulmonary disease. European Respiratory Journal 23(4): 629–636. doi:10.1183/09031936.04.00016804
23. Dodd S, Dean O, Copolov DL, Malhi GS, Berk M (2008) N-acetylcysteine for antioxidant therapy: pharmacology and. Expert Opin Biol Ther 8(12): 1955–1962
24. Earnshaw WC (1999) A cellular poison cupboard. Nature vol. 397, no. 6718, pp. 387–389.
25. Bleske BE, Clark MM, Wu AH, Dorsch MP (2013) The Effect of Continuous Infusion Loop Diuretics in Patients With Acute Decompensated Heart Failure With Hypoalbuminemia. Journal of cardiovascular pharmacology and therapeutics 00(0): 1–4.
26. Fang C, Shen L, Dong L, Liu M, Shi S, et al. (2013) Reduced urinary corin levels in patients with chronic kidney disease. Clinical Science 124(12): 709–717.
27. Warburg O (1956) On the origin of cancer cells. Science vol. 123, no. 3191, pp. 309–14
28. Mathupala SP, Rempel A, Pedersen PL (1997) Aberrant glycolytic metabolism of cancer cells: a remarkable coordination of genetic, transcriptional, post-translational, and mutational events that lead to a critical role for type II hexokinase. J Bioenerg Biomembr vol. 29, no. 4, pp. 339–43
29. Casellas P, Galiegue S, Basile AS (2002) Peripheral benzodiazepine receptors and mitochondrial function. Neurochem Int vol. 40, no, 6, pp. 475–86.
30. Green DR, Kroemer G (2004) The pathophysiology of mitochondrial cell death. Science vol. 305, no. 5684, pp. 626–9.
31. Foster BC, Arnason JT, Briggs CJ (2005) Natural health products and drug disposition. Annual review of pharmacology and toxicology 45: 203–26. doi:10.1146/annurev.pharmtox.45.120403.095950
32. Chandra VB, et al. (2014) Metabolic profiling of Piper species by direct analysis using real time mass spectrometry combined with principal component analysis. Analytical Methods, 4234–4239.

19

Running for Exercise Mitigates Age-Related Deterioration of Walking Economy

Justus D. Ortega[1]*, Owen N. Beck[1,2], Jaclyn M. Roby[2], Aria L. Turney[1], Rodger Kram[2]

1 Department of Kinesiology & Recreation Administration, Humboldt State University, Arcata, California, United States of America, 2 Department of Integrative Physiology, University of Colorado, Boulder, Colorado, United States of America

Abstract

Introduction: Impaired walking performance is a key predictor of morbidity among older adults. A distinctive characteristic of impaired walking performance among older adults is a greater metabolic cost (worse economy) compared to young adults. However, older adults who consistently run have been shown to retain a similar running economy as young runners. Unfortunately, those running studies did not measure the metabolic cost of walking. Thus, it is unclear if running exercise can prevent the deterioration of walking economy.

Purpose: To determine if and how regular walking vs. running exercise affects the economy of locomotion in older adults.

Methods: 15 older adults (69±3 years) who walk ≥30 min, 3x/week for exercise, "walkers" and 15 older adults (69±5 years) who run ≥30 min, 3x/week, "runners" walked on a force-instrumented treadmill at three speeds (0.75, 1.25, and 1.75 m/s). We determined walking economy using expired gas analysis and walking mechanics via ground reaction forces during the last 2 minutes of each 5 minute trial. We compared walking economy between the two groups and to non-aerobically trained young and older adults from a prior study.

Results: Older runners had a 7–10% better walking economy than older walkers over the range of speeds tested (p = .016) and had walking economy similar to young sedentary adults over a similar range of speeds (p = .237). We found no substantial biomechanical differences between older walkers and runners. In contrast to older runners, older walkers had similar walking economy as older sedentary adults (p = .461) and ~26% worse walking economy than young adults (p<.0001).

Conclusion: Running mitigates the age-related deterioration of walking economy whereas walking for exercise appears to have minimal effect on the age-related deterioration in walking economy.

Editor: Yuri P. Ivanenko, Scientific Institute Foundation Santa Lucia, Italy

Funding: Support was provided by the California State University Program for Education and Research in Biotechnology New Investigator [Grant #: HM531] (http://www.calstate.edu/csuperb/grants/). This funder had no role in study design, data collection and analysis, decision to publish, or preparation of the manuscript. Support was also provided by the National Institutes of Health Clinical and Translational Science Award [Grant #: UL1 TR000154] (http://www.ncats.nih.gov/research/cts/ctsa/funding/funding.html). This funder provided the facility where the data was collected.

Competing Interests: The authors have declared that no competing interests exist.

* Email: Justus.Ortega@humboldt.edu

Introduction

Walking performance typically deteriorates with advanced age [1], and impaired walking performance is a key predictor of morbidity among older adults [2]. A distinctive characteristic of impaired walking performance among older adults is a 15–20% greater metabolic cost for walking (worse economy) compared to young adults [3–5]. Several factors are known to determine the metabolic cost of walking in humans across all ages. These major biomechanical factors include the costs associated with: supporting body weight, performing mechanical work, leg swing and balance [5–8]. Studies investigating age-related biomechanical determinants of walking cost have found that older adults have a similar cost of balance and perform a similar amount, or even less, external mechanical work during walking as young adults [5,9,10].

Despite these similarities, other studies suggest that a decrease in muscular efficiency and an increase in antagonist leg muscle co-activation, contribute to the greater cost of walking in both healthy sedentary and active older adults [3,5,7,10,11]. Yet, no study has found a sole mechanical determinant that accounts for the 15–20% greater metabolic cost of walking in older adults. Therefore, interventions for improving walking economy in older age have been elusive.

Recent studies by Thomas et al. [12] and Malatesta et al. [13] show that vigorous walking interval training effectively reduces the metabolic cost of walking in older adults by as much as 20%. Yet, the mechanisms for the decreases were not elucidated. Conversely, a generalized year-long training program that included resistance, aerobic and balance exercises had no effect on post-training walking economy in older adults [14]. The different effects of these exercise interventions, high intensity aerobic versus generalized exercise with only a moderate aerobic component, suggest higher intensity aerobic activities may mitigate the typical age-related decrease in walking economy, and consequently, preserve mobility into older age.

In contrast, running economy does not exhibit the same age-related trend as walking economy. Two studies have reported that adults (45–61 years) who consistently participated in running exercise retain a similar metabolic economy of running as young runners (23–27 years) [15,16]. Although these results seem to support the hypothesis that vigorous aerobic exercise mitigates the decline in locomotion economy, i.e. metabolic cost of running and walking, it is also possible that a decline in running economy does not occur until late into the 6th decade of life, as observed with walking economy [17]. Perhaps the subjects in these studies [15,16] were not "old" enough to exhibit declines in locomotion economy. Another possible explanation is that running economy, unlike walking economy, is simply not affected by age. However, since these running studies did not measure walking economy, it remains unclear if regular participation in running exercise mitigates the typical age-related deterioration of walking economy.

Our purpose was to determine if and how regular participation in walking or running exercise affects the metabolic cost and biomechanics of walking in older adults. We hypothesized that older runners would consume less metabolic energy for walking than older walkers. Further, we also investigated whether the two groups demonstrate different walking biomechanics. We measured metabolic rates, ground reaction forces and spatio-temporal stride variables of two groups, older walkers and older runners, while they walked on a dual-belt, force-sensing treadmill at three speeds.

Methods

Subjects

Thirty healthy older adults (15 males and 15 female) who either walk (4 Male, 11 Female) or run (10 Male, 5 Female) regularly for exercise volunteered. Table 1 summarizes the anthropometric characteristics of the subjects. We recruited subjects with a minimum age of 65 years, which is in accordance with prior studies reporting age-related impairments of walking performance become most apparent at this age [3,18–20]. All subjects were free of neurological, orthopedic and cardiovascular disorders. Walkers self-reported walking for exercise three or more times per week for at least 30 minutes per bout and for at least six months prior to the study. Runners self-reported running for exercise three or more times per week for at least 30 minutes per bout and for at least six months prior to the study. The experiment was performed in accordance with the ethical standards of the 1964 Declaration of Helsinki and was approved by the Humboldt State University and University of Colorado Institutional Review Boards. All subjects gave written informed consent prior to participation in the study.

Protocol

Subjects completed three sessions. In the first session, subjects underwent a physician's examination to determine neurological, orthopedic and cardiovascular health, a body composition test (DXA) to determine percent body fat and lean tissue mass and a VO_2 max treadmill test to determine maximal aerobic capacity. In the second session, at least five days following the first session, we measured standing metabolic rate and familiarized the subjects to treadmill walking. For the treadmill familiarization, subjects walked on a dual-belt, force-instrumented treadmill (FIT, Bertec Corporation, Columbus, OH, USA) at three speeds (0.75, 1.25 and 1.75 m/s) for at least 7 minutes at each speed. These speeds

correspond to 1.67, 2.80, 3.91 MPH. Thus, subjects completed a minimum of 21 minutes total of walking familiarization. This familiarization period is over double the recommended minimum treadmill habituation time of 10 minutes [21,22]. In the third session, at least two days following familiarization, we measured each subject's metabolic rate during quiet standing and while walking on the treadmill at three speeds (0.75, 1.25 and 1.75 m/s) in random order. All trials were five minutes in duration with at least five minutes of rest between trials. Throughout each trial, we measured the rates of oxygen consumption (VO_2) and carbon dioxide production (VCO_2) in order to determine metabolic rate. We calculated the average VO_2 and VCO_2 for the last two minutes of each trial. We also measured ground reaction forces (GRFs) from the force-instrumented treadmill for 1 minute during the last 2.5 minutes of each trial to determine kinetics and spatio-temporal stride variables.

Metabolic Power Consumption

We measured VO_2 and VCO_2 using an open-circuit expired gas analysis system (TrueOne 2400, ParvoMedic, Sandy, UT, USA). We calculated average gross metabolic power per kilogram body mass (W/kg) [23] using the average VO_2 (mlO$_2$/min) and VCO_2 (mlCO$_2$/min) for the last two minutes of each trial, when VO_2 and respiratory exchange ratio reached steady state ensuring that each subject was working sub-maximally and oxidative metabolism was the main metabolic pathway. We then divided gross metabolic power by speed to calculate gross metabolic cost of transport (CoT) (J/kg/m) for walking.

Ground Reaction Forces and Spatio-temporal Stride Variables

For each walking trial, we collected the ground reaction forces (vertical and horizontal components) of each leg from the force-sensing treadmill at 2000 Hz for a 1 minute period during the last 2.5 minutes of each trial. A custom MATLAB script (Math Work Inc., Natick, Mass) was then used to process all force data. The

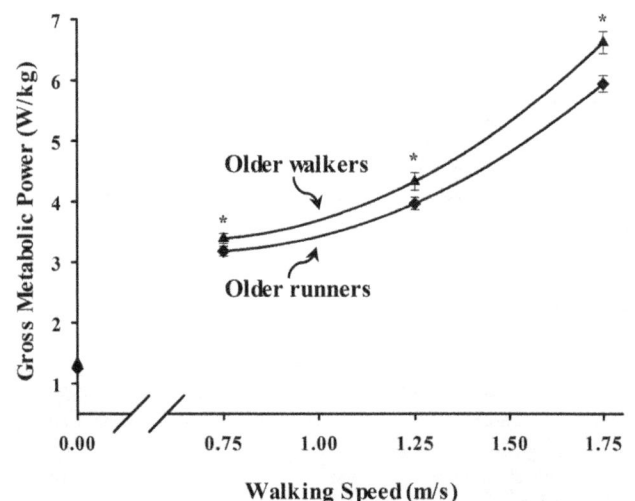

Figure 1. Mean (SE) gross metabolic power as a function of walking speed in older walkers (▲) and older runners (◆) walkers (▲). Lines represent least square regression for older walkers ($y = 2.709x^2 - 3.539x + 4.523$, $r^2 = 0.86$) and older runners ($y = 2.382x^2 - 3.189x + 4.233$, $r^2 = 0.89$). Symbols shown on vertical axis represent standing metabolic rate of both groups. Asterisks (*) indicate significant differences between older runners and walkers (p<0.05).

Table 1. Subject characteristics (Mean ±SD) with statistics for older walkers and older runners.

	Older Walkers (n = 15; 4M, 11 F)	Older Runners (n = 15; 10M, 5 F)
Age, years	68.9±3.0	68.9±4.7
Height, m	1.61±0.09	1.70±0.09*
Leg length, m	0.83±0.06	0.88±0.06
Body mass, kg	61.7±11.0	66.5± 13.0
Lean tissue mass, kg	39.2±7.1	48.6±9.2*
Body fat, % body mass	31.5±9.6	23.4±6.0*
VO_2 Max, mlO2/kg/min	27.7±3.6	37.3±5.3*
Standing metabolic rate, W/kg	1.34±0.21	1.26±0.14
0.75 m/s, gross metabolic power, W/kg	3.39±0.33	3.18±0.31*
1.25 m/s, gross metabolic power, W/kg	4.33±0.56	3.97±0.40*
1.75 m/s, gross metabolic power, W/kg	6.33±0.71	5.95±0.52*

Asterisk indicates the only significant group difference (p<.05).

GRF data were filtered with a 4th order zero-lag low pass Butterworth filter with a cutoff frequency of 30 Hz. For each trial, we calculated vertical and horizontal peak GRFs across all 10 strides. Using the filtered GRF data, we determined gait cycle events and spatio-temporal stride variables (stride frequency, stance time, and duty factor as percent of the gait cycle) for 10 strides of each trial (10 steps per each leg).

Statistical Analyses

We used a repeated-measures ANOVA (p<.05) to determine statistical differences due to exercise group (walkers vs. runners) and walking speed, as well as, the exercise group-walking speed interaction. When a significant main effect of exercise group was found, we performed independent-samples t-tests with Bonferroni correction to determine at which speed(s) the differences occurred. To determine if difference in metabolic cost and GRF was related to sex differences in our runner and walker groups, we examined differences in metabolic cost, ground reaction forces and spatio-temporal stride variables due to sex among each group and analyzed difference in metabolic cost, GRFs and spatio-temproal

stride variables using sex as a covariate. We found no effect of sex on any dependent variable and differences between runners and walkers were not affected by sex. We performed all statistical analyses using SPSS 21.0 (SPSS, Inc.) software. In addition to our comparison between older walkers and runners, we used a mixed-model repeated-measures ANOVA (p<.05) to make further post-hoc comparisons of gross metabolic cost in walkers and runners collected in the present study to data for young and older sedentary adults previously collected in our lab at similar speeds [5]. To make these comparison between exercise/age group (old walkers, old runners, old sedentary and young sedentary) using a linear mixed model, walking speed squared $(m/s)^2$ was used as the repeated measure.

Results

In support of our hypothesis, older runners consumed 7–10% less metabolic energy for walking than older walkers across the range of speeds tested (Fig. 1; p = .016). Gross metabolic power consumption increased significantly across the range of walking speeds tested in both older runners and walkers, (p<.0001). Compared to walking at the slowest speed of 0.75 m/s, gross metabolic power increased by 95% to walk at 1.75 m/s in older walkers but only 86% in older runners (speed X group interaction, p = .009). Mass-specific standing metabolic rates were similar between older runners and walkers (p = .250; Table 1).

Following from the metabolic rate data, the older runners had an average of 7–10% lower gross metabolic cost of transport compared to the older walkers. Older walkers and runners exhibited similar U-shape relations between gross CoT and walking speed (Fig. 2). Between the three speeds, gross CoT was significantly lower at the intermediate speed of 1.25 m/s as compared to the faster and slower walking speeds in both the older walkers (3.49±0.09 J/kg/m, p<.0001) and older runners (3.18±0.08 J/kg/m, p<.0001). Although there were a greater number of male runners in the study, our statistical analysis showed that the difference in metabolic cost between runners and walkers was not due to sex or any other anthropometric variable.

Despite the substantial differences in walking economy, older walkers and runners exhibited nearly identical spatio-temporal stride variables and kinetics across the range of speeds (Table 2). Among spatio-temporal gait characteristics, we found no significant differences between older walkers and older runners in

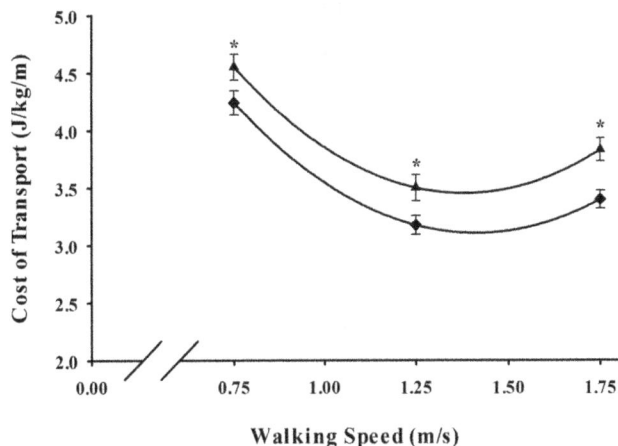

Figure 2. Mean (SE) gross metabolic cost of transport as a function of speed in older walkers (▲) and older runners (◆). Asterisks (*) indicate significant differences between older walkers and runners (p<.05).

regards to stride time, stride frequency (p = .879), single leg stance time (p = .126) or duty factor (p = .126). However, older runners walked with slightly (6%) shorter strides in relation to their leg length compared to older walkers (p = .033). This difference remained nearly constant across the range of speeds. With regards to ground reaction forces, older walkers and runners exhibited similar first (p = .838) and second (p = .282) peak vertical ground reaction force (Figure 3). Additionally, peak anterior-posterior braking (p = .182) and propulsive (p = .056) ground reaction forces were similar for both exercise groups.

We also compared gross metabolic cost of walking for older walkers and older runners to data from young and older sedentary adults collected in our lab from a prior study over a similar range of speeds [5]. The speeds used in these two studies were slightly different. Thus, in order to statistically make this comparison using a linear mixed model repeated measures ANOVA, we determined gross metabolic power as a function of speed squared (Fig. 4). The results of this analysis showed that across the range of speeds, older

walkers consume metabolic energy at a similar rate as sedentary older adults (p = .461) and 14–22% faster than young sedentary adults (p<.0001). In contrast, older runners consume metabolic energy at a slower rate compared to older sedentary adults (p = .016). However, our most striking finding was that older runners consumed metabolic energy at a similar rate as young sedentary adults across the range of walking speeds (p = .237).

Discussion and Conclusions

In this study, we distinguished the effects of regular walking vs. running exercise on the metabolic cost and biomechanics of walking in older adults. In support of our hypothesis, older runners consumed less metabolic energy for walking than older walkers. Although the older runners consumed less metabolic energy for walking than the older walkers, the two groups had almost identical walking biomechanics.

Given that there were virtually no differences in walking biomechanics between the older walkers and runners, other factors

Table 2. Spatio-temporal stride variables and ground reaction force data (Mean ±SD) with statistics for older walkers and older runners.

	Older Walkers (n = 15)	Older Runners (n = 15)
Speed 0.75 m/s		
Stride Time, sec	1.26±0.11	1.19±0.08
Stance Time, % of stride	65±2	66±1
Swing Time, % of stride	35±2	34±1
Stride Frequency, Hz	0.80±0.07	0.84±0.06
Stride Length, Leg Length	1.14±0.08	1.02±0.10*
First Peak VGRF, BW%	104±3	104±3
Second Peak VGRF, BW%	101±3	100±2
Braking HGRF, BW%	−8±1	−8±1
Propulsive HGRF, BW%	11±2	10±1
Speed 1.25 m/s		
Stride Time, sec	1.04±0.07	1.05±0.06
Stance Time, % of stride	63±2	64±2
Swing Time, % of stride	37±2	37±2
Stride Frequency, Hz	0.97±0.07	0.95±0.06
Stride Length, Leg Length	1.57±0.08	1.49±0.10*
First Peak VGRF, BW%	110±5	108±4
Second Peak VGRF, BW%	106±5	105±3
Braking HGRF, BW%	−17±2	−16±2
Propulsive HGRF, BW%	19±0.02	17±2
Speed 1.75 m/s		
Stride Time, sec	0.92±0.05	0.93±0.04
Stance Time, % of stride	61± 1	63±2
Swing Time, % of stride	39±1	38±2
Stride Frequency, Hz	1.09±0.06	1.08±0.05
Stride Length, Leg Length	1.88±0.10	1.83±0.10
First Peak VGRF, BW%	134±12	129±4
Second Peak VGRF, BW%	110±9	119±7
Braking HGRF, BW%	−28±2	−26±5
Propulsive HGRF, BW%	26±3	25±3

Peak vertical ground reaction forces (VGRF) and horizontal ground reaction forces (HGRF) are represented as % body weight (BW). Asterisk indicates significant group difference (p<.05).

A

B

Gait Cycle (%)

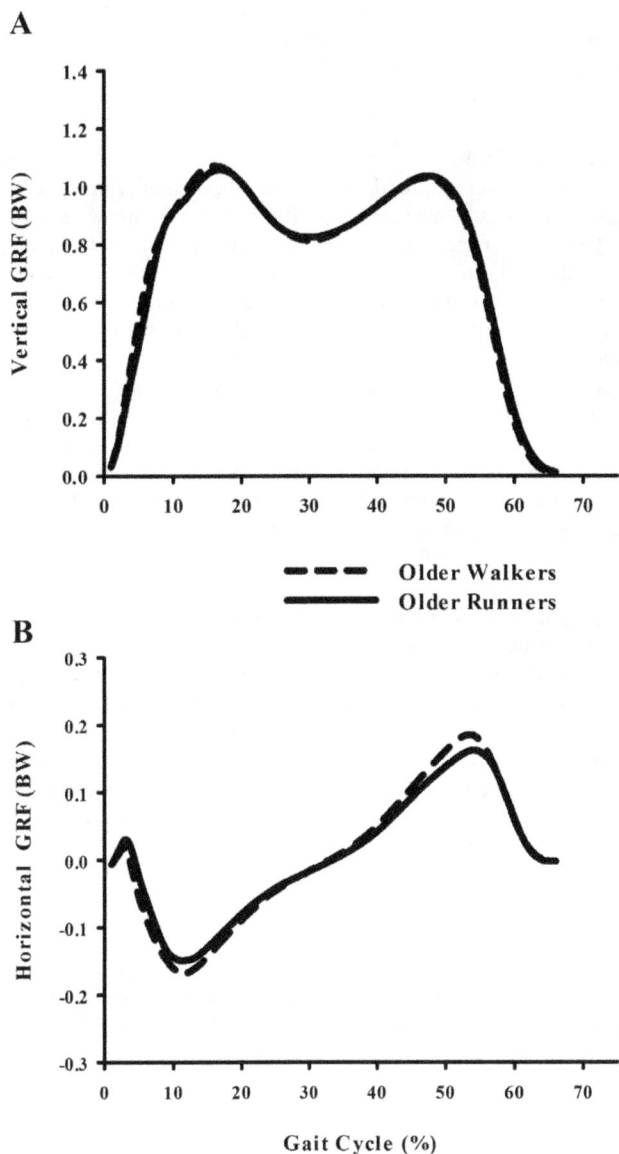

Figure 3. Average individual leg vertical (A) and horizontal (B) ground reaction force for older walkers (dashed lines) and older runners (solid lines) at the intermediate walking speed of 1.25 m/s.

muscle strength and reduce co-activation. However, a decrease in co-activation associated with running that is similar in magnitude to the decrease observed after strength training is likely not sufficient to explain the 7–10% difference in metabolic cost of walking. It is also possible that other neuromuscular factors such as widening of EMG/motoneuronal bursts [27] may also help to explain the difference in metabolic cost between older runners and walkers.

Better muscular efficiency may also help explain why older runners have a lower metabolic cost of walking than older walkers. Aging has been associated with reduced muscular efficiency [10,28]. More specifically, mitochondrial dysfunction associated with the uncoupling of oxidative phosphorylation (reduced ATP synthesis per O_2 uptake) effectively reduces muscular efficiency and increased the metabolic cost of muscle activation [28]. Interestingly, recent evidence suggests that aerobic exercise training may ameliorate mitochondrial uncoupling and improve muscular efficiency in older adults [29].

Perhaps studies of cycling efficiency in older adults can provide insight. In contrast to the effects of running we have observed, the muscular efficiency of cycling declines with age despite regular cycling exercise [30]. More recently, Brisswalter et al. [31] measured the cycling efficiency of active triathletes (who regularly swim, bike, and run for exercise) across age-groups and found a decline in cycling efficiency past the 5th decade. These data suggest that older cyclist and triathletes are unable to maintain muscular efficiency with age. However, Peiffer et al. [32] found no difference in cycling efficiency between their youngest age group (39±3 years) and their oldest (65±4 years). Intriguingly, their oldest training group cycled 58 km more per week (359 km per week) than the youngest group. Possibly the greater quantity of aerobic cycling exercise mitigated the decrease in muscular efficiency with age.

Alternatively, the intensity of exercise may hold the key to maintaining or improving muscular efficiency. Two prior studies have found that 6–7 weeks of vigorous aerobic exercise (fast walking) that elicits a heart rate close to the ventilatory threshold can improve walking economy by 8–20% [12,13]. More vigorous aerobic exercise such as walking uphill, fast walking or running may be required to elicit improvement in walking economy. Clinicians and others who work with older adults to improve their fitness may need to prescribe more vigorous, more prolonged and/ or more frequent aerobic exercise to prevent the decline in walking performance. To test this hypothesis and help guide clinicians, a future study should investigate the effects of different intensity aerobic exercises on muscular efficiency and more specifically, the economy of walking.

Limitations

One limitation of the current study is the cross-sectional design. It is possible that older runners may not be economical walkers because of the effect of running exercise but rather they run because they are more economical in their locomotion. To better address this issue, a future study might quantify the longitudinal effects of a running training program. One such study conducted by Trappe et al. [16] on the longitudinal effect of running exercise on running economy spanned 22 years. In that study, Trappe et al. [16] showed that running economy did not decline in older adults who maintained their health and fitness over the 22 year period, whereas runners who became unfit had worse running economy. Although these results suggest that running may help to prevent a decline in running economy, Trappe et al. [16] did not measure walking economy.

must underlie the lower cost of walking observed for the older runners. One factor may be muscle co-activation. Older adults, both sedentary and active walkers, use 30–50% greater co-activation of antagonist leg muscles compared to young adults [6,10,24]. It has been suggested that older adults may use greater co-activation to increase joint stiffness and the stabilization of the body, thus reducing the risk of walking related falls [25]. Yet, increased co-activation has been associated with increased metabolic cost of walking in older adults [6,10]. It is possible that older runners are able to maintain a lower metabolic cost of walking compared to older walkers because they use less antagonist leg muscle activation. Some research shows that older adults who participated in a lower limb strength training program reduce leg muscle co-activation by 5–10% [26]. Perhaps, by regularly running three or more times per week for 30 minutes per bout, older runners are able to maintain or even increase leg

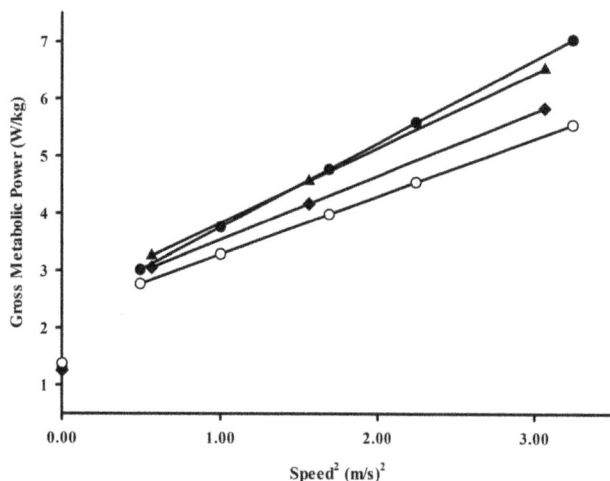

Figure 4. Gross metabolic power as a function of speed² in older sedentary adults (•), older walkers (▲), older runners (♦), and young sedentary adults (○). Lines denote least square regression within each group (older sedentary: $y = 1.46x + 2.30$, $r^2 = 0.91$; older walkers: $y = 1.31x + 2.52$, $r^2 = 0.86$; older runners: $y = 1.12x + 2.42$, $r^2 = 0.88$; young sedentary: $y = 1.01x + 2.27$, $r^2 = 0.87$). Symbols on vertical axis represent standing metabolic rate of each group.

Another potential limitation of the current study is the different numbers of male and female participants in each group. Although the sex difference may have influenced the difference in anthropometrics between runners and walker, our results showed no main effect of sex on walking economy ($p = .211$) and no sex difference in walking economy among older runners ($p = .131$) or older walkers ($p = .331$). Based on post-hoc power analysis, it is clear that we did not have sufficient statistical power to detect sex differences that might exist but that would require ~300 subjects. However, when treated as covariates, sex and anthropometrics did not statistically account for the difference in walking economy

between runners and walkers. Thus, while it would have been preferable to have a larger sample size with more similar sex and anthropometric matched cohorts, it would not have changed our overall conclusion.

Future Studies

Based on the results of this study and others, future studies of the effect of age and exercise on walking economy are warranted. Although the average age of our runners and walkers was 69 years, a future study might look to see if running exercise continues to prevent or slow the decline in walking economy in even older runners (over the age of 80 years). It seems plausible that at some age that exercise may not be able to sufficiently offset the normal decline in muscular efficiency and walking economy associated with aging. It is also not known whether there is an intensity threshold of aerobic exercise that is needed to prevent the decline in walking economy. Thus, it would be beneficial for future studies to investigate the relative effect of exercises with different levels of aerobic intensity on walking economy.

Conclusions

In conclusion, older runners mitigate the age-related deterioration of walking economy. However, older walkers are unable to forestall the decline of walking economy as they require the same metabolic consumption as sedentary older adults. The difference in walking economy between older runners and older walkers remains unexplained due to no substantial differences found in either the kinetic or spatio-temporal data between the groups. Other factors such as decreased muscle co-activation and/or increased muscular efficiency may contribute to the superior walking economy exhibited by the older runners.

Author Contributions

Conceived and designed the experiments: JO OB JR RK. Performed the experiments: JO OB JR AT RK. Analyzed the data: JO OB JR AT RK. Contributed reagents/materials/analysis tools: JO RK. Wrote the paper: JO OB JR AT RK.

References

1. Himann JE, Cunningham DA, Rechnitzer PA, Paterson DH (1988) Age-Related-Changes in Speed of Walking. Med Sci Sports Exerc 20: 161–166.
2. Studenski S, Perera S, Patel K, Rosano C, Faulkner K, et al. (2011) Gait speed and survival in older adults. JAMA 305: 50–58.
3. Martin PE, Rothstein DE, Larish DD (1992) Effects of age and physical activity status on the speed-aerobic demand relationship of walking. J Appl Physiol 73: 200–206.
4. Waters RL, Lunsford BR, Perry J, Byrd R (1988) Energy-speed relationship of walking: standard tables. J Orthop Res 6: 215–222.
5. Ortega JD, Farley CT (2007) Individual limb work does not explain the greater metabolic cost of walking in elderly adults. J Appl Physiol 102: 2266–2273.
6. Ortega JD, Farley CT (2014) Effects of aging on mechanical efficiency and muscle activation during level and uphill walking. J Electromyogr Kinesiol 0.
7. Ortega JD, Fehlman LA, Farley CT (2008) Effects of aging and arm swing on the metabolic cost of stability in human walking. J Biomech 41: 3303–3308. Epub 2008 Sep 3323.
8. Gottschall JS, Kram R (2005) Energy cost and muscular activity required for leg swing during walking. J Appl Physiol 99: 23–30.
9. Franz JR, Lyddon NE, Kram R (2012) Mechanical work performed by the individual legs during uphill and downhill walking. J Biomech 45: 257–262.
10. Mian OS, Thom JM, Ardigo LP, Narici MV, Minetti AE (2006) Metabolic cost, mechanical work, and efficiency during walking in young and older men. Acta Physiol Scand 186: 127–139.
11. Malatesta D, Simar D, Dauvilliers Y, Candau R, Borrani F, et al. (2003) Energy cost of walking and gait instability in healthy 65- and 80-yr-olds. J Appl Physiol 95: 2248–2256.
12. Thomas EE, Vito GD, Macaluso A (2007) Speed training with body weight unloading improves walking energy cost and maximal speed in 75- to 85-year-old healthy women. J Appl Physiol 103: 1598–1603.
13. Malatesta D, Simar D, Ben Saad H, Prefaut C, Caillaud C (2010) Effect of an overground walking training on gait performance in healthy 65- to 80-year-olds. Exp Gerontol 45: 427–434.
14. Mian OS, Thom JM, Ardigo LP, Morse CI, Narici MV, et al. (2007) Effect of a 12-month physical conditioning programme on the metabolic cost of walking in healthy older adults. Eur J Appl Physiol 100: 499–505.
15. Quinn TJ, Manley MJ, Aziz J, Padham JL, MacKenzie AM (2011) Aging and factors related to running economy. Journal of strength and conditioning research/National Strength & Conditioning Association 25: 2971–2979.
16. Trappe SW, Costill DL, Vukovich MD, Jones J, Melham T (1996) Aging among elite distance runners: a 22-yr longitudinal study. J Appl Physiol 80: 285–290.
17. Prince F, Corriveau H, Hebert R, Winter DA (1997) Gait in the elderly. Gait Posture 5: 128-135.
18. Studenski S, Perera S, Patel K, Rosano C, Faulkner K, et al. (2011) Gait speed and survival in older adults. JAMA: the journal of the American Medical Association 305: 50–58.
19. Himann JE, Cunningham DA, Rechnitzer PA, Paterson DH (1988) Age-related changes in speed of walking. Medicine and science in sports and exercise 20: 161–166.
20. Murray MP, Kory RC, Clarkson BH (1969) Walking patterns in healthy old men. J Gerontol 24: 169–178.
21. Wall JC, Charteris J (1981) A kinematic study of long-term habituation to treadmill walking. Ergonomics 24: 531–542.
22. Van de Putte M, Hagemeister N, St-Onge N, Parent G, de Guise JA (2006) Habituation to treadmill walking. Biomed Mater Eng 16: 43–52.
23. Brockway JM (1987) Derivation of formulae used to calculate energy expenditure in man. Hum Nutr Clin Nutr 41: 463–471.
24. Franz JR, Kram R (2012) How does age affect leg muscle activity/coactivity during uphill and downhill walking? Gait Posture 37: 378–384.

25. Finley JM, Dhaher YY, Perreault EJ (2012) Contributions of feed-forward and feedback strategies at the human ankle during control of unstable loads. Exp Brain Res 217: 53–66.

26. Hakkinen K, Kallinen M, Izquierdo M, Jokelainen K, Lassila H, et al. (1998) Changes in agonist-antagonist EMG, muscle CSA, and force during strength training in middle-aged and older people. J Appl Physiol 84: 1341–1349.

27. Monaco V, Ghionzoli A, Micera S (2010) Age-related modifications of muscle synergies and spinal cord activity during locomotion. J Neurophysiol 104: 2092–2102.

28. Amara CE, Shankland EG, Jubrias SA, Marcinek DJ, Kushmerick MJ, et al. (2007) Mild mitochondrial uncoupling impacts cellular aging in human muscles in vivo. Proc Natl Acad Sci U S A 104: 1057–1062.

29. Conley KE, Jubrias SA, Amara CE, Marcinek DJ (2007) Mitochondrial dysfunction: impact on exercise performance and cellular aging. Exerc Sport Sci Rev 35: 43–49.

30. Sacchetti M, Lenti M, Di Palumbo AS, De Vito G (2010) Different effect of cadence on cycling efficiency between young and older cyclists. Med Sci Sports Exerc 42: 2128–2133.

31. Brisswalter J, Wu SX, Sultana F, Bernard T, Abbiss C (2014) Age difference in efficiency of locomotion and maximal power output in well-trained triathletes. Eur J Appl Physiol: 1–8.

32. Peiffer JJ, Abbiss CR, Chapman D, Laursen PB, Parker DL (2008) Physiological characteristics of masters-level cyclists. J Strength Cond Res 22: 1434–1440.

PERMISSIONS

LIST OF CONTRIBUTORS

Xuan Zheng
Department of Neuroscience and Regenerative Medicine, Georgia Regents University, Augusta, Georgia, United States of America
Center for Gene Diagnosis, Zhongnan Hospital, Wuhan University, Wuhan, China

William S. Dynan
Department of Neuroscience and Regenerative Medicine, Georgia Regents University, Augusta, Georgia, United States of America
Departments of Radiation Oncology and Biochemistry, Emory University, Atlanta, Georgia, United States of America

Xinyan Zhang
Department of Biostatistics, University of Alabama at Birmingham, Birmingham, Alabama, United States of America

Lingling Ding
Department of Anatomy and Embryology, Wuhan University School of Medicine, Wuhan, China

Jeffrey R. Lee
Department of Pathology, Georgia Regents University, Augusta, Georgia, United States of America

Paul M. Weinberger
Department of Otolaryngology and Center for Biotechnology & Genomic Medicine, Georgia Regents University, Augusta, Georgia, United States of America

Leon M. Straker
School of Physiotherapy and Exercise Science, Curtin University, GPO Box U1987, Perth, Western Australia, 6845, Australia

Erin K. Howie, Kyla L. Smith, Rebecca A. Abbott and Anne J. Smith
School of Physiotherapy and Exercise Science, Curtin University, Perth, Australia

Ashley A. Fenner
chool of Psychology and Speech Pathology, Curtin University, Perth, Australia

Deborah A. Kerr
School of Public Health, Curtin University, Perth, Australia

Tim S. Olds
Health and Use of Time (HUT) Group, University of South Australia, Adelaide, Australia

Hsu-Liang Hsieh
Institute of Plant Biology, National Taiwan University, Taipei, 116, Taiwan, ROC

Ya-Wen Hsu
Institute of Plant Biology, National Taiwan University, Taipei, 116, Taiwan, ROC
Institute of Plant and Microbial Biology, Academia Sinica, Nankang, Taipei, 11529, Taiwan, ROC

Huei-Jing Wang and Ming-Hsiun Hsieh
Institute of Plant and Microbial Biology, Academia Sinica, Nankang, Taipei, 11529, Taiwan, ROC

Guang-Yuh Jauh
Institute of Plant and Microbial Biology, Academia Sinica, Nankang, Taipei, 11529, Taiwan, ROC
Biotechnology Center, National Chung-Hsing University, Taichung, 402, Taiwan, ROC

Lei Jiang, Hao Wang, Chunli Shi, Ke Liu, Meidong Liu, Nian Wang, Kangkai Wang, Huali Zhang, Guiliang Wang and Xianzhong Xiao
Department of Pathophysiology, Xiangya School of Medicine, Central South University, 110 Xiangya Road, Changsha, Hunan 410078, P. R. China

Akira Kohsaka, Partha Das, Izumi Hashimoto, Tomomi Nakao, Yoko Deguchi, Sabine S. Gouraud, Masanobu Maeda and Hidefumi Waki
Department of Physiology, Wakayama Medical University School of Medicine, Wakayama, Japan

Yasuteru Muragaki
First Department of Pathology, Wakayama Medical University School of Medicine, Wakayama, Japan

Kaya Yoshida
Department of Oral Healthcare Education, Institute of Health Biosciences, University of Tokushima Graduate School, Tokushima, Japan

Masami Yoshioka
Department of Oral Health Science and Social Welfare, Institute of Health Biosciences, University of Tokushima Graduate School, Tokushima, Japan

Hirohiko Okamura
Department of Histology and Oral Histology, Institute of Health Biosciences, University of Tokushima Graduate School, Tokushima, Japan

Satomi Moriyama and Daisuke Hinode
Department of Hygiene and Oral Health Science, Institute of Health Biosciences, University of Tokushima Graduate School, Tokushima, Japan

Kazuyoshi Kawazoe
Department of Clinical Pharmacy, Institute of Health Biosciences, University of Tokushima Graduate School, Tokushima, Japan

Daniel Grenier
Oral Ecology Research Group, Faculty of Dentistry, Laval University, Quebec City, QC, Canada

Daisuke Sunaga, Masaya Tanno, Atsushi Kuno, Satoko Ishikawa, Makoto Ogasawara, Toshiyuki Yano, Takayuki Miki and Tetsuji Miura
Department of Cardiovascular, Renal and Metabolic Medicine, Sapporo Medical University School of Medicine, Sapporo, Japan

José Julián Garde
SaBio, Instituto de Investigación en Recursos Cinegéticos (IREC), CSIC-UCLM-JCCM, Campus Universitario, Albacete, Spain

JoséLuis Ros-Santaella
SaBio, Instituto de Investigación en Recursos Cinegéticos (IREC), CSIC-UCLM-JCCM, Campus Universitario, Albacete, Spain
Department of Animal Science and Food Processing, Faculty of Tropical AgriSciences, Czech University of Life Sciences, Prague, Czech Republic

Álvaro Efrén Domínguez-Rebolledo
Instituto Nacional de Investigaciones Forestales, Agrícolasy Pecuarias (INIFAP), Mocochá, Yucatán, México

Mei-pian Chen, Shing Chan, Godfrey Chi-fung Chan and Yiu-fai Cheung
Department of Pediatrics and Adolescent Medicine, The University of Hong Kong, Hong Kong, China

Z. Ioav Cabantchik
Department of Biological Chemistry, Alexander Silberman Institute of Life Sciences, Hebrew University of Jerusalem, Safra Campus at Givat Ram, Jerusalem, Israel

Jennifer Turner and Drew Titmarsh
Tissue Engineering and Microfluidics Laboratory, The Australian Institute for Bioengineering and Nanotechnology, The University of Queensland, St Lucia, Queensland, Australia
Stem Cell Engineering Group, The Australian Institute for Bioengineering and Nanotechnology, The University of Queensland, St Lucia, Queensland, Australia

Justin Cooper-White
Tissue Engineering and Microfluidics Laboratory, The Australian Institute for Bioengineering and Nanotechnology, The University of Queensland, St Lucia, Queensland, Australia
School of Chemical Engineering, The University of Queensland, St Lucia, Queensland, Australia

Li-Pin Kao and Ernst Wolvetang
Stem Cell Engineering Group, The Australian Institute for Bioengineering and Nanotechnology, The University of Queensland, St Lucia, Queensland, Australia

Lake-Ee Quek
Centre for Systems and Synthetic Biotechnology, The Australian Institute for Bioengineering and Nanotechnology, The University of Queensland, St Lucia, Queensland, Australia

Jens O. Krömer and Lars Nielsen
Centre for Systems and Synthetic Biotechnology, The Australian Institute for Bioengineering and Nanotechnology, The University of Queensland, St Lucia, Queensland, Australia
Metabolomics Australia, The Australian Institute for Bioengineering and Nanotechnology, The University of Queensland, St Lucia, Queensland, Australia

Blanca Arango-Gonzalez, Dragana Trifunović, Ayse Sahaboglu, Marius Ueffing and François Paquet-Durand
Institute for Ophthalmic Research, University of Tuebingen, Tuebingen, Germany

Pietro Farinelli
Institute for Ophthalmic Research, University of Tuebingen, Tuebingen, Germany
Division of Ophthalmology, Department of Clinical Sciences, University of Lund, Lund, Sweden

Eberhart Zrenner and Thomas Euler
Institute for Ophthalmic Research, University of Tuebingen, Tuebingen, Germany
Centre for Integrative Neuroscience, University of Tuebingen, Tuebingen, Germany

Katharina Kranz, Ulrike Janssen-Bienhold and Karin Dedek
Department of Neurobiology, University of Oldenburg, Oldenburg, Germany

Stylianos Michalakis, Susanne Koch, Fred Koch and Martin Biel
Center for Integrated Protein Science Munich and Department of Pharmacy - Center for Drug Research, Ludwig-Maximilians-University Munich, Munich, Germany

Per Ekström
Division of Ophthalmology, Department of Clinical Sciences, University of Lund, Lund, Sweden

Sandra Cottet
Institute for Research in Ophthalmology, Sion, Switzerland

Mei-Ling Cheng
Department of Biomedical Sciences, College of Medicine, Chang Gung University, Tao-Yuan, Taiwan
Healthy Aging Research Center, Chang Gung University, Tao- Yuan, Taiwan
Metabolomics Core Laboratory, Chang Gung University, Tao-Yuan, Taiwan

Shiue-Fen Weng and Chih-Hao Kuo
Department of Medical Biotechnology and Laboratory Science, College of Medicine, Chang Gung University, Tao-Yuan, Taiwan

Hung-Yao Ho
Healthy Aging Research Center, Chang Gung University, Tao- Yuan, Taiwan
Department of Medical Biotechnology and Laboratory Science, College of Medicine, Chang Gung University, Tao-Yuan, Taiwan
Office of Research and Development, Chang Gung Memorial Hospital, Tao-Yuan, Taiwan

Esther Roselló-Lletí, Estefanía Tarazón, Ana Ortega, Maria Micaela Molina-Navarro and Miguel Rivera
Cardiocirculatory Unit, Health Research Institute Hospital La Fe, Valencia, Spain

María G. Barderas
Department of Vascular Physiopathology, Hospital Nacional de Parapléjicos, SESCAM, Toledo, Spain

Manuel Otero, Francisca Lago and Jose Ramón González-Juanatey
Cellular and Molecular Cardiology Research Unit, Department of Cardiology and Institute of Biomedical Research, University Clinical Hospital, Santiago de Compostela, Spain

Antonio Salvador
Cardiology Service, Hospital La Fe, Valencia, Spain

Manuel Portolés
Cell Biology and Pathology Unit, Health Research Institute Hospital La Fe, Valencia, Spain

Kim Vancampenhout and Claudia Spits
Research Group Reproduction and Genetics (REGE), Vrije Universiteit Brussel (VUB), Brussels, Belgium

Katrien Stouffs, Willy Lissens and Sara Seneca
Research Group Reproduction and Genetics (REGE), Vrije Universiteit Brussel (VUB), Brussels, Belgium
Center for Medical Genetics, UZ Brussel, Vrije Universiteit Brussel (VUB), Brussels, Belgium

Meirleir
Research Group Reproduction and Genetics (REGE), Vrije Universiteit Brussel (VUB), Brussels, Belgium
Department of Pediatric Neurology, UZ Brussel, Vrije Universiteit Brussel (VUB), Brussels, Belgium

Ben Caljon
Center for Medical Genetics, UZ Brussel, Vrije Universiteit Brussel (VUB), Brussels, Belgium

An Jonckheere
Department of Pediatric Neurology, UZ Brussel, Vrije Universiteit Brussel (VUB), Brussels, Belgium

Linda De Arnaud Vanlander, Joél Smet, Boel De Paepe and Rudy Van Coster
Department of Pediatrics, Division of Pediatric Neurology and Metabolism, University Hospital Ghent, Ghent University, Ghent, Belgium

Anderson De Andrade
Department of Translational Neurodegeneration, German Centre for Neurodegenerative Diseases (DZNE), Munich, Germany

Julius Bruch, Hong Xu and Günter Höglinger
Department of Translational Neurodegeneration, German Centre for Neurodegenerative Diseases (DZNE), Munich, Germany
Department of Neurology, Technische Universität München, Munich, Germany

Hojun Lee and Keisuke Kawata
Department of Kinesiology, Temple University, Philadelphia, Pennsylvania, United States of America

Boa Kim and Joon-Young Park
Department of Kinesiology, Temple University, Philadelphia, Pennsylvania, United States of America
Cardiovascular Research Center, Temple University, Philadelphia, Pennsylvania, United States of America

Tridib Kumar Sinha, Ratan Kumar Basak and Basudam Adhikari
Materials Science Centre, Indian Institute of Technology Kharagpur, West Bengal, India

Sougata Roy Chowdhury
Materials Science Centre, Indian Institute of Technology Kharagpur, West Bengal, India
Immunology lab, Department of Zoology, University of Calcutta, West Bengal, India

Suman Sengupta, Subir Biswas and Arindam Bhattacharyya
Immunology lab, Department of Zoology, University of Calcutta, West Bengal, India

Ramkrishna Sen
Department of Biotechnology, Indian Institute of Technology Kharagpur, West Bengal, India

Pamela Ovadje, Dennis Ma, Phillip Tremblay, Alessia Roma, Matthew Steckle and Siyaram Pandey
Department of Chemistry & Biochemistry, University of Windsor, Windsor, ON, Canada

Jose-Antonio Guerrero and John Thor Arnason
Department of Biology, University of Ottawa, Ottawa, ON, Canada

Justus D. Ortega, Aria L. Turney
Department of Kinesiology & Recreation Administration, Humboldt State University, Arcata, California, United States of America

Owen N. Beck
Department of Kinesiology & Recreation Administration, Humboldt State University, Arcata, California, United States of America
Department of Integrative Physiology, University of Colorado, Boulder, Colorado, United States of America

Jaclyn M. Roby and Rodger Kram
Department of Integrative Physiology, University of Colorado, Boulder, Colorado, United States of America

Index

www.ingramcontent.com/pod-product-compliance
Lightning Source LLC
Chambersburg PA
CBHW080518200326
41458CB00012B/4259